WIRELESS BODY AREA NETWORKS

edited by

Mehmet R. Yuce
Jamil Y. Khan

WIRELESS BODY AREA NETWORKS

Technology, Implementation, and Applications

Pan Stanford Publishing

Published by

Pan Stanford Publishing Pte. Ltd.
Penthouse Level, Suntec Tower 3
8 Temasek Boulevard
Singapore 038988

Email: editorial@panstanford.com
Web: www.panstanford.com

British Library Cataloguing-in-Publication Data
A catalogue record for this book is available from the British Library.

Wireless Body Area Networks: Technology, Implementation, and Applications

ISBN 978-981-4316-71-2 (Hardcover)
ISBN 978-981-4241-57-1 (eBook)

Contents

Preface

Sensor-based measurements and monitoring techniques have been widely used in electronic patient care systems for a long time. The concept of wireless-sensor-based patient monitoring using wireless body area network (WBAN) will bring revolutionary changes in health care systems. WBAN allows flexibility in providing location-independent and seamless patient monitoring without affecting the lifestyle of patients. A WBAN system can be deployed at care centers for elderly people or at homes to look after elderly citizens without affecting their mobility or activities, thus improving the quality of life. The WBAN technology has advanced significantly in the last decade and gradually moved from the research laboratories to clinical trials and test environments. In future, WBAN applications will not remain restricted to the field of medicine. They will register their presence in other areas such as sports and training, military applications, and human safety. This book covers a range of topics on the WBAN technology, introduced in a manner that will be suitable for a broad range of readers. The book will be a key resource for medical ICT (information and communication technology) professionals, biomedical engineers, and graduate and senior undergraduate students in computer, electronic, and biomedical engineering.

This book discusses the current state of the art by focusing on the latest research and new design and development methodologies related to the WBAN technology. The book systematically introduces basic concepts, hardware, software and system design techniques, and various WBAN algorithms for wearable and implantable sensor applications. It also discusses some potential applications of WBAN in the field of e-health. Low-cost electronics and off-the-shelf consumer electronics-based wireless embedded systems are introduced, which can be used to design basic low-cost wireless body area networks.

The book offers a comprehensive review of WBAN design, development, and deployment techniques. All chapters are written by leading experts in their fields. The contributions by the authors focus on applications of wireless body area networks, implementation, channel modeling, signal transmission around and in the body, antenna design, in-body communication, and hardware implementation of body sensors and nodes. Chapter 1 starts with an introduction to the WBAN to help readers understand the topic and the associated basic techniques. The next three chapters concentrate on the implementation of WBAN in clinical environments. Chapter 2 discusses clinical applications of WBAN, followed by discussions on the cardiac arrhythmias monitoring in Chapter 3. Human bio-kinematic monitoring using body area networks is described in Chapter 4. Ambulatory health monitoring applications are presented in Chapter 5. Hardware design, developments, and architectures are given in Chapter 6. Chapter 7 presents an overview of hardware implementation for the ambulatory health monitoring systems.

The design of sensor front-end circuits to detect biological signals is described in detail in Chapter 8. Network architecture and medium access control techniques used to design sensor networks are discussed in Chapter 9. Chapter 10 discusses the power management techniques for wireless body area networks. Chapter 11 concentrates on the radio transmission channel modeling for body-centric wireless communications. Chapter 12 focuses on antenna design and signal propagation techniques for WBAN applications. The effect of interference from other electronic systems such as Wi-Fi and microwave ovens could influence the performance of WBAN; these issues are discussed in Chapter 13. Implanted communication systems have been dealt with in Chapter 14. Chapter 15 describes the design of implantable devices that could form a sensor network by connecting with the on-body wireless nodes. The ultra-wideband (UWB) technology for both wearable and implantable WBAN applications is studied in the last chapter, Chapter 16.

We hope that this book will be a key resource for researchers and students who are working in this emerging area of the medical technology. Considering the rapid progress in the area, we firmly believe that the wireless body area network technology will play important roles in future health care and associated areas. Finally we would like to thank all the authors for their excellent contributions, which enabled us to develop a key book on wireless body area network. We also thank the publisher for delivering an important book in a timely manner for one of the most important technologies in this century.

Mehmet R. Yuce
Jamil Y. Khan

Chapter 1

Introduction to Wireless Body Area Network

Mehmet Rasit Yuce[a] and Jamil Y. Khan[b]

[a] *Department of Electrical and Computer Systems Engineering, Monash University, Clayton, VIC 3800, Australia*
[b] *School of Electrical Engineering and Computer Science, The University of Newcastle, Callaghan, NSW 2308, Australia*
Mehmet.Yuce@monash.edu; Jamil.Khan@newcastle.edu.au

1.1 Introduction

Recent advances in wireless technologies and ICT (information and communication technology) systems are enabling the health care sector to efficiently administer and deliver a range of health care services. Advanced ICT systems will be able to deliver health care services to patients not only in hospitals and medical centers, but also in their homes and workplaces, thus offering cost savings and improving the quality of life of patients. For example, Internet is currently available almost everywhere in different forms, using either cable or wireless networking technologies. With the advancement of mobile and satellite communication technologies along with broadband communication techniques, e-health services can be delivered anywhere at any time. Especially, when wireless devices are

Wireless Body Area Networks: Technology, Implementation, and Applications
Edited by Mehmet R. Yuce and Jamil Y. Khan

ISBN 978-981-4316-71-2 (Hardcover), ISBN 978-981-4241-57-1 (eBook)
www.panstanford.com

integrated with sensors, it is possible to acquire and monitor human signals at any environment at any time. Hence, Internet can be used as a major tool to deliver e-health services to both developing and developed countries. E-health services can take advantages of wireless body area network (WBAN), which can act as an enabling technology. WBAN systems could offer great advancement for a ubiquitous health care, which has the potential to improve many aspects of everyday living leading to improved quality of life of many patients. Telemedicine and e-health services could offer several significant advantages, including speedup of diagnosis, therapeutic care for emergencies, offering specialist services to remote and rural locations, and supporting patients' mobility and lifestyles [1].

A WBAN monitors physiological signals from some tiny sensors with wireless transmission capability placed either inside or around a person's body, which are used to collect important health data of a person during a particular activity — medical or sport or training-related activities. These nodes form a network between the sensors and a control device. Figure 1.1 shows a generic WBAN application scenario. Basically, a WBAN system consists of a number of tiny sensor nodes and a gateway node used to connect them to remote locations (i.e., hospital, call center) as shown in Fig. 1.1. In literature, a number of different terminologies or names have been used for the gateway device; mainly terminologies such as body control unit (BCU) or central control unit (CCU) or personnel control unit (PCU) are used. The gateway device can be a smart phone or any portable device that can aggregate collected sensor data and forward them to remote stations. The gateway node can connect the sensor nodes to a range of communication networks. These communication networks can be either a standard telecommunication network, mobile/wireless network, a dedicated medical center/hospital LAN (local area network) or a public WLAN (wireless local area network) hotspot, commonly known as the Wi-Fi.

A WBAN allows a user to store collected data in his/her PDA (personal digital assistant) or iPod or any other portable devices and then transfer those information to a suitable computer when a communication link is available. Future applications of WBAN could

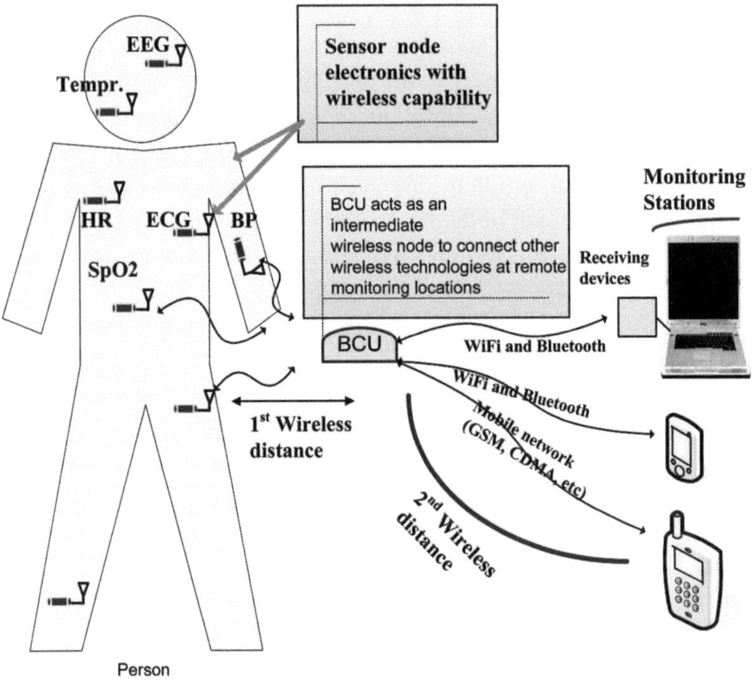

Figure 1.1. A wireless body area network scheme.

introduce numerous possibilities to improve health care and sports training facilities. In recent years, the WBAN concept has attracted the attention of medical and ICT researchers. Standard ICT systems are already in use in medical areas mostly related to patient record keeping and scheduling tasks. Some ICT systems or equipment are used to collect patient physiological data for treatments mainly using on-site wired equipment. The main drawback of the current system is the location specific nature of the system due to the use of fixed/wired systems. Introduction of WBANs opens up new possibilities in patient care and monitoring using e-health and tele-health systems. Another major area of health care where the WBAN will find its application is the aged care, where the quality of life of elderly people can be significantly improved if the health of older people is invasively monitored and appropriate health care is provided.

1.2 Applications

With the widespread use of portable devices such as mobile phones, pagers, PDAs, MP3 players, etc., the concept of collecting personal data in those devices becomes more realistic. In 1996, T. G. Zimmerman studied how such electronic devices operate on and near the human body. He used the term wireless personal area network (PAN) [2]. He characterized the human body and used it as a communication channel for intra-body communications. Later around 2001, the term PAN was modified to body area network (BAN) to represent all the applications and communications on, in, and near the body [3]. One of the most attractive applications of BAN is in medical environment to monitor physiological signals from patients [4].

In recent years, interest in the application of WBAN has grown significantly. According to ABIresearch (Allied Business Intelligence Inc.), "*The market for wireless devices that monitor patients' condition and report that data to health care providers is on the verge of explosive growth, according to a new study from ABI Research. Over the next few years, it will show a remarkable 77% compound annual growth rate (CAGR) resulting in global revenue of almost $950 million in 2014*" [5]. A WBAN based on low-cost wireless sensor network technology can significantly enhance patient monitoring systems in hospital, residential, and work environments. A WBAN system allows easy internetworking with other devices and networks, thus offering health care workers easy access to a patient's critical as well as non-critical data without or minimal manual intervention.

Currently many service providers use very basic manual/semi-automatic remote patient monitoring systems. Most of the current systems use the standard telephone network where a patient could send some medical data by manually connecting a medical sensor to its body that sends information via telephone networks. Various companies or service providers offer these services. AMAC (American Medical Alert Corp.) provides a system called health buddy, which allows patients to send medical data to health care professionals remotely [6]. Such a system requires manual intervention. According to the leading technology company Frost & Sullivan, it is expected that the remote patient monitoring technology will be deployed in the near future to provide the following medical services [7]:

- Cardiac care
- Diabetes
- Pulmonary diseases
- Pharmaceutical compliance
- Mental health
- Co-morbidities and others

Various medical equipment vendors have started to produce remote tele-health equipment to support remote patient monitoring services. One of the examples of such equipment is the TeleStation produced by Philips [8]. The TeleStation unit acts as a hub that connects to other physiological data collection devices such as weighting scale, blood pressure and pulse cuff, ECG/rhythm strip recorder and the pulse oximeter. The TeleStation collects all the measurements from different units and sends those data over the telephone line to a Philips database server. The TeleStation can also receive instructions from physicians and/or health care workers. Physicians or health care workers can request more data from patients to study or analyze their condition. The TeleStation can be seen as a gateway of a basic WBAN, which comprises measurement devices. The TeleStation solution is still not a true WBAN because in this case, patients have to carry out manual measurements using standard equipment, whereas in a true WBAN we would expect sensors attached to a human body to automatically collect physiological data and transmit them over communication interfaces. Also, the TeleStation is only capable of transmitting data over a phone line only and doesn't support mobility because of the use of bulky measurement equipment.

Recently a European vendor Broek op Langedijk of the Netherlands has developed the RS TechMedic, a portable cardiac monitor [9]. The unit has an internal GSM mobile phone interface as well as Bluetooth class I and USB connections. Such devices and systems are yet to conform to WBAN requirements; however, the trend in the tele-health/e-health sector is moving towards WBAN applications.

Aged care is one of the prime areas of WBAN applications. Aging population, shortage of medical staff, and high demand of hospital resources are problems faced in many countries around the world. A wireless medical sensor network facilitates remote monitoring system, which allows the medical and health staff to detect early symptoms of any illness or even they can use those medical data to alter medications or care pattern. Timely medical intervention significantly improves a patient's chances of recovery. WBAN-based solution will provide invasive, mobile, and cost effective services for the above applications.

A WBAN-based monitoring system can also be extended to monitor athletes' performance to assist them in their training activities. Using sensor nodes, a trainer can get both on and off the track performance data. For example, for a cricketer or a tennis player, movement of arms and body postures are very important for their success. In this case, a trainer can obtain data from a player via a WBAN and store those data in computers for further analysis.

Another exciting application of WBAN is in space and military applications. With wireless and wearable sensors, the conditions and status of astronauts and soldiers can also be monitored. Military applications of WBAN have different operating environments. In the case of military application, a WBAN will be formed around a soldier's body by interconnecting various sensors, equipment, and perhaps with different communication devices [10]. In future, deployment of WBAN may not only be limited to medical applications but may also be extended to many other applications, including training, industrial safety, military and other logistic applications. Applications of WBANs to help disabled persons could be another major application. It is worthwhile to mention here that a WBAN is a special purpose sensor network; hence, it is possible for a WBAN to accomplish a range of tasks. Further research and development work is necessary to develop future WBAN applications.

1.3 Wireless Personal Area Network (WPAN)/Wireless Local Area Network (WLAN)

Wireless personal and local area networks are low power, short range wireless networks used to establish small size networks

Table 1.1. Wireless technologies for short-range data communications

	Frequency Band	DataRate	Range	Standard	Transmission Power
Bluetooth	2.4 GHz	Up to 1 Mbps	1–100 m	IEEE 802. 15.1, WPAN	1–100 mW
ZigBee	2.4 GHz Worldwide 868/915 Europe/US	Up to 250 kbps	0–10 m	IEEE 802. 15.4 WPAN	1–10 mW
Wi-Fi	2.4 GHz , 5 GHz	Up to 400 Mbps	300 m	IEEE802.11(b/g/n)	250–1000 mW
6LoWPAN	2.4 GHz	250 kbps	$\sim$ 30m	IEEE802.15.4/IETF	$\sim$ 10mW

to exchange data. Using these technologies, it is possible to develop low-cost, small-size communication networks to support many applications, including WBAN applications. Table 1.1 lists some of the current available wireless technologies that could be used to develop small-scale wireless sensor networks for various applications. Bluetooth, ZigBee, and 6LoWPAN systems have mainly been optimized for short range (10 m) communication transmitting at low power to connect sensors and control nodes. Wi-Fi devices are generally used for long-range applications and thus consume more energy due to higher transmission power. In addition, board dimensions of the wireless platform for Wi-Fi links are larger.

Most of the current WBAN systems tend to use commercially available WPAN platforms such as ZigBee and Bluetooth as sensor nodes of a WBAN system. Bluetooth technology can be used to communicate with new smart phones. It will be important for future systems to evaluate their designs for multi-patient environment, especially devices using a wireless platform at 2.4 GHz ISM band due to the massive presence of other wireless devices and equipment operating in the same frequency band. Interference from other devices operating in the ISM band could reduce the reliability of WBANs.

With the arrival of new smart phones (e.g., iPhones), PDAs , and other pervasive wireless devices (iPods, iPads), it is now possible to use many wireless connections to transmit data from anywhere to the IT unit of a medical center. Smart phones generally have multiple wireless links, which could be either 2G/3G mobile links or satellite links or Wi-Fi connections to transfer data. Smart phones can select the most appropriate link based on availability, quality, or user preferences.

1.4 Wireless Body Area Network

A body area network is a collection of wireless sensors (i.e., sensor nodes) placed around or in a human body that are used to exchange important information from a human body to remote stations. Although WPAN devices have been used for WBAN applications, WBAN is a very small-scale network requiring a communication distance of a few meters between sensors and the control unit. For most applications, a WBAN node requires a low data transmission rate as each sensor node will transmit only one physiological signal. Physiological signals usually occupy small frequency components. Table 1.2 lists the biological signals that will most likely be the source of sensors for WBAN applications. Current WPAN standards are optimized for industrial applications. As an example, ZigBee is optimized for sensor network applications and Wi-Fi is optimized for data network with longer transmission ranges. Thus WBAN systems should be based on a different standard, which should be developed and optimized specifically for a short-range low power sensor network around a body [11]. Figure 1.2 shows the standing of WBAN among other wireless standards.

In addition to wearable vital signal monitoring sensors (ECG, temperature, heart rate, etc.), the use of implanted sensors is also

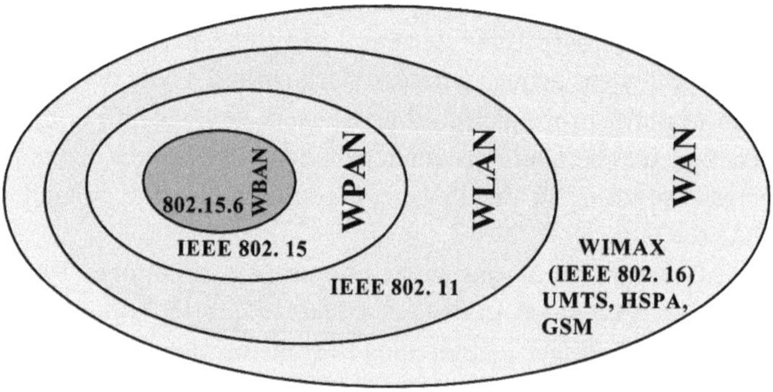

Figure 1.2. Wireless standards. See also Color Insert.

Table 1.2. List of biosensors

Wearable Sensors	Implantable
Electrocardiogram (ECG)	Pacemaker
Heart Rate	Cochlear Implants
Electromyography (EMG)	Implantable defibrillators
Electroencephalogram (EEG)	Wireless capsule endoscope (Electronic Pill)
Temperature	Electronic pill for drug delivery
Pulse oximeter	Deep brain stimulator
Blood pressure	Retina implants
Oxygen, pH value	
Glucose sensor	
Movement (accelerometer)	

increasing. Some examples are wireless capsule endoscope, cochlear implant, and implantable defibrillators. Among the biosensors listed in Table 1.2, except for retina implants and electronic pills, the remaining sensors require low data transmission rates. The electronic pill requires the highest data transmission rate, around 10 Mbps, to transmit good quality video from inside the body to a monitoring device.

In order to monitor the inner organs as well as the status of medical implants such as pacemakers and defibrillators, a frequency around 400 MHz has been used as a popular transmission band for recent systems [12]. To treat a large number of patients wearing implanted systems in the same environment (e.g., hospital), a reliable wireless networking is required to monitor and differentiate each individual implanted device and patient. Thus implanted wireless nodes in a patient's body should form a wireless body area network so that one or more implanted devices inserted in the bodies of a number of patients in a hospital environment can be controlled with minimum complexity. It is quite possible that for some applications both implantable and wearable sensors form a WBAN system.

A WBAN is mostly likely to incorporate wearable and implantable nodes operating in two different frequencies. An implantable node is most likely to operate at 400 MHz using the MICS (Medical Implantable Communication Service) band, whereas the wearable node could operate in ISM/UWB (Instrumentation Scientific Medical/Ultra Wide Band) or some other specific bands [13].

Both implantable and wearable sensor nodes require the use of a gateway-body control unit (BCU) to communicate with remote monitoring stations as depicted in Fig. 1.1.

A WBAN system can interface with or accommodate different wireless platforms. As an example, a multilink WBAN can have multiple wireless links. Multiple wireless links could consist of the standard ZigBee or Bluetooth to cover WPAN areas and a 802.11 based Wi-Fi link to cover a larger transmission range possibly to connect to a remote station or a database server [14, 15]. Different configurations of WBAN nodes are illustrated in Fig. 1.3.

Using WBAN nodes after obtaining raw data from a human body, sensor nodes transmit those data to the control device — BCU

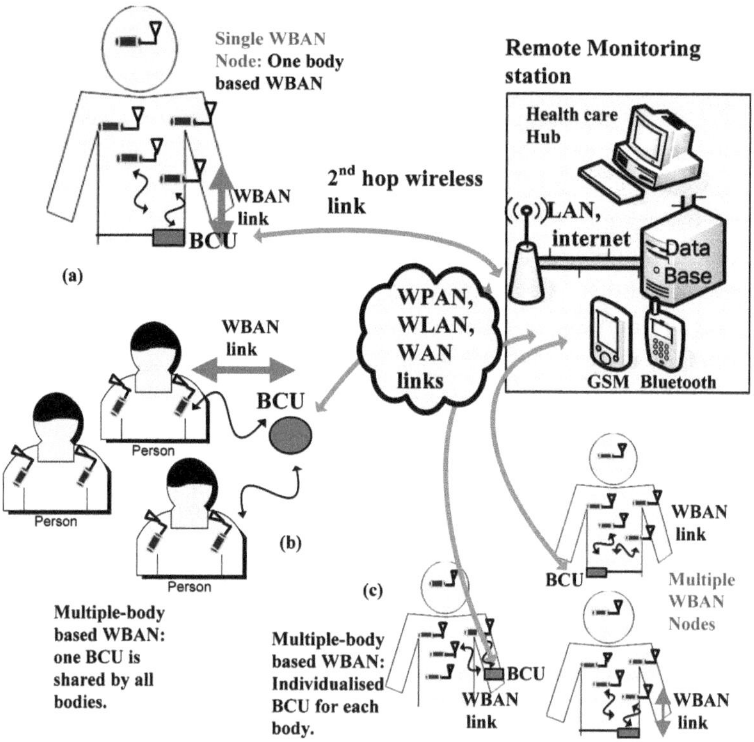

Figure 1.3. Wireless body area network implementation scenarios: (a) single WBAN node, (b) a WBAN node with multiple bodies, and (c) multiple WBAN nodes. See also Color Insert.

— over a short-distance wirelessly using one of the short-range wireless systems. The control device can be placed on the body like a mobile phone as shown in Fig. 1.3a, or it can be placed at an accessible location. Most suitable technologies for this link are ZigBee, Bluetooth, 6loWPAN, and WMTS. The main task of the BCU is to transfer data to a PC or to a smart phone. Wireless technologies used on this segment (BCU to PC or to an Internet device) could be a mobile communication network, a satellite link, or a Wi-Fi link. When the Internet is used, the data collected at this PC can be transferred to remote stations in remote medical centers across the network.

As described in Fig. 1.3, the collected sensor data can easily be transferred to remote stations (i.e., medical centers) with the existing wireless and information technology infrastructures such as satellite, mobile communication system, Internet, etc. Accessing the medical data of injured people through Internet is an effective solution at the moment, which will allow medical professionals at the hospital to collect and evaluate data while patients are being transferred to the hospital in an emergency vehicle. If provisioned, these data can also be accessed outside the emergency areas as they will be made available online.

The control device will be similar to smart phones we use in our daily life to receive and monitor the data obtained from sensors. They will be like minicomputers, which will most likely be connected via a wireless technology such as Bluetooth, Wi-Fi, 3G/2G networks or the satellite (VSAT: very small aperture terminal). These technologies offer flexible communication links that can be configured to send data from medical sensors to remote medical centers (second wireless link in Fig. 1.1 and Fig. 1.3), which can be accessed by medical professions at any time. The BCU can be attached to the body as a wristwatch or around belt.

The first scenario sharing a single WBAN node can be used for home care where there is need for only one body monitoring. Scenarios presented in Fig. 1.3a,b can be used at a disaster area, in emergency rooms in hospitals, and in ambulance while patients are taken to the hospital.

Above discussion shows that WBAN is a specialized sensor network with definite application requirements. For the commercial deployment of WBANs, it is necessary to develop an industrial standard by considering its different application scenarios and requirements.

1.5 Design Requirements

Design of miniaturized, low-power, reliable, and wearable sensor node devices is the key requirement of a WBAN design. A WBAN is a special purpose wireless sensor network designed to transmit data within a very short distance. The quality of service (QoS) of a WBAN will depend on applications, which are primarily medical applications. As discussed earlier, WBANs could be used in other applications such as sports training, rehabilitation purposes, or military applications where the QoS will be dictated by the nature of those applications. For medical applications where the source of information is physiological signals such as heart rate, blood pressure, ECG, etc., the system generally demands low latency and high reliability. Also, most of the physiological signal sources produce shorter data bursts at a regular sampling frequency. Hence, a WBAN transmission data rate requirement is low to medium for each connection. However, one should keep in mind that a WBAN will comprise multiple sensor nodes where data transmission should be coordinated by using a reliable and efficient medium access control protocol [4]. Whatever may be the application of a WBAN, the main design requirement of a WBAN is to develop a short-range and reliable wireless sensor network.

A WBAN may need to send data over a longer distance or may need to support mobility. The range of a WBAN can be extended by using a multi-hop network using a gateway or a router, which could connect a WBAN to external networks. The range of a WBAN network should not affect the sensor node designs because the gateway will be responsible for seamless long distance services; hence, a modular design approach should be taken when a long distance WBAN application design is developed. Similarly, the mobility feature of a WBAN will be taken care by the gateway or a router, thus reducing the design complexity of sensor nodes. The gateway should be able to detect movement of a WBAN or a person and connect itself to available external networks to exchange information with the WBAN. The mobility speed support will be determined by the external network. For example, a Wi-Fi can only support very low speed mobility up to 5 km/h, whereas a 2G/3G network could support mobility up to 240 km/h. If we are looking for mobility within a hospital or inside a home, then a Wi-Fi-type connection could be sufficient, whereas for a sport training application, it may be necessary to use a 2G/3G network to obtain data from the sensors located at an athlete's body. So the WBAN design requirements can be classified into basic and advanced requirements. Advance design requirements will be influenced by the design requirements of the applications that a WBAN is used. The basic WBAN design attributes are summarized below:

- WBAN sensor nodes should be able to transfer data over a distance of a few meters using a single hop connection.
- Sensor nodes should be miniaturized so that they can be easily wearable.
- Suitable frequency bands should be selected to reduce interference and thus increase the coexistence possibility of sensor node devices with other network devices available at medical centers or in homes.
- WBAN nodes should consume low power so that battery could last for a very long time. Also, WBAN nodes should use small size batteries so that light-weight nodes can be used.
- A WBAN should be designed in a fail-safe manner so that failure of a node can be automatically detected or failure of a node should not affect the operation of the network.
- A WBAN should be scalable so that health care workers can increase or decrease the number of nodes on a patient's body without any manual intervention of IT personnel.

To achieve the above design attributes, WBAN designers should consider the following points in their design work:

Wireless sensors: A transmitter circuit can be designed with a few components, which may consume extremely low power when designed with an integrated circuit technology. Appropriate sleep and wake-up cycles should be incorporated with the designed hardware to reduce the power consumption to a minimum level.

Reliable data communication: Reliable, error free, and robust information should be received from sensors. A WBAN will use a wireless channel to transmit data, which is inherently unreliable. Error checking and correction mechanisms should be incorporated so that the unreliability of the transmission channels can be countered. The wireless technologies used in a medical environment should operate in the frequency bands that are **immune to interference and thus increase the coexistence of** sensor node devices with other network devices available at the same location. The wireless technology used should have less interference effect on other medical equipment. Any network outage should be automatically detected, and the sensor data should be delivered in a fail-safe manner, which could be a critical requirement for a patient monitoring system.

Wireless network security and privacy: Key software components should be defined and developed to accommodate secure and effective wireless networking. Protocols should be designed in such a manner that a WBAN data cannot be collected by intruders. A WBAN will use a wireless transmission channel, which opens up the possibility of external intrusion while transmitting data. Data encryption techniques could be used, but designers should keep in their mind that undue complexities should not be introduced at the sensor node level to avoid higher battery power and larger physical size.

Handover mechanism: Handover mechanism should be integrated in a WBAN using the gateway or a router. Handover features should not overload the sensor node design.

Miniaturized antenna: Unobtrusive small antenna design should be used, which will operate at high frequencies. Directional or narrow beam antenna design could be considered for specific medical applications.

Gateway devices: A WBAN node may have a wearable or implantable node. Gateway devices should be developed to interface with the existing wireless networks used in health care systems. Gateway devices should implement advanced algorithms, which are more power hungry so that sensor node design and requirements remain simple.

Alarm option: An alarm option should be included when an outage occurs or a sensor node fails.

Comfort: Sensor node electronics could be designed using flexible and stretchable technology so that sensor nodes can easily be embedded in textiles (i.e., patient's clothes). It can be attached to the human body using a plaster to eliminate movements.

1.6 Scope of the Book

This book addresses the applications, hardware, and software design of WBANs. It informs readers about some of the current and possible future applications of WBAN-based monitoring systems. A number of chapters have been devoted to WBAN applications as mentioned in the preface of the book. Following the application chapters, readers are presented with hardware and software design techniques, which can be utilized to develop WBAN systems for medical applications. Hardware design techniques and system level developments are discussed in detail. An important component of the WBAN is the antenna design, which is also introduced in a separate chapter. Protocol and network design techniques are discussed in several chapters, which will provide readers clear understanding of WBAN network requirements. It is important to understand protocol and network design issues because the efficiency and the performance of a WBAN system could be largely determined by them. The book also introduces various current industrial standards, which will inform readers about possible choice of hardware and software system selections. As mentioned in this chapter, a number of industrial wireless standards exist, which could be followed to develop a WBAN by procuring and configuring them in a suitable manner. Signal transmission around and in-body, in-body communication techniques, and the design of implantable sensor nodes are also described in individual chapters. The book also presents a chapter on coexistence issues. This is an important issue because nowadays many other wireless devices are used in many locations in hospitals and homes; hence, it is necessary to understand the impact of other wireless devices and choice of appropriate frequency bands. In conclusion, the book provides a comprehensive coverage of WBAN applications, design, and deployment techniques.

References

1. ETHEL, http://www.e-wwg.com/Publications/eHealth/EHTEL%20Briefing%20Paper%20Sustainable%20Telemedicine.pdf.
2. Zimmerman, T. G. (1996) Personal Area Networks: Near-Field Intrabody Communication, *IBM Systems Journal*, 35(3–4), 609–617.
3. Dam, K. V., Pitchers, S., and Barnard, M. (May, 10–11, 2001) From PAN to BAN: Why Body Area Networks?, in *Proceedings of the Wireless World Research Forum (WWRF) Second Meeting*, Nokia Research Centre, Helsinki, Finland.
4. Khan, J. Y., and Yuce, M. R. (2010). Wireless Body Area Network (WBAN) for Medical Applications, New Developments, in *Biomedical Engineering* (ed. Domenico Campolo), ISBN: 978-953-7619-57-2, InTech.
5. http://www.abiresearch.com/press/3244-Healthcare.
6. Health Buddy, http://www.amac.com/remote_patient_monitoring.cfm.
7. http://www.frost.com/prod/servlet/market-insight-top.pag?docid=18086184.
8. TeleStation, http://www.healthcare.philips.com/in/products/telehealth/products/telestation.wpd.
9. RSTechMedic, http://www.rstechmedic.com/.
10. Cotton, S. L., Scanlon, W. G. and Madahar, B.K. (2009) Millimeter-Wave Soldier-to-Soldier Communications for Covert Battlefield Operations, *IEEE Communications Magazine*, 47(10), 72–81.
11. http://www.ieee802.org/15/pub/TG6.html.
12. Bradley, P. D. (2007) Implantable Ultralow-Power Radio Chip Facilitates In-Body Communications, *RF Design*, 20–24.

13. Hanna, S. (2009) Regulations and Standards for Wireless Medical Applications, *ISMICT*.
14. Istepanian, R. S. H., Jovanov, E., and Zhang, Y.T., (December 2004) Guest Editorial Introduction to the Special Section on M-Health: Beyond Seamless Mobility and Global Wireless Health-Care Connectivity, *IEEE Transactions on Information Technology in Biomedicine*, 8(4), 405–414.
15. Jovanov, E., Milenkovic, A., Otto, C., and de Groen, P.C., (March 2005) A Wireless Body Area Network of Intelligent Motion Sensors for Computer Assisted Physical Rehabilitation, *Journal of NeuroEngineering and Rehabilitation*, 2 (6), doi: 10.1186/1743-0003-2-6.

Chapter 2

Wireless Patient Monitoring in a Clinical Setting

Esteban J. Pino[a], Dorothy Curtis[b], Tom O. Stair[c], John V. Guttag[b], and Lucila Ohno-Machado[d]

[a]*Electrical Engineering Department, Universidad de Concepción, Concepción, Chile*
[b]*Computer Science and Artificial Intelligence Laboratory, Massachusetts Institute of Technology, Cambridge, Massachusetts, USA*
[c]*Department of Emergency Medicine, Brigham and Womens Hospital, Boston, Massachusetts, USA*
[d]*Division of Biomedical Informatics, Department of Medicine, University of California San Diego, La Jolla, California, USA*
estebanpino@udec.cl

This chapter presents the technology, implementation, and application of a wireless patient monitoring system for triage support. The SMART project was developed jointly at Brigham and Women's Hospital in Boston, MA and at the Massachusetts Institute of Technology, Cambridge, MA. It was deployed and evaluated at the Brigham and Women's Hospital over an 18-month period. We present a discussion of similar projects, all involving wireless patient monitoring, and then describe the SMART system design and the results of our experience with its deployment in the waiting area of the Emergency Department of the Brigham and Women's Hospital.

Wireless Body Area Networks: Technology, Implementation, and Applications
Edited by Mehmet R. Yuce and Jamil Y. Khan

ISBN 978-981-4316-71-2 (Hardcover), ISBN 978-981-4241-57-1 (eBook)
www.panstanford.com

2.1 Introduction

According to World Health Organization data, globally there are about 1.4 physicians per 1000 inhabitants [1]. In most developed countries, the ratio is only slightly better: 2 to 4 physicians per 1000 inhabitants. It is only natural that patients far outnumber care providers, particularly during peak demands. In large overcrowded settings, patient oversight becomes increasingly difficult, such as in waiting rooms during seasonal outbreaks, at improvised first-care centers or in other nonstandard settings. To manage a large concentration of patients, a monitoring system is key for ensuring high quality of care.

In health care settings, a wireless body area network (WBAN) can be used to collect patient-relevant information instead of having a caregiver constantly monitoring a group of patients. A wireless, automated data collection system provides several benefits:

- Ease of data collection with minimum patient discomfort
- Scalability to a large number of patients
- Real-time monitoring and assessment of changing physiological conditions
- Higher patient-to-caregiver ratio possible, while maintaining the same quality of care (better human resource utilization)

In current scenarios, caregivers are forced to collect patient data on a case-by-case basis. Bedside data collection is only used after patients have been admitted to the hospital. Collecting data is essential to support good medical decisions. In emergency departments, the first set of data is collected during triage, where an initial assessment of the patient status is performed. However, a single snapshot of the patient vitals may lead to misprioritization. Furthermore, it is very difficult to notice any changes in patient status after the triage process has been completed.

In terms of scalability, caregiver–centered systems can easily collapse during peak demand. A patient–centered system has obvious advantages. Such a system scales naturally as more nodes are added to monitor more patients. The only constraint is an eventual overload at the network layer or at the monitoring station. To this end, as much computation as possible should be programmed into the WBAN components.

In order to assist in the monitoring of multiple patients, we developed SMART (Scalable Medical Alert and Response Technology). [2–4] SMART implements a patient-based wireless monitoring system as an aid for multiple patient scenarios. It can help first responders in emergency situations by providing a fast on–line triage system requiring little set-up time. In overloaded waiting rooms, it can help in re-prioritization of patients. In nursing homes, it can facilitate the supervision of a large number of people, without degrading their quality of life. SMART provides a steady flow of data from the patients and a real-time alarm system to prompt re-triage in the case where a patient's condition deteriorates.

Physiological signals such as electrocardiogram (ECG) and pulse oximetry (SpO_2) are simultaneously acquired from different patients and processed at a central server where a single operator can oversee a large number of patients. Location information is also collected wirelessly to facilitate locating a particular patient who needs attention.

Similar projects originate from the military. The Artemis project [5] uses a WBAN to evaluate the physiological state of soldiers in the field. A fuzzy logic based algorithm [6] gathers information from the SpO_2 sensor to trigger a distress signal or it can be initiated by a fellow team member. The distress signal is forwarded to medical personnel, along with location information provided by GPS, to guide the caregivers to the injured soldier. Konoske et al. present an evaluation of a Mobile Medical Monitor (M3) [7] for soldier care. The mobile unit is very complete, incorporating an assortment of physiological sensors. However, its main drawback and complaint among users is its inability to supervise multiple patients and save the data in relational databases, features SMART can support.

There are many proposals from academic institutions [8] for pervasive, wireless response systems. They all implement solutions for multiple patient monitoring. In the WIISARD project, a different solution is proposed for mass casualty incident response [9], with a clear focus on logistics. Mobile caregivers are equipped with wireless PDAs to evaluate patients in the field and relay that information to a central command post. Patients only had basic online physiological monitoring, unlike SMART, preventing adequate detection of adverse events. An intelligent triage tag is proposed for WIISARD [10] as a replacement for the paper tag, mainly for notification and prioritization in the field. The wireless infrastructure [11] is provided using commercially available equipment.

AID-N project, [12] an extension of CodeBlue, [13] presents a mobile patient monitoring prototype for an electronic triage system. The sensors are similar to SMART: SpO_2, blood pressure, ECG and GPS/MoteTrack for outdoors/indoors location. The primary goal is triage management for mass casualty events. [14, 15] To this end, a central station collects data from mobile units, caregivers carry wireless PDAs to communicate with the server, and patients are equipped with physiological sensors and visual and audible alarms to alert caregivers to emergent conditions. A survey conducted after preliminary usability tests shows that the most important feature for caregivers is triaging support. SMART also provides triaging support in the main server, alerts and location information. A difference between AID-N and SMART is that our system is easily extensible to regular settings, such as a waiting room, helping ease the learning curve involved in using a new system.

Another application of WBAN systems is concerned with patient monitoring during transport [16] among units in a hospital. In this case, ECG and SpO_2 sensors are used, but no location information is necessary. The patient equipment is similar to SMART, based on an iPaq H5450. Another pre-hospital patient monitoring system [17] shows that body sensor networks can also prevent over- or under-triage of trauma patients. The authors propose an iPaq–based system that records eight different vital signs during transport to improve triage. They finally conclude that SpO_2 and ECG heart rate (HR) are a better predictors of the need for abdominal surgery than reports from pre-hospital clinicians. A low SpO_2 and high HR combination is a good predictor for internal bleeding injury. A smart vest can also be used as support for WBAN. [18] In this case, sensors are embedded into a vest or T-shirt. Wires are woven into the fabric and data and power are transmitted to and from a main processing hardware with wireless capability. Sensors used in the smart vest acquire ECG, photoplethysmogram (PPG), blood pressure (BP), skin temperature, galvanic skin response (GSR), and ECG–derived HR.

A project conducted at Pennsylvania Hospital [19] evaluated an ECG-based alarm system. Using a commercially available wireless ECG, the authors were able to assess the validity of ventricular fibrillation, asystole, tachycardia, or bradycardia alarms. The authors reported high patient and clinician satisfaction, even though there were more false alarms than useful alarms, in line with SMART's findings. Their system did not provide an automated location system, although the authors acknowledge its importance.

A complete survey of wearable medical monitors is discussed by Raskovic et al. [20] Issues, current proposed systems, and a general description of different systems and mobile monitoring implementations are presented by these authors.

2.2 Smart System

SMART is designed as a mass casualty response system, but is also easily used in everyday environments. The rationale is that a system that is used regularly by physicians and nurses will have a much easier transition into use in a disaster situation. In SMART, the wireless component allows the subject to move freely, and has little negative impact on their regular behavior. Furthermore, the patients perceive a higher sense of care as shown by our post-study survey. In terms of medical performance, the system helped detect several cases that would have otherwise gone undetected.

The SMART patient monitoring system was deployed in the waiting room of the Emergency Department at the Brigham and Women's Hospital in Boston, MA. The system was operational from June 2006 to December 2007, collecting 6815 minutes of data from 145 patients.

2.2.1 *Architecture*

2.2.1.1 Hardware

On the patient, physiological and location data are collected and wirelessly sent to a central server. The physiological sensors measure ECG and SpO_2 and are wired to a PDA that acts as a sensor hub and transmits the data via standard 802.11b Wi-Fi network. Location is implemented via ultrasound tags and detectors that can pinpoint a particular patient to a predetermined area. Privacy issues are considered and patient-sensitive data is encrypted prior to transmission to caregiver PDAs. Physiological data transmitted from the patients to the server is not encrypted because it contains no identifying information.

The ECG sensor's printed circuit board is the only custom-made hardware in the project. It implements an instrumentation amplifier with fixed gain and filters. An 8-bit analog-to-digital converter (ADC) samples the signal at 200 Hz and sends the data using RS-232 serial communication. For SpO_2 sensing, a commercial sensor from Nonin Medical Inc. [21] is used. This sensor reports status, SpO_2 level, and heart rate every second, also via serial interface. As shown in Fig. 2.1, both sensors arrive at a sensor box, connected to a standard PDA acting as a wireless node. The HP iPaq 5500 is chosen because it provides Wi-Fi connectivity, extended battery life, and serial ports. Having a PDA as the basic building block was useful in configuration and debugging tasks.

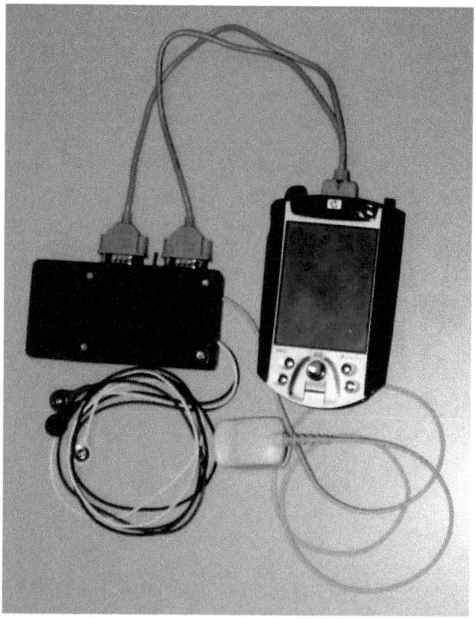

Figure 2.1. Patient PDA with ECG leads and SpO_2 sensor. See also Color Insert.

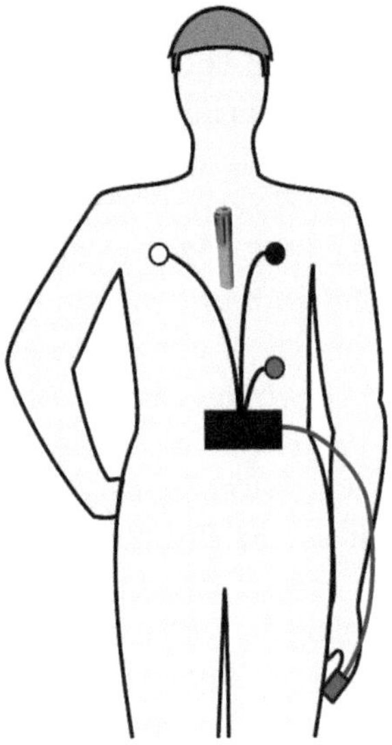

Figure 2.2. Patient with ECG, SpO_2 and location sensor.

Location data is provided by an ultrasonic (US) system from Sonitor Technologies. [22] US tags attached to patients transmit a unique code that is read by US receivers mounted on the walls. A simple signal strength algorithm assigns a tag to its closest receiver. The advantage of the US system is that the signal is confined by walls. This way, there is absolute certainty that a patient is in a given room. Large rooms can be subdivided by properly placing the receivers. The tags are pen-like and are usually worn on a lanyard around the neck. Figure 2.2 shows a diagram of the complete gear worn by the monitored patients.

Fourteen US detectors are installed covering the complete waiting area and the overflow area. By having detectors outside the waiting area, we can infer when a patient has left the waiting room and their general direction.

SMART also implements a caregiver wireless node. Caregivers can be equipped with a hand-held PDA and a location sensor. This way they can receive and respond to alarms forwarded to their PDA, check the status of a patient, and the server can know their location.

The server hardware is a standard PC with dual monitors, wired to a wireless router. ECG and SpO_2 data arrive wirelessly from the patients to the router. Location information is also received from the US detectors via wireless adapters. Only three detectors are wired to the router, due to their close proximity to the server. All wireless communications use the IEEE 802.11b standard.

The maximum bandwidth provided by the 802.11b network is 11 Mbps, which was enough for our needs. A simple calculation sets the requirements for 10 wireless nodes (patients and caregivers) at under 0.5 Mbps. [4] There are some other technical aspects to consider when using wireless networks. First, the available bandwidth is reduced as the distance to the wireless router or access point increases. Proper placement of the wireless router can minimize the number of nodes with reduced data rate, while serving the majority of the nodes at higher data rates. Second, wiring the server to the router reduces by half the number of wireless data packets on the network. Finally, radio interference can be avoided by proper coordination of channel usage in the hospital. In our particular case, our wireless network used a different channel than the existing network, and we did not experience any problems on either network.

2.2.1.2 Software

Different programming languages are used in SMART according to the task and platform, as shown in Fig. 2.3 For the mobile nodes, the

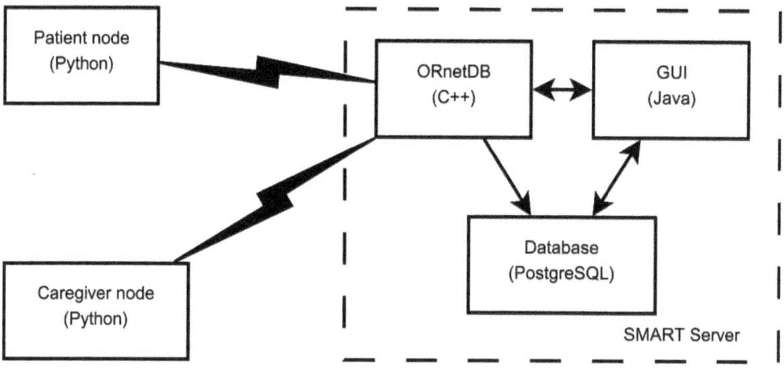

Figure 2.3. Software and data flow in SMART.

language of choice was Python. The PDA was running a linux-based OS, Familiar Linux, that provided better control over the hardware, and an open source platform. There are two Python programs, the data-collecting and forwarding program running on the patient PDA, and the data-fetching program running on the caregiver PDA. The patient PDA program is not intended to be used by patients. It provides a simple configuration screen to specify serial port usage and a "live" data screen used in the early stages of the project for debugging. Data are identified as ECG or SpO_2, timestamped and sent via TCP to the server. To reduce the wireless traffic, 20 samples of ECG data are packed together and sent every 0.1 s. Individual patient data streams are identified according to their originating PDA identification number. The caregiver PDA program displays a roster of patients being monitored and their main vital signs. It also has the ability to query the server for data for a particular patient and show live or retrospective ECG and SpO_2 data. This communication is encrypted to protect patient privacy, since at this point patient names are associated with the data. Both programs present a simple user interface (UI) programmed using Glade and GTK+ for configuration and monitoring.

At the server, data first arrive at ORnetDB. ORnetDB was developed in a previous project as an operating room streaming database. [23] This database is able to log and perform computations on live feeds of incoming data. Raw data and computed data such as QRS positions are then saved into a standard relational database, implemented in PostgreSQL. [24] Finally, the main UI for the operator is programmed in Java. The operator has an overview of all patients being monitored and can select one in particular to view detailed information (Fig. 2.4). There is also a list of issued alarms for all patients being monitored. The patient roster on the top of the screen shows: patient number, in/out status, name, SpO_2 and HR from the finger sensor, calculated HR from the ECG data, the Emergency Severity Index (ESI) [25, 26] assigned upon admission, the algorithm's proposed diagnosis, location and current alarm. Patient management (enrolling in monitoring, and completion of monitoring) is handled via web page forms. Location information is shown on a secondary monitor (Fig. 2.5) with colored tags for the patients and the caregivers, displayed over a map of the waiting area.

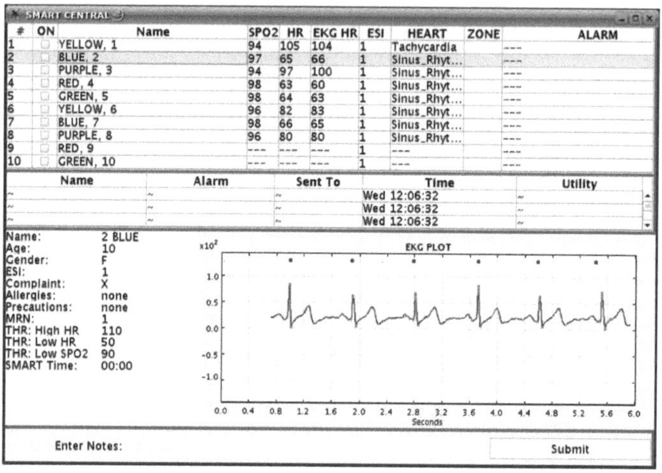

Figure 2.4. SMART central main GUI. See also Color Insert.

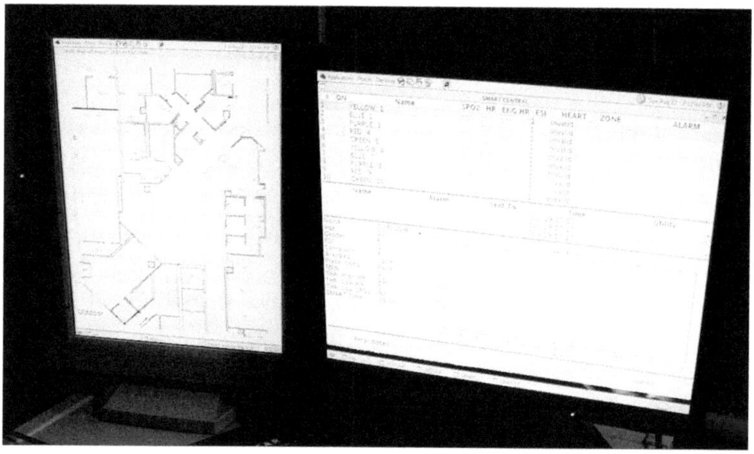

Figure 2.5. SMART dual monitor display. Left: location data. Right: main GUI.

2.2.2 *Clinical Implementation*

The goal of SMART is to enhance response in the presence of a high number of patients. Introducing a large number of wireless monitoring nodes requires that the system assist in the detection of critical cases. This approach allows scalability with minimal requirements of extra-trained personnel. To this end, SMART implements a set of parameter computations and a decision tree to assist in alarm generation and diagnosis suggestion. To produce a diagnosis algorithm that will feed the alarming system, several steps were necessary. [3] We selected an ECG processing algorithm based on its ability to handle noisy data. 2) The algorithm was tested against standard ECG databases publicly available from Physionet. [27] We conducted tests on healthy volunteers to gain experience handling live and untethered patients. 4) We conducted tests on patient simulators to get feedback on how the system responded to different health conditions.

The final algorithm is programmed based on all the tests and preliminary results. [3] The ECG algorithm is based on the SQRS algorithm available on the Physionet website. The modifications introduced include an auto-calibration interval to account for differences in electrode placement and patient variability, an adaptive threshold to handle noisy segments, and a "no beat" output after 3 s of inactivity. The diagnosis suggestion is based on detected QRS positions and raw ECG statistical properties, sensor integration with SpO_2, and noise detection. All possible diagnoses and alarms generated are shown in Table 2.1.

Sinus rhythm, bradycardia, and tachycardia are determined by simple HR comparison to patient specific thresholds. By default, the low HR threshold is 60 beats per minute (BPM) and the

Table 2.1. Diagnosis suggestions and alarm generation

Diagnosis	Source	Validation	Alarm
Sinus rhythm	ECG or SpO_2	-	None
Bradycardia	ECG or SpO_2	-	Medical
Tachycardia	ECG or SpO_2	-	Medical
Ventricular tachycardia	ECG	-	Medical
Ventricular fibrillation	ECG	SpO_2	Medical
Irregular rhythm	ECG	-	Medical
Asystole	ECG	SpO_2	Medical
Leads off	ECG	-	Technical
Noise	ECG	-	Technical
Mismatch	ECG & SpO_2	-	Technical

high HR threshold is 100 BPM. The HR is either calculated from the detected QRS waves in the ECG or reported by the SpO_2 sensor. Ventricular tachycardia, fibrillation, and irregular rhythm can only be computed from the ECG data, since they depend on QRS width, ECG waveform, and QRS periodicity, respectively. Asystole status is determined when no QRS complexes can be found in the ECG.

Technical alarms are generated when an abnormal situation most likely related to the PDA or sensors (and not the patient) is detected. Leads off is easily detected from the ECG waveform. Noise is indicated by the QRS algorithm when a series of QRS-like peaks are detected in a short period. Finally, the mismatch technical alarm is issued when the ECG and SpO_2 data are contradictory.

The final decision tree programmed in SMART is shown in Fig. 2.6 The diagnosis with the most votes in the preceding 15 s is used as the final suggested diagnosis and alarms are issued for abnormal conditions.

Sensor integration is fundamental for alarm generation. While SpO_2 data proved to be extremely stable, it can't detect certain conditions of interest. ECG provides a quicker response time and better diagnosis capability. However, ECG is highly affected by noise from electrode movements and muscle activity. The solution is to validate preliminary ECG findings with SpO_2 information. For instance, an ECG finding of asystole is not possible with a normal O_2 saturation and reported HR greater than zero.

No major problems were found during the clinical implementation. However, it is clear that if the number of patients were to increase considerably, the load on the main server might become unsustainable. One proposed solution is to push most of the load to the mobile units. This was tested in the early stages of the project but later discarded in favor of a simpler, centralized system. However, all signal processing and alarm generation can run on the patient PDAs. This way, the main server is responsible for only minor tasks. Alternatively, SMART Central could be implemented using multiple computers that share the load as the number of patient nodes increases.

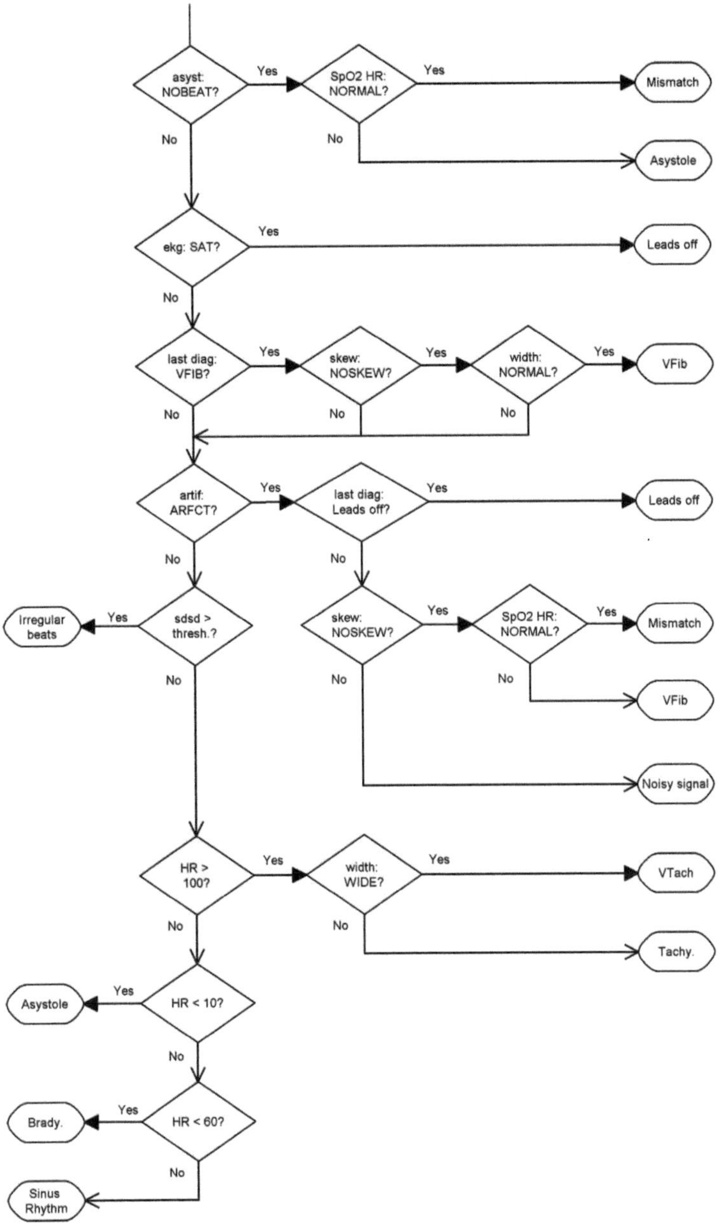

Figure 2.6. Decision tree programmed in SMART for diagnosis suggestion and alarm generation.

2.3 Results

During the implementation in the Emergency Department (ED), some important lessons were learned. Signal quality was worse than in the healthy volunteer training session data. Wireless, nonintrusive systems must be prepared to deal with bad quality data. Despite this, several patients and the system as a whole benefited from SMART. It was possible to detect unnoticed medical conditions and to dismiss a medical complaint sooner by using the data collected from patients in the waiting room. The critical cases were detected by SMART Central and reported to the SMART Operator (whose presence was required by the hospital's Institutional Review Board) and ED personnel were alerted when necessary.

2.3.1 *Medical Usefulness*

ECG, SpO_2, and location data were collected from 145 patients during 18 months. The amount of data collected per patient depends how long the patient was in the waiting room, prior to being seen by a physician. This time ranges from 5 min to slightly over 3 h. Only patients presenting with symptoms of shortness of breath and chest pain were approached for enrollment in the study. This selection was based on which symptoms might deteriorate quickly, resulting in a negative outcome for the patient, which determined the set of physiological sensors available in SMART.

The physiological data collected show that in normal conditions the system works. However, there is an important difference between real patients and simulators or healthy volunteers. The noise level encountered on real patients is much higher than expected, but relatively short-lived. This produces a significant number of false alarms that quickly become annoying to the operator. This usually happens on particular patients for different reasons, the most recurring one being restlessness. However, most patients present a normal, quiet ECG tracing with minimal number of alarms. Setting up and removing the equipment from the patients also causes false alarms. To alleviate this problem, an option is provided in the main GUI to disable alarms for particular patients.

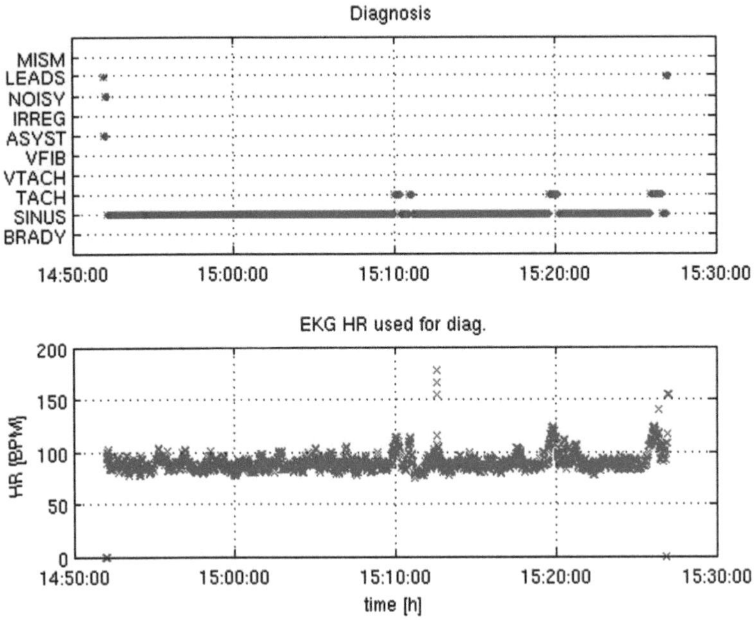

Figure 2.7. Diagnosis estimation and ECG-computed HR of an average patient. See also Color Insert.

This way, the alarm system is not enabled until the patient has settled. Figure 2.7 shows a typical tracing and its accompanying diagnosis suggestion.

During the evaluation period, three patients presented health problems detected by SMART that required re-prioritization. The first case was a patient with premature ventricular contractions that SMART detected as irregular rhythm and tachycardia. The second case presented a severe bradycardia. In the third case, the ECG-reported HR was significantly different than the SpO_2- reported HR. Upon closer examination of the live ECG acquired by SMART, the patient was admitted sooner to the ED, presenting with junctional tachycardia.

The last case worth mentioning is a patient who complained that her pacemaker wasn't working properly. Figure 2.8 shows the ECG HR and SpO_2 HR and the ECG tracing on the bottom. Every time the HR dropped below 60 BPM, the pacemaker is engaged and produces

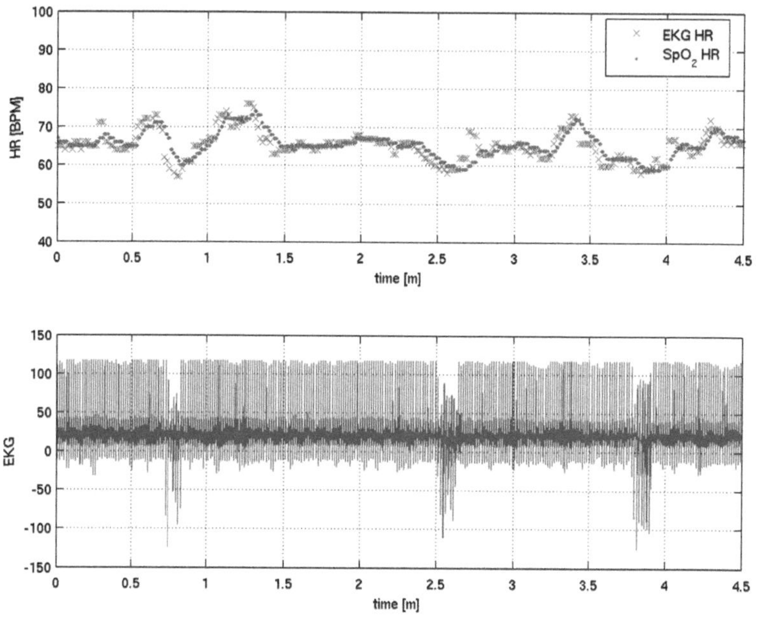

Figure 2.8. ECG – SpO_2 HR comparison and ECG pacemaker activity when patient HR dropped below 60 BPM. See also Color Insert.

a different tracing on the ECG. An added benefit of SMART is that the physicians are presented with valuable information by the time they get to see the patient for the first time, helping their diagnosis.

2.3.2 *User Acceptance*

Acceptance among the patients was very good. Most patients felt safer as they knew they were being monitored. For them, it was a large improvement over just waiting to be seen by a physician. Knowing that their location was known also helped build confidence in case they needed prompt care.

A survey was conducted among the patients right after they were admitted into the ED. Seventy patients completed the survey that evaluated their perception of the system. Figure 2.9 shows that 51% of the patients would definitively wear the system again. Another 41% would be inclined to do so. Only 7% are against or did not respond to the question. To the question "Did the monitoring system

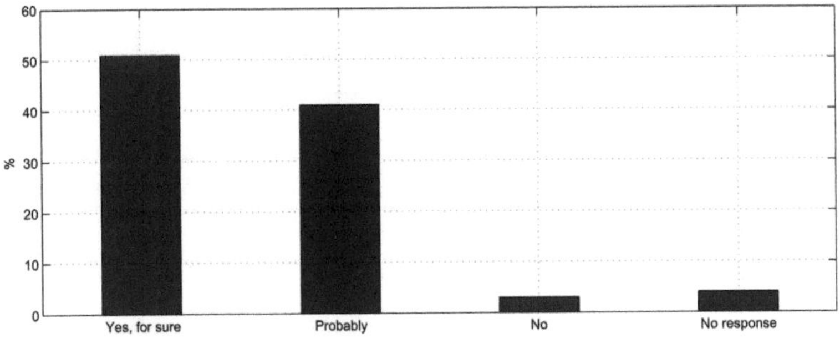

Figure 2.9. Response to survey question: Would you wear a SMART pouch again?

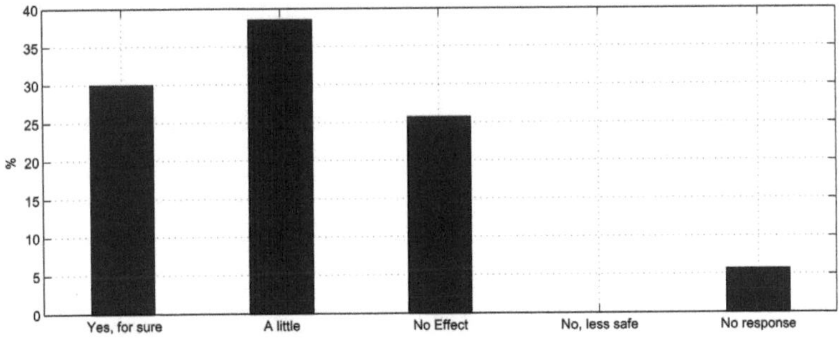

Figure 2.10. Response to survey question: Did the monitoring system make you feel safer?

make you feel safer?," over 69% of the patients answered positively, as shown in Fig. 2.10

Even though the patients who participated in this study were told that the system would not change the attention provided by the hospital, Table 2.2 shows that most of them felt that they were getting a higher level of care.

We also asked the patients about their perception of the value of having their location and their vital signs being monitored. We were concerned that they might find the location sensor too intrusive. However, 84.3% considered the location information valuable and 94.3% considered vital sign monitoring valuable (Table 2.3).

Finally, a majority of the patients found the system acceptably comfortable to wear (Table 2.4).

Table 2.2. Perceived effect on care

Effect of SMART monitoring on care	Preferences
Improved it a lot	17%
Improved it a little	24%
No effect	49%
Impaired it a little	0%
Impaired it a lot	0%

Table 2.3. Perceived value of monitoring

Category	Location	Vitals
Very important	68.6%	78.6%
A little	15.7%	15.7%
Not important	12.9%	5.7%
Not important and actually annoying	0.0%	0.0%
No response	2.8%	0.0%

Table 2.4. Patient evaluation of system comfort

Was the monitoring system comfortable?	Preferences
Didn't bother at all	82.9%
A little uncomfortable	14.3%
Very uncomfortable	1.4%
No response	1.4%

2.4 Conclusion

The SMART system facilitates the collection, storage, review, and detection of events that otherwise may go unnoticed regarding the patients' medical conditions. To accomplish this, the set of sensors on each patient in conjunction with the patient's PDA send their data wirelessly to a central station where individual care needs are evaluated.

SMART addresses the need to provide better care to multiple patients. It can extend care to nonstandard settings such as nursing homes for the elderly, improvised care facilities in case of mass casualty events, locations designed for patients such as hospital waiting rooms, and homes or offices for people at risk during their daily activities. As cited, there is concern among health care facilities regarding patients in unmonitored areas. [19] In standard care centers, a wireless monitoring system that provides alarms for critical cases can reduce the workload on the triage staff and provide a better quality of care for the patients. Initial triage assessment time can be reduced, and by the time a physician first contacts the patient, a considerable amount of physiological data is available.

One of the main advantages of wireless networks in medicine is the ability to provide an enhanced level of care without interfering with the patient's quality of life. Even though in SMART the sensors are wired to the PDA hub on the patient, the wireless transmission of the data to a central station allows patients be untethered. The survey conducted among patients after wearing the device shows with high satisfaction levels, both in terms of care and comfort.

Acknowledgements

This work was supported in part by the National Library of Medicine, NIH, N01LM33509.

References

1. World Health Organization data and statistics (April 2010), URL: http://www.who.int/research/en/.
2. J. Waterman, D. Curtis, M. Goraczko, E. Shih, P. Sarin, E. Pino, L. Ohno-Machado, R. Greenes, J. Guttag, and T. Stair (2005) Demonstration of SMART (Scalable Medical Alert Response Technology), in *AMIA 2005 Annual Symposium*, Washington, DC, American Medical Informatics Association, pp. 1182–1183.
3. E. Pino, L. Ohno-Machado, E. Wiechmann, and D. Curtis (2005) Real-Time ECG Algorithms for Ambulatory Patient Monitoring, in *Proceedings of the AMIA Annual Symposium*, pp. 604–608.
4. D. W. Curtis, E. J. Pino, J. M. Bailey, E. I. Shih, J. Waterman, S. A. Vinterbo, T. O. Stair, J. V. Guttag, R. A. Greenes, and L. Ohno-Machado (2008) SMART — An Integrated Wireless System for Monitoring Unattended Patients,

Journal of the American Medical Informatics Association, **15**(1), pp. 44–53. doi: 10.1197/jamia.M2016.

5. S. P. McGrath, E. Grigg, S. Wendelken, G. Blike, M. D. Rosa, A. Fiske, and R. Gray (2003) Artemis: A Vision for Remote Triage and Emergency Management Information Integration, URL: <http://citeseerx.ist.psu.edu/viewdoc/summary? doi>= 10.1.1.128.4959.
6. S. M. Wendelken, S. P. McGrath, and G. T. Blike (2003) A Medical Assessment Algorithm for Automated Remote Triage, in *Proceedings of the 25th Annual Conference of the Engineering in Medicine and Biology Society,* Cancun, pp. 123–135.
7. P. Konoske, W. Deniston, R. Barker, and D. Moses (1998) Evaluation of the Mobile Medical Monitor (m3) in a Field Environment, in *Proceedings of the Medical Technology Symposium,* Pacific, pp. 83–92, doi: 10.1109/PACMED.1998.767929.
8. C. Orwat, A. Graefe, and T. Faulwasser (2008) Towards Pervasive Computing in Health Care — a Literature Review, *BMC Medical Informatics and Decision Making*, **8**(1), pp 26, ISSN 1472-6947, doi: 10.1186/1472-6947-8-26, URL: http://www.biomedcentral.com/1472-6947/8/26.
9. J. P. Killeen, T. C. Chan, C. Buono, W. G. Griswold, and L. A. Lenert (2006) A Wireless First Responder Hand-Held Device for Rapid Triage, Patient Assessment and Documentation During Mass Casualty Incidents, in *Proceedings of the AMIA Annual Symposium*, pp. 429–433.
10. L. Lenert, D. Palmer, T. Chan, and R. Rao (2005) An Intelligent 802.11 Triage Tag for Medical Response To Disasters, in *Proceedings of the AMIA Annual Symposium,* pp. 440–444.
11. M. Arisoylu, R. Mishra, R. Rao, and L. A. Lenert (2005) 802.11 Wireless Infrastructure To Enhance Medical Response to Disasters, in *Proceedings of the AMIA Annual Symposium*, pp. 1–5.
12. T. Massey, T. Gao, M. Welsh, J. H. Sharp, and M. Sarrafzadeh (2006) The Design of a Decentralized Electronic Triage System, in *Proceedings of the AMIA Annual Symposium*, pp. 544–548.
13. D. Malan, T. Fulford-Jones, M. Welsh, and S. Moulton (2004) CodeBlue: An Ad Hoc Sensor Network Infrastructure for Emergency Medical Care, in *International Workshop on Wearable and Implantable Body Sensor Networks.*
14. T. Gao, M. I. Kim, D. White, and A. M. Alm (2006) Iterative User-Centered Design of a Next Generation Patient Monitoring System for Emergency Medical Response, in *Proceedings of the AMIA Annual Symposium,* pp. 284–288.

15. T. Gao and D. White (August 2006) A Next Generation Electronic Triage To Aid Mass Casualty Emergency Medical Response, in *28th IEEE Annual International Conference of the Engineering in Medicine and Biology Society,* pp. 6501–6504, doi: 10.1109/IEMBS.2006.260881.
16. Y.-H. Lin, I.-C. Jan, P. C.-I. Ko, Y.-Y. Chen, J.-M. Wong, and G.-J. Jan (2004) A Wireless PDA-Based Physiological Monitoring System for Patient Transport, *IEEE Transactions on Information Technology in Biomedicine.* **8**(4), 439–447, ISSN 1089-7771.
17. C. F. Mackenzie, P. Hu, A. Sen, R. Dutton, S. Seebode, D. Floccare, and T. Scalea (2008) Automatic Pre-Hospital Vital Signs Waveform and Trend Data Capture Fills Quality Management, Triage and Outcome Prediction Gaps, in *Proceedings of the AMIA Annual Symposium,* pp. 318–322, URL: http://www.pubmedcentral.nih.gov/articlerender.fcgi? artid=2656089.
18. P. Pandian, K. Mohanavelu, K. Safeer, T. Kotresh, D. Shakunthala, P. Gopal, and V. Padaki (2008) Smart Vest: Wearable Multi-Parameter Remote Physiological Monitoring System, *Medical Engineering and Physics,* **30**(4), pp. 466 - 477, ISSN 1350-4533, doi: DOI:10.1016/j.medengphy.2007.05.014, URL: http://www.sciencedirect. com/science/article/B6T9K-4PNF2X4-1/2/733ffbd1f1553e04ed4fc73ef13 31c40.
19. C. V. Pollack (March 2009) Wireless Cardiac Event Alert Monitoring Is Feasible and Effective in the Emergency Department and Adjacent Waiting Areas, *Crit Pathw Cardiol,* **8**(1), pp. 7–11, doi: 10.1097/HPC.0b013e3181980f8b, URL: http://dx.doi.org/10.1097/HPC.0b013e 3181980f8b.
20. D. Raskovic, T. Martin, and E. Jovanov (2004) Medical Monitoring Applications for Wearable Computing, *The Computer Journal,* **47**(4), pp. 495–504, doi: 10.1093/comjnl/47.4.495, URL: http://comjnl.oxfordjournals.org/cgi/content/abstract/47/4/495.
21. Nonin Medical Inc. Nonin Medical Inc., http://www.nonin.com/.
22. Sonitor R home page, Sonitor R Technologies Inc., URL: http://www.sonitor.com. Accessed April 30, 2010.
23. D. W. Curtis, ORNetDB software, Available on request from dcurtis@csail.mit.edu.
24. PostgreSQL open source database, URL: http://www.postgresql.org/.
25. P. Tanabe, D. Travers, N. Gilboy, A. Rosenau, G. Sierzega, V. Rupp, Z. Martinovich, and J. G. Adams (2005) Refining Emergency Severity Index Triage Criteria, *Academic Emergency Medicine,* **12**(6), pp. 497–501.

26. R. C. Wuerz, L. W. Milne, D. R. Eitel, D. Travers, and N. Gilboy (2000) Reliability and Validity of a New Five-Level Triage Instrument, *Academic Emergency Medicine*, **7**(3), pp. 236–242, doi: 10.1111/j.1553-2712.2000.tb01066.x.

27. A. L. Goldberger, L. A. N. Amaral, L. Glass, J. M. Hausdorff, P. C. Ivanov, R. G. Mark, J. E. Mietus, G. B. Moody, C. -K. Peng, and H. E. Stanley (2000) PhysioBank, PhysioToolkit, and PhysioNet: Components of a New Research Resource for Complex Physiologic Signals, *Circulation*, **101**(23), pp. 215–220, doi: http://circ.ahajournals.org/cgi/reprint/101/23/e215.pdf.

Chapter 3

Real-Time Cardiac Arrhythmias Monitoring for Pervasive Health Care

Zhou Haiying,[a] Hou Kun-Mean,[b] de Vaulx Christophe,[c] and Li Jian[d]

[a]*School of Computer Science and Technology, Harbin Institute of Technology, N.92, Xi DaZhi Jie, Harbin, China*
[b]*Laboratoire LIMOS UMR 6158 CNRS, ISIMA, University of Blaise Pascal Clermont-Ferrand II, France*
[c]*Polytech'Clermont-Ferrand, CUST, University of Blaise Pascal Clermont-Ferrand II, France*
[d]*School of software, Harbin Institute of Technology, N.92, Xi DaZhi Jie, Harbin, China*
haiyingzhou@hit.edu.cn; kun-mean.hou@isima.fr;
christophe.devaulx@cust.univ-bpclermont.fr; lijian_susan@hit.edu.cn

Pervasive health care (PHC) is a new health care model that enables patient mobility, continuous health monitoring, and timely detection of anomalies. Comparing with the ideal PHC services, the current available (commercial) remote patient monitoring services such as portable monitoring equipment and prototype systems need to be improved in real-time capability, effectiveness, and reliability of context-sensitive anomaly detection. This chapter addresses the PHC system architecture and anomaly detection techniques for chronic diseases; a pervasive cardiac care (PCC) prototype that supports real-time indoor and outdoor continuous

Wireless Body Area Networks: Technology, Implementation, and Applications
Edited by Mehmet R. Yuce and Jamil Y. Khan

ISBN 978-981-4316-71-2 (Hardcover), ISBN 978-981-4241-57-1 (eBook)
www.panstanford.com

cardiac arrhythmias monitoring service health care models is presented. Moreover, this chapter explores key PHC technologies such as architecture model and system, including context-aware recognition, abnormal detection, and adaptive environment, etc.

3.1 Introduction

Health care, as an essential part of modern society, which contributes to social stability and government reputation, claims a remarkable percentage of national budgets. Due to global population aging in the 21st century, health care system worldwide is facing a serious shortage of financial and human resources [1, 2].

World Health Organization (WHO) reports that the world's aging population over the age of 65 is reaching 761 million in 2025. [3, 4] The white paper of "The Development of China's Undertakings for the Aged" declares that China is entering the aging society [5]. By the end of 2005, the aging population over 60 in China was more than 1.44 million, and will reach 400 million in 2037, ranking first in the world [6, 7]. Furthermore, China is suffering from a serious shortage and an extremely unbalanced distribution of health resources: China takes up 22% of world population, yet only 2% of world total health resources, among which the 80% of the health resources in China is concentrated in urban areas, and the 80% of the urban resources is concentrated in handful large hospitals. Therefore, in China, the severe contradiction between supply and demand of health care services is prominent [8, 9].

Many studies have shown that more than 80% elderly people have chronic diseases as well as lifestyle-related diseases, and more than 30% of them suffer from two or more types of chronic diseases [10, 11]. Further studies reveal that long-term continuous health care observations, rather than clinical care or treatment, are essential for patients suffering from several chronic diseases. Also according to epidemiological data and related research findings, we discover that successful prevention and reduction of acute attack of diseases depends on the improvement of health care quality, proactive management of chronic diseases, detection of abnormal symptoms, and healthy out-of-hospital lifestyles [12].

To cope with the rapid rise of medical cost and the growing shortage of health care capabilities, a personalized out-of-hospital healthcare model, which supports independent lifestyle, is highly demanded [3, 13]. The model is characterized by its: (1) excellent chronic disease management in 7 × 24 h, by assisting medical experts with technologies of pervasive biomedical sensing, computing and communication; (2) excellent abnormality detection ability, which effectively helps avoid emergencies (sudden death); (3) timely and accurate health care services and emergency handling capability, available anytime anywhere [14, 15].

PHC is designed to provide high-quality health care services for patients or elderly people with more than one chronic disease. By integrating and balancing health care resources and thus reducing medical cost, PHC is a strong complement for the existing health care system in China and developed countries. With the help of PHC model, health care providers can largely be relieved from physical and mental pressures, the quality of collaborate-treatment can be improved and the cost be lowered, and eventually a better and healthier out-of-hospital lifestyle can be established for the patients [8].

Patient health care service is a billion-dollar industry [16]. Yet, research on the key technologies of PHC system model and service provision is still rare [17]. Hence, the major research purposes of this chapter are: (1) to provide the designing rules of PHC services and to explore PHC service model; (2) to study the key technologies of PHC for chronic diseases, including context-aware computing, abnormal detection and environment adaptive technologies, etc. Health care is crucial not only to people's lives and physical health, but also to national security and harmony. PHC is the technical foundation of future health care system, particularly in China. The application of PHC service will help to solve the conflict between resources and needs, and eventually facilitate social stability. By adopting pervasive service and IT technology and integrating pervasive computing concept into health care services, this chapter proposes an innovative service model for public and private PHC services provision with no limitation in time and place.

3.2 History of PHC Research

PHC technology, as a cutting edge of its kind, demonstrates significant differences in research content, methods, and techniques from its counterparts such as biomedical engineering technology, medical informatics, and pervasive computing technology. Due to the complexity, diversity, and variability of PHC applications, how to build a ubiquitous ICT-based embedded biomedical technology health care environment, is a complicated interdisciplinary research issue [18]. This chapter provides a review of the state-of-the-art of research and development in the PHC field.

Today the progress in areas of wireless communication technology and medical telemetry equipment has set forth the rapid advancement of patients' clinical care and out-of-hospital care, and a variety of patient monitoring systems have thus been developed and applied. Thanks to the technology development in pervasive computing, medical engineering, medical telemetry, and other associated disciplines, it is now possible to extract, record, analyze, and transmit data concerning patients health symptoms (in the form of physical signal characterization) in a timely way via medical equipment worn by patients to the computer or PDA device of medical service providers. In this way, the time required for medical diagnosis and treatment is largely reduced [15, 19]. Lately, research institutions and organizations involved in health care studies have made a remarkable achievement in developing patient monitoring prototype system for accurate monitoring of patients' vital signs and timely detection of patients' abnormal signs [20]. Recognizing the importance of the aged health care market, a number of world-renowned medical equipment companies, including General Electric, Hewlett-Packard, Honeywell, and Intel, jointly set up Research Center for Aging Services Technology (CAST) in 2002 in Washington, so as to promote the application of wireless and remote patient monitoring service model and to encourage the joint efforts in technology development and cooperation of elderly people service.

The development of a health care prototype system has gone through three typical phases [17]. According to the patient monitoring scene, the prototype can be divided into three different types: hospital clinical care, home care as well as outdoor mobile monitoring [21]. The first generation of monitoring systems, such as Micropaq [22], collects and transmits multi-parameter monitoring information via short-range wireless medium (such as Bluetooth) to the network infrastructures within the building framework in the hospital. The second generation of surveillance systems, such as Medtronic [23], supports patient monitoring at home and is able to make time-appointed collection and transmission of patients' information. Based on television and cable networks, Motiva system provides a secure, personalized health care communication platform, which supports the interaction between patients at home and remote service providers, so as to achieve monitoring and management of chronic diseases and to improve patients' quality of life [24]. Other second-generation systems, such as CardioNet [25] and Biotronik [26], provide short-term care for patients at home (7–14 days). The new generation of surveillance systems is designed for mobile patients with continuous collection and transmission of patients' vital signs via infrastructure-based wireless networks (WLAN, cellular PCS, satellite, etc.). Relative research includes smart wearable health care research [27], medical telemetry devices [28], PDA Mobile Gateway [29], smart clothing care [30] and so on. References [31–34] present recent researches of the medical decision-making research for PHC service. There are also some typical PHC systems for elderly people such as Gator Tech Smart Room [35], perception Room [36], and the elite care [37].

By now, a large number of health care prototype systems have been designed, but there is still not much study in the architecture model of PHC system. The existing literature also mainly addresses the requirements and challenges in PHC system design, but not the actual system modeling. Kafeza et al. [38] analyzed the demand model of warning messages delivery. From the point of view of ICT infrastructure, Haux et al. [39] defined four types of health care architectures: people-centered, family-centered, remote health care service-centered, service provider-centered. The main functions of health care service are classified as: emergency state detection and alarm, disease management, health status feedback and suggestions. It also proposes the main components of a 24 h medical care environment: indoor home monitoring system, remote monitoring and location-based outdoor activities, emergency rescue system, and service support system. Koch et al. [40] investigated the relevant researches in the health care field in recent five years and classify them into five categories: decision support information systems, consumer health informatics, and remote home healthcare, emergency information technology, and informatics methods. According to the service controllability and application location, Doukas et al. [41] divided health care into two categories: health controlled environment (such as health care units and hospitals) and immediate health care services unavailable environment (such as at home or outdoor). Sneha et al. [17] proposed a component-based system framework of pervasive health care service and explore the design methods of major system components (sensing, reasoning, and transferring function). Based on the wireless sensor network technology, Daramolar et al. [42] presented a grid-based framework for PHC system. By analyzing the existing pervasive health care applications, Salvador et al. [43] attempted to define a general PHC system framework, which is compatible with its software and classify the PHC application into three types of typical environment: home, grid space, and mobile space.

Although some remarkable achievements in the field of patient care have been made, there are still some major limitations and constraints: 1) the majority of proposed solutions is not dedicated to PHC services, and there is no in-depth study of the application characteristics of PHC, such as patient mobility, continuous monitoring, and timely detection of anomalies; 2) most of the solutions do not make full use of mobile computing platforms for biomedical data analysis, which not only produces large amount of raw data that leads to high network traffics and more bandwidth requirements, but also brings huge mental and physical pressure on medical professionals when performing signal analysis and abnormalities monitoring.

Between the ideal PHC services and the current remote patient monitoring services with commercially available portable monitoring equipment and prototype systems, a huge gap still exists, especially in their real-time capability, effectiveness and reliability of context-sensitive anomaly detection. Partly because the existing solutions are lack of in-depth understanding, exploration of the complicated processing mode, the relevant system parameters as well as reasoning and decision-making mechanisms of the PHC system.

To solve the above problems, this chapter addresses the PHC system architecture and anomaly detection techniques for chronic diseases. The objective of this chapter is 1) to provide timely anomaly detection, prevent sudden attack of illness, and reduce avoidable medical costs; 2) to provide a PHC service, which is based upon the mobile computing platform, with the capabilities of biomedical data analysis and medical intervention, and is able to reduce network traffics and bandwidth requirements. This chapter presents a WSN-based pervasive cardiac care prototype that supports clinical, home, and outdoor models to provide continuous cardiac arrhythmias monitoring service [44, 45]. Based on this prototype system, this chapter explores the system architecture technology [46, 47], context-aware recognition and abnormal detection technology [48, 49], adaptive environment technology [50], and so on.

3.3 Overview of PCC System

Heart disease ranks as the top threat to human health. Elderly people and middle-aged person in long-term sub-health state are among the high risk population of cardiac sudden death (CSD). Clinical experiences prove that the most efficient way to prevent CSD is to detect the cardiac abnormalities in time by making long-term regular cardiac monitoring. Traditional health care systems are not suitable for pervasive cardiac monitoring (PCC) due to the time and space constraints in health care applications, as well as the inability of real-time continuous ECG monitoring. In recent years, a new generation of cardiac monitoring system named pervasive cardiac monitoring is booming. By introducing ubiquitous computing into traditional healthcare, the PCC system provides a real-time continuous cardiac monitoring service for different subjects with no constraints of time or space.

3.3.1 *PCC System Architecture*

The PCC system integrates the advanced technologies of wireless communication, embedded system, automatic ECG diagnosis (AED),

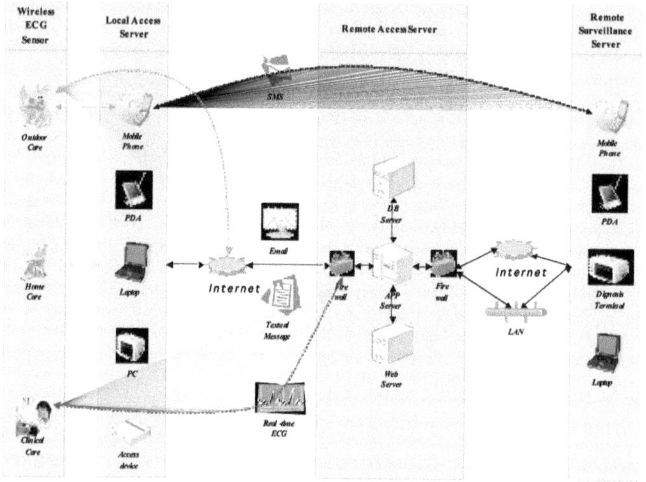

Figure 3.1. Architecture of PCC system. See also Color Insert.

and telemedicine, which consists of four main functional components: wireless ECG sensor (WES), local access server, remote access server, and remote surveillance terminal. Figure 3.1 shows the architecture and operation fashions of the PCC system.

3.3.1.1 Wireless ECG sensor

To minimize cost and to correspond to the latest AHA recommendations [51], an energy-efficient compact ECG collection device, named WES, is implemented. WES conforms to the basic characteristics of wireless sensor device: tiny resource, tiny power-consumption, and associated tiny cost. Furthermore, the basic embedded software such as a distributed real-time fault tolerant microkernel [52], dedicated hardware and firmware [53], and a TCP/IP protocol stack [54] are implemented and ported into WES.

The WES prototype (Fig. 3.2) is a real-time wireless embedded portable sensor (size = 70 mm × 100 mm) based on TI MSP430 microcontroller. The key features of WES are

- Gain: 1000
- CMMR (min): 120 dB
- Bandwidth: 0.05 Hz to 125 Hz

Figure 3.2. Wireless ECG sensor. See also Color Insert.

- Programmable sample frequency more than 500 Hz
- Analogue to digital converter (ADC): 12 bit
- Leakage current: 10 μA

WES enables the capture of 4-leads ECG signals sampled at 500 Hz (sample frequency) in real-time. Note that the sample frequency is reprogrammable (100 Hz to 2000 Hz). The sample data is sent to the local server over a wireless medium such as Wi-Fi or Bluetooth. In offline mode, ECG signals can be stored in the flash memory card. The signal store duration depends on the flash card capacity, the sample frequency, and the number of ECG leads. For example, a 128 megabyte flash memory card can store 24 h data of continuous 4-leads ECG signals sampled at 500 Hz. In this way, the WES works as Holter or RTEST device.

3.3.1.2 Local access server

Local access server may be implemented on a standard PC, a PDA, a mobile phone, or a dedicated network access device. It provides two kinds of network connection services: connection with WES via a local wireless medium (Wi-Fi or Bluetooth) and connection with remote system via infrastructure network, such as cable modem, ADSL, 2G/3G, etc.

In view of the difference of network mediums, local servers, and transmission speeds caused by network traffics, the PCC system provides an adaptable communication mechanism to guarantee a reliable data transmission service between local server and remote server. Furthermore, in order to provide a real-time data transmission, it is important to minimize the amount of transmission data to reduce network traffics over low-bandwidth connection links. A lossless ECG signals compression algorithm is thus implemented.

If the local access server is a high-performance machine, the ECG diagnostic module can be migrated from WES into the local server and the cost of WES will thus be reduced. In addition, patients' video information is necessary as an aided method for online diagnosis. Hence, a "webcam" can be installed in the local server to provide patients' images.

3.3.1.3 Remote access server

The remote access server provides the capability of network connection between local access server and remote surveillance server. In view of network types, it can support two types of access servers: PPP server and WAP server. The PPP server establishes connections between patients and cardiologists via PSTN, while the WAP server establishes the connections via wide wireless network. In addition, if the patients and the cardiologists are located in a same area (e.g., in hospital or at clinic) and share local network infrastructures (e.g., high-speed LAN), or they utilize the global network connection services provided by commercial ISPs (Internet service providers), this component is no more necessary and can be removed from the PCC system architecture.

Several function modules can be loaded in the remote access server, including database service and Web service. The medical history records of patients are important for the diagnosis of heart diseases. In the remote access server, the medical records with multimedia formats are stored in the patient database system, which include the ECG signals sequences, cardiologists' diagnostic reports, patients' video, voice and individual profile information, etc. The Web service provides the capacity of Internet access to the patients' database system.

3.3.1.4 Remote surveillance server

The remote surveillance server has an interactive visualization graphical user interface (GUI, see Fig. 3.3) which enables cardiologists to diagnose cardiac arrhythmias events in real-time and

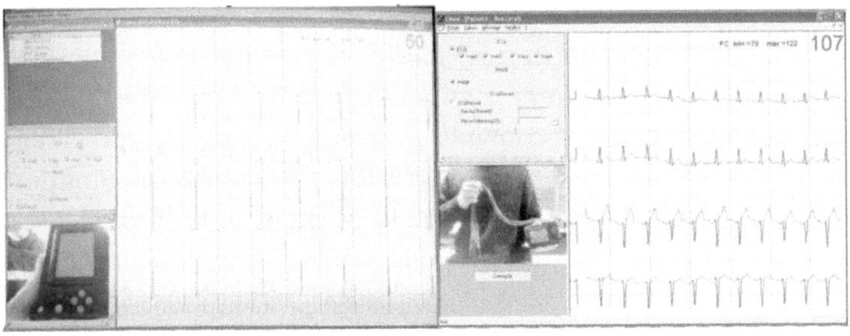

Figure 3.3. Remote surveillance server. See also Color Insert.

to respond to alarm messages by monitoring the ECG signals sequence and patients' images. This server supports multi-patients surveillances and one patient on-line diagnosis (by cardiologists) at one time. The 4-lead ECG signals and related diagnostic results (by the AED algorithm) are displayed on the screen and stored into data files with the format of WaveForm DataBase (WFDB) [55]. The diagnostic reports can be produced automatically and be printed with the ECG signals sequence and the related statistic results.

When a patient's number of a PCC application is few, e.g., an application of small clinic, the cardiologist of the remote surveillance server can directly connect with the patient in the local access server via the Internet. In this way, the database service and Web service can be implemented directly in the remote surveillance server.

3.3.2 *PCC Operation Modes*

This PCC system enables four operation modes which work together to make the system always adaptable to different application scenarios and requirements. Cardiologists can reset the operation mode by taking into account patients' physical status and network medium access bandwidth. The key features of the four operation modes are illuminated as follows:

- *Real-time continuous ECG signal.* For the sake of remote real-time displaying and diagnosing, the data including continuous ECG signals acquisition and its diagnosis report will be sent in real-time to the remote system. This operation mode has the highest alarm level that enables on-line diagnosis in real-time. Note that this mode does not fit to monitor a large number of patients due to the limitations of network bandwidth, system resources and medical resources, but it has remarkable performance in monitoring the CSD high risk patient. In practice, each cardiologist can survey approximately four patients simultaneously. In this mode, to assure reliable cardiac arrhythmias diagnosis, a patient's image is required.

- *ECG signal sequence.* In order to satisfy remote real-time multi-patient detection and monitoring, WES is configured to send automatically a sequence of ECG signals (pre- and post-abnormality) to the remote system when a cardiac arrhythmia event defined by cardiologists is detected. This operation mode is suitable for long-term multi-patient (lower risk of sudden death than the previous class of patients) cardiac arrhythmias events surveillance.
- *Textual emergency message.* In this mode, only a short textual emergency message will be sent to cardiologists when a cardiac arrhythmia event is detected. According to the gravity of the symptom, the cardiologists can decide to intervene immediately or later. This mode may be operated on any access medium (wire or wireless).
- *Diagnosis report email.* It is the lowest level operation mode. The local server will send periodically a report (like HOLTER report) attached to an email to the remote server. The cardiologists predefined the reporting period. This mode is suitable to monitor a large number of patients.

It is to be noted that cardiologists can remotely reconfigure the operation mode to adapt to the changes of a patient's status and environments.

3.4 Key Technologies of PCC System

In this chapter, we present the key technologies developed for the PCC system, including the lossless signal compression technique, adaptive real-time communication mechanism, and the adaptive ECG diagnosis (AED) algorithm.

3.4.1 *Lossless ECG Signal Compression*

Minimizing the amount of the data transmission is the most effective technique to reduce network traffics, and it is thus necessary to compress the ECG signals before data transferring. The 500 Hz sampling frequency in WES offers high quality ECG signals for the ECG diagnosis, but also increases network traffics for the data transmission. For example, a 4-lead ECG signals with 500 Hz sampling frequency and 12 bit sample data (ADC resolution), the size of its 5 s frame is 20 KB (4 leads * 500 Hz * 5 s * 2 byte). In order to support real-time data transmission through classical modem, the network bandwidth must be superior to 32 Kbps (20,000 * 8 bit/5 s) when PCC works in the level-I operation mode. This speed is obviously unfitted for the low bandwidth networks. The ECG signals with the high sampling frequency (500 Hz in WES) are absolutely necessary to guarantee the accuracy of the AED algorithm. Note that ECG signals with the low sampling rate (often 128 Hz) are acceptable for the purpose of the ECG observation in the remote visualization system. In general the ECG signals issued from HOLTER devices are sampling at 128 Hz. Hence, the sampling frequency of ECG signals can be sub-sampled to reduce network traffics.

On the other hand, the morphological features of ECG signals show that the signals have huge data redundancies. A good data compression algorithm can thus reduce network traffics. In fact, there are many high ratio data compression techniques such as MPEG3/4, etc., but they are lossless and CPU time consuming. Thus, to minimize energy consumption and ease VLSI integration of WES, we adopt a very simple no-loss compression algorithm: only the difference value between two consecutive samples is sent to the remote server. Since the potential amplitudes of the ECG signals in the cardiac cycle duration are almost equal to the zero isoelectric value, the difference value between the samples data of the cardiac cycle duration is thus almost equal to zero. According to the data sizes of the difference values, the sample compression format is defined in Fig. 3.4.

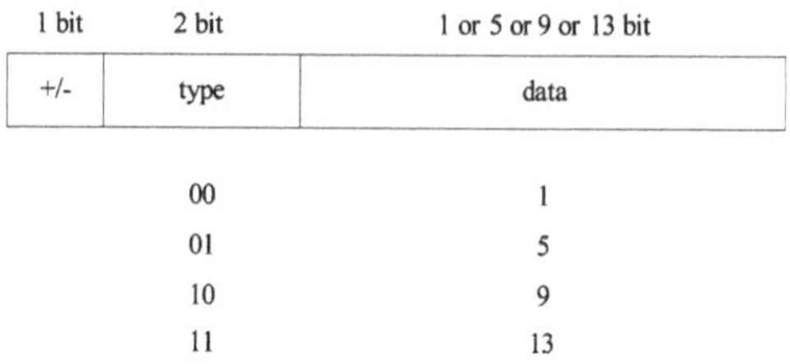

Figure 3.4. Data format of signal compression algorithm.

Where the " + /−" field indicates the positive/negative property of the difference value: 1 — positive and 0 — negative; the "type" field indicates the type of data length, and the "data" field stores the difference data. The statistical results show that this compression algorithm has 50%–60% compression ratios. Hence, in case of a 4-lead ECG signals at 500 Hz sampling frequency and 12 bit ADC resolution, when adopting the sub-sampling frequency technique (to 125 Hz) and the signal compression algorithm, PCC can reduce 87.5% $\sim$ 90% network traffics.

3.4.2 *Adaptive Communication Mechanism*

In order to support a real-time data transmission, the UDP protocol is adopted in PCC to transmit ECG signals. Because the UDP protocol does not offer a guaranteed datagram delivery service, a reliable data transmission mechanism must be implemented in the application layer. In fact, the network bandwidth of the PCC communication system is normally fluctuated and affected by network traffics and some interference factors, an application layer communication mechanism dedicated to PCC is thus implemented, which enables reliable real-time data transmission in various healthcare environments.

3.4.2.1 PCC data frame

The transfer data unit between remote and local peers in PCC is named a *PCC frame*, which is used to establish/terminate connection, to deliver data (ECG signals or images), and to configure operation modes or other system parameters. Each frame consists

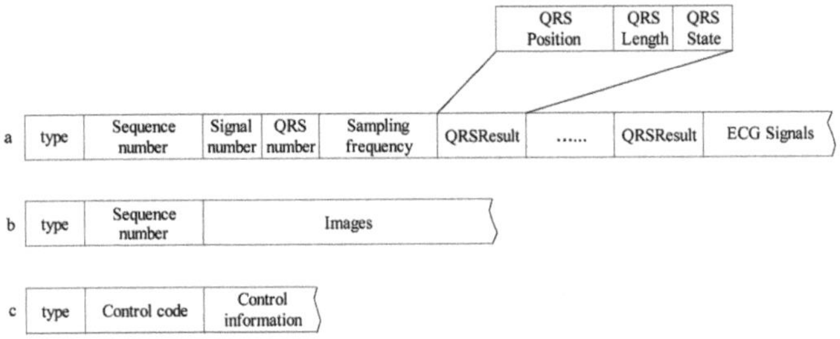

Figure 3.5. Frame format of PCC.

of two parts: a frame header followed by data. Figure 3.4 shows the PCC frame format.

The type field is used to identify the type of PCC frame. Three frame types are implemented in PCC: a value of 0×01 indicates a system control frame; a value of 0×02 indicates an ECG signal frame; and a value of 0x03 indicates an image frame. Each type of PCC frames has a unique system priority identified by the type value (1 to 3, from high to low).

Figure 3.5a shows the format of an ECG signal frame. Each ECG frame has a unique identifier specified by the sequence number. Every time when the local server sends an ECG frame, the sequence number increments automatically by one for next frame. Basing upon this value, PCC calculates the surveillance time of heart monitoring. The value of signal number field indicates the channel number of ECG signals and the default channel number is four in WES. The value of QRS number fields represents the number of QRS complexes detected by the AED algorithm in the ECG frame. The default sampling frequency of WES is 500 Hz, which is identified in the sampling frequency field. This value is alterable by the sub-sampling frequency operation because the signal compression algorithm changes the size of original ECG frame. Hence this frequency value is useful to define the length of the uncompressed original frame in the remote server.

The following area of the ECG frame stores the diagnostic results of the AED algorithm. It is a QRS structure queue, where the QRS member number is indicated in the QRS number field. Each QRS member named QRSResult consists of three elements: QRS position,

QRS length, and QRS state. The QRS position indicates the onset position of a QRS complex in the uncompressed ECG frame. The heart rate of each beat can be calculated basing on the positions of two consecutive QRS. The QRS length represents the time interval of a QRS. The QRS state shows the type of heart rhythm and its related heart statue classified by the AED algorithm. The ECG signals field stores the compressed ECG signals. In the PCC system, each ECG frame contains 5 s of ECG signals.

Figure 3.5b shows the format of the image frame. The sequence number field is unused for an image frame. The image field contains an image in the jpeg format. Figure 3.5c shows the format of the system control information frame. The value of control code field indicates the type of control code. The related control information is stored in the following control information field.

3.4.2.2 PCC communication mechanisms

Three kinds of UDP connections are established in PCC. They are responsible for the system control (udp_CMD), the ECG signal transmission (udp_SIG), and the image transmission (udp_IMG).

3.4.2.2.1 PCC system control PCC control frames are responsible for the system remote configuration, the patient online/offline notifications, and the ECG frames retransmission management. The control frame is transferred via the UDP_CMD connection. PCC defines nine types of control frames:

- REQ_Connect /ACK_Connect; REQ_Terminate /ACK_Terminate
- REQ_Configure /ACK_Configure; REQ_Restra
- ACK_5Frames; ACK_Image.

The control frames with the code of REQ_Connect and ACK_Connect are responsible for the connection establishment between patients and cardiologists. The control frames with the code of REQ_Terminate and ACK_Terminate are responsible for the connection termination. Both patients and cardiologists have a right to open and close the UDP_CMD connection. The control frames with the code of REQ_Configure and ACK_Configure are responsible for the system configuration. Currently, PCC provides the configurations of the operation mode and the sampling frequency. The other control frames with the code of REQ_Restra, ACK_5Frames, and ACK_Image will be introduced in the following subsections.

3.4.2.2.2 Signal retransmission mechanism PCC guarantees a reliable ECG signal delivery service by providing a signal retransmission mechanism. Two ECG frame queues are defined, respectively, for the local access server (Hold_Queue) and the remote surveillance server peer (Wait_Queue). The standard length of the two queues is set to five. The retransmission mechanism is described in Fig. 3.6.

When an ECG frame with the sequence number k is lost during network transmission, the remote surveillance server sends a retransmission requirement frame with a control code of REQ_Restra

```
If (an ECG frame is recieved)
    If(Seq_cur == Seq_exp)
            Insert current frame to wait_queue; The position of current
            frame in wait_queue is decided by its Seq_cur;
            Seq_exp++;
    else if (Seq_cur < Seq_exp)
            drop it;
            Send Control Frame REQ_Restra to the local server;
    else if (Seq_cur > Seq_exp)
            Add current frame to the tail of wait_queue;
    end

    if (continuous Five frame is received)
            Send Control Frame ACK_5Frames to the local server;
    end
end
```

```
If(ACK_5Frames frame is received)
    Release the ECG frames whose Seq_fra <= Seq_ack from the
    hold_queue;
else if (REQ_Restra frame is received )
    Retransmit an ECG frame of Seq_req to the remote surveillance
    server;
    Release the ECG frames whose Seq_fra < Seq_req from the
    hold_queue;
end

if (an ECG frame is ready)
    Send an ECG frame of Seq_cur to remote surveillance
    server;
    Seq_cur++;
end
```

Figure 3.6. Signal Retransmission Mechanism: Remote Surveillance Peer (U) and Local Access Peer (D).

and the local access server responses to this requirement by retransmitting the ECG frame k. When five consecutive ECG frames are received by the remote surveillance server, a control frame containing the code of ACK_5Frames will be sent to the local server to inform the release of the remainder ECG frames in Hold_Queue.

3.4.2.2.3 Data competition mechanism Considering the network quality and communication cost, PCC employs a data competition mechanism to guarantee those important data (i.e., ECG signals) being transferred in a higher priority, so that in the remote surveillance peer, cardiologists can detect and diagnose the ECG signals in real time.

As mentioned above, PCC has three types of data frames. Each kind of data frame has a unique transmission priority. The system control frame has the highest priority in this system. Furthermore, since the ECG signals are more important than the images data for the diagnosis of cardiologists, the ECG frame has higher priority than the image frame. In order to guarantee a real-time ECG signal transmission service, a data competition mechanism is implemented in the PCC system, described in Fig. 3.7.

When the queue length of Wait_Queue is equal to or greater than 5, it means that there is at least 25 s of ECG signals stored in the remote surveillance server. Hence, it is acceptable to allow the local server to transmit images. Whereas, when the queue length of Wait_Queue is smaller than 5, it means that the network speed begins to fluctuate and the network quality is lowered; hence, the system will stop image transmission so as to reduce network traffics.

3.4.3 *AED Algorithm*

In recent years, the study of automatic ECG diagnosis (AED) techniques addressed mainly clinical applications, but rarely PHC applications, due to two reasons: 1) to be applied on various scenes, i.e., different objectives and environments, the system signals are ambulatory, easily being disturbed by interferences; 2) to meet the needs of portable, the system node has strict resource constraints, traditional AED techniques are resource-consumption in views

```
If (an Image frame is received)
    If(the length of wait_queue >= 5)   //25 seconds delays
        Send a Control frame ACK_Image to  the local server;
        Set the Image_Enable = TRUE;
    else
        Set the Image_Enable = FALSE;
    end
end

if ( an ECG frame is received)
    If(the length of wait_queue > 5 && Image_Enable == FALSE )
        Send a Control frame ACK_Image to  the local server;
        Set the Image_Enable = TRUE;
    end
end
```

```
If (Image_Enable == TRUE)
    Send an Image frame to  the remote surveillance server;
    Set the Image_Enable = FALSE;
end

if ( a Control frame ACK_Image is received)
    Set the Image_Enable = TRUE;
end
```

Figure 3.7. System competition mechanism: remote surveillance peer (U) and local access peer (D).

of PCC services. A novel AED algorithm dedicated to the PCC applications has thus been developed. By offering the PCC service, this system can reduce the risk of cardiac sudden death by detecting cardiac arrhythmia events in time.

3.4.3.1 Signal preprocessing and conditioning

Due to the nonstationary and easy-to-be-disturbed features, the ambulatory ECG signals must be de-noised before decision-making. Most of artifacts, such as baseline drifts, electrical noises, and muscle tremor interferences, can be eliminated or restrained by adopting suitable filters.

3.4.3.1.1 ECG time series Three ECG signal series, i.e., $R(t)$, $AD(t)$, and $RC(t)$, are adopted in the AED algorithm. The $R(t)$ series is raw ECG signals acquired from electrodes. It is generally contaminated by different kinds of noises. The $AD(t)$ series is the adaptive differential signals with the processing of the differential filter and the adaptive filter. The inferences of the baseline drift and the motion artifacts can be eliminated in the $AD(t)$ series; hence, this series is used to detect and localize the QRS complexes. The $RC(t)$ series is the de-noised ECG signals with the operations of the band-pass filter and the linear amplifier. Since electrical noises and muscle tremors have been removed from the $RC(t)$ series, the $RC(t)$ series is used to extract the characteristics of the QRS complexes.

3.4.3.1.2 Adaptive filter Classical filters for the ECG series, e.g., notch filter, low-pass filter, and high-pass filter, can effectively remove or reduce most of interferences. But for motion artifacts, because of their irregular occurrences and irregular morphological attributes, these filters cannot eliminate these disturbances. These artifacts can cause much trouble in QRS detection when encountering QRS-like artifacts. This algorithm adopts an adaptive filter (AT) to reduce motion artifacts. The resultant signal series, named $A(t)$, is generated by performing AT operation in the raw series $R(t)$. The AT expression is

$$\begin{cases} Aecg(0) = R(0) \\ & 0 < \alpha < 1,\ t = 1 \cdots N \\ Aecg(t) = \alpha * Aecg(t-1) + (1-\alpha) * R(t) \end{cases} \tag{3.1}$$

where α is the balance coefficient of AT, which is a key factor to influence the performance of the AT filter. If the value of α is augmented, the previous estimated value $Aecg(t - 1)$ then gives more proportion in the estimation of current $Aecg(t)$. Thereby, the achieved $A(t)$ is stable but cannot reveal the changes of ECG signals. In contrast, if the value of α is diminished, the original value $R(t)$ then gives more proportion and the achieved $A(t)$ is thus more dynamic and adaptive to the changes of ECG signals. Hence, the value of α should be cautiously selected to make $A(t)$ both stable and adaptive to the changes in $R(t)$. The value of α is set to 0.95 by default in views of the ECG sample frequency in PCC (500 Hz). Figure 3.8 shows the QRS complex waveform after the adaptive filter operation with different impact factors.

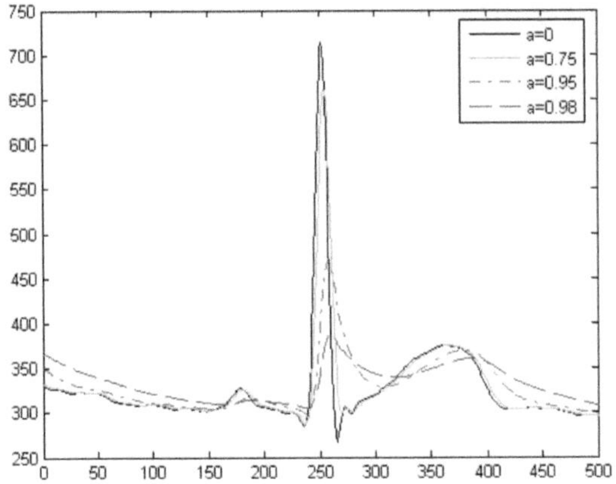

Figure 3.8. QRS complex waveform after different impact factor in the adaptive filters. See also Color Insert.

Figure 3.9 shows different ECG series: (a) is the raw signals $R(t)$, which are seriously polluted by noises. (b) represents the reconstructed series $RC(t)$ when filtering the $R(t)$ series by traditional filters, i.e., notch filter, low-pass filter, and high-pass filter. The $RC(t)$ series still contain the interferences generally caused by baseline wandering and motion artifacts. (c) is the adaptive filter signal $A(t)$ when filtering the $R(t)$ series by AT, which has

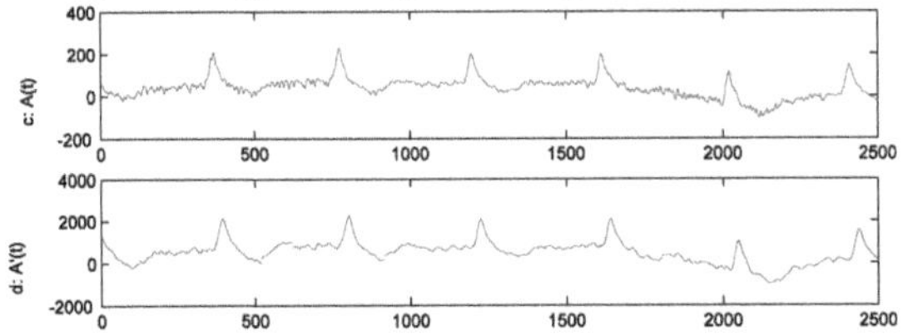

Figure 3.9. ECG series after different filters.

better signal quality than $RC(t)$. (d) is the reconstructed signal $RC^*(t)$ based on the adaptive filter signal $A(t)$. Obviously, in contrast to the previous reconstructed signal $RC(t)$, the signal $RC^*(t)$ has better signal quality in which the motion artifacts are effectively eliminated.

3.4.3.2 QRS complex detection

A new QRS detector has been developed to eliminate noises and artifacts by exploiting a self-adaptive threshold (SAT) method and designing state transition recognition (STR) procedure. The SAT method is used to estimate the peaks of ECG subsegments and the means of contextual thresholds, which allow estimating the optimum thresholds in a segment space. The STR procedure traces the waveform changes of signal series and identifies QRS complexes based on the optimum thresholds and the rules of state transition.

3.4.3.2.1 Diagnostic segment window (DSW) A short-term redundant data (default 5 s) is important in QRS detection. First, this short-term segment enables the complex contextual correlative analysis and reduces the interferences of baseline drift. In view of the low-frequency baseline drift, a short-term segment has fewer disturbances caused by baseline wandering than long-term signals. The redundant data enable the QRS detector to identify current QRS complex by comparing with fore-and-aft QRS complexes. Furthermore, in view of the unpredictability and variability of network quality, redundancy is necessary for data retransmission and network communication.

3.4.3.2.2 Self-adaptive threshold The QRS complexes of AECG have rapid changes and high potential amplitudes so that the differential series $D(t)$ can exactly represent the changes. The QRS signals have higher absolute amplitudes in a cardiac cycle of the $D(t)$ series. The solution to QRS detection is to search the optimum pair-peak for each QRS complex, i.e., the positive and negative peaks in a cardiac cycle. In Diagnostic Segment Window (DSW), there are generally multiple pair-peaks because several heartbeats will occur within 5 s. These pair-peaks make up of a pair-peak series in a DSW. Based on the pair-threshold extracting from the

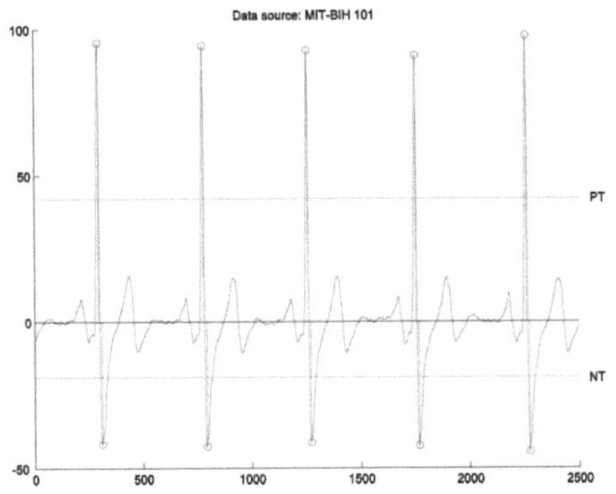

Figure 3.10. Mean of pair-peak series in a diagnostic window.

pair-peaks series in a DSW, the STR procedure is then able to locate QRS complexes. The absolute amplitude of each peak is generally greater than the associated absolute threshold in $D(t)$. Furthermore, since the position offset between the $D(t)$ series and the $A(t)$ series is constant, we can thus obtain the positions of QRS complexes in $A(t)$ by locating the complexes in $D(t)$.

The SAT method aims to determine the optimum pair-threshold, which is estimated from two aspects: the mean of the pair-peak series in a DSW and the pair-threshold of the previous DSW. The pair-threshold results from the means of the negative and positive pair-peaks series in a DSW. In order to accurately estimate these pair-peaks, the diagnostic segment window is divided into five subsegments with the length of 1 s (see Fig. 3.10). Because the normal heart rate of a healthy adult is 60–100 bmp, each subsegment thus contains one heartbeat. Since the differential signals of QRS complex have the maximum absolute amplitudes in a cardiac cycle, a pair-peak will indicate a QRS complex and then can be used to estimate the thresholds. Furthermore, the shorter the subsegment, the less interference of baseline drift the subsegment has. A subsegment with the length of 1 s can thus be regarded as a stationary series.

3.4.3.2.3 QRS location: State transition recognition In view of the QRS morphology properties in the $D(t)$ series, the complexes are categorized into two groups: positive and negative. Therefore,

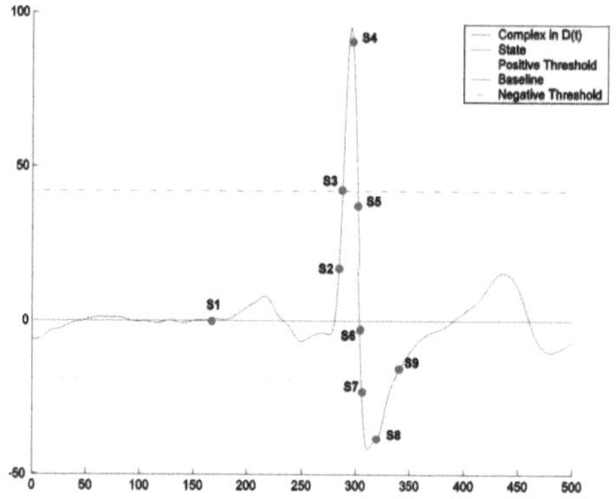

Figure 3.11. Positive states of a QRS complex in $D(t)$. See also Color Insert.

different states are defined to outline the phases of QRS complex in $D(t)$. $S2\sim S9$ represent the positive states of a QRS complex (see Fig. 3.11); corresponding $S20\sim S29$ represent negative states. An adaptive and self-corrected procedure, named STR, is developed to automatically track the changes of signal series, to correct error detection, and to record detected complexes. The states transitions are based on three basic reference lines: the baseline, the positive threshold, and the negative threshold.

3.4.3.2.4 Feature extraction: geometric analysis method QRS complex has triangular-alike or triangular-component morphological characteristics (see Fig. 3.12). This chapter thus adopts the geometric analysis method (GAM) to extract the features of QRS complexes. GAM has simple operations and low resource consumption, being able to predict and estimate the key points of QRS complexes under noisy situations, such as R wave peak, end point of Q wave (Q_t), and onset point of S wave (S_i). Therein, R wave peak can be obtained from $T_{peak\ 1}$ or $T_{peak\ 2}$, and it has mono-peak or poly-peaks. The measurement and detection phases of Q_t and S_i points are illuminated as follows.

- Defining two-level thresholds for left and right sides of R wave ($LH = (1/4) \times V_{peak\ 1}$, $LL = (3/4) \times V_{peak\ 1}$, $RH = (1/4) \times V_{peak\ 2}$, $RL = (3/4) \times V_{peak\ 2}$).

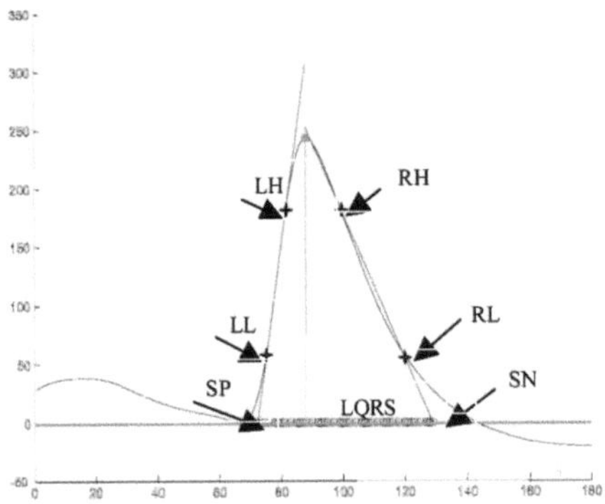

Figure 3.12. Illumination of geometric analysis method. See also Color Insert.

- Calculating the intersection points between the threshold values and complex signals. The slopes of two approaching lines represent two characteristics of QRS complex: SP (positive slope) and SN (negative slope).
- Obtaining the duration length of QRS (LQRS) which is the distance of two intersection points between the baseline and two approaching lines.

3.4.3.3 AED performance analysis

The AED algorithm has been evaluated on two ECG databases: MIT-BIH arrhythmia database [56] and CSD database (Clinic *STAR* Database). The former contains 48 half-hour excepts of two-channel ambulatory ECG recordings, and the latter is obtained from 30 patients of C.H.R.U. of Gabriel Montpied's hospital (Clermont-Ferrand, France) by using the PCC system named *STAR* [57]. The CSD signals are recorded by using WFDB format as the MIT-BIH Database one.

Dotsinsky et al. [58] defined four performance parameters to assess the algorithm efficiency (Se: sensitivity and Sp: specificity): TP (true positive), FP (false positive), FN (false negative), and shifted SH beats, shown as follows:

$$\mathrm{Se} = \frac{\mathrm{TP}}{\mathrm{TP}+\mathrm{FN}+\mathrm{SH}} \quad \mathrm{Sp} = 1 - \frac{\mathrm{FP}}{\mathrm{TP}+\mathrm{FP}} = \frac{\mathrm{TP}}{\mathrm{TP}+\mathrm{FP}} \tag{3.2}$$

Table 3.1. Performance evaluations of QRS detection algorithms

		Se (%)		Sp (%)	
Afonso *et al.* [59]		99.59		99.56	
Poli *et al.* [60]		99.60		99.51	
Dotsinsky *et al.* [58]		99.04		99.62	
Kaiser *et al.* [61]		99.68		99.72	
Datex-Ohmeda Corp. [62]		99.86		99.88	
Millet *et al.* [63]	Alg 1	94.6		98.0	
	Alg 2	97.3		98.0	
Our algorithm		99.43	99.25	98.55	97.94
		MIT	CSD	MIT	CSD

Comparing with the performance results of other algorithms listed in Table 3.1, the overall results of our detection algorithm, 99.37% sensitivity and 99.68% specificity on MIT-BIH database, 99.67% sensitivity and 99.74% specificity on CSD database, show the high sensitivity and specificity. This detection algorithm has minimal beat detection latency, low computational consumption, and fast detection ability.

3.5 Conclusion

Currently, the PCC system has been evaluated on 30 patients who had acute cardiac arrhythmia disturbances at the CHRU of Gabriel Montpied's hospital (Clermont-Ferrand, France), shown in Fig. 3.13. Since patients and cardiologists are located at the same area (at the hospital) in this application, the PCC system is thus configured as *clinical care* state. The local access server is a laptop on which the real-time cardiac arrhythmias program is installed. The remote access server and the remote surveillance server are installed in the same laptop or a standard PC, which connect with the local server via the local high-speed network. This system supports continuous remote cardiac arrhythmias monitoring, and the 4-lead ECG signals are sent directly to the remote surveillance terminal in real time with *real-time continuous ECG signal* operation mode.

The system has also been utilized to evaluate athletes' cardiac status during physical exercises, shown in Fig. 3.13. The obtained

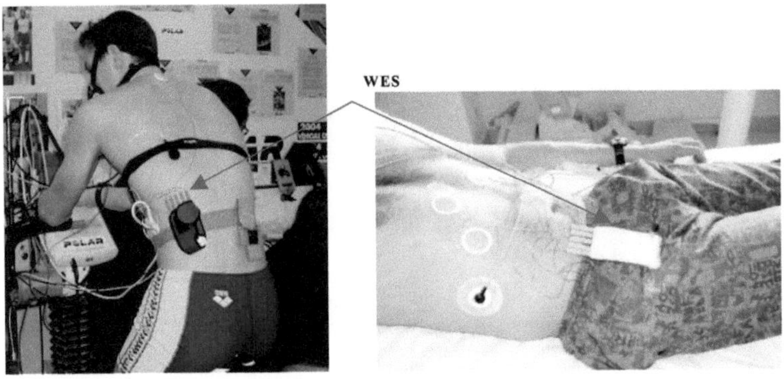

Figure 3.13. Application example of PCC system. See also Color Insert.

results show this system still provides high quality ECG signals and accurate QRS detection PCC is an efficient system for diagnosing cardiac arrhythmias. It provides a new clinical approach adapted to monitor accurately and efficiently large scale high-risk patients. Furthermore, it may be used by cardiologists to remotely monitor patients, to evaluate the efficiency of drugs and to discuss difficult cardiac pathology cases with other colleagues.

In order to compare with other system, the HP telemetry system has been applied during PCC evaluation. For each patient, the evaluation duration is 30 min. The evaluation results obtained by the PCC system and the HP telemetry system are compared, as shown in Fig. 3.14. Due to the higher sampling frequency, the ECG signals of our system have better quality than HP ones. Concerning the clinical off-line cardiac arrhythmia detection, the two systems have similar results. The evaluation results prove that the PCC system meets the requirements of real-time cardiac monitoring and diagnosing application.

In order to improve this system, the following techniques should be developed: (i) Embedded RTOS (i.e., HEROS [64]) adopts a full "modularity" design fashion. The system primitives and tasks of HEROS will be defined as a set of actions. Thus, it may be configured according to different applications. (ii) AED algorithm has been ported into the local access unit and the remote center server, but it can be integrated as an ECG diagnostic chip. We are working on the

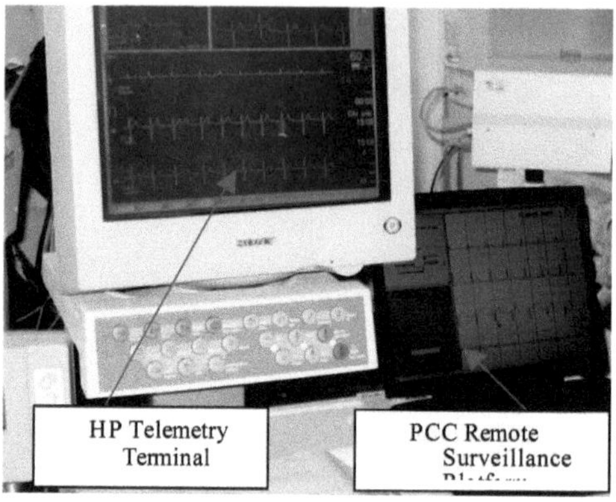

Figure 3.14. PCC test and comparision with HP telemetry.

implementation of an intelligent wireless ECG sensor (IWES) [65] by integrating the algorithm into a VLSI chip. This chip is currently under evaluation and test on an ALTERA FPGA board.

Acknowledgment

We would like to thank the MENRT, the ANVAR and the Conseil Regional d'AUVERGNE for their support, and also the colleagues of the SMIR group of LIMOS UMR 6158 CNRS for their help.

References

1. Goldberg, S. and Wickramasinghe N. (2003) 21st Century Healthcare: the Wireless Panacea in *Proceedings of the 36th Hawaii International Conference on Systems Sciences.*
2. Varshney, U. (2005) Pervasive Healthcare: Applications, Challenges and Wireless Solutions *Communications of the Association of Information Systems,* 16(3).
3. Tablado, A. Illarramendi, A., Bermudez, J., and Goni, A. (2003) Intelligent Monitoring of Elderly People, in *Proceedings of the 4th Annual IEEE Conference on Information Technology Applications in Biomedicine.*
4. World Health Organization (1999) Death by Cause, Sex and Mortality Stratum in WHO Regions, *WHO General Report.*

5. Chinese National Aging office (2006) Forecast Study of the Development Trend of China's Population Aging Issue, *General Report* (in Chinese).
6. Zou, N. J. (2007) Old-Age Security, Old-Age Home and Community Support: New Option for Old-Age Models, *Jiangsu Social Sciences*, 4 (in Chinese).
7. Chinese National Bureau of Statistics (2006) *China's Demographic Yearbook2006*, pp. 289 (in Chinese).
8. Zhu, Z. M. (2010) Health Ubiquitous Service: Exploring Pervasive Computing in Health/Medical Applications, http://www.bjkp.gov.cn/bjkpzc/kjqy/it/lsywl/ 250584.shtml (in Chinese).
9. Yang, T. Li, Z. G. and Shi, Y. C. (2007) Study Elderly Care Services by Integration With the Community Health Services in Rural China, *General Report* (in Chinese).
10. Yan, J. L. and Wei, S. (2009) The Health Education Problems and Countermeasures for Urban Elderly in Community, *Modern Nursing*, 35(6), pp. 122 (in Chinese).
11. Chronic Care Improvement (2004) A Product of the E-Health Committee, *ITAA E-Health White Paper*, pp. 5.
12. Sneha, S. (2008) Patient Monitoring via Mobile Ad Hoc Networks: Maximizing Reliability While Minimizing Power Usage and Delays, *Doctoral Dissertation*, Georgia State University.
13. Website (2010) US Administration on Aging Report on Demographic Changes, http://www.aoa.dhhs.gov/aoa /stats/aging21 /demography.html.
14. Chiasson, M. Davidson, E. Kaplan, B., et al. (2004) Strangers in a Strange Land: Can IS Meet the Challenges and Opportunities of Research in Healthcare? in P*roceedings of the 10th Americas Conference in Information Systems.*
15. Gouaux, F. Simon-Chautemps, L. Adami, S., et al. (2003) Smart Devices for the Early Detection and Interpretation of Cardiological Syndromes, in *Proceedings of the 4th International IEEE EMBS Conference on Information Technology Applications in Biomedicine.*
16. Website (2010) US Patient Monitoring Industry Overlook., http://www.frost.com/prod/servlet/report-homepage.pag ?repid=A369-01-00-00-00.
17. Sneha, S. and Varshney, U. (2009) Enabling Ubiquitous Patient Monitoring: Model, Decision Protocols, Opportunities and Challenges, *Decision Support Systems*, **46**, pp. 606–619.

18. Bardram, J. E. (2008) Pervasive Health Ccare as a Scientific Discipline, *Methods Information Medicine*, **47(3)**, pp. 178–185.
19. Hung, K. and Zhang, Y. (2003) Implementation of a WAP-Based Telemedicine System for Patient Monitoring, *IEEE Transactions on Information Technology in Biomedicine* **7(2)**, pp. 101–107.
20. Sneha, S. and Varshney, U. (2005) Wireless ECG Monitoring System for Pervasive Health Care, in *Proceedings of the 11th Americas Conference on Information Sys*tems.
21. Zhou, H. Y. and Hou, K. M. (2010) Pervasive Cardiac Monitoring System for Remote Continuous Heart Care, in *Proceedings of International Conference on Bioinformatics and Biomedical Engineering.*
22. Website (2010) Patient Monitoring by Welch Allyn, http://www.monitoring. welchallyn. com/products/wireless.
23. Mendoza, G. G. and Tran, B. Q. (2002) In-Home Wireless Monitoring of Physiological Data for Heart Failure Patients, in *Proceedings of the 24th Joint IEEE EMBS/BMES Annual Conference.*
24. Website (2010) TV-Based Monitoring by Philips http://www.medical.philips.com/main/ news/content/file_630.html.
25. Website (2010) Cardionet, http://www.cardionet.com/.
26. Website (2010) Biotronik, http://www.biotronik.com/.
27. Bonato, P. (2003) Wearable Sensors/Systems and Their Impact on Biomedical Engineering, *IEEE Engineering Medicine and Biology Magazine* **22(3)**, pp. 18–20.
28. Gieras, I. A. (2003) The Proliferation of Patient-Worn Wireless Telemetry Technologies Within the US Healthcare Environment, in *Proceedings of the 4th International IEEE EMBS Special Topic Conference on Information Technology Applications in Biomedicine.*
29. Jovanov, E. O'Donnell Lords, A. Raskovic, D., et al. (2003) Stress Monitoring Using a Distributed Wireless Intelligent Sensor System, *IEEE Engineering in Medicine and Biology Magazine* **22(3)**, pp. 49–55.
30. Website (2010) Smart Shirt, http://www.smartshirt.gatech.edu.
31. Brahnam, S. Chuang, C. F., Sexton, R. S. and Shih, F. Y. (2007) Machine Assessment of Neonatal Facial Expressions of Acute Pain, *Decision Support Systems* **43(4)**, pp. 1242–1254.
32. Hu, P. J. H., Wei, C. P. Cheng, T. H. and Chen, J. X. (2007) Predicting Adequacy of Vancomycin Regimens: a Learning-Based Classification Approach to Improving Clinical DecisionMaking, *Decision Support Systems* **43(4)**, pp. 1226–1241.

33. Lin, L. J-HHu, P. and Sheng, O. R. L. (2006) A Decision Support System for Lower Back Pain Diagnosis: Uncertainty Management and Clinical Evaluations, *Decision Support Systems*, **42(2)**, pp. 1152–1169.
34. Varshney, U. (2008) A Framework for Supporting Emergency Messages in Wireless Patient Monitoring, *Decision Support Systems*, **45(4)**, pp. 981–996.
35. Website (2010) Gator Tech Smart House, http://www.icta.ufl.edu/gt.html.
36. Website (2010) Aware Home Project, http://www.awarehome.gatech.edu/projects/ index.html.
37. Stanford, V. (2002) Using Pervasive Computing to Deliver Elder Care, *IEEE Pervasive Computing Magazine* **1(1)**, pp. 10–13.
38. Kafeza, E., Chiu, D. K. W., Cheung, S. C. and Kafeza, M. (2004) Alerts Immobile Health Care Applications: Requirements and Pilot Study, *IEEE Transactions on Information Technology in Biomedicine* **8(2)**, pp. 173–181.
39. Haux, R. Howe, J. Marschollek, M., Plischke M. and Wolf, K. H. (2008) Health-Enabling Technologies for Pervasive Health Care: on Services and ICT Architecture Paradigms, *Informatics for Health and Social Care* **33(2)**, pp. 77–89.
40. Koch, S. and Hägglund, M. (2009) Health Informatics and the Delivery of Care to Older People, *Maturitas* **63(3)**, pp. 195–199.
41. Doukas C. and Maglogiannis, I. (2008) Intelligent Pervasive Healthcare Systems, *Advanced Computational Intelligence Paradigms in Healthcare* (Springer Press), **107**, pp. 95–115.
42. Daramolar, J. O. Osamor, V. C., and Oluwagbemi, O. O. (2008) A Grid-based Framework for Pervasive Healthcare Using Wireless Sensor Networks: a Case for Developing Nations, *Asian Journal of Information Technology*, **7(6)**, pp. 260–267.
43. Salvador, Z. Larrea M. and Lafuente, A. (2007) Infrastructural Software Requirements of Pervasive Health Care, in *Proceedings of IADIS International Conference on Applied Computing*, pp. 557–562.
44. Zhou, H. Y. (2004) Wireless Sensor Networks Dedicated to Remote Continuous Real-Time Cardiac Arrhythmias Detection and Diagnosis, Ph.D Thesis, University of Blaise Pascal Clermont-Ferrand II (France).
45. Zhou, H. Y. et al. (2006) A New System Dedicated to Real-Time Cardiac Arrhythmias Tele-Assistance and Monitoring, *Special Issue of Journal of Universal Computer Science*, **12(1)**, pp. 30–44.

46. Zhou, H. Y. et al. (2004) Remote Continuous Cardiac Arrhythmias Detection and Monitoring, *Transformation of Healthcare With Information Technologies, Studies in Health Technology and Informatics* (IOS Press), **105**, pp. 112–120.
47. Zhou, H. Y. et al. (2005) A Real-Time Continuous Cardiac Arrhythmias Detection System: RECAD, in *Proceedings of the 27th Annual International Conference of the IEEE Engineering in Medicine and Biology* **(EMBC05)**, pp. 875–881.
48. Zhou, H. Y. Hou, K. -M. and Zuo, D. C. (2009) Real-Time Automatic ECG Diagnosis Method Dedicated to Pervasive Cardiac Care, *Wireless Sensor Network*, **1(4)**, pp. 276–283.
49. Zhou H. Y. and Hou, K. -M. (2008) Embedded Real-Time QRS Detection Algorithm for Pervasive Cardiac Care System, in *Proceedings of the 9th International Conference on Signal Processing*, pp. 2150–2153.
50. De Sousa, G. Zhou, H. Y., Hou, K. -M. et al. (2007) Adaptive System for Wireless Sensor Networks Applications, *Journal of Harbin Institute of Technology* **39**, pp. 153–157.
51. Bailey, J. J. Berson, A. S. Garson A. et al (1990) Recommendations for Standardization and Specifications in Automated Electrocardiography: Bandwidth and Digital Signal Processing, *Circulation* American Heart Association **81(2)**, pp. 730–739.
52. De Vaulx, C. and Hou, K. -M. (2002) DREAM: un micro noyau réparti, temps réel orienté pour la tolérance aux fautes, *Revue Informatique et santé*, Télémédecine et e-santé, **13**, pp. 63–69.
53. Gineste, L (2002) Plate-forme de suivi à distance d'arythmie cardiaques, *DEA CSTI report*, University of Blaise Pascal Clermont-Ferrand II, France.
54. Palau, P. Hou, K. -M. and Ponsonnaille, J. (2002) TelmedTCP: Protocole TCP/IP temps réel dédié à la télémédecine, *Revue Informatique et santé*, Télémédecine et e-santé, **13**, pp. 53–60.
55. PhysioNet (2002) The WFDB Software Package Software for Viewing, Analyzing, and Creating Recordings of Physiologic Signals, http://www.physionet.org/physiotools/ wfdb.shtml.
56. Moody G. B. and Mark, R. G. (1990) The MIT-BIH Arrhythmia Database on CD-ROM and Software for Use With It, *Computers in Cardiology* **17**, pp. 185–188.
57. Zhou, H. Y. Hou, K. -M. Ponsonnaille, J. et al. (2004) Remote Continuous Cardiac Arrhythmias Detection and Monitoring, in *Proceedings of the 2nd International Conference on E-health in Common Europe*.

58. Dotsinsky I. A., and Stoyanov, T. V. (2004) Ventricular Beat Detection in Single Channel Electrocardiograms, *BioMedical Engineering OnLine* 3(3), doi:10.1186/1475-925X-33.
59. Afonso, V. X. Tompkins, W. J. Nguyen T. Q. and Luo S. (1999) ECG Beat Detection Using Filter Banks, *IEEE Transactions on Biomedical Engineering*, **46**, pp. 192–202.
60. Poli, R. Cagnoni, S. and Valli G. (1995) Genetic Design of Optimum Linear and Nonlinear QRS Detectors, *IEEE Transactions on Biomedical Engineering* **42**, pp. 1137–1141.
61. Kaiser W. and Findeis M. (2000) Novel Signal Processing Methods for Exercise ECG, Special Issue on Electrocardiography in Ischemic Heart Disease, in *Proceedings of IJBEM.*
62. Datex-Ohmeda Corp. (2002) Bedside Arrhythmia Monitoring Quick Guide, Internal Web Journal for Medical Professionals, http://www.clinicalwindow.com.
63. Millet, J. Perez, M. Joseph, G. Mocholi A. and Chorro J. (1997) Previous Identification of QRS Onset and Offset Is Not Essential for Classifying QRS Complex in a Single Lead, *Computers in Cardiology*, **24**, pp. 299–302.
64. Zhou, H. Y. Hou, K. -M. de Vaulx C. and Zuo, D. C. (2009) A Hybrid Embedded Real-Time Operating System for Wireless Sensor Networks, *Journal of Networks*, **4(6)**, pp. 428–435.
65. Ding, H. Hou, K. M. Lecoq J. et al. (2009) Toward a Low Cost and Single Chip Holter: SoC-Holter, *in Proceedings of the 3rd International Conference on New Technology, Mobility, and Security-2009.*

Chapter 4

Human Bio-Kinematic Monitoring with Body Area Networks

Roozbeh Jafari, Hassan Ghasemzadeh, Eric Guenterberg, Vitali Loseu, and Sarah Ostadabas

Embedded Systems and Signal Processing Laboratory, Department of Electrical Engineering, University of Texas at Dallas
Richardson, TX, 75080, USA
rjafari@utdallas.edu; h.ghasemzadeh@utdallas.edu; mavpion@gmail.com; vitali.loseu@utdallas.edu; and sarahostad@student.utdallas.edu

Body Area Networks (BANs), known as enabling technology for many biomedical applications, are composed of body-worn sensor devices that can provide mobile and continuous monitoring of the human body. This chapter presents an overview of platform design strategies for BANs with applications in physical movement monitoring. First, an introduction to several compelling applications is given, which shows BAN versatility in both medical and recreational fields. Applications are important in the sense that they help in understanding design requirements of the BAN platform. An architecture of the system, including hardware and software components, is then described. It is followed by a description of typical signal processing for movement monitoring applications. While this type of signal processing flow can be generally used in movement monitoring applications, in order to take the most

Wireless Body Area Networks: Technology, Implementation, and Applications
Edited by Mehmet R. Yuce and Jamil Y. Khan

ISBN 978-981-4316-71-2 (Hardcover), ISBN 978-981-4241-57-1 (eBook)
www.panstanford.com

advantage of the system, the signal processing needs to be custom tailored for each individual application. An example of this process is shown based on the Hidden Markov Model (HMM) movement annotation applications. Finally, the chapter is concluded with discussion of possible BAN system optimizations. In particular, it is shown that the energy consumption of the system can be reduced by using buffers to decrease the number of transmissions.

4.1 Physical Movement Monitoring

A BAN system uses several sensor units each equipped with motion sensors, processing units, and wireless and memory components. The system aims to collect sensor readings and transform them into useful information. In order for a wearable system to be successful, it is required to be comfortable and not introduce any additional movement constraints. The main deployment objective of any BAN is to improve its wearability and the ease of use. This idea has a few consequences. First, it forces sensor units to be powered by a battery. Using a large energy source would hinder mobility and, therefore, is not acceptable. Second, wearability concerns rule out a wired communication scheme because a set of wires all over the body is not natural and may constrain some human movements. Finally, reducing the form factor of sensor units is crucial for wearability. It can be done via reducing the size of the sensor unit's components such as the processor, memories, sensors, and the battery. While technology made a wide leap in reducing the size of microprocessors, memories, and sensors, the progress in battery size reduction has not been as fertile [1]. It suggests that currently the battery size dominates the form factor of the wearable sensor units. The requirement of the extreme battery efficiency motivates the need for light-weight and highly efficient signal processing algorithms and optimization techniques. The signal processing, however, needs to exhibit sufficient reliability and sensitivity in extracting the relevant parameters.

BAN systems are desirable because they provide objective, quantitative measurements while not restricted to a laboratory environment. Furthermore, depending on the number and locations of the sensing units, BAN systems can vary greatly in the scope of possible tasks from a general action recognition to extracting a very specific detail about a movement. This property makes BAN systems extremely useful in a large set of applications. In particular, they find applications in rehabilitation [2], sports medicine [3], geriatric care [4], gait analysis [5], balance evaluation, [6] and sports training [7, 8].

4.2 Applications

BANs can collect a wide range of information that monitor different aspects of human life from movement monitoring and emotion recognition [9, 10] to sports training and leisure applications. Applications can be divided into two categories, including medical and nonmedical applications. With traditional health care, patients are either observed by doctors who rely on personal experience to identify symptoms and the severity of a condition or have to be examined in a laboratory environment. With wearable mobile sensors, however, remote and continuous monitoring the patients is available. Sports training is another important application area of BANs. Quantitative feedback provided by wearable sensors can be used to provide atheletes with better instruction on improving their game without an involved participation from a human expert trainer.

4.2.1 *Medical Applications*

An emerging application for BANs involves their use in medical application. BAN area revolutionizes the health care system by allowing inexpensive, continuous, and ambulatory health monitoring with real-time capabilities. Using human bio-kinematic monitoring that provides sensor data as a biometric of body functionality is of increasing interest since it is noninvasive and can be measured without subject contact or knowledge. In clinical application, as therapists increase their emphasis on evidence-based practice, they must increase the use of objective, quantitative methods to demonstrate efficacy of their method. For instant, gait analysis is an proper method for demonstrating change from treatment or from disease progression. Gait analysis by means of BANs for the person with Parkinsons disease (PD) can be used as a tool for determining the treatment efficacy of pharmacologic, surgical, or physical therapeutic interventions. In this section, we review some of the existence methods for gait analysis and Parkinson's disease assessment.

4.2.1.1 Gait analysis

In the human action recognition, gait analysis is the systematic measurement, description, and assessment of those quantities thought to characterize human locomotion. Through gait analysis, kinematic and kinetic data are acquired and analyzed to provide information, which describes fundamental gait characteristics and which is ultimately interpreted by clinicians to form a medical assessment.

For the purpose of gait analysis, wearable sensors of the BAN can be attached at the externor side of the thigh [11]. The hip angle is defined as the angle between the thigh and gravity direction. The swing velocity (angular velocity) of the thigh is defined as $v = \frac{d\theta}{dt}$. Kalman filter is applied to estimate θ and v, which are key features of the gait cycle. Another proposed method for gait analysis is a generic algorithm for temporal parameter extraction called the hidden Markov event model based on HMMs. This method constrains the state structure to facilitate location of key events of gait [12].

In the context of signal processing, gait recognition experiments using spectral features in terms of magnitude, phase, and phase-weighted magnitude show that both magnitude and phase spectra are effective gait signatures. In [13], authors proposed a gait recognition approach using spectral features of horizontal and vertical movement of ankles in a normal walk. They used an integration of magnitude and phase spectra for gait recognition using AdaBoost classifier. At each round, a weak classifier evaluates each magnitude and phase spectra of a motion signal as dependent subfeatures, then classification results of each subfeature are normalized and summed for the final hypothesis output. Authors in [14] investigate an ear- worn sensor for the development of a gait analysis framework. Instead of explicitly defining gait features that indicate injury or impairment, an automatic method of feature extraction and selection is proposed.

4.2.1.2 Parkinson's disease assessment systems

Many efforts have been made to come up with objective measures for motor impairment and disability assessment in Parkinson's disease. Hoehn and Yahr Clinical staging [15], Webster Rating scale [16], Columbia University Rating scale [17], the short Parkinson's evaluation scale (SPES) [18] and the Unified Parkinson's Disease Rating Scale (UPDRS) [19] are a few of them. The UPDRS has become the gold standard of rating scales for PD detection and assessment [20], as it is one of the most evaluated and reliable, bearing close correlation to other widely accepted standards at the same time [21].

Systems have been designed for PD assessment as described in [22] and [23] albeit for specific activities like recording tremor and measuring sleep disorder [24]. In [25], Homann et al. examined a method to measure bradykinesia and akinesia using finger-tap test on a computer. Rajaraman et al. [26] demonstrated a 3D tremor measurement system for comparing tremor across patients and in measuring the efficacy of therapeutic interventions. Their system employed 3D electromagnetic position sensors to measure the actual and cumulative displacement of the tremoring finger. The tremor frequency and amplitude are estimated using spectral analysis on the data after low-pass filtering at 15 Hz. Systems for clinical measurement of tremor have also been demonstrated in [27] and [28], but unrelated to the assessment of any disease in particular.

The methods used for the measurements have varied from electromyography (EMG) [29, 30] to accelerometers [23, 31], electromagnetic sensors, [26] and custom transducers [22]. Hassan et al. [32] examines light-weight sensor devices which enable on-body and mobile health-care monitoring. Goetze et al. [20] employed an at-home testing device for testing motor impairment in PD, which consisted of a dedicated apparatus for conducting the tests. The device consisted of a testing panel on the base that contains a two-key keyboard for finger tapping, two buttons for reaction time/movement time and repetitive hand tapping assessments, an eight-peg pegboard, control buttons, and a docking station for an actiwatch device measuring tremor.

4.2.2 *Sports Training Application*

Sport training represents the body's adaptation to conditions of certain exercises. One can achieve considerable progress in a sport with the aid of appropriate exercises and training methods [33]. The biggest challenges of sports training is the evaluation of a movement and an ability to provide feedback regarding movement's quality. Traditionally, these two tasks are performed by a human coach who uses personal expertise and experience to evaluate exercises and provide feedback. However, personal coaches are expensive and, to make training more affordable, often coach multiple students at the same time, thus diverting attention from detailed diagnosis of problems of individual students. An automated system that can assess the overall performance of a learner and pinpoint problem areas in the learners movements would facilitate performance assessment, increasing the effectiveness of unsupervised practice.

Traditionally, mechanical training systems [34, 35] have been used to provide movement feedback. However, these types of systems lack fine grain details of movements. Authors in [36] use a motion capture system to record and analyze tennis movements. They show that alternative forehand and backhand movements outperform discrete forehand or backhand practices due to the inertia of the trunk rotation movements between subsequent strokes. The approach in [37] tracks the movements in rugby football using close-range or long-range photography. Shoulder orientation and trunk flexion are demonstrated to be significantly different in experienced players and novices and, therefore, can be used to measure players' performance and training quality. A motion capture system is also used in [38] to design a virtual baseball training system. In the system, a batter swings a bat toward the virtual ball, and the trajectory of the swing is used to calculate the quality of the swing. While these systems tend to perform well, they require an expensive laboratory setup and, therefore, are not desirable. A BAN-based solution to sports training is either based on a solution where sensors are placed in the environment [39] or embedded within the sports equipment [40].

Overall, there are two areas of interest in sports training. Training can focus on a precise execution of each movement individually. Training can also focus on coordination in execution of series of movements. While these two components are connected, they can be considered individually. Next two sections demonstrate how BAN systems can address both of these cases.

4.2.2.1 Golf swing training

The popular sport of golf requires a complicated sequence of motions to swing the golf club properly with the primary goal of propelling the golf ball a certain distance in a desired direction. A repeatable and consistent golf swing can dramatically improve a golfer's score. However, this single movement, which has such a major impact on the player's overall game, is difficult to master and execute consistently for players who are new to the sport or have little experience.

A full swing is a complex motion of the body aimed at accelerating the club at great speed. The motion starts at an initial position, referred to as the address position [41], followed by the swing. A golf swing can be divided into smaller segments. A general golf swing model considers a full swing composed of four major segments: takeaway, backswing, downswing, and follow-through [42]. *Takeaway* starts as the first movement after the address position and ends when the club is approximately parallel to the target line and at waist level. The *backswing* follows the takeaway and continues until the golf club is lifted to its highest point behind the player. Following this is the *downswing* in which the club is brought back down to hit the ball. After impact with the ball, the *follow-through* motion brings the club to its stopping point in front of the player. This model enables more precise analysis of the actions by reporting the quality of each individual part of the swing.

The goal in achieving a perfect swing is to hit the ball squarely and straight [43]. This would also give the golfer maximum distance. Consequently, it is important to investigate actions that prevent development of a perfect swing. According to the literature, there are two kinds of common mistakes new players make, resulting in a poor shot [41]: *wrist rotation* and *out-of-plane* movements. To simplify the system, a virtual coach can address each one of the issues individually. For example, the system can focus on evaluating golf swing in terms of the angle of wrist rotation. This information can be obtained by placing sensing units on the upper body and arms to capture significant motions during the swing [44] and placing two sensing units on the golf club to capture its movements [7].

4.2.2.2 Baseball swing training

Baseball batting involves hitting a thrown ball with the primary objective of transferring maximum force to propel the ball as far as possible in a desired direction. Numerous baseball players and coaches have suggested methods for successful batting. A good swing is the result of a sequence of rotational movements including foot, knees, hips, shoulder, and hands movements. Generally, the action of the batter starts in the lower body and moves upward. Properly performed motions executed at the right time maximize the power of the swing. Major components of a good swing include bat speed, bat swing plane, and timing. Common mistakes include late rotation of lower body, back shoulder dip, and drifting of the front foot. Late movement of the foot and hips impairs the swing timing. Dropping the back shoulder affects the bat plane so as the bat does not pass through the strike zone horizontally, decreasing the chance of a successful hit. Drifting refers to improper weight transfer from the back foot to the front foot. One consequence is losing power in the hips, which decreases the bat speed at impact. Therefore, proper weight transfer necessitates coordination between different body segments during the swing.

The considered swing model emphasizes three major events: 1) Rotation of the lower body (feet, knees, hips) toward the pitcher, 2) rotation of the upper body into the swing, and 3) the swing of the arms and hands toward the pitcher. These key events should be executed in a specific and overlapping sequence. The coordination is extremely important as it ensures that the maximum power from arms, shoulders, and hips is delivered exactly as the bat crosses the plate [45]. In order to recognize movement coordination, sensing units placed on the ankle, hip, upper body, and the bat have to identify body movements and their timings. It can be achieved with detailed body, choreography modeling with motion transcripts introduced in [8].

4.3 Hardware and Software Architecture

The purpose of an action recognition system is to classify transitional movements into pre-defined actions. Given a set of movements, the system must distinguish between every pair of motions. These sensors capture inertial information from physical movements. An example platform used to generate data and achieve results in this chapter is shown Fig. 4.1. The system consists of several sensor units; each has a tri-axial accelerometer, a bi-axial gyroscope, a microcontroller, and a radio. In this specific setup, the accelerometers are LIS3LV02DQ with 1024 LSb/g sensitivity. The IDG-300 gyroscopes have 2 mV/°/s sensitivity. The processing unit of each node, or mote, can sample sensor readings at certain rate and transmit the data wirelessly to a base station. A common protocol

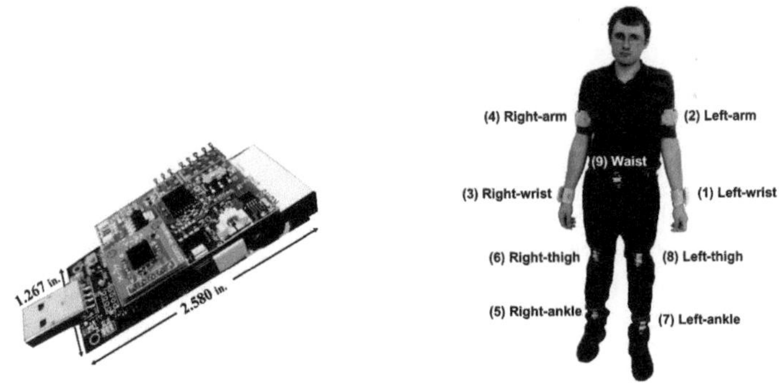

(a) Sensor node (b) Subject

Figure 4.1. (a) A sensor node composed of processing unit and custom-designed sensor board. The motion sensor board has a triaxial accelerometer and a biaxial gyroscope Sensor node and subject (b) A subject wearing nine sensor nodes. See also Color Insert.

for this transmission is Time Division Multiple Access (TDMA). For the purpose of action recognition, authors of [46] use the TelosB motes [47], which are commercially available from XBow®. The sensor board they use is predominantly a custom designed with an integrated Li-Ion battery that powers both sensor board and mote as shown in Fig. 4.1a. The sensor nodes can be placed on different locations on the body to capture movements of their subjects. Figure 4.1b shows a subject with nine nodes placed on different body segments.

Signal processing for body sensor network usually comprises multiple levels of data abstraction, from raw sensor data to data calculated from processing steps. SPINE (Signal Processing in Node Environment) [48] is a software framework that aims to enable efficient implementations of signal processing on sensor nodes in BAN. SPINE is a framework for distributed signal processing founded on the following principles:

Open source: SPINE is developed as an open source project to establish a broad community of users and developers that contributes to extend the framework with new capabilities and applications.

Interoperability through APIs: SPINE provides local and remote applications with lightweight Java APIs that can be used by local and remote applications to manage the sensor nodes or issue service requests, and are easily portable to devices of various capabilities, such as PCs or mobile phones, that can be used as BAN coordinator.

High-level abstractions: SPINE provides libraries of protocols, utilities, and processing functions; hence, it simplifies the task of application developers by raising the level of abstraction. The layer defined by the SPINE libraries allows designers to focus on application-specific issues and program at a higher level of abstraction than TinyOS.

Distributed implementations of classification algorithms: SPINE simplifies the development of applications that require complex signal processing algorithms and classifiers. For example, SPINE supports distributed implementation of classification algorithms to reduce the amount of data to be transmitted and save energy.

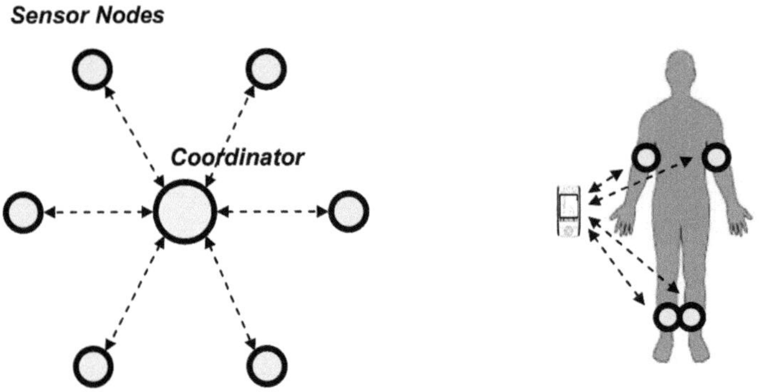

Figure 4.2. SPINE 1.0 Network Architecture [48].

SPINE relies on a BAN architecture with star topology, including one or more sensor nodes and a BAN coordinator node. The coordinator typically manages the BAN, collects and analyzes the data received from the sensor nodes, and acts as a gateway to connect the BAN with wide area networks for remote data access. Figure 4.2 visualizes a typical architecture with star topology.

4.4 Signal Processing for Body Area Networks

Figure 4.3 shows several processing tasks typically used for signal processing and action recognition in BANs. Each processing block is described as follows.

Sensor data collection: Data is collected from all of the sensors on each of the nodes at a specified frequency. The sampling rate can be empirically chosen to provide sufficient resolution while compensating for bandwidth constraints of the system [49], or it can be determined to satisfy the Nyquist criterion [50]. Usually, a 20 samples per seconds would provide fine details of human movements [51].

Preprocessing: Data is filtered with a small window moving average to remove high frequency noise. The number of points used to average the signal can be chosen by examining the power spectral density of the signals. The filter is required to remove unnecessary motions (e.g., tremors) while maintaining significant data. With

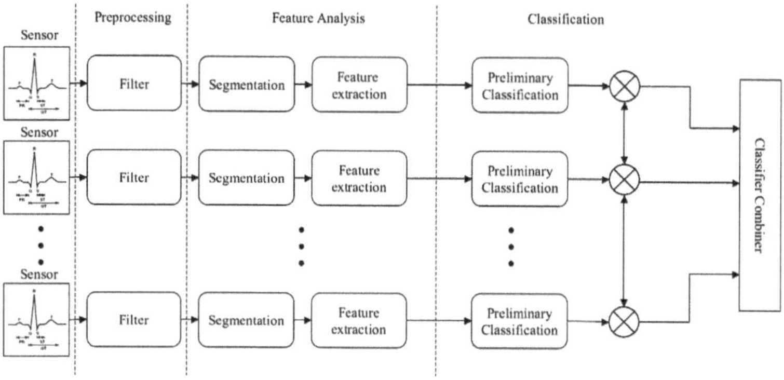

Figure 4.3. Signal processing flow.

these objectives, authors in [52] test several moving average filters with varying window sizes and choose the filter that best satisfies the aforementioned requirements.

Segmentation and annotation: Segmentation algorithms divide continuous data streams into discrete time intervals of the type expected by the information processing step, while annotation algorithms locate and label specific events. To enable real-time movement monitoring, an automated method is required.

Feature extraction: Statistical and morphological features are extracted from the signal segment. For example, sensor readings can be transformed into a set of informative attributes, including mean, start-to-end, standard deviation, peak-to-peak amplitude, RMS power, median, and maximum value.

Per-node classification: Each node uses the feature vector generated during feature extraction to determine the most likely action. Due to its simplicity and scalability, k-Nearest Neighbor (k-NN) [53] is a widely used classifier [54].

Final classification: The final decision can be made using either a data fusion or a decision fusion scheme. In the data fusion, features from all sensor nodes are fed into a central classifier. The classifier then combines the features to form a higher dimensional feature space and classifies movements using the obtained features. In the decision fusion, however, each sensor node makes a local classification and transmits the result to a central classifier where a final decision is made according to the received labels.

In Section 4.5, we focus on the human movement annotation. Many movement models divide an action into several parts. In sports, many swings can be divided into phases. The golf swing is separated into takeaway, backswing, downswing, and follow-through portions [55]. In walking, the human stride is marked by several events such as initial stance (the foot placed on the ground), mid-stance, initial swing (the foot has just been lifted from the ground), and mid-swing, which repeat indefinitely [56]. These divisions can be used directly. For instance, high standard deviation of stride time during walking is indicative of Parkinson's disease or Huntington's disease and can be used to assess the risk of falling [57, 58]. We introduce a model based on HMMs that divides walking into aforementioned events. Inertial sensors provide movement data directly; therefore, these events can be found by looking for patterns in the sensor data.

4.5 An Automatic Parameter Extraction Method Based on HMM

Human movement models often divide movements into parts. These parts are often divided based on key events, also called temporal parameters. When analyzing a movement, it is important to correctly locate these key events, and so automated techniques are needed. There exist many methods for dividing specific actions using data from specific sensors, but for new sensors or sensing positions, new techniques must be developed. To address this problem, this section introduces a generic method for temporal parameter extraction called the Hidden Markov Event Model based on HMMs. This method can be quickly adapted to new movements and new sensors/sensor placements. This method is validated on a walking dataset using inertial sensors placed on various locations on a human body. The technique is designed to be computationally complex for training, but computationally simple at runtime to allow deployment on resource-constrained sensor nodes.

4.5.1 *HMEM Training and Use*

The Hidden Markov Event Model (HMEM) is the name of the introduced key event labeling system, which uses an HMM with a specific state structure and a modified training procedure designed to find key events. The model also adds a feature selection and model parametrization system based on Genetic Algorithms (GAs). The HMEM makes several assumptions about the underlying data: 1) there are a number of different event types, 2) the events always occur in a specific order and for cyclical movements they repeat, 3) every single event type is represented in every action, and 4) there are a number of unlabeled samples between two adjacent events.

A traditional pattern recognition technique used for time-varying signals is the Hidden Markov Model. A basic HMM describes a discrete-time Markov process. At a particular moment the process is in just one state. At fixed time intervals, the process produces an output and then transitions to another state (or remains in the current state). The transitions and outputs are probabilistic and based exclusively on the present state. The process generates a sequence of outputs, and a corresponding sequence of states. The states cannot be directly observed, and are thus hidden. An HMM is completely described by initial state probabilities, state transition probabilities, and output probabilities. Algorithms exist to 1) build an HMM to describe a given set of output sequences, 2) choose which of several models best describe an output sequence, 3) find the most likely current state of an HMM given the output sequence up until now, and 4) the most likely state sequence associated with a particular output sequence for a specified HMM [59, 60].

For a general HMM, it is possible for any state to transition to any other state. This is called an ergodic model. Another variant is to enforce a specific ordering for the states: each state can only transition to itself or state to the "right" of it in the ordering. This is called a left-right model [60].

Each key event can be represented by a unique state. Ideal events occur at a specific time but have no duration. However, given the idea that the key event might be associated with unique features in the

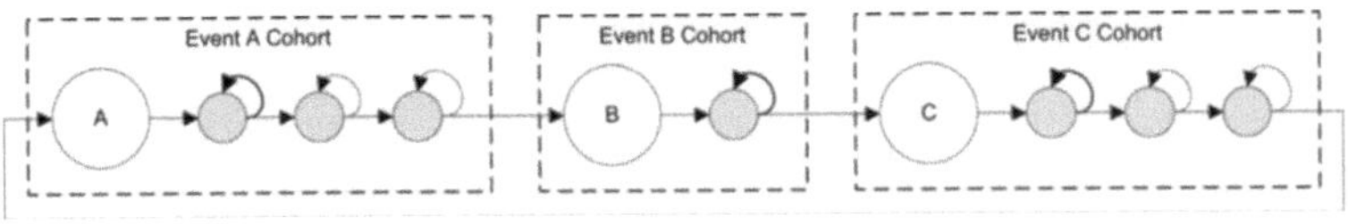

Figure 4.4. HMEM model and structure. See also Color Insert.

observation sequence, the key event state should have a one-sample duration. The HMEM encodes this concept into an HMM by removing the self-transition from states associated with key events, forcing a transition after one sample. The samples between key events are represented by transition states which support both self-transitions and forward-transitions, as seen in Fig. 4.4 states are grouped into cohorts, which start with a key event state and end with the last transition state before the next key event state. For any observation sequence in the training data, the positions of the key event states are known. This means that training each cohort independently is identical to training the whole system at once.

4.5.2 *Overview*

There are several stages required to train the HMEM as shown in Fig. 4.5.

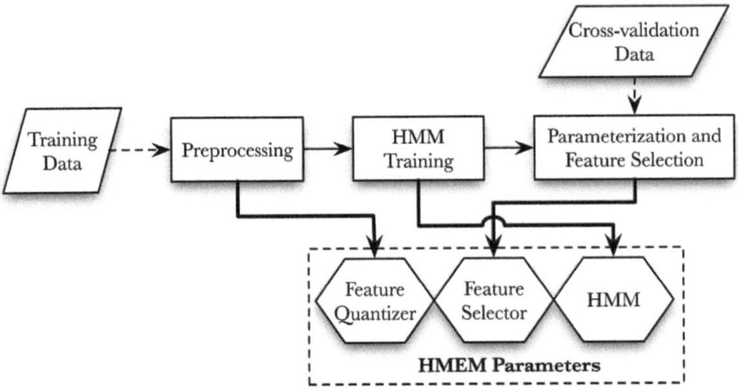

Figure 4.5. HMEM training procedure.

4.5.2.1 Preprocessing and feature extraction

The signal is filtered with a moving average filter to remove high frequency noise. Then it is normalized by subtracting a large-window mean and dividing by a large-window standard deviation. Several parameters representing the action data inside the signal, referred to as features, are extracted at each sample. These features are further quantized with a 10-level uniform quantizer based on the range of the features in the training data [10].

4.5.2.2 HMM training

The HMM is effectively a finite state machine with probabilistic transitions and certain emission probabilities. These probabilities must be specified as part of the mathematical model defining an HMM. The exact locations of all key event states and what observations they emit are known from the annotations in the training data. However, the number of transition states and their transition and emission probabilities are unknown and must be trained. There are several well-known techniques for training HMMs, including Baum-Welch and Viterbi Path Counting [59, 61].

The training data is segmented using the canonical annotations. Each cohort is trained independently using a set of segments that start with a sample that should be labeled with the cohorts event and end just before the next labeled event. According to this model, the first state must be the cohort's event state, and the last sample must be associated with the last transition state in the cohort. During training, it is important to make sure that all considered paths meet this constraint. Viterbi Path Counting produces a single path for each event that can be edited to meet the constraints if necessary, while Baum-Welch can also be constrained in this way, VPC is much faster, which is important given the already high training times. This feature is one of the primary reasons for choosing VPC over Baum-Welch. The details of the training process are discussed in Section 4.5.3.

4.5.2.3 Parametrization and feature selection

HMMs are trained to represent a process, not to minimize segmentation error. It is possible to explicitly attempt to increase classification accuracy by choosing model parameters with that goal in mind. A genetic algorithm with uniform crossover is used to train the model. The population fitness is evaluated using the training model, and then at the end, the model that gives the best results for the cross-validation data is selected. Further discussion follows in Section 4.5.4.

4.5.3 *HMM Training and the Viterbi Algorithm*

A Markov process has N states $S = \{s_1, s_2, \cdots, s_N\}$, and can emit M observations $X = \{x_1, x_2, \cdots, x_M\}$. For a given observation sequence $O_T = (o_1, o_2, \cdots, o_T)$ with T observations, there is a corresponding state sequence $Q_{(T)} = (q_1, q_2, \cdots, q_T)$. The HMM $\lambda = \{\pi, A, B\}$ is defined by three sets of probabilities: initial state probabilities $\pi = \{\pi_i | \pi_i = P(q_o = s_i)\}$, state transition probabilities $A = \{a_{ij} | a_{ij} = P(q = s_j | q_{prev} = s_i)\}$, and observation probabilities $B = \{b_j(k) | b_j(k) = P(o = x_k | q = s_j)\}$.

The most common training algorithm is the Baum-Welch algorithm [59], however a newer algorithm, the Viterbi Path Counting (VPC) algorithm is more appropriate for this work [61]. Both algorithms follow the training procedure shown in Algorithm 4.1. They start with a fixed number of states and an initial set of model parameters, then extract probabilistically weighted state sequences

---

Algorithm 4.1 HMM Training Procedure

Require: $\lambda_0, \hat{} = \{O^1_{(T_1)}, O^2_{(T_2)}, \cdots, O^Y_{(T_Y)}\}$

$\lambda \leftarrow \lambda_0$

for $i \leftarrow 1$ to K **do** {estimates from $\tilde{}$}

  $\tilde{} \leftarrow \emptyset$

  for all $O_{(T)} \in \hat{}$ **do**

    $Q_{(T)} \leftarrow$ collect weighted state sequences using λ

    $\tilde{} \leftarrow \tilde{} \cup Q_{(T)}$

  end for

  $\bar{\pi}_i = \dfrac{\text{number of times in state } s_i \text{ at time } 1}{\text{number of times at time } 1}$

  $\bar{a}_{ij} = \dfrac{\text{number of transitions from state } s_i \text{ to state } s_j}{\text{number of transitions from state } s_i}$

  $\bar{b}_j(k) = \dfrac{\text{number of times in state } s_j \text{ observing symbol } x_k}{\text{number of times in state } s_j}$

  $\lambda \leftarrow \{\bar{\pi}, \bar{A}, \bar{B}\}$

end for

---

using the model parameters. Next, the transition and emission probabilities are updated based on the transitions and observations associated with each state sequence. The initial model is updated with these new probabilities. This process repeats until some desired level of convergence is reached. It will implicitly converge to a local minima.

The key to VPC is extracting the most likely state sequence.

$$Q_{(T)\max} = \arg \max_{Q_{(T)} \in S^T} P\left(Q_{(T)}, O_{(T)}\right) \tag{4.1}$$

A dynamic programing algorithm, called the Viterbi algorithm [59], solves this problem. Using the most likely state sequence extracted using the Viterbi algorithm, the transition and emission probabilities are found simply by counting the occurrences in all the most likely state sequences for each observation sequence in the training set. Since every conhort is trained individually, each observation sequence is a sequence starting on the key event and ending right before the next key event.

Maximizing the probability is equivalent to maximizing the log probability; therefore, we use log probabilities to prevent numerical underflow and facilitate faster computing. The order of the Viterbi algorithm for a left-right model with independent features is $O\left(\text{Viterbi}\right) = O\left(TxN\right) \cdot O\left(Pr_{est}\right)$, where $O\left(Pr_{est}\right)$ is the order of algorithm required to estimate probability. This order is constant time for the state transition probability, but based on the number of features for the observation probability estimation. This means $O\left(\text{Viterbi}\right) = O\left(TxNx|\vec{F}|\right)$.

4.5.4 *Feature Selection and Model Parametrization Using Genetic Algorithms*

The choice of whether or not to include each feature and the choice of the number of transition states for the cohorts are all tunable parameters of the HMEM. The feature selection $\Psi = \{\psi_k | \psi_k \in \{0,1\}, i = 1, \ldots, |F|\}$ represents a choice of which features out of an exhaustive list are to be included and which are to be discarded. The number of selected features is $|\Psi| = \sum_k \psi_k$. Feature k is selected if $\psi_k = 1$ and is discarded if $\psi_k = 0$. The other parameter is the number of transitions states for each cohort, $\Omega = \{\omega_e | \omega_e \in \{1, \ldots, 5\}, e = 1 \ldots E\}$, where E is the number of key events. The full HMEM model is represented by $\lambda_{HMEM} = \{\lambda_{HMM}, \Psi, \Omega\}$.

In essence, parametrization consists of choosing several good models based on one or more objective functions applied to the training set. These models are then compared against each other using the same objective function(s) applied to the cross-validation set. The best model on the cross-validation set is chosen. Because the cross-validation set exclusively contains data from subjects not in the training set, generalizability of the models to new subjects is improved. Genetic algorithms are used to generate the list of "good" models.

4.5.5 *HMEM Application Procedure*

After the HMEM is trained, it can be used to find key events in a data stream for the movement it has been trained on. The data flow for the algorithm is shown in Fig. 4.6. First, the data is filtered using the procedure described above, then features are extracted and quantized. Next, the feature selection is applied, and finally the most likely state sequence is extracted using the Viterbi algorithm. The annotation converter finds all the event states in this sequence and outputs an ordered set where each element consists of a time and event label.

4.5.6 *Experimental Analysis*

The results discussed in this section are based on an experiment reported in [62] and the node placement as described in Section 4.3.

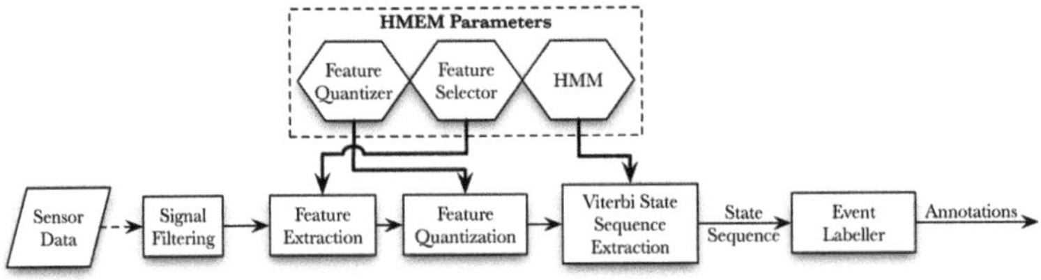

Figure 4.6. Application of the HMEM as described in Section 1.5.5.

The results are reported with precision (P), recall (R), Quality (RMSE). The first and last annotations were ignored because the algorithm needs context to determine annotations, and we are interested in the steady-state performance only. We show error for each mote using just the accelerometer readings, look at per-subject error for a poor performing sensor node, and a well-performing sensor node.

4.5.6.1 Examination of per-subject error

One of the goals of the HMEM system is good generalization to new subjects. Table 4.1 shows per-subject error for the sensor on the right thigh. Initially subjects two to four were in training, five to seven in cross validation, and eight and nine in test. However, subjects five and seven have walking patterns that differ significantly from the others, but are similar to each other. Therefore, subjects four and five were exchanged. It is likely that with a larger dataset the system could generalize better to such subjects. All the results are shown from the portion of the subjects' data that was in the test dataset.

The sensor on the right thigh, as shown in Table 4.1, performs well. Subjects eight and nine perform a little worse than subjects in the training and cross-validation sets. The worst per-subject error comes from subjects five and seven. The reason for this is not entirely clear. The use of the GA does not significantly reduce the discrepancy in per-subject error. Since the final selection criteria

Table 4.1. Subject error for R thigh with Accel and TP

	Default (132 Features)			GA (38 Features)		
Subject	P	R	RMSE	P	R	RMSE
Sub 2 Train	100	100	1.44	100	100	1.62
Sub 3 Train	100	100	1.33	100	100	1.11
Sub 4 Cross	100	100	1.36	100	100	1.38
Sub 5 Train	100	100	3.54	100	100	3.43
Sub 6 Cross	100	100	1.03	100	100	1.02
Sub 7 Cross	100	100	3.25	100	100	3.09
Sub 8 Test	100	100	1.74	100	100	1.76
Sub 9 Test	100	100	1.55	100	100	1.66

Table 4.2. Cross-validated subject error (R thigh)

Subject	P	R	RMSE	Fsel
Sub 2	100	100	1.47	6.7
Sub 3	100	100	1.38	8.7
Sub 4	100	100	1.68	4.7
Sub 5	99.8	99.8	3.55	4.7
Sub 6	100	100	1.20	6.7
Sub 7	100	100	3.09	28.7
Sub 8	100	100	1.52	10.7
Sub 9	100	100	1.59	16.0

for the solution is minimum total error, not minimum worst-case subject error, this is not surprising.

Manual partitioning of the data into training, cross-validation, and testing sets can artificially bias the results. Therefore, we performed an experiment for each subject where the subject was placed exclusively in the testing set, and the training and cross-validation sets were selected randomly from the remaining subjects. The results shown in Table 4.2 are the average of three tests after the GA. These results demonstrate that the model has good generalization to many subjects, but performs poorly on some.

4.6 System Optimizations

In Section 4.1 we discussed some of the constraints of the BAN systems. It was pointed out that energy constraint dominates the wearability of BAN systems. In general, energy can be used for local computation and communication. Previous studies of embedded sensor nodes have shown that data communication is expensive in terms of energy consumption, whereas data processing is relatively inexpensive [63].

The energy cost of the system can be improved in multiple ways. First, communication protocols used in embedded devices tend to be very general. Protocols, such as ZigBee [64], do not take advantage of the BAN specific features such as node proximity, data requirements, and signal processing flow. For example, due to proximity of the sensor units on the body, it may not be required for the radio to operate at full range. Sensors nodes can be optimized to operate at the minimum required range to save energy. Second, not all of the sensor nodes may be required to contribute to a given computation. If that is the case, then the system can be optimized to select only required nodes at any given time. Finally, the signal processing technique can explicitly address the energy problem by selecting less expansive operations. In the following section, we describe how in the case of applications that do not have hard data deadlines it is possible to optimize energy expenditure of the communication system.

4.6.1 *Burst Communication*

Most BAN applications have specific signal processing requirements as can be seen in Section 4.2. The typical signal processing flow involves several processing blocks, which may require collaboration among the sensing units. Intensive signal processing may require a lot of inter-node communication. Frequent inter-node communication leads to increased energy consumption. Therefore, if an application does not have a strict deadline, transmissions in short bursts at a high bit rate can help minimize transmission energy per bit [65].

4.6.1.1 Task graph

The signal processing model described earlier is usually implemented in a distributed manner over a BAN. Information flow across the network can be represented with a task graph model. In a system of n sensor nodes, task graph is composed of n subgraphs connected through inter-node links. Each subgraph represents static information dependencies within a node as shown in Fig. 4.7, where u_i^R, u_i^S, u_i^F, and u_i^C denote sensor reading and

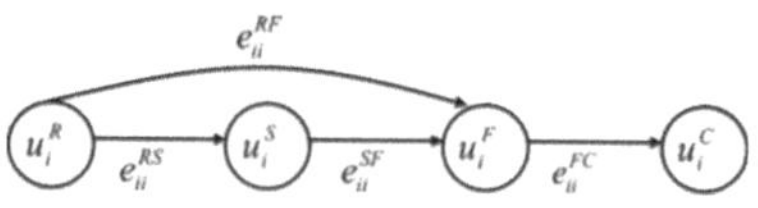

Figure 4.7. Intra-node task subgraph.

preprocessing, segmentation, feature extraction, and classification blocks, respectively.

Given a set of n sensor nodes, *intra-node task subgraph* for node s_i is defined by $G_i = (V_i, E_i)$, where V_i is the set of four vertices and E_i is the set of four edges. Each vertex, denoted by u_i^{μ}, corresponds to a processing unit, and each edge is denoted by $e_{ii}^{\mu\eta}$ $(\mu, \eta \in \{R, S, F, C\})$ representing intra-node dependencies. Inter-node links that represent dependencies across sensor nodes are used to connect subgraphs and form a task graph. An inter-node link is determined according to dependencies induced by application.

Given a set of n sensor nodes $s_1, \ldots, s_n$ and inter-node dependencies, *task graph* $G = (V, E)$ is formed by connecting n intra-node task subgraphs $G_1, \ldots, G_n$ through dependency links E_b defined by application criteria. The set of edges E is given by

$$E = E_w \cup E_b \tag{4.2}$$

where the set of intra-node edges E_w is given by

$$E_w = \bigcup_{i=1}^{n} E_i \qquad i = 1, \ldots, n \tag{4.3}$$

and the set of vertices V is defined by

$$V = \bigcup_{i=1}^{n} V_i \qquad i = 1, \ldots, n \tag{4.4}$$

where each V_i is the set of vertices within subgraph G_i.

As an abstract example, Fig. 4.8 shows a task graph for a network of two sensor nodes where processing units are connected through five inter-node dependency links.

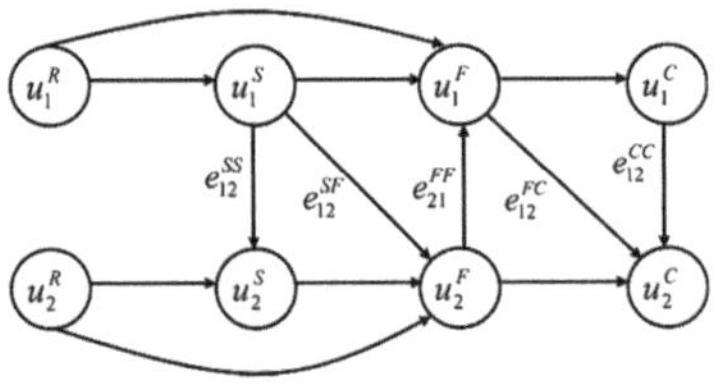

Figure 4.8. Sample task graph with five internode links.

In the context of the task graph, the idea behind using buffers is to transmit the maximum amount of data on the inter-node links in short time intervals. The large number of data blocks produced by each processing unit is stored locally and is transmitted using available bandwidth. This would conform to real situations of health care systems where physicians are interested in receiving reports on daily activities rather than immediate reports. By assuming no immediate deadlines in the system, individual buffers on each link are maintained and the data blocks are transmitted separately for each inter-node link. The optimization problem can be defined as follows:

Optimization Problem: Given task graph G, each inter-node link $e_{ij}^{\mu\eta}$ is associated with a number $x_{ij}^{\mu\eta}$ denoting the number of actions for which data blocks produced by the source unit u_i^{μ} are buffered prior to every transmission. The objective is to find values $x_{ij}^{\mu\eta}$ that minimize the number of transmissions subject to memory constraints on nodes.

4.6.1.2 Problem formulation

Given the task graph G as described earlier, let $a_{ij}^{\mu\eta}$ be a binary that represents existence of inter-node links and is given by

$$a_{ij}^{\mu\eta} = \begin{cases} 1, & \text{if } s_i \text{ and } s_j \text{ are dependent through } u_i^{\mu} \text{ and } u_j^{\eta} \\ 0, & \text{otherwise} \end{cases} \tag{4.5}$$

and let d_i^{μ} be another binary denoting whether or not a processing unit is slower than its predecessor.

$$d_i^{\mu} = \begin{cases} 1, & \text{if } r_i^{\mu} < r_i^{\mu-1} \\ 0, & \text{otherwise} \end{cases} \tag{4.6}$$

The number of actions A occurred at action rate k during time period T is given by

$$A = k \times T \tag{4.7}$$

and the number of transmissions can be calculated by

$$Z = \sum_{i,j} A \frac{a_{ij}^{\mu\eta}}{x_{ij}^{\mu\eta}} \tag{4.8}$$

The total size of the intra-node buffers W_i for node s_i is given by

$$W_i = \sum_{\mu} b_i^{\mu} \tag{4.9}$$

and the total size of the inter-node buffers for node s_i is determined by

$$B_i = \sum_{\mu} (\max_{j,\eta}(a_{ij}^{\mu\eta} b_i^{\mu} x_{ij}^{\mu\eta}) + \max_{j,\eta}(a_{ji}^{\eta\mu} b_j^{\eta} x_{ji}^{\eta\mu}) + \max_{j,\eta}(a_{ji}^{\eta\mu} b_j^{\mu-1} x_{ji}^{\eta\mu})) \tag{4.10}$$

and the total size of delay buffers for node s_i is given by

$$D_i = T \sum_{\mu} d_i^{\mu}((r_i^{\mu-1} - r_i^{\mu}) b_i^{\mu-1} + \sum_{j,\eta} a_{ji}^{\eta\mu} b_j^{\eta}) \tag{4.11}$$

Let M_i be the size of the memory on node s_i. The problem of minimizing the number of transmissions can be formulated as a convex optimization problem as follows:

$$Min\ Z \tag{4.12}$$

subject to

$$W_i + B_i + D_i \leq M_i \qquad \forall i \in \{1, ..., n\} \tag{4.13}$$

$$x_{ij}^{\mu\eta} \in Z^{+} \quad \forall i, j, \mu, \eta \tag{4.14}$$

The non-linear constraints due to the *max* functions in Eq. (4.10) can be transformed into linear equations by expanding every function over all values taken by the function. The integrality condition (4.14) can be relaxed to solve the problem using common convex programming tools.

$$x_{ij}^{\mu\eta} > 0 \quad \forall i, j, \mu, \eta \tag{4.15}$$

The solution obtained due to integrality relaxation will not carry the optimality condition, but it is possible to find a lower bound on the size of memory for which the result is optimal. For each sensor node s_i, the integer relaxed convex optimization finds optimal solutions for memories of size

$$M_i - \sum_{\mu} (1 - \varepsilon) \alpha_i^{\mu} \beta_i^{\mu} \gamma_i^{\mu} \tag{4.16}$$

where

$$\begin{aligned}\alpha_i^{\mu} &= a_{ij}^{\mu\eta} b_i^{\mu} \quad \text{s.t.} \quad j,\eta = \underset{j,\eta}{\arg\max}(a_{ij}^{\mu\eta} b_i^{\mu} x_{ij}^{\mu\eta}) \\ \beta_i^{\mu} &= a_{ji}^{\eta\mu} b_j^{\eta} \quad \text{s.t.} \quad j,\eta = \underset{j,\eta}{\arg\max}(a_{ji}^{\eta\mu} b_j^{\eta} x_{ji}^{\eta\mu}) \\ \gamma_i^{\mu} &= a_{ji}^{\eta\mu} b_j^{\mu-1} \quad \text{s.t.} \quad j,\eta = \underset{j,\eta}{\arg\max}(a_{ji}^{\eta\mu} b_j^{\mu-1} x_{ji}^{\eta\mu})\end{aligned} \tag{4.17}$$

Inter-node buffers assigned to each processing unit can have at most three components as given in Eq. (4.10). Integer relaxation would increase size of each optimal buffer associated with $x_{ij}^{\mu\eta}$ by factor $1-\varepsilon$ of the unit data $(\alpha_i^{\mu}, \beta_i^{\mu}, \gamma_i^{\mu})$.

4.6.1.3 Experimental results

The results discussed in this section are based on the experiment reported in [66]. During the experiment three communication networks are considered. Each network is based on a particular signal processing requirements, which define specific inter- and intra-connection networks. For each network, Table 4.3 shows the number of network transmissions required, including the number of packets exchanged with and without using the buffer approach and highlights the improvement of applying buffers.

Based on the results in Table 4.3, a few conclusions can be made. In addition to communication simplification, the approach can reduce the communication coordination and enhance the system lifetime. Since the considered metric is based on the number of packets, it is independent of the communication protocol. However, the choice of packet size is important for the system. A larger

Table 4.3. Number of transmissions for different communication networks

Configuration	Action Rate (Hz)	# Packets		Improvement
		Without Buffer	With Buffer	
1	1.8	185760	74485	59%
	0.0167	1440	80	94%
2	1.8	46440	31226	32%
	0.0167	1440	153	89%
3	1.8	69660	37297	46%
	0.0167	1440	17	98%

packet size would enhance energy per bit transmitted ratio. Finally, the discussed solution assumes that buffers are preallocated for a particular implementation. An approach that would dynamically update the buffer size to accommodate for the system requirements at any given time would be potentially more beneficial.

References

1. S. Park, A. Savvides, and M. Srivastava (2001) Battery capacity measurement and analysis using lithium coin cell battery, in *Proceedings of the 2001 international symposium on Low power electronics and design*, pp. 387, ACM.
2. V. Jones, R. Bults, D. Konstantas, and P. Vierhout (2001) Healthcare PANs: personal area networks for trauma care and home care, in *Proceedings of the Fourth International Symposium on Wireless Personal Multimedia Communications (WPMC)*, pp. 9–12.
3. O. Bott, M. Marschollek, J. Bergmann, K. Wolf, U. Tegtbur, and R. Haux (2007) Sensor-enhanced health information system architectures for home and telecare: concept and prototype, in *European Conference on eHealth*, **118**, pp. 193–203.
4. S. Po, G. Dagang, M. Hapipi, N. Naing, W. Shen, A. Ongkodjojo, and F. Hock (2006) Overview of MEMSWear II-incorporating MEMS technology into smart shirt for geriatric care, *Journal of Physics: Conference Series*, **34**, pp. 1079–1085, Institute of Physics Publishing.
5. E. Jovanov, A. Milenkovic, C. Otto, and P. De Groen (2005) A wireless body area network of intelligent motion sensors for computer assisted physical rehabilitation, *Journal of NeuroEngineering and Rehabilitation*, **2(1)**, pp. 6.
6. R. Ramachandran, L. Ramanna, H. Ghasemzadeh, G. Pradhan, R. Jafari, and B. Prabhakaran (2008) Body sensor networks to evaluate standing balance: interpreting muscular activities based on inertial sensors, in*Proceedings of the second International Workshop on Systems and Networking Support for Health Care and Assisted Living Environments*, pp. 4, ACM.
7. H. Ghasemzadeh and R. Jafari (2009) sport training using body sensor networks: a statistical approach to measure wrist rotation for golf swing, in *The Fourth International Conference on Body Area Networks (BodyNets 09)*, Los Angeles, CA.

8. H. Ghasemzadeh and R. Jafari (2010) Body sensor networks for baseball swing training: coordination analysis of human movements using motion transcripts, *in Proceedings of the Eight IEEE International Conference on Pervasive Computing and Communications Workshops*, pp. 792–795.
9. A. Van Halteren, R. Bults, K. Wac, D. Konstantas, I. Widya, N. Dokovsky, G. Koprinkov, V. Jones, and R. Herzog (2004) Mobile patient monitoring: The mobihealth system, *The Journal on Information Technology in Healthcare*, **2**(5), pp. 365–373.
10. K. Kim, S. Bang, and S. Kim (2004) Emotion recognition system using short-term monitoring of physiological signals, *Medical and Biological Engineering and Computing*, **42**(3), pp. 419–427.
11. L. Dong, J. Wu, X. Bao, and W. Xiao (2006) Extraction of gait features using a wireless body sensor network (BSN), in *Proceedings of the Sixth International Conference on ITS Telecommunications*, pp. 987–991.
12. E. Guenterberg, A. Yang, H. Ghasemzadeh, R. Jafari, R. Bajcsy, and S. Sastry (2009) A method for extracting temporal parameters based on hidden markov models in body sensor networks with inertial sensors, *IEEE Transactions on Information Technology in Biomedicine*, **13(6)**, pp. 1019, (2009).
13. A. Lie, R. Shimomoto, S. Sakaguchi, T. Ishimura, S. Enokida, T. Wada, and T. Ejima (2005) Gait recognition using spectral features of foot motion, *Lecture Notes in Computer Science*, **3546**, pp. 767–776.
14. L. Atallah, O. Aziz, B. Lo, and G. Yang (2009) Detecting walking gait impairment with an ear-worn sensor, in *Proceedings of the 2009 Sixth International Workshop on Wearable and Implantable Body Sensor Networks 00*, pp. 175–180, IEEE Computer Society.
15. M. Hoehn and M. Yahr (1967) Parkinsonism: onset, progression and mortality, *Neurology*, **17**(5), pp. 427–442.
16. D. Webster (1968) Critical analysis of the disability in Parkinson's disease, *Modern Treatment*, **5(2)**, pp. 257.
17. G. Montgomery, N. Reynolds Jr., and R. Warren (1985) Qualitative assessment of parkinson's disease: study of reliability and data reduction with an abbreviated columbia scale., *Clinical Neuropharmacology*, **8(1)**, pp. 83.
18. J. Rabey, H. Bass, U. Bonuccelli, D. Brooks, P. Klotz, A. Korczyn, P. Kraus, P. Martinez-Martin, P. Morrish, W. Van Sauten, et al. (1997) Evaluation of the short parkinson's evaluation scale: a new friendly scale for

the evaluation of Parkinson's disease in clinical drug trials, *Clinical Neuropharmacology*, **20**(4), pp. 322–337.

19. S. Fahn and R. Elton, Members of the UPDRS Development Committee (1987) Unified Parkinson's disease rating scale, *Recent Developments in Parkinson's Disease*, **2**, pp. 153–163.
20. C. Goetz, G. Stebbins, D. Wolff, W. DeLeeuw, H. Bronte-Stewart, R. Elble, M. Hallett, J. Nutt, L. Ramig, T. Sanger, et al. (2009) Testing objective measures of motor impairment in early Parkinson's disease: Feasibility study of an at-home testing device Potential conflict of interest: Nothing to report., *Movement Disorders*, **24(4)**, pp. 551–556.
21. C. Ramaker, J. Marinus, A. Stiggelbout, and B. Van Hilten (2002) Systematic evaluation of rating scales for impairment and disability in Parkinson's disease, *Movement Disorders*, **17(5)**, pp. 867–876.
22. J. Ackmann, A. Sances, S. Larson, and J. Baker (1977) Quantitative evaluation of long-term Parkinson tremor, *IEEE Transactions on Biomedical Engineering*, pp. 49–56.
23. E. Van Someren, B. Vonk, W. Thijssen, J. Speelman, P. Schuurman, M. Mirmiran, and D. Swaab (1998) A new actigraph for long-term registration of the duration and intensity of tremor and movement, *IEEE Transactions on Biomedical Engineering*, **45(3)**, pp. 386–395.
24. B. Van Hilten, J. Hoff, H. Middelkoop, E. Van der Velde, G. Kerkhof, A. Wauquier, H. Kamphuisen, and R. Roos (1994) Sleep disruption in Parkinson's disease. Assessment by continuous activity monitoring, *Archives of Neurology*, **51**(9), pp. 922–928.
25. C. Homann, K. Suppan, K. Wenzel, G. Giovannoni, G. Ivanic, S. Horner, E. Ott, and H. Hartung (2000) The bradykinesia akinesia incoordination test (BRAIN TEST©), an objective and user-friendly means to evaluate patients with Parkinsonism, *Movement Disorders*, **15(4)**, pp. 641–647.
26. V. Rajaraman, D. Jack, S. Adamovich, W. Hening, J. Sage, and H. Poizner, (2000) A novel quantitative method for 3D measurement of Parkinsonian tremor, *Clinical Neurophysiology*, **111**(2), pp. 338–343.
27. P. Bain, (1998) Clinical measurement of tremor, *Movement Disorders*, **13**, pp. 77–80.
28. J. Matsumoto, D. Dodick, L. Stevens, R. Newman, P. Caskey, and W. Fjerstad (1999) Three-dimensional measurement of essential tremor, *Movement Disorders*, **14(2)**, pp. 288-294.
29. G. Deuschl, P. Bain, and M. Brin (1998) Consensus statement of the Movement Disorder Society on tremor, *Movement Disorders*, **13**, pp. 2–23.

30. D. Wright, K. Nakamura, T. Maeda, K. Kutsuzawa, K. Miyawaki, and K. Nagata (2008) Research and development of a portable device to quantify muscle tone in patients with Parkinson's disease, in *Proceedings of the 30th IEEE Annual International Conference of the Engineering in Medicine and Biology Society*, pp. 2825–2827.
31. P. Bain, L. Findley, P. Atchison, M. Behari, M. Vidailhet, M. Gresty, J. Rothwell, P. Thompson, and C. Marsden (1993) Assessing tremor severity, *British Medical Journal*, **56(8)**, pp. 868–873.
32. H. Ghasemzadeh, E. Guenterberg, and R. Jafari (2009) Energy-efficient information-driven coverage for physical movement monitoring in body sensor networks, *IEEE Journal on Selected Areas in Communications*, **27(1)**, pp, 2.
33. A. Viru and M. Viru (2001) *Biochemical Monitoring of Sport Training*, Human Kinetics.
34. M. Rizzo (2001) Softball/baseball training machine, United States Patent, US 6,305,366 Bl (45).
35. Kuster (20 November 2003) Golf swing training device, US Patent App. 10/717,841.
36. Y. Yamamoto (2004) An alternative approach to the acquisition of a complex motor skill: Multiple movement training on tennis strokes, *International Journal of Sport and Health Science*, **2**, pp. 169–179.
37. A. Chong (March 2009) A photogrammetric application in virtual sport training, *The Photogrammetric Record*, **24**, pp. 51–65(15).
38. T. Komura, A. Kuroda, and Y. Shinagawa (2002) NiceMeetVR: facing professional baseball pitchers in the virtual batting cage, in *Proceedings of the 2002 ACM Symposium on Applied Computing*, pp. 1065. ACM.
39. I. Karliga and J. Hwang (2006) Analyzing human body 3-d motion of golf swing from single camera video sequences, in *Proceedings of the IEEE International Conference on Acoustics, Speech and Signal Processing (ICASSP 2006)*.
40. A. Baca and P. Kornfeind (2006) Rapid feedback systems for elite sports training, *IEEE Pervasive Computing*, **5**(4), pp. 70–76.
41. A. Cochran, J. Stobbs, and Golf Society of Great Britain (1968) *Search for the Perfect Swing*, Triumph Books.
42. P. A. Hume, J. Keogh, and D. Reid (2005) The role of biomechanics in maximising distance and accuracy of golf shots, *Sports Medicine*, **35**, pp. 429–449(21).

43. J. Tartagni and P. Schmid (6 January 2004) Putting stroke training aid, US Patent 6,672,974.
44. M. Bellagamba (18 June 1991) Athletic swing practice apparatus, US Patent 5,024,443.
45. C. Welch, S. Banks, F. Cook, and P. Draovitch (1995) Hitting a baseball: a biomechanical description, *The Journal of Orthopaedic and Sports Physical Therapy*, **22(5)**, pp. 193.
46. H. Ghasemzadeh, E. Guenterberg, and R. Jafari (2009) Energy-Efficient Information-Driven Coverage for Physical Movement Monitoring in Body Sensor Networks, *IEEE Journal on Selected Areas in Communications*, **27**, pp. 58–69.
47. J. Polastre, R. Szewczyk, and D. Culler (April 2005) Telos: enabling ultra-low power wireless research, in *Fourth International Symposium on Information Processing in Sensor Networks (IPSN 2005)*, pp. 364–369, doi: 10.1109/IPSN.2005.1440950.
48. R. Gravina, A. Guerrieri, G. Fortino, F. Bellifemine, R. Giannantonio, and M. Sgroi (2008) Development of body sensor network applications using SPINE, in *Proceedings of the IEEE International Conference on Systems, Man, and Cybernetics (SMC 2008)*, Singapore.
49. H. Ghasemzadeh, J. Barnes, E. Guenterberg, and R. Jafari (2008) A phonological expression for physical movement monitoring in body sensor networks, in *Fifth IEEE International Conference on Mobile Ad Hoc and Sensor Systems (MASS 2008)*, pp. 58–68.
50. N. Stergiou, (2003) *Innovative Analyses of Human Movement: Analytical Tools for Human Movement Research*, Human Kinetics.
51. H. Ghasemzadeh, E. Guenterberg, K. Gilani, and R. Jafari (2008) Action coverage formulation for power optimization in body sensor networks, in *Design Automation Conference, 2008 (ASPDAC 2008)*, Asia and South Pacific, pp. 446–451.
52. H. Ghasemzadeh, V. Loseu, and R. Jafari (2009) Structural action recognition in body sensor networks: Distributed classification based on string matching, *IEEE Transactions on Information Technology in Biomedicine*, **14(2)**, pp. 425 - 435.
53. R. Duda, P. Hart, and D. Stork (1973) *Pattern Classification and Scene Analysis*, Wiley, New York.
54. R. Jafari, R. Bajcsy, S. Glaser, B. Gnade, M. Sgroi, and S. Sastry (June 2007) Platform design for health care monitoring applications, in *Joint Workshop on High Confidence Medical Devices, Software, and Systems and Medical Device Plug-and-Play Interoperability*, pp. 88–94.

55. S. Newell and E. Els (2001) *The Golf Instruction Manual*, Dorling Kindersley.
56. J. Perry (1992) *Gait analysis: normal and pathological function*, 1st edn. SLACK incorporated.
57. J. Hausdorff, M. Cudkowicz, R. Firtion, J. Wei, and A. Goldberger (1998) Gait variability and basal ganglia disorders: stride-to-stride variations of gait cycle timing in Parkinson's disease and Huntington's disease, *Movement Disorders*, **13(3)**, pp. 428–437.
58. J. Hausdorff, D. Rios, and H. Edelberg (2001) Gait variability and fall risk in community-living older adults: a 1-year prospective study, *Archives of Physical Medicine and Rehabilitation*, **82(8)**, pp. 1050–1056.
59. L. Rabiner and B. Juang (1986) An introduction to hidden Markov models, *IEEE ASSP Magazine*, **3**(1 Part 1), pp. 4–16.
60. L. Rabiner (1989) A tutorial on hidden Markov models and selected applications in speech recognition, in *Proceedings of the IEEE*, **77(2)**, pp. 257–286.
61. N. Liu, B. Lovell, and P. Kootsookos (2003) Evaluation of HMM training algorithms for letter hand gesture recognition, in *Proceedings of the Third IEEE International Symposium on Signal Processing and Information Technology (ISSPIT 2003)*, pp. 648–651.
62. E. Guenterberg, H. Ghasemzadeh, and R. Jafari (2009) A distributed hidden Markov model for fine-grained annotation in body sensor networks, in *Proceedings of the 2009 Sixth International Workshop on Wearable and Implantable Body Sensor Networks*, 00, pp. 339–344, IEEE Computer Society.
63. V. Raghunathan, C. Schurgers, S. Park, and M. Srivastava (March 2002) Energy-aware wireless microsensor networks, *Signal Processing Magazine, IEEE*, **19(2)**, pp. 40–50, ISSN 1053-5888. doi: 10.1109/79.985679.
64. Z. Alliance, ZigBee specification, ZigBee Document 053474r06, Version **1**, 2005.
65. J. M. Kahn, R. H. Katz, and K. S. J. Pister (1999) Next century challenges: mobile networking for "smart dust," in *Proceedings of the Fifth Annual ACM/IEEE International Conference on Mobile Computing and Networking (MobiCom 1999)*, pp. 271–278, ISBN 1-58113-142-9.
66. H. Ghasemzadeh, N. Jain, M. Sgroi, and R. Jafari (2009) Communication minimization for in-network processing in body sensor networks: a buffer assignment technique, in *Proceedings of the ACM/IEEE Design, Automation and Test in Europe (DATE09)*.

Chapter 5

Signal Processing In-Node Frameworks for Wireless Body Area Networks: From Low-Level to High-Level Approaches

Francesco Aiello, Giancarlo Fortino, Stefano Galzarano, Raffaele Gravina, and Antonio Guerrieri

Department of Electronics, Informatics, and Systems (DEIS), University of Calabria, Via P. Bucci, cubo 41C, 87036 Rende (CS), Italy

faiello@si.deis.unical.it; g.fortino@unical.it; galzarano@si.deis.unical.it; rgravina@deis.unical.it; aguerrieri@deis.unical.it

To develop applications based on wireless body area networks (WBANs), either low-level or high-level programming approaches can be adopted. Low-level approaches are based on low-level APIs (application programming interfaces) offered by sensor platforms, whereas high-level approaches are based on high-level programming models and frameworks that facilitate programming and increase productivity. This chapter proposes a high-level approach based on the agent-oriented programming model to flexibly design and efficiently implement signal processing in-node environments supporting WBAN applications. The approach is exemplified through a case study concerning a real-time human

Wireless Body Area Networks: Technology, Implementation, and Applications
Edited by Mehmet R. Yuce and Jamil Y. Khan

ISBN 978-981-4316-71-2 (Hardcover), ISBN 978-981-4241-57-1 (eBook)
www.panstanford.com

activity monitoring system, which is developed through two different agent-based frameworks: MAPS (Mobile Agent Platform for Sun SPOT) and AFME (Agent Factory Micro Edition). A comparison of the effectiveness and efficiency of the developed systems is finally presented.

5.1 Introduction

Wireless sensor networks (WSNs) are currently emerging as one of the most disruptive technologies enabling and supporting next generation ubiquitous and pervasive computing scenarios [24]. WSNs are capable of supporting a broad array of high-impact applications in several domains such as disaster/crime prevention, military, environment, logistics, health care, and building/home automation. WSNs applied to human body monitoring are usually called wireless body area networks (WBANs) [29]. WBANs are conveying notable attention as their real-world applications aim at improving the quality of life of human beings by enabling continuous and real-time, non-invasive medical assistance at low cost. Health care applications where WBANs could be greatly useful include early detection or prevention of diseases, elderly assistance at home, e-fitness, rehabilitation after surgeries, motion and gestures detection, cognitive and emotional recognition, medical assistance in disaster events, etc.

However, programming WBAN applications is a complex task due to the hard constraints of wearable devices in terms of limited resources (computing power, memory, and communications) and due to the lack of proper and effective software abstractions. To deal with these issues, several software frameworks have been developed such as CodeBlue [18], Titan [16], and SPINE [12, 9]. They aim at decreasing development time and improving interoperability among signal processing intensive applications based on WBANs while fulfilling efficiency requirements. In particular, they basically rely on a star-based network architecture, which is organized into a coordinator node and a set of sensor nodes. Moreover, they are developed in TinyOS at the sensor node side and in Java at the coordinator node side.

Apart from the adoption of software frameworks, we believe that the exploitation of the agent-oriented programming paradigm to develop WSN applications could provide more effectiveness as demonstrated by the application of agent technology in several key application domains [8, 17]. In this chapter, we, therefore, propose high-level, agent-oriented design and agent-based implementation to support the development of WBAN applications. In particular, the approach is applied in the context of a real-time human activity monitoring system through two different Java-based agent platforms running on the Sun SPOT sensor platform: MAPS (Mobile Agent Platform for Sun SPOTs) [4, 2] and AFME (Agent Factory Micro Edition) framework [21]. The system architecture relies on a typical star-based WBAN composed of a coordinator node and two sensor nodes, which are located on the waist and the thigh of the monitored human being, respectively. The coordinator is based on an enhancement of the Java-based SPINE coordinator [12, 27] and allows configuring the sensing process, receiving sensed data features, and recognizing predefined human activities. Each sensor node executes either a MAPS-based agent or an AFME-based agent that performs sensing of the on-board 3-axial accelerometer sensor, computation of significant features on the acquired data, and aggregation and transmission of features to the coordinator. Finally, a comparison between the MAPS-based and the AFME-based implementations is discussed.

This chapter is organized as follows. Section 5.2 describes a reference architecture for WBAN applications from network and functional perspectives. Section 5.3 details and compares the state-of-the-art on signal processing in-node frameworks and applications based on TinyOS for the development of WBAN applications: CodeBlue, Titan, and two versions of SPINE. In Section 5.4, the most known available agent-oriented platforms for developing WSN-based applications are described. Section 5.5 presents an agent-oriented approach for the construction of WBAN applications, whereas Section 5.6 exemplifies it by describing the implementation and evaluation of an agent-based, real-time human activity monitoring system through aforementioned different agent platforms. Finally conclusions are drawn and future research delineated.

5.2 A WBAN Reference Architecture

The network architecture of the reference WBAN for signal processing is organized into multiple sensor nodes and one coordinator node according to the scheme reported in Fig. 5.1. The coordinator manages the network, collects, stores, and analyzes the data received from the sensor nodes, and also can act as a gateway to connect the WBAN with other networks (e.g., Internet) for remote data access. Sensor nodes measure local physical parameters and send raw or preprocessed data to the coordinator. In this system, sensor nodes only communicate with the coordinator according to the star network topology.

The software architecture of the system consists of two main components installed, respectively, on the coordinator (e.g., a PC or a smartphone) and on the sensor nodes. Figure 5.2 shows a schema of the software architecture from a functional point of view.

At the coordinator side, an interface to the WBAN, which is placed between user applications and the hardware and software host platform, is made available. User applications manage the WBAN through a *system API*. The top level of the software architecture at the coordinator allows registered applications being notified of the following events generated by the WBAN: discovery of new nodes, sensor data communication, node alarms, and system messages such as low battery warnings. Commands issued by user applications and network-generated events are, respectively, coded in lower-level messages and decoded in higher-level information by

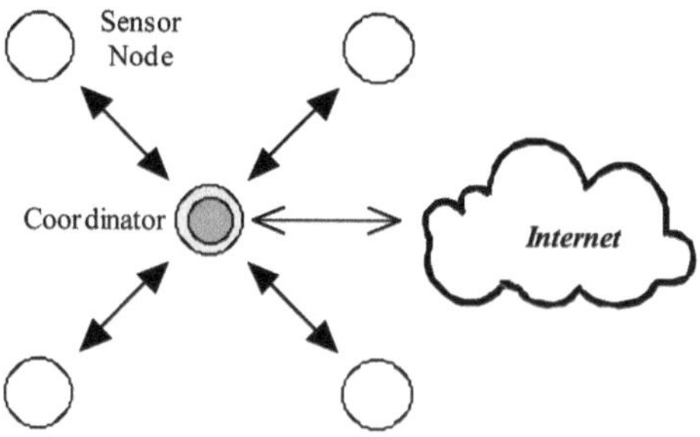

Figure 5.1. WBAN reference network architecture.

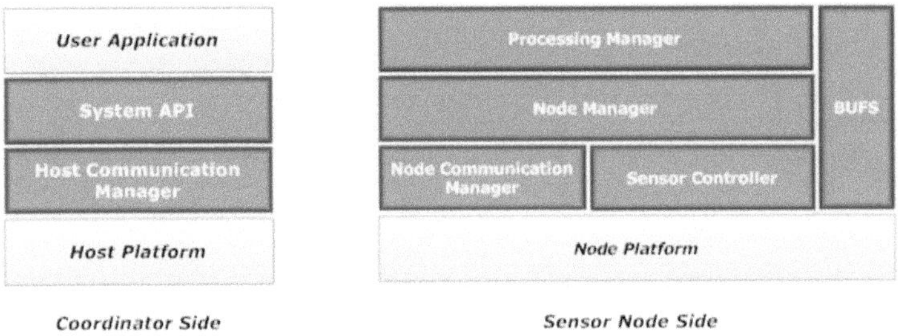

Figure 5.2. WBAN reference software architecture.

the *host communication manager* according to a specific over-the-air protocol. This component handles the generation and reception of messages and is interfaced with specific software components of the host platform to access the physical radio module to transmit/receive messages to/from the sensor nodes.

At the sensor node side, the software architecture provides: (i) abstractions of hardware resources such as sensors and the radio, (ii) a default set of ready-to-use common signal processing functions, and (iii) a flexible and modular architecture to be customized and extended to support new physical platforms and sensors and introduce new signal processing services. In particular, the *Node Communication Manager* acts as the counterpart of the *Host Communication Manager*. The *Sensor Controller* manages and abstracts the sensors on the node platform, providing a standard interface to the diverse sensor drivers. Specifically, it is responsible for sampling the sensors and storing the sensed data in properly defined *Buffers*. The *Node Manager* is the central component of the sensor node side, responsible of recognizing the coordinator requests and dispatching them to the proper components. Finally, the *Processing Manager* offers core processing services, which can be enriched by user-defined services.

5.3 Software Frameworks for Programming WBANs

WBAN applications can be mainly implemented according to the following approaches: (i) *application-specific code*, which aims at

Table 5.1. Comparison among different approaches for WBAN application development

Approaches/ Features	Application-specific code	Domain-specific framework	General-purpose middleware
Code Reusability		•	•
Rapid prototyping		•	•
Ease of debugging		•	•
Code efficiency	•	•	
System interoperability		•	•
Specific support to flexible sensing operations		•	
Specific support to in-node signal processing		•	

developing a WBAN application by using the low-level API provided by the sensor node platform; (ii) *general-purpose middleware*, which provides a high-level distributed framework for programming WSN applications for distributed signal processing; (iii) *domain-specific frameworks*, which aims at supporting specific development of WBAN applications according to a reference architecture. Table 5.1 reports the characteristics of the three approaches: code reusability, rapid prototyping, ease of debugging, code efficiency, system interoperability, specific support to flexible sensing operations, and specific support to in-node signal processing. In the following, specific applications and frameworks belonging to such approaches are described.

Most of the previous research efforts on WBANs were focused on proof-of-concept applications with the aim to demonstrate the feasibility of new context-aware algorithms and techniques, e.g., for recognition of physical activity or prompt detection of heart diseases, considering issues like power consumption and also radio channel usage but not taking into account code reusability and modularity. In reference [22], a method for physical activity monitoring is presented, which is able to detect body postures and periods of walking in elderly persons using one kinematic sensor attached to the chest. In reference [13], wearable motion sensors are used to guide post-stroke rehabilitation by models to predict clinical scores of motor abilities. In reference [23], activity recognition is improved by integrating wearable sensors with ambient blob-based vision sensor data. Another physical activity recognition system based on wearable devices is presented in reference [19]. In reference [15], it is discussed in detail the interesting issue of developing a personal activity recognition system, which is based on data coming from a single body location regardless of the specific sensor location and able to work with different individuals. The system can be further personalized to enhance the activity recognition accuracy. Finally, the most exhaustive effort on activity recognition based on wearable sensors is presented in reference [5]. All these research efforts have been focused on efficient application-specific solutions rather than on the definition of reusable frameworks facilitating the development of WBAN applications.

The first important solution provided to define a general platform able to support various WBAN applications is CodeBlue [18]. CodeBlue is a framework running on TinyOS specifically designed for integrating wireless medical sensor nodes and other devices. CodeBlue allows these devices to discover each other, report events, and establish communications. CodeBlue is based on a publish/subscribe-based data routing framework in which sensors publish relevant data to a specific channel and end-user devices subscribe to channels of interest. It includes a naming scheme, a multi-hop communication protocol, authentication and encryption capabilities, location tracking, and in-network filtering and aggregation. CodeBlue provides end-user devices with a query interface for retrieving data from previously discovered sensor nodes. Although CodeBlue provides a sensor driver abstraction architecture, which allows an easy integration of new sensors within the system, selection of sensor types or physical node identifiers as data sources, tuning of the data rate, and definition of threshold-based filters to avoid unnecessary data being transmitted, it does not allow inserting complex signal processing functionalities into the sensor nodes. It supports just simple threshold-based triggers on the sensor readings that do not give enough flexibility for the variety of requirements of the WBAN applications.

A higher-level approach is adopted by Titan [16]. Titan is a general-purpose middleware implemented in TinyOS that supports implementation and execution of context recognition algorithms in dynamic WSN environments. Titan represents data processing by a data flow from sensors to recognize result. The data are processed by tasks, which implement elementary computations. Tasks and their data flow interconnections define a task network, which runs on the sensor network as a whole. Tasks are mapped onto each sensor node according to the sensors and the processing resources they provide. Titan dynamically reprograms the WSN to exchange context recognition algorithms, handle defective nodes, variations in available processing power, or broken communication links. The architecture of Titan is composed of several software components, which enhance modularity. Although Titan raises the programming abstraction level by offering a middleware for effectively developing signal processing applications in WSNs, it is based on too generic mechanisms for providing efficient solutions in the specific BAN application domain. In fact, the programming of a feature extraction operation on the sensor node, which is often carried out in a BAN application, requires the creation of at least five tasks (sampling, buffering, loading, feature calculation, and transmission). Moreover, some overhead can be introduced due to the connections of the output and input ports among tasks through which data are exchanged.

The first example of domain-specific framework in the WBAN area is represented by SPINE (Signal Processing in Node Environment) [11, 12, 27]. SPINE is a software framework for the design of collaborative WBAN applications. It provides programming abstractions, APIs and libraries of protocols, utilities and data processing functions, which simplify development of distributed signal processing algorithms for the analysis and the classification of sensor data. SPINE [27] is distributed in open source under the LGPL (Lesser General Public License) license to facilitate establishing a broad community of users and developers that contribute to the scientific evolution of the framework with new capabilities and applications. The SPINE framework is composed of two distinctive parts: a node side runtime system residing on the sensor nodes and a Java application, the coordinator, residing on a PC and having functionalities such as nodes configuration and control, data gathering and data analysis. To date, two releases of SPINE are available:

- The TinyOS release (version 1.3), which supports different kinds of sensor platforms running the TinyOS [28] operating system (supported platforms are TelosB, MicaZ, Shimmer).
- The Z-Stack release (version 1.0), which allows the development of WBAN applications on the Z-Stack platform according to the ZigBee standard [30]. In particular, Z-Stack is the implementation of the ZigBee stack carried out by Texas Instruments.

The SPINE2 framework [10, 9] is an evolution of SPINE based on the C language for reaching a very high platform independency for C-like programmable sensor platforms (e.g., TinyOS, EmberZNet, Z-Stack) so raising the level of the provided programming abstractions from platform-specific to platform-independent. SPINE2 is founded on a software layering approach based on the C language, which is the language used for programming the majority of embedded systems. It embodies the following features: (i) execution on commercial resource-constrained sensor platforms each one having a different operating system; (ii) minimization of the amount of code that should be replicated for each specific implementation; (iii) enabling C developers (eventually C++) to extend the SPINE2 framework without having to learn low-level details of specific sensor platforms or without having to learn new programming languages; (iv) enabling, compiling, and simulating the code by using normal ANSI C tools.

While SPINE is centered on a programming model based on functions, SPINE2 is based on a task-oriented programming model in order to best fit the requirements of collaborative distributed applications in resource-constrained environments. Distributed and collaborative applications can then be programmed as a dynamically schedulable and reconfigurable set of tasks. Different tasks can be assigned to each node of the network, and tasks can be controlled at execution time via proper message exchange; in this way the network can overall adapt to changes in context, in overall goals, in the state of each single node, and it can better balance load and task types between each element of the network. Dynamic distribution of tasks also allows preprocessing of sensed data directly on the node, a significant reduction of data transmission and battery consumption, and an overall increase in the network lifetime. Due to task-oriented programming, application developers do not need to program in tiny environments but only configure tasks on the WBAN coordinator.

To summarize, we promote the use of the domain-specific framework approach, which stands in the middle between application-specific code and general-purpose middleware approaches. It specifically addresses and standardizes the core challenges of WSN development within a particular application domain. While maintaining high efficiency, it allows for a more effective development of customized applications with little or no additional hardware configuration and with the provision of high-level programming abstractions tailored for the reference application domain. In particular, the SPINE and SPINE2 frameworks are domain-specific frameworks in the context of signal processing in-node intensive WBAN applications, whereas CodeBlue and Titan can be seen as general-purpose middleware solutions for WBAN applications. SPINE allows for both code efficiency similarly to an application-specific code approach, and code reusability, rapid prototyping, easy debugging, and system interoperability as CodeBlue and Titan. Moreover, it specifically supports functionalities for flexible sensing operations and easily programmable in-node signal processing. SPINE2 extends the functionalities of SPINE with the development of WBAN applications on heterogeneous sensor platforms based on C-like languages.

5.4 Agent-Oriented Platforms for Wireless Sensor Networks

In the context of WSNs, it is challenging to develop agent platforms (APs) for supporting (mobile) agent-based programming [8, 3]. Due to the currently available resource-constrained sensor nodes and related operating systems, building flexible and efficient agent platforms is a very complex task. Very few APs for WSNs have been to date proposed and actually implemented. An AP architecture (see Fig. 5.3) relies on the services offered by the hosting sensor node OS and support execution of agents by managing agent

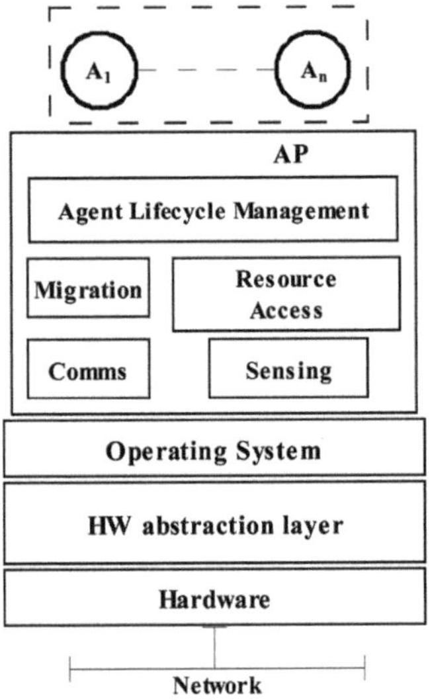

Figure 5.3. Reference architecture of a sensor node agent platform.

lifecycle, migration, communication, sensing capabilities, and sensor resource access. The most representative APs developed to date are SensorWare [7], Agilla [8], actorNet [14], MAPS [4], and AFME [21].

SensorWare [7] is a general middleware framework based on agent technology, where the mobile agent concept is exploited. Mobile control scripts in Tcl model network participants' functionalities and behaviors, and routing mechanisms to destination areas. Agents migrate to destination areas performing data aggregation reliably. The script can be very complex, and diffusion gets slower when it reaches destination areas. The replication and migration of such scripts in several sensor nodes allows the dynamic deployment of distributed algorithms into the network. SensorWare defines, creates, dynamically deploys, and supports such scripts. SensorWare is designed for iPAQ devices with megabytes of RAM. The verbose program representation and on-node Tcl interpreter can be acceptable overheads; however, they are not yet on a sensor node.

Agilla [8] is an agent-based middleware where each node supports multiple agents and maintains a tuple space and neighbors' list. The tuple space is local and shared by the agents residing on the node. Special instructions allow agents to remotely access another node's tuple space. The neighbors' list contains the addresses of all one-hop nodes. Agents can migrate carrying their code and state, but do not carry their own tuple spaces. Agilla is currently implemented on MICA2, MICAZ, and TelosB motes.

While both Agilla and SensorWare rely on mobile agents, they employ a different communication model: Agilla's agent interaction is based on local tuple spaces, whereas SensorWare's agent interaction is based on direct communication based on network messages. In reference [26], another mobile agent framework is proposed. The framework is implemented on Crossbow MICA2DOT motes. In particular, it provides agent migration and agent interaction based both on locally shared memory and network messages. In reference [25], the authors propose an extension of Agilla to support direct communication based on messages. In particular, to establish direct communications, agents are mediated by a middle component (named landmark) that interact with agents through zone-based registration and discovery protocols.

In reference [14], actorNet, a mobile agent platform for WSNs based on the actor model, is proposed. In particular, it provides services such as virtual memory, context switching, and multitasking to support a highly expressive yet efficient agent functional language. Currently, the sensor node actorNet platform is specifically designed for TinyOS on MICA2 sensors.

The aforementioned mobile agent systems for WSNs are all implemented for TinyOS-based sensor platforms and use ad-hoc languages for agent programming (Agilla uses a micro-programming language, whereas actorNet employs a function-oriented language). Although some supported operations (e.g., migration) are very efficient, programming complex tasks is not so straightforward and, moreover, developers need to learn another very specific language. In the following, the available Java-based agent platforms are introduced.

The Mobile Agent Platform for Sun SPOT framework (MAPS) [4, 20] is an innovative Java-based framework for wireless sensor networks based on Sun SPOT technology, which enables agent-oriented programming of WSN applications. The MAPS architecture is based on components that interact through events. Each component offers a minimal set of services to mobile agents, which are modelled as multi-plane state machines driven by ECA rules. In particular, the offered services include message transmission, agent creation, agent cloning, agent migration, timer handling, and easy access to the sensor node resources (sensors, actuators, input switches, flash memory, and battery).

The Agent Factory Micro Edition (AFME) [21, 1] is a lightweight version of the Agent Factory framework purposely designed for wireless pervasive systems and implemented in J2ME, recently ported onto Sun SPOT and purposely used for implementing agent communication and migration in WSNs. The AFME agents are based on a BDI (beliefs-desires-intentions)-like architecture and are programmed through a high-level logic-based language. AFME also provides a translator, which converts high-level agent specifications into Java code.

However, AFME was not specifically designed for WSNs and, particularly, for Java Sun SPOT. MAPS is conversely specifically designed for WSNs, and it fully exploits the release 5.0 red of the Sun SPOT library to provide advanced functionality of communication, migration, sensing/actuation, timing, and flash memory storage. Moreover, it allows developers to program agent-based applications in Java according to the rules of the MAPS framework so that no translator and/or interpreter need to be developed and no new language has to be learnt.

5.5 An Agent-Oriented Design of Signal Processing In-Node Environments

The exploitation of the agent-oriented programming paradigm to develop WBAN applications could provide more effectiveness as demonstrated by the application of agent technology in several key application domains [17]. In this section, we describe an agent-oriented design of SPINE (see Section 5.3), named ASPINE [6], which extends the functionalities provided by SPINE and allows for a more rapid development of signal processing intensive WBAN applications in terms of agent-based systems. The reference architecture defined in Section 5.2 was designed by using an

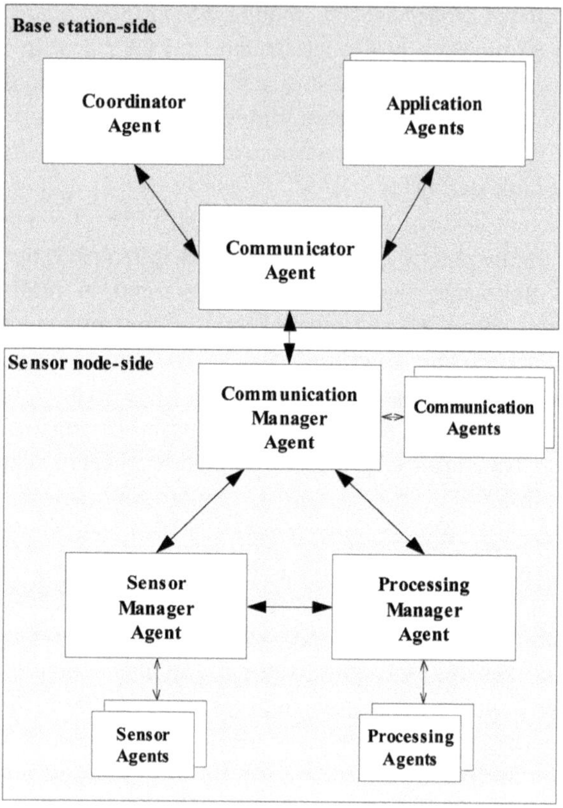

Figure 5.4. The ASPINE high-level architecture.

agent-based approach. Figure 5.4 shows the high-level architecture of ASPINE through a class diagram.

In particular, the following core agents are defined:

Base station-side

- The CoordinatorAgent is responsible for managing the set of nodes of the sensor network under control. Management involves configuring and monitoring nodes.
- The ApplicationAgents are agents implementing application-specific or domain-specific logics.

- The CommunicatorAgent allows the CoordinatorAgent and the ApplicationAgents to interact with the sensor nodes through an efficient over-the-air application-level protocol.

Sensor node-side

- The SensorManagerAgent manages the sensor/actuator resources of the node through specific SensorAgents able to interact with specific sensors (temperature, light, accelerometer, etc.), actuators (LEDs, actuation devices, etc.), and the battery.
- The CommunicationManagerAgent manages the communication with the communicator agent and with the other CommunicationManagerAgents, located at different sensor nodes, by means of specific CommunicationAgents. Moreover, such agent manages the radio settings (e.g., duty cycling, transmission power, etc.).
- The ProcessingManagerAgent supports one or more local processing tasks or parts of global processing tasks (spanning multiple sensor nodes) through processing agents. They are able to perform computation on sensed data (e.g., feature extraction, threshold functions, classification algorithms, etc.) and data aggregation.

An example of in-node signal processor is portrayed in Fig. 5.5 by means of a data-flow model based on tasks. The in-node signal processor aims at sampling the on-board 3-axial accelerometer, computing selected features on sample windows, and sending the results to the coordinator. In particular, the sensed data periodically produced by the sensing task acting on the three channels (X, Y, Z) of the accelerometer are split for the computation of the features Mean on all three axes (XYZ), and Min and Max on axis X. Each triple

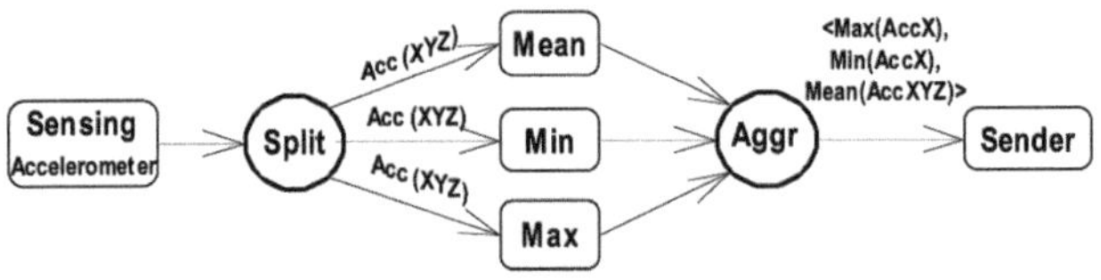

Figure 5.5. Data-flow-based model of an in-node signal processing task.

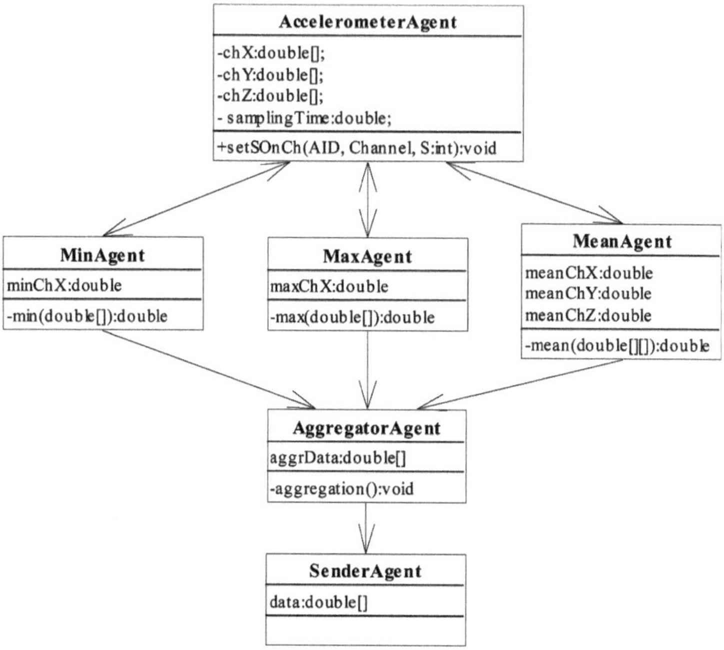

Figure 5.6. ASPINE-based design of the example in-node signal processing task.

of computed features (<Mean(AccXYZ), Min(AccX), Max(AccX)>) are aggregated by the aggregation task (Aggr) and sent to the coordinator node by the data transmission task (Sender) as soon as the aggregated data is available.

The ASPINE design of the data-flow-based model of Fig. 5.5 is shown in Fig. 5.6. The AccelerometerAgent interacts with the accelerometer sensor and, according to the set sampling time (samplingTime), acquires one sample per channel (X, Y, Z), and stores them into the corresponding buffers chX, chY, chZ. Once *S* samples are acquired, the AccelerometerAgent passes them to the MinAgent, MaxAgent, and MeanAgent. These relationships are created through the method setSOnCh(AID, Channel, *S*), where AID is the agent identifier, Channel refers to the channels to be considered, and *S* is the number of samples. In this case, all agents are based on the same *S* but on different channels. In particular,

MinAgent, MaxAgent, and MeanAgent receive the last acquired *S* samples, respectively, from the chX buffer, from the chX buffer, and from the chX, chY, and chZ buffers. Upon reception of such data, the agents compute their respective functions and pass the results to the AggregatorAgent. This waits for the aggregation of the data triple aggrData=<minChX, maxChX, <meanChX, meanChY, meanChZ≫ and passes it to the SenderAgent, a specific CommunicationAgent, which, in turn, transmits it to the CommunicatorAgent located at the base station.

5.6 An Analysis of Agent-Oriented Implementations of In-Node Signal Processors

A WBAN-based activity human monitoring system aims at recognizing postures and movements of assisted livings [22]. The developed prototype is able to recognize some postures (lying down, sitting, and standing still) and a movement (walking). The architecture of the system, shown in Fig. 5.7, is organized into a coordinator and two sensor nodes according to the reference WBAN architecture described in Section 5.2.

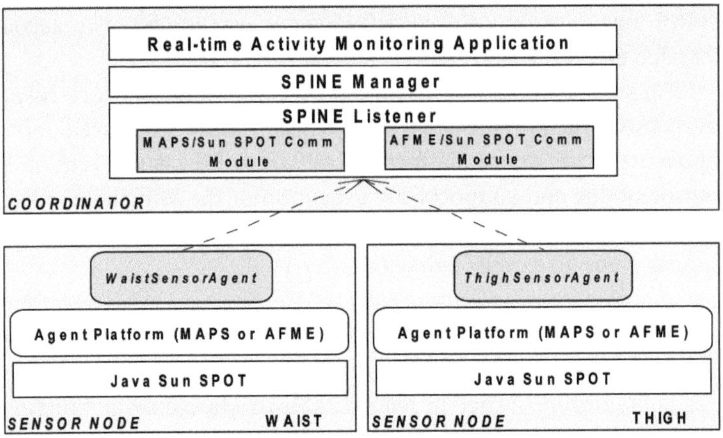

Figure 5.7. Architecture of the WBAN-based real-time activity monitoring system.

The coordinator is based on the Java-based SPINE coordinator developed in the context of the SPINE project [27]. In particular, the SPINE Manager is used by end-user applications (e.g., real-time activity monitoring application) for sending commands to the sensor nodes. Moreover, the SPINE Manager is responsible for capturing low-level messages and node events through the SPINE Listener, which integrates several sensor platform-specific SPINE communication modules (e.g., TinyOS, Z-Stack, etc.) to notify registered applications with higher-level events and message content. A SPINE communication module is composed of a send/receive interface and some components that implement such interface according to the specific sensor platform and that formalize the high-level SPINE messages in sensor platform-specific messages. The SPINE Listener is enhanced with MAPS/Sun SPOT and AFME/Sun SPOT communication modules to configure and communicate with MAPS-based and AFME-based sensor nodes, respectively. Such modules translate high-level SPINE messages formatted according to the SPINE OTA (over-the-air) protocol [27] into lower-level MAPS or AFME/Sun SPOT messages through its transmitter component and vice versa through its receiver component. Such modules also integrate an application-specific logic for synchronizing the operations of the two sensors. The SPINE-based real-time activity monitoring application was thus completely reused as well as the SPINE Manager; only the SPINE Listener was modified to account for such enhancement.

The sensor nodes are based on Java Sun SPOT sensors respectively positioned on the waist and the thigh of the monitored person. In particular, an agent platform (MAPS or AFME) is resident on the sensor nodes and supports the execution of the WaistSensorAgent and the ThighSensorAgent. Such sensor agents have the following similar step-wise cyclic behavior:

1. *Sensing* the 3-axial accelerometer sensor according to a given sampling time (ST).
2. *Computation* of specific features (Mean, Max, and Min functions) on the acquired raw data according to the window (W) and shift (S) parameters. In particular, W is the sample size on which features are computed, whereas S is the number of new

acquired sample data for calculating a new feature. Usually S is set to 50% of W.
3. Features *aggregation* and *transmission* to the coordinator.
4. Goto 1.

The agents differ in the specific computed features even though the W and S parameters are equally set. In particular, while the waist sensor agent computes the mean values for data sensed on the XYZ axes, the min and max values for data sensed on the X axis, the ThighSensorAgent calculates the min value for data sensed on the X axis.

In the following subsections, we describe the implementation of the WaistSensorAgent carried out with MAPS and AFME and, then, compare the implemented solutions.

5.6.1 *MAPS-Based and AFME-Based Implementation of Sensor Agents*

The architecture of the MAPS-based sensor agent (Fig. 5.8) is composed of two planes (*sensing* and *feature calculation and transmission*) modeled as finite state machines. Each plane is able to handle specific events, which are instances of system and user-defined event types. The *sensing plane* (see Fig. 5.9a) specifies the sensing process: the MSG.START event initiates the sensing of the three channels of the accelerometer; after the sensed data are returned through the ACC_CURRENT_ALL_AXES event, the sampling timer is scheduled for the next acquisition; if a MSG.RESYNCH, formalizing a re-synchronization operation among the sensor agents sent by the coordinator, is received, the sensing process is paused until a MSG.RESTART event arrives; finally, the MSG.STOP halts the sensing process. The *feature calculation and transmission* plane (see Fig. 5.9b) specifies the feature extraction process and the transmission of the computed features to the coordinator, which are triggered by the MSG.COMPFEATURES event.
The architecture of the AFME-based sensor agent (Fig. 5.10) is composed of:

- *Two perceptors.* MTSPerceptor checks for the arrival of a new message and SensingPerceptor checks for the feature

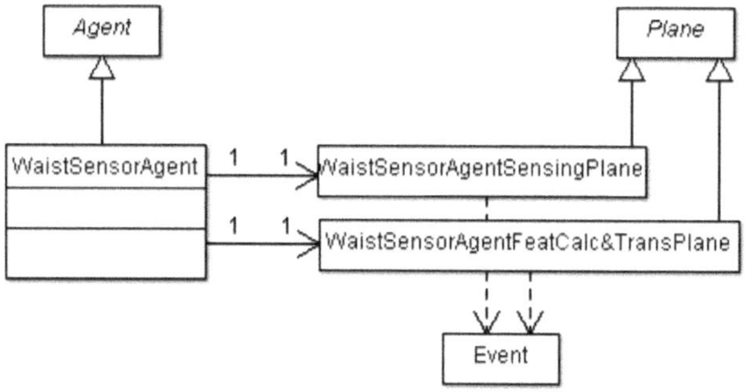

Figure 5.8. MAPS-based sensor agent.

computation and send the computation results to the coordinator if such results are available.

- *Three actuators.* ActivateSensorActuator allows activating the sensing operation, ResetActuator resets the data buffer after the reception of the resynch message, and RequestActuator is used to send a request message to the coordinator.
- *Rules.* TerImplication contains the rules of the agent behavior (see below).

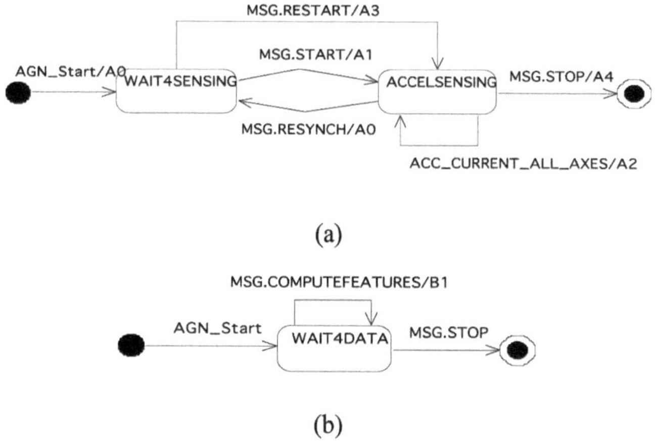

Figure 5.9. Planes of the MAPS-based sensor agent: (a) sensing plane; (b) feature calculation and transmission plane.

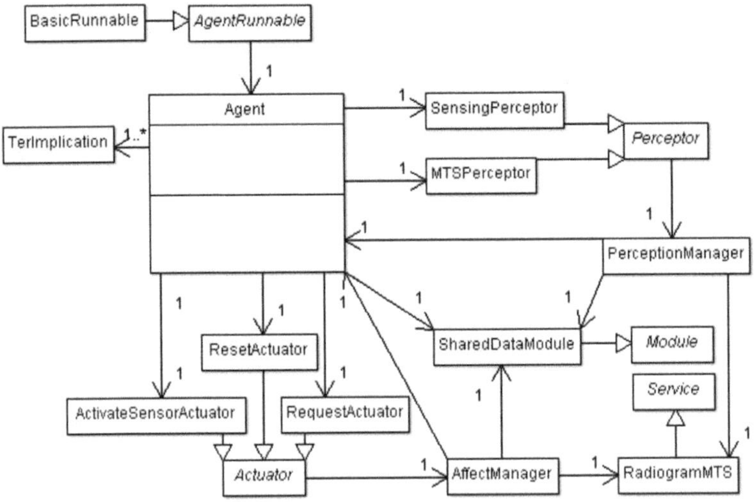

Figure 5.10. AFME-based sensor agent.

- *One Module.* SharedDataModule contains buffers storing the data sensed from the three accelerometer channels, the feature extraction parameters, the flag variable (*activation*) enabling the sensing process.
- *One Service.* RadiogramMTS represents the transport service for data transmission to the coordinator.

The agent behavior is modeled through the following three basic rules:

1. message(inform, sender(BaseStation, addresses(BSAddress)), begin) > activateSensors(1);
2. sense(?val), !message(inform, sender(BaseStation, addresses (BSAddress)), resynch) > request(agentID(BaseStation, addresses ("radiogram://"+BSAddress)), ?val);
3. message(inform, sender(BaseStation, addresses(BSAddress)), resynch) > reset;

Rule (1) enables the sensor reading operation on the sensor node. In particular, this rule states that the message belief, which is generated upon the reception of an inform message with "begin" content sent by the coordinator with address BSAddress, triggers the action ActivateSensors(1), which will then set the *activation* flag on.

Rule (2) regulates the agent behavior during the sensing phase. If a new sense belief having the sensed data is generated and a message belief generated upon reception of the resynch message sent by the coordinator does not exist, the request action is carried out so that the new message containing the feature calculation is sent to the coordinator.

Finally, rule (3) allows re-synchronizing the sensor agent. In particular, it states that if the belief message is generated upon reception of the resynch message, the action reset, which re-initializes all the data structures in SharedDataModule, is executed.

5.6.2 *Agent Implementation Comparison*

The comparison of the two different agent implementations are carried out according to the following two aspects: (i) effectiveness in prototyping the agent-based solutions; (ii) timing and synchronization of the real-time monitoring.

The development of MAPS and AFME agents is based on different approaches. MAPS uses state machines to model the agent behavior and directly the Java language to program guards and actions. AFME uses a more complex model centered on perceptors, actuators, rules, modules, and services that define the agent behavior. They are both effective in modeling agent behavior even though MAPS is more straightforward as it relies on a programming style based on state machines widely known by programmers of embedded systems.

Two important issues to deal with are the timing of the sensing process in terms of admissible sampling rate and the synchronization between the operations of the two agents, which is to be maintained within a maximum skew for not affecting real-time monitoring. If such skew is overtaken, the two agents are to be re-synchronized. Indeed such two aspects are strictly correlated. In particular, as the sensor agents compute a different number of features, when the sampling rate is high, the agent computing more features (i.e., the WaistSensorAgent) takes more time to complete its operations for each S sample acquisition than the ThighSensorAgent. Re-synchronization is driven by the synchronization logic included in the developed MAPS/Sun SPOT and AFME/Sun SPOT communication modules, which sends a re-synchronization message as soon as it detects that the synchronization skew is greater than a given threshold. Detection is based on the skew time between the receptions of two messages sent by the agents that contain features referring to the same interval of S sample acquisition: if $skew \geq P \times S \times ST$, then *resynchronize*, where P is a percentage, S = 0.5 W is the shift, W is the data sample window, ST is the sampling time. Thus, the evaluation aims at analyzing the synchronization of the sensor agents and their monitoring continuity. The defined metrics are:

- The *packet pair average time* (PPAT), which is the average reception time between two consecutive pairs of synchronized packets, having the same logical timestamp and containing the computed features sent by the sensor agents.
- The *synchronization packet percentage* (SPP), which is the percentage of resynchronization packets, which are sent by the coordinator for re-synchronizing the sensor agents, calculated with respect to the total number of received feature packets.

PPAT should be ideally equals to ST*S, i.e., the packet pair arrives each monitoring period and so there is no de-synchronization in the average. SPP should be as much as possible close to 0, i.e., a few or no re-synchronizations are carried out and so the monitoring can be continuous as a resynch operation usually takes 600 ms. Thus, system parameters (W, ST, and P) should be carefully set to optimize the monitoring process. To this purpose, a set of experiments was set up to tune the system parameters. In particular, the experiments were carried out by fixing ST (ms) = [25, 50, 100], W (samples) = [40, 20], and P (%) = [5, 15, 25]. Figures 5.11 and 5.12 show the obtained results.

As can be noticed, the system cannot support an ST = 25 ms because PPAT is always greater than the ideal value and SPP is too high. This leads to a non-continuous monitoring due to very frequent

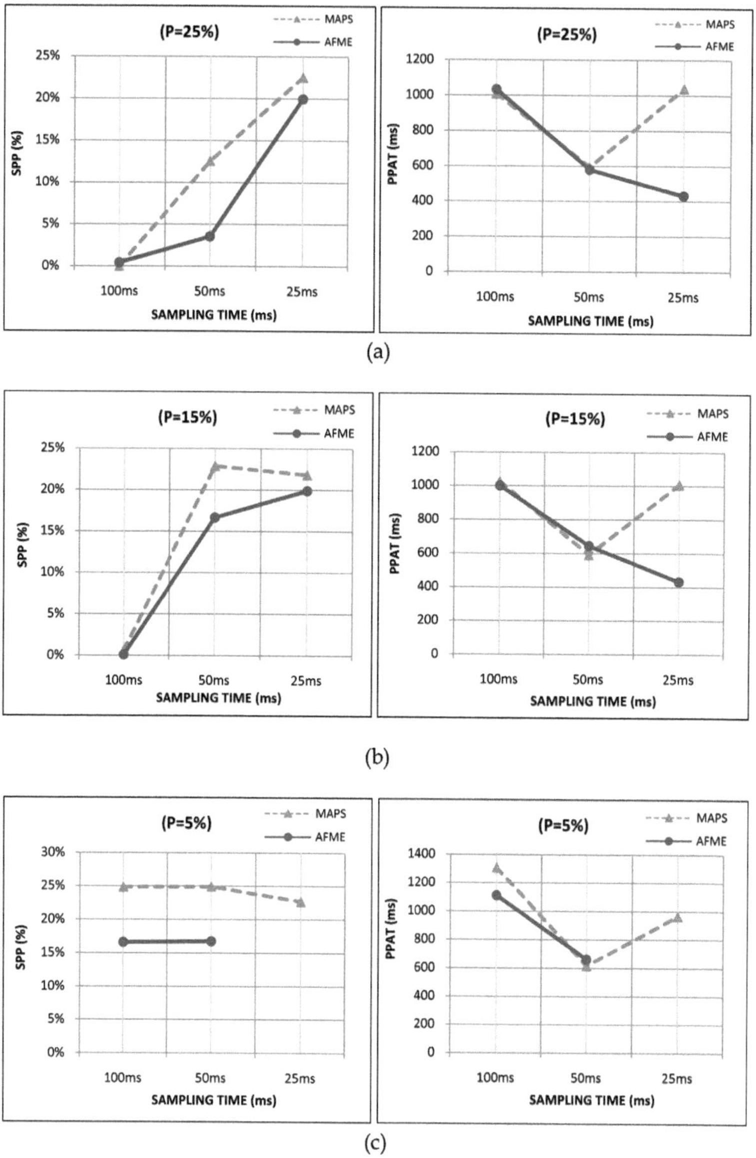

Figure 5.11. Analysis of PPAT and SPP of the sensor agents with W = 20, S = 10, and P = (a) 25%, (b) 15%, (c) 5%. See also Color Insert.

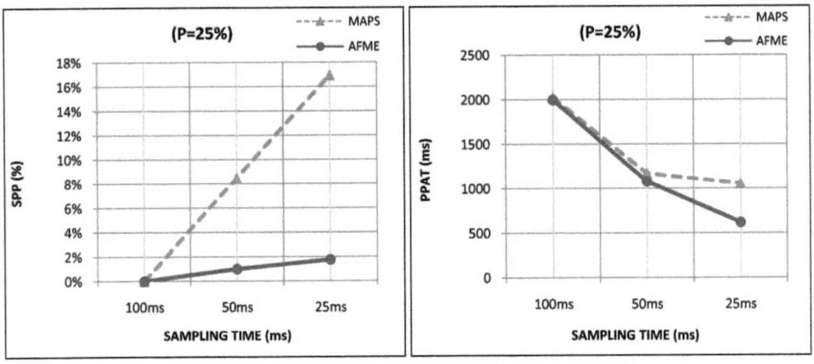

(a)

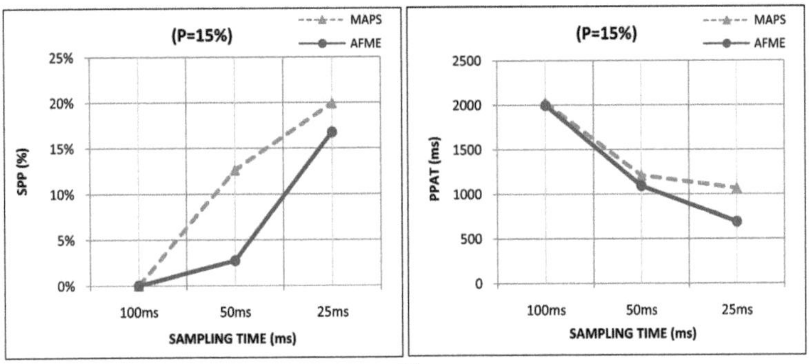

(b)

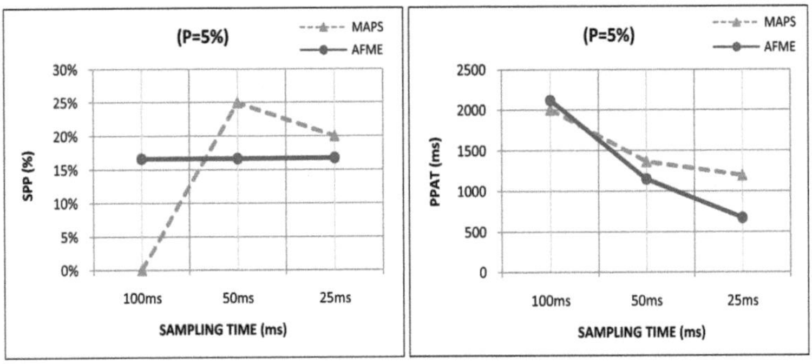

(c)

Figure 5.12. Analysis of PPAT and SPP of the sensor agents with W = 40, S = 20, and P = (a) 25%, (b) 15%, (c) 5%. See also Color Insert.

re-synchronizations (SPP >= 20% for W = 20 and S = 10). The best results are obtained with ST = 100 ms, P = 25% or 15% and W = 20; they guarantee monitoring continuity due to an SPP $\approx$ 0% and regularity as the experimented PPAT $\approx$ the ideal PPAT for W = 20. If W = 20 and P = 5%, ST = 100 ms is not a good value either because an out-of-limits skew frequently occurs. Although the AFME implementation performs better than the MAPS implementation, the AFME implementation collapses in the case W = 20, S = 10, and P = 5%. It is worth noting that even though a lower ST would allow a finer monitoring, the considered human activities can be well captured by an ST = 100 ms as demonstrated by the good accuracy results (95% in the average) obtained from the carried-out real-time human activity monitoring [12].

5.7 Conclusions and Future Work

In this chapter, we have presented an agent-oriented approach for high-level programming of WBAN applications. The agent approach is not only effective during the design of a WBAN application but also during the implementation phase. In particular, the presented frameworks MAPS and AFME allow for a more rapid prototyping of sensor node code than low-level APIs, which can be effectively used only by sensor node skilled programmers having knowledge of sensor drivers, data buffers, radio and energy mechanisms. The higher level software abstractions provided by MAPS and AFME are suitable for the development of real-time monitoring systems as demonstrated by the performance evaluation of the proposed case study concerning a real-time human activity monitoring system. Ongoing work is aimed at implementing ASPINE through MAPS to provide a full-fledged agent-based signal processing in-node environment.

References

1. *Agent Factory Micro Edition (AFME)*, documentation and software at http://sourceforge.net/projects/agentfactory/files/ (2011).

2. Aiello, F., Bellifemine, F., Fortino, G., Gravina, R., and Guerrieri, A. (10–14 May 2010) An agent-based signal processing in-node environment for real-time human activity monitoring based on wireless body sensor networks, in *Proceedings of the 1st International Workshop on Infrastructures and Tools for Multiagent Systems (ITMAS-2010)*, jointly held with the9th *International Joint Conference on Autonomous Agents and Multiagent Systems (AAMAS-2010)*, Toronto, Canada.
3. Aiello, F., Fortino, G., and Guerrieri, A. (25–31 August 2008) Using mobile agents as an effective technology for wireless sensor networks, in *Proceedings of the 2nd IEEE/IARIA International Conference on Sensor Technologies and Applications (SENSORCOMM 2008)*, Cap Esterel, France.
4. Aiello, F., Fortino, G., Gravina, R., and Guerrieri, A. (2011) A Java-based agent platform for programming wireless sensor networks, *The Computer Journal*, 54(3), pp. 439–454.
5. Bao, L., and Intille, S. S. (2004) Activity recognition from user-annotated acceleration data, in *Proceedings of the 2nd International Conference on Pervasive Computing (PERVASIVE)*, pp. 1–17.
6. Bellifemine, F., and Fortino, G. (9–10 July 2009) ASPINE: an agent-oriented design of SPINE, in *Proceedings of the Workshop on Objects and Agents (WOA'09)*, Parma, Italy.
7. Boulis, A., Han, C. -C., and Srivastava, M. B. (2003) Design and implementation of a framework for efficient and programmable sensor networks, in *Proceedings of the 1st International Conference on Mobile systems, Applications and Services (MobiSys)*, pp. 1 87–200.
8. Fok, C. -L., Roman, G. -C., and Lu, C. (6–10 June 2005) Rapid development and flexible deployment of adaptive wireless sensor network applications, in *Proceedings of the 24th International Conference on Distributed Computing Systems (ICDCS'05)*, Columbus, Ohio, pp. 653–662.
9. Fortino, G., Guerrieri, A., Giannantonio, R., and Bellifemine, F. (11–14 October 2009) Platform-independent development of collaborative WBSN applications: SPINE2, in *Proceedings of the IEEE International Conference on Systems, Man, and Cybernetics (SMC 2009)*, San Antonio, Texas, USA.
10. Fortino, G., Galzarano, S., Giannantonio, R., Gravina, R., and Guerrieri, A. (2010) SPINE-based application development on heterogeneous wireless body sensor networks, *International Journal of Computing*, 9(1), pp. 80–89.

11. Gravina, R., Andreoli, A., Salmeri, A., Buondonno, L., Raveendranathank, N., Loseuk, V., Giannantonio, R., Seto, E., and Fortino, G. (7–9 June 2010) Enabling multiple BSN applications using the SPINE framework, in *Proceedings of the International Conference on Body Sensor Networks (BSN 2010)*, Biopolis, Singapore.
12. Gravina, R., Guerrieri, A., Fortino, G., Bellifemine, F., Giannantonio, R., and Sgroi, M. (12–15 October 2008) Development of body sensor network applications using SPINE, in *Proceedings of the IEEE International Conference on Systems, Man, and Cybernetics (SMC 2008)*, Singapore.
13. Hester, T., Hughes, R., Sherrill, D. M., Knorr, B., Akay, M., Stein, J., and Bonato, P. (3–5 April 2006) Using wearable sensors to measure motor abilities following stroke, in *Proceedings of the 3rd International Workshop on Wearable and Implantable Body Sensor Networks, 2006 (BSN 2006)*, MIT, Boston (MA), USA, pp. 5–8.
14. Kwon, Y., Sundresh, S., Mechitov, K., and Agha, G. (2006) ActorNet: An actor platform for wireless sensor networks, in *Proceedings of the 5th International Joint Conference on Autonomous Agents and Multiagent Systems (AAMAS)*, pp. 1297–1300.
15. Lester, J., Choudhury, T., Borriello, G. (2006) A practical approach to recognizing physical activities, in *International Conference on Pervasive Computing (PERVASIVE)*, pp. 1–16.
16. Lombriser, C., Roggen, D., Stager, M., and Troster, G. (26 February–2 March 2007) Titan: a tiny task network for dynamically reconfigurable heterogeneous sensor Networks, in *Verteilten Systemen (KiVS 2007)*, Bern, Switzerland.
17. Luck, M., McBurney, P., and Preist, C. (2004) A manifesto for agent technology: towards next generation computing, *Autonomous Agents and Multi-Agent Systems*, 9(3), pp. 203–252.
18. Malan, D., Fulford-Jones, T., Welsh, M., and Moulton, S. (June 2004) CodeBlue: An ad hoc sensor network infrastructure for emergency medical care, in *Proceedings of MobiSys 2004 Workshop on Applications of Mobile Embedded Systems (WAMES 2004)*.
19. Maurer, U., Smailagic, A., Siewiorek, D. P., and Deisher, M. (2006) Activity recognition and monitoring using multiple sensors on different body positions, in *Proceedings of the 3rd International Workshop on Wearable and Implantable Body Sensor Networks (BSN 2006)*, MIT, Boston (MA), USA.
20. *Mobile Agent Platform for Sun SPOT (MAPS)*, documentation and software at http://maps.deis.unical.it/ (2011).

21. Muldoon, C., O'Hare, G. M. P., O'Grady, M. J., and Tynan, R. (2008) Agent migration and communication in WSNs, in *Proceedings of the 9th International Conference on Parallel and Distributed Computing, Applications, and Technologies.*

22. Najafi, B., Aminian, K., Ionescu, A., Loew, F., Büla, C. J., and Robert, P. (June 2003) Ambulatory system for human motion analysis using a kinematic sensor: monitoring of daily physical activity in the elderly, *IEEE Transactions on Biomedical Engineering*, 50(6), pp. 711–723.

23. Pansiot, J., Stoyanov, D., McIlwraith, D., Lo, B. P. L., and Yang, G. Z. (26–28 March 2007) Ambient and wearable sensor fusion for activity recognition in healthcare monitoring systems, in *Proceedings of the 4th International Workshop on Wearable and Implantable Body Sensor Networks (BSN 2007)*, RWTH Aachen University, Germany.

24. Sohraby, K., Minoli, D., and Znati, T. (2007) *Wireless Sensor Networks: Technology, Protocols, and Applications*, Wiley.

25. Suenaga, S., and Honiden, S. (14 May 2007) Enabling direct communication between mobile agents in wireless sensor networks, in *Proceedings of the 1st International Workshop on Agent Technology for Sensor Networks (ATSN 2007)*, jointly held with the6th *International Joint Conference on Autonomous Agents and Multiagent Systems (AAMAS 2007)*, Honolulu, Hawaii.

26. Szumel, L., LeBrun, J., Owens, J. D. (30–31 May 2005) Towards a mobile agent framework for sensor networks, in *Proceedings of the 2nd IEEE Workshop on Embedded Networked Sensors (EmNetS-TT)*, Sydney, Australia.

27. *Signal Processing In-Node Environment (SPINE)*, documentation and software at http://spine.tilab.com (2011).

28. *TinyOS website*, www.tinyos.net (2011).

29. Yang, G. -Z. (2006) *Body Sensor Networks*, Springer.

30. *ZigBee alliance*, technical documents and standard specifications at http://www.zigbee.org (2011).

Chapter 6

Hardware Development and Systems for Wireless Body Area Networks

Mehmet Rasit Yuce
Department of Electrical and Computer Systems Engineering, Monash University, Clayton, VIC 3800, Australia
mehmet.r.yuce@ieee.org

Miniaturized and wearable sensor nodes are required to form a wireless body area network (WBAN) to monitor physiological parameters from human bodies. In this part of the book, we will present the existing hardware solutions to realize a small wireless sensor for use in WBAN applications. Implementation issues are discussed, and details of techniques for the design of wireless sensors in WBAN applications are presented.

6.1 Introduction

A WBAN network aims to provide efficient and optimized wireless link for physiological signal monitoring from single or multiple human bodies. As shown in Fig. 6.1, the future WBAN-based sensor network requires miniaturized and wearable sensor nodes that can communicate with the receiving device. A WBAN system consists of individual wireless sensor nodes that can transfer a person's

Wireless Body Area Networks: Technology, Implementation, and Applications
Edited by Mehmet R. Yuce and Jamil Y. Khan

ISBN 978-981-4316-71-2 (Hardcover), ISBN 978-981-4241-57-1 (eBook)
www.panstanford.com

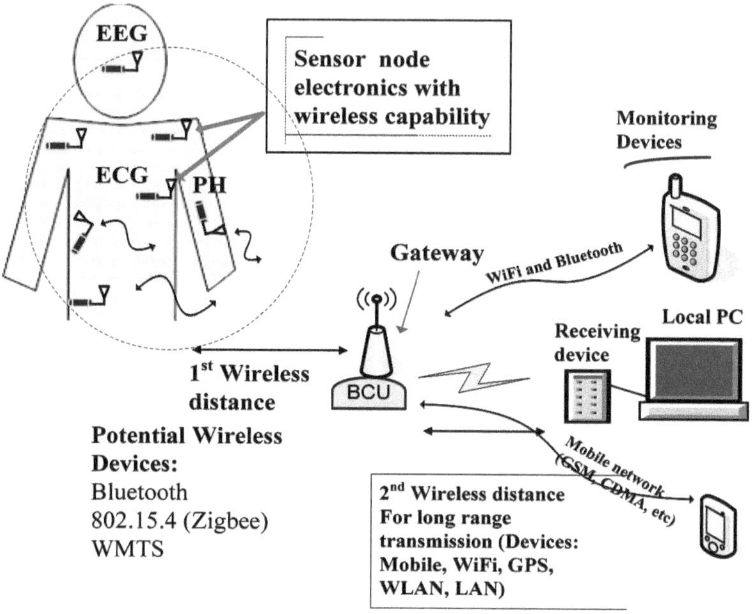

Figure 6.1. A wireless sensor network system detecting and transmitting signals from a human body.

physiological data such as heart rate, blood pressure, and ECG via a wireless link, without the need of any wired connection. Each sensor will have wireless capability and its design will be optimized in terms of the physical characteristic of the physiological signal. Having individual wireless nodes are also very beneficial since not all patients require all the physiological parameters for diagnosis. Each sensor can easily be plugged in and plugged out from a human body.

6.2 Wireless Body Sensors

Figure 6.1 shows a WBAN-based telemedicine system for collecting medical data from a human body using wireless sensors. This telemedicine system comprises sensor nodes, a body control unit (BCU) that transmits data to a receiver station (i.e., remote PC) that can have communication with remote stations at a medical center [1].

The BCU can be a smart phone or any portable device that will act as a gateway device to pass on the collected data from sensors to remote stations. The control device will be similar to smart phones we use in our daily life, which can also be used to monitor the data obtained from sensors. Such portable electronic devices will mostly contain Bluetooth, Wi-Fi, and GPS technologies built-in, which can be connected to medical sensors and to remote medical centers so that data can be accessed by medical professionals at any time. The BCU can be carried on the body as a wristwatch or can be placed on the belt.

For collecting data from a human body, usually short-range wireless technologies operating at MICS (medical implant communication service), WMTS (wireless medical telemetry service), 2.4 GHz ISM (industrial, scientific, and medical) bands have been developed by designers. However, it is necessary to connect the sensor nodes to remote stations using Wi-Fi and mobile communication links for longer range data transfer. In order to provide long-range remote monitoring, several gateway devices should be developed to interface with the existing ICT (information and communication technology) infrastructure in emergency vehicles or in medical centers. These gateway devices will mainly be used to provide communication between BCUs and remote computers or mobile devices.

6.2.1 *Sensor Nodes and Hardware Designs*

Low-power wireless technologies should be used to form small, light-weight, and wearable sensor nodes to collect and sense physiological data from a human body. Especially, when the number of sensors used on a single body increases, the weight of each sensor node becomes an important issue. Wireless sensor nodes are being developed by biomedical companies and researchers to make them safe, secure, and light weight so that they can be wearable. Key parameters of a wireless sensor node can be as follows:

- The wireless sensor nodes should be able to transfer data over a distance of a few meters.
- Sensor nodes should be miniaturized so that they can be easily wearable or attachable to a patient body.

Table 6.1. Physiological parameter range and signal frequencies

	Parameter	Range of parameter	Signal frequency
Continuous Signals	ECG signal	0.5–4 mV	0.01–250 Hz
	EMG (Electromyogram)	10–15 mV	10–500 Hz
	EEG	3–300 μV	0.5–70 Hz
Discrete Signal*	Body Temperature	32–40° C	0– 0.1 Hz
	Blood Pressure (BP)	10–400 mm Hg	0–50 Hz
	GSR (Galvanic Skin Reflex)	30–3 mV	0.03–20 Hz
	Respiratory rate	2–50 breaths/min	0.1–10 Hz

*Here the term "discrete" indicates large time intervals.

Alternatively, they can also be inserted in clothes using textile technology to improve the patient's comfort.

Wireless sensor nodes are developed specifically to detect and transmit the physiological signals listed in Table 6.1. These signals can be put in two categories: discrete and continuous. ECG/EEG signals are periodic signals, which require continuous monitoring. Body temperature is an example of discrete physiological data since it is taken usually at discrete times, e.g., every hour.

Most physiological signals are low frequency in nature and occupy a small information bandwidth. At such low frequencies and low amplitudes, some problems inherent to circuits need additional attention. For reliable information transfer, it is necessary that the interface electronic in the sensor nodes detect physiological signals in the presence of noise and increase the signal-to-noise ratio (SNR) of the detected signal for processing by the subsequent blocks of the sensor nodes.

Sensor nodes are designed to be small and power efficient so that their battery lasts for a long time. They collect weak, raw signals from a human body. Signals from a human body are usually weak and coupled with noise, and thus they need to go through amplification/filtering process to increase the signal strength and to remove unwanted signals and noise. Then an analog-to-digital conversion (ADC) stage is employed to convert the analog body signals into digital for digital processing. The digitized signal is processed and stored in the microprocessor. The microprocessor (i.e., microcontroller) will then pack the data and transmit over the

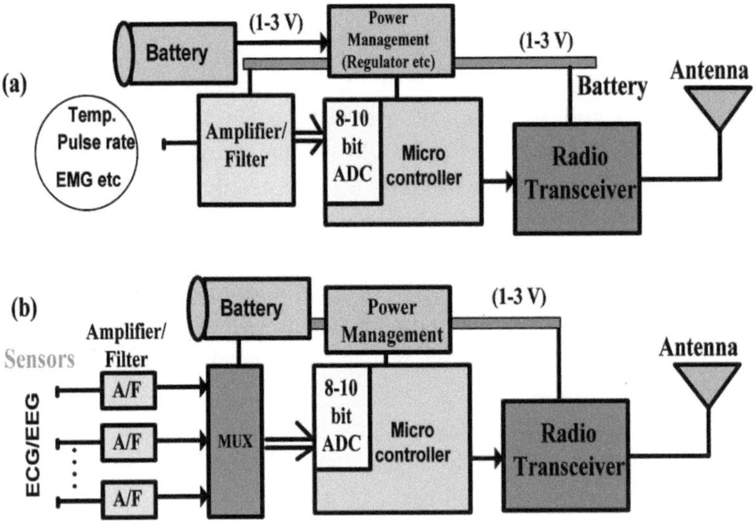

Figure 6.2. An example of sensor node hardware: (a) single wireless sensor node, (b) multi-channel wireless sensor node. See also Color Insert.

air via a wireless transceiver. Figure 6.2 shows two commonly used hardware implementations of sensor nodes.

A sensor node consists of three main blocks: the sensing front end, microcontroller, and the radio receiver to undertake the tasks mentioned above. In addition to these blocks, there is also a battery and its power management circuitry. The power management circuit is usually a regulator chip used to distribute the power source to the individual blocks. It is advised to keep all the sensor blocks' power supply level the same (e.g., all at 3 V) so that the regulator will not consume a large power.

Considering the signal attributes given in Table 6.1, a sampling rate of around 200–1000 Hz will be necessary for the ADC in the microcontroller (the sampling rate should be a minimum of twice the highest frequency in the signal that is digitized). Unlike other wireless sensor network applications, in a medical scenario, each sensor node has a different frequency and thus should be optimized according to its frequency band. The trade-off between the reduction in sampling rate and the total power consumption of the ADC is thus determined by the choice of the specific physiological parameter used in the sensor node.

In case of emergency in a medical application, usually the monitoring and analyzing of critical physiological signals such as ECG and blood pressure (BP) will be the top priority. Thus, it is important if the signals in Table 6.1 are grouped into critical and noncritical data before attaching sensors on the injured people in a medical area. The wireless communication from sensors to the BCU should be designed such that priority should be given to the critical data. A WBAN-based telemedicine system will result in better data gathering process if data are prioritized according to the vital sign and patient situation.

The front end of a sensor node (i.e., interface electronics) is the first unit connected to sensors and electrodes. It is an analog electronics circuit that consists of amplifiers and filters. In case of detecting a continuous physiological signal such as EEG, ECG, or EMG, the front end should use an instrumental amplifier. An instrumental amplifier helps to amplify and detect noisy body signals at the presence of common mode noise (DC noise) at the input. By looking at Table 6.1 for amplitude and frequency information on continuous physiological signals, parameter values could be similar or different in some cases. It mainly depends on the quality of electrodes used to detect these signals. The same analog front end can be used to detect these continuous physiological signals. On the other hand, the required amplification and filtering can easily be adjusted for a specific signal by modifying capacitors and resistors in the front-end circuit.

Figure 6.3 shows a front-end circuit that can be used to detect ECG, EEG, and EMG signals. This circuit utilizes an instrumentation amplifier (INA321) to detect noisy body signals and a low-pass filter with an op-amp (LTC6081) [2]. The final stage is a 100 Hz LPF added for antialiasing. Typically the useful spectrums of ECG and EEG signals have amplitude of less than 500 μV with a frequency less than 100 Hz. In this example, the input medical signals are amplified by 60 dB; INA321 produces a gain of 14 dB, and the second amplification stage with low-pass filter characteristic-LTC6081 provides a gain of 46 dB with a cutoff frequency at 100 Hz. The circuit is running from a 3.3 V source with virtual ground set

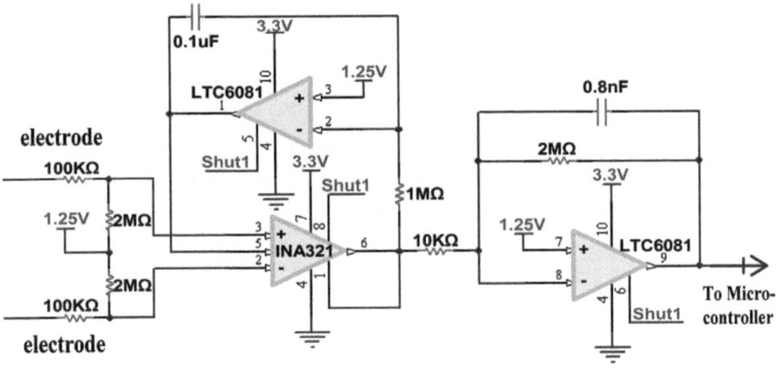

Figure 6.3. Schematic of analog front end for continuous physiological parameter detection.

at 1.25 V. Both components possess high common mode rejection ratio (CMRR) and low current consumption properties. INA321 has a CMRR of 94 dB, while LTC6081 has a CMRR of 100 dB. High CMRR values help obtain better quality signal. The active current consumption is 40 μA, and when operating in shutdown mode, it consumes less than 1 μA. The INA321 not only has extremely good CMRR and meets the low power need, but also has the ability to shut down when not in use. This is directly controlled by the microcontroller by connecting the shutdown pin to a CMOS I/O pin. The INA321 instrumental amplifier has 100 nV/√Hz, @ 1 kHz, input voltage noise and 3 fA/√Hz @ 1 kHz input current noise.

To eliminate DC drift caused by ICs, an integrator op-amp circuit is used as a negative feedback to the reference pin of the INA321. Any DC buildup is inverted and fed back into the system to cancel it out. This is only necessary for high gain circuits since any DC drift is multiplied just like any other signal. With a gain of 1000, a shift of 1 mV will cause the output to jump 1 V, which is close to saturation in this case. Apart from the integrator feedback, this circuit is relatively straight forward and has many common features to any other signal processing circuits.

The 2 MΩ and 100 kΩ resistors are used to limit the current to provide protection for the human body. In addition to this, it is also advised to use a safe power supply. Using a portable battery during

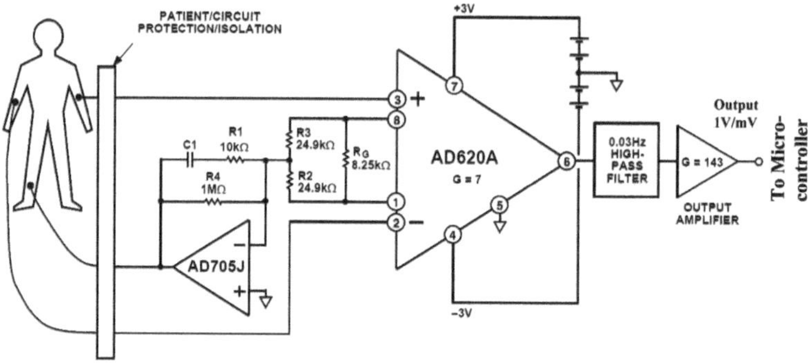

Figure 6.4. ECG detection with the AD620 instrumentation amplifier.

the design of this circuit is a better idea rather than using a direct source from a power line.

The AD620 is another well known low cost, high accuracy instrumentation amplifier (Analog Devices, USA), which can be used to detect body signals. It has 9 nV/√Hz, @ 1 kHz, input voltage noise and 0.1 pA/√Hz @ 1 kHz input current noise. Similar to INA321, the AD620 can be configured to have a CMRR value of 100 dB. The details of circuit configuration as a front end to detect sensor signals can be found in references [3] and [4]. Figure 6.4 shows a schematic for ECG detection designed with AD620. As explained before, the additional filter and a gain stage are added to make the detected signal ready for digitization at the micro-controller.

Another noise exists in detecting body signals, known as the power-line interference at 50 Hz (or 60 Hz in some countries). A notch filter (a combination of low-pass filter and high pass filter) can be designed at the front end to remove this noise. Alternatively, this can be done digitally at the microcontroller or at the remote computer after the continuous signal is received wirelessly [4]. An analogue notch filter works for most applications, but it requires more power and increases the board size and is, thus, avoided. Figure 6.5 shows an ECG signal obtained using the circuit given in Fig. 6.3. Figure 6.5b shows a clear ECG signal after using a notch filter to remove the 50/60 Hz noise.

Employing a digital-notch filter at the computer will make the sensor node less complex. It is also possible to use a combination

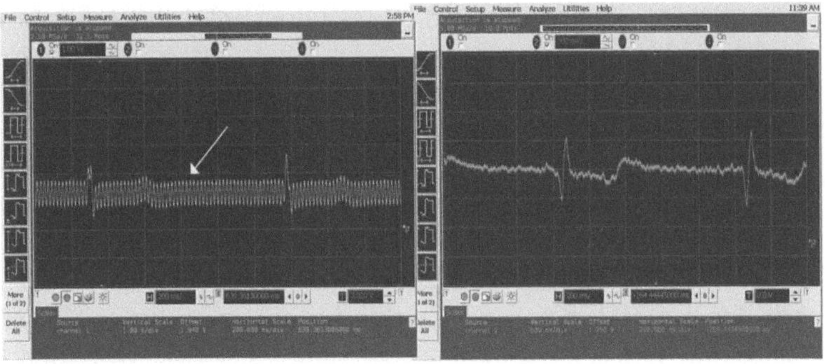

Figure 6.5. ECG with 60 Hz (or 50 Hz) noise (a) and without noise (b). See also Color Insert.

of an analog notch filter at the front end and then digitally via computation programs like MATLAB. The latter approach will increase the clarity of the signal further.

Some ECG projects use a "right leg drive" circuit, which is simply the inverse of the 50/60 Hz injected back into the patient to cancel the power line noise out. Similar to the above approach, this would require more op-amps and an extra electrode and is preferred not to be used. This method also requires a small amount of current, approximately 1–3 μA, to be passed through the patient.

In case of detecting discrete physiological signals such as body temperature and respiratory rate, a simple active low-pass filter (a filter together with a gain amplifier) is sufficient to bring the signal detected by a sensor to a level to be digitalized by a microcontroller.

Figure 6.2b shows a sensor node based on multichannel detection. A multiplexer is an array of analog switches operated with a clock that allows all the input signals to a single output to be digitized at the microcontroller. A multichannel sensor node requires the use of wires connected to electrodes for a single sensor node which should be avoided if possible for a WBAN application. The existing of wires will restrict human movements and thus comfort.

Having said that, for some clinical applications, a multichannel sensor node may be required for certain diagnostic tools. For example, a multichannel sensor node is used for EEG signals for brain-computer interface applications [5]. Multichannel signal detection will increase the data rate requirement. As an example, if we use 200 Hz sampling frequency for 10 channel EEG signal, the required data rate is calculated as 200 sample/s per channel × 10 channels × 8–10 bits/sample = 16–20 Kbit/s.

One of the requirements of body area network (BAN) is to have individual wireless sensor nodes. Even if we integrate a multichannel EEG signal into a BAN system and considering the additional preamble bits in a data package, the required data rate for a WBAN-based physiological monitoring system is still sufficiently low. A lot of available wireless transceiver will provide enough bandwidth to handle such data rates. When the number of sensor nodes increases, the required data rate will depend on the multi-access communication used. In case of TDMA (Time Division Multiple Access) and CSMA (Carrier Sense Multiple Access), the bandwidth requirement is low. For frequency division multiplexing techniques (e.g., FDM, FHSS, OFDMA) and CDMA schemes, a wideband will be required for data transmission when the number of users (here sensors) is increased. Probably this is one of the main reasons why the WPAN (wireless personal area network) systems such as ZigBee use CSMA as a multi-access communication. It can have a higher number of nodes with available low bandwidth. For more information about multi-access communication techniques in WBAN, refer to the related chapters of the book.

6.2.2 *Wireless Systems and Platforms*

In addition to the sensing front end, a sensor node and the BCU require a microcontroller and a wireless transceiver to coordinate the transfer of data. Numerous frequency bands can be used in medical monitoring system. So far there is no available standard to define operation frequencies specifically targeting a WBAN. It is important to select a proper and safe (interference free) wireless band. Especially for medical data transmission, it is crucial for the patient's safety to monitor accurate information. Table 6.2 presents the unlicensed frequency bands that can be used for WBAN applications.

Low ISM bands, especially the 13.56 MHz frequency, are widely used for radio-frequency identification (RFID) applications,

Table 6.2. Available frequency bands

Model	Country	Frequency	BW	RF power (dBm)
UWB	Worldwide	3.1–10.6 GHz	>500 MHz	−41
MICS	US, Australia Europe, Japan Worldwide	402–405 MHz	300 kHz, 10 channels	−16(25 μW)
High ISM	Worldwide	2.4–2.5 GHz, 5.725–5.875 GHz	20 MHz, 40 MHz	> 0
Mid ISM	Europe US Canada, Australia	433.05–434.79 MHz 865– 868 MHz 902–928 MHz	kHz range , 200–500 kHz 10–15 channels MHz	Up to 15
Low ISM	Worldwide	6.765–6.795 MHz 13.55–13.567 26.95–27.283 40.66–40.70	kHz range, 14 kHz for 13 MHz	0
WMTS	US, Canada	608–614 1395–1400 1427–1432	1.5 MHz 5–6 MHz	>10

inductive links for implantable systems, and smart cards for security systems [6, 7]. Most probably designers will continue to use these low-frequency ISM bands for inductive link based implantable systems as they present better performance at low frequencies. For medium ISM (Mid-ISM) range, countries in Europe, New Zeeland, and Honk Kong use 865–868 MHz. Australia, Korea, Taiwan, Hong Kong, and Singapore use a frequency range within and the US and Canada use the whole band between 902–928 MHz. Japan has 950–956 MHz ISM band frequencies. These frequencies have widely been used for cordless headphones and microphones.

In addition to unlicensed ISM bands, there are medical bands such as MICS and WMTS, which are specifically regulated for medical monitoring by communication commissions around the world. The MICS is an ultra-low power, unlicensed, mobile radio service for transmitting data to support diagnostic or therapeutic functions associated with implanted medical devices [8]. In order to monitor medical implant devices and status of inner organs, a frequency around 400 MHz have been a popular transmission band. In fact this frequency is an optimal frequency to provide a fully integrated circuit approach for a wireless transceiver design and causes relatively insignificant penetration loss (10 dB with 10 mm penetration) [9]. Using a higher frequency causes higher penetration loss. Meantime high-level integration becomes difficult at lower frequency. In addition, a small antenna design is difficult at lower frequency (less than 400 MHz).

In addition to medical implants, metrological aid service has primary allocation at 402–405 MHz band and the Earth exploration-satellite service together with metrological-satellite service has secondary allocation at 402–403 MHz. Thus transmission in the MICS has some regulated parameters in terms of channel spacing and transmission power level. At the moment in some countries, the MICS is being extended from 402–405 to 401–406 MHz, introducing additional 2 MHz wider band, which can increase the population of monitoring patients in hospital environment.

WMTS is used in the United States and Canada but not in Europe. In Japan, WMTS frequencies are 420–429 MHz and 440–449 MHz. WMTS has long been used for wireless patient monitoring system in the United States and Japan [10].

The most common wireless systems such as ZigBee (IEEE 802.15.4), WLANs, and Bluetooth (IEEE 802.15.1) operate at 2.4 GHz ISM band. The technologies operating at 2.4 GHz ISM band may suffer from interference issues when they are located in the same environment. In addition, there may be a variety of other sources such as microwave and cordless phones operating in 2.4 GHz band. Latency sensitive WBAN nodes like an ECG node can suffer significance performance loss from this mutual radio-frequency interference as a correct submission of a sensor will be delayed.

The electromagnetic interference is another source of interference that may limit the use of wireless in certain environments, especially in medical applications. Electromagnetic energy from a device can interfere with another when they are operating in the same environment. There have been already incidents where telemetry devices have affected the available medical systems in the hospitals [11].

In addition to the above wireless bands, the ultra-wide band (UWB) can also be used for monitoring of physiological signals. High data rate transmission, as commonly known, is not the only unique property of an UWB-based monitoring system. Some other advantages of a wide band technology are its low transmitter power, the physical size which can be extremely small because of the design at GHz range, the simplicity of transmitter design, and the band is not crowded as compared to other available bands [12]. The maximum transmission power is regulated as -41 dBm. Such a low-signal level will have insignificant RFI/EMI (radio frequency interference/electromagnetic interference) effect on medical equipment. The current drawback of UWB technology is that not enough devices available in the market that can be used as a complete solution.

The MICS band has also low emission power (25 μW, comparable to UWB) and lower power consumption, and will thus provide one of the most suitable medical sensor nodes. It is envisioned that the sensor nodes of a WBAN will use a specified wireless link, which will most likely be a narrow ISM band considering the current wireless technology developments. The UWB band can be used when interference and EMI are an issue for the environment. For implantable body area networks, the MICS band may still continue to be the preferable choice.

6.2.2.1 Wireless transceivers and microcontrollers

Most common wireless chips used in low power sensor network applications are listed in Table 6.3. The chips that exhibit low power and have small size are Zarlink's MICS chip, Nordic's nRF24E1 device, and CC2420 of TI. Zarlink provides one of the lowest power wireless chips available today. The low power is achieved by reducing the supply voltage to a value as low as 1.2 V, while most of the other transceiver chips operate with a nominal voltage around 3 V. The power figures shown in Table 6.3 are for the maximum transmission power levels and a nominal supply voltage, which is 3 V. Although these chips are advised to operate at 3 V, their supply voltage in practice can be reduced to as low as 2 V. This way the power consumption will be lower than what is shown in the table. The Zarlink chip has been designed for implantable sensor nodes. The Nordic device nRF24E1 consumes a transmit current of 10.5 mA for an output signal level of −5 dBm and 18 mA receiver current for

Table 6.3. Some suitable wireless chips for potential use as a wireless medical sensor node

						Power consumption	
Model	Company	Frequency	Data rate	RF power (dBm)	Physical dimension	**Tx**	**Rx**
CC1010	TI[1]	300 to 1000 MHz	76.8 Kbps	−20 to +10	12 × 12 mm^2 (chip)	26.6 mA	9.1 mA
CC24xx	TI[1]	2.4 GHz	1 Mbps	−25 to 0	7.1 × 7.1 mm^2 (chip)	19 mA	17 mA
XE1205	Xemics[2] (Semtech)	433, 868, 915 MHz	304 Kbps	Up to 15	8 × 8 mm^2 (chip)	62 mA	14 mA
CX72303	Conexant[3]	2.4 GHz	1 Mbps	−10 to +2	—	34 mW	43 mW
nRF24E1	Nordic[4]	2.4 GHz	2 Mbps	−20 to 0	6 × 6 mm	11 mA	18 mA
AMIS52100	AMIS[5]	402–405 MHz	16 Kbps	−3 to 12	7.5 × 7.8 mm^2 (chip)	25 mA	7.5 mA
ZL70250	Zarlink[6]	402–405 MHz, 433 MHz ISM	800 Kbps	< 0	7 × 7 mm^2 (chip)	5 mA, TX/ RX	

[1]http://www.ti.com/.
[2]http://www.semtech.com.
[3]http://www.conexant.com.
[4]http://www.nordicsemi.com.
[5]http://www.amis.com.
[6]http://www.zarlink.com/.

a data rate of 250 Kbps. The physical dimensions shown in the table are based on the packages used by the companies.

The Nordic has another transceiver -nRF401 that operates in the 433 MHz ISM frequency band. It uses frequency shift keying (FSK) modulation and presents a data rate up to 20 kbit/s. The current consumptions for both transmitter and receiver are 11 mA and 8 mA, respectively.

The CC1010 transceiver chip from TI (Chipcon) has the capability to transmit anywhere within 300 and 1000 MHz. It can be used to generate signals for MICS, Mid-ISM bands, and WMTS frequencies. The AMIS52100 chip from AMI Semiconductor requires a crystal oscillator at 12.56 MHz to generate MICS band. A crystal with any frequency can be used with CC1010 to generate medical and ISM band frequencies. The effective frequency of the device is arranged by the external VCO inductor in addition to the crystal oscillator. For example, an inductor value of 12 nH gives a frequency range of 565–730 MHz. The PLL in chip is able to quickly lock onto WMTS frequency of 610 MHz. The frequency registers of the device allow for jumps of 250 Hz, making this device suitable for multiple channel selection and taking advantage of the available bandwidth by using FDMA or frequency hopping.

The Semtech's XE1205 chip can be used for the 433, 868, and 915 MHz license-free ISM frequencies. Similar to CC1010, it can also operate in other frequency bands in the 180–1000 MHz range. CX72303 from Conexant operates at 2.4 GHz. This technology is specifically designed for Bluetooth system solutions

Another Chipcon's chip CC2420 is a 2.4 GHz IEEE 802.15.4/ ZigBee compliant RF transceiver. It is based on digital direct sequence spread spectrum (DSSS) baseband modem providing a spreading with data rate of 250 Kbps. This chip has been widely used in sensor nodes because of its low-power design and its operation for ZigBee and Bluetooth applications.

AMIS52100 and TI's cc10xx chip series can provide MICS and WMTS signals. Both chips' output can go as low as −20 dBm meeting the requirement of −16 dBm transmission power for the MICS band. Zarlink's ZL70250, CC1010, and Semtech XE1205 uses FSK data modulation, AMIS52100 amplitude shift key/on-off key (ASK/OOK) modulation, while CC2420 uses offset quadrature phase- shift keying (O-QPSK) with half-sine chip shaping, which is equivalent to MSK modulation. The newer version transceiver from AMI Semiconductor (AMIS53000) can operate using one of the following modulation schemes: OOK, FSK, or GFSK. The nRF24E1 operates at 2.4 GHz with GFSK modulation. The transceivers can work based on both Manchester and NRZ bit coding schemes; however, the data rate is half in the case of Manchester coding.

6.2.2.2 Existing sensor boards

The unit responsible for processing data, controlling the functionality of the components of the sensor node, and coding the sensor data for the wireless transceiver is the microcontroller. The task of a microcontroller for a sensor node in WBAN applications is relatively simple due to the low frequency of biological signals. Thus a less complex and small size microcontroller can usually be utilized. The commercially available microcontrollers have a 10-bit and 8-bit ADC built-in, which eliminates the need for an external off-shelf ADC.

Considering that the sensor node board is required to be light and small in order to be wearable, a wireless transceiver chip containing a microcontroller would be advantageous. As an example, CC1010 from TI has a microcontroller builtin which has a memory of 2048 + 128, 10-bit ADC, and 22.7 kHz sampling frequency. A small board can easily be designed, similar to that of Mica2DOT shown in Fig. 6.6. The Nordic's nRF24LE1 chip also contains a microcontroller unit and an ADC. The microcontroller is an 8051 compatible microcontroller, and the ADC is 10 bit with 100 Kbps.

Figure 6.6. A Mica2DOT board (taken from www.xbow.com). See also Color Insert.

The sensor board with these devices will require only a few external components: the needed sensing front end given in Fig. 6.3, regulator IC for power supply arrangement and matching network together with antenna.

Some of the commonly used wireless sensor platforms (as a complete sensor boards) are shown in Table 6.4. Crossbow's Mica2DOT and T-node are the smallest sensor nodes available in commercial domain. Another widely used sensor node is Tmote Sky node [13]. It uses 250 Kbps 2.4 GHz IEEE 802.15.4 Chipcon Wireless Transceiver (CC2420) and a separate 16-bit 8 MHz Texas InstrumentsMSP430 microcontroller (10 K RAM, 48 K Flash).

MicaZ node also uses CC2420 transceiver and 10-bit ATmega 128L microcontroller, which has 4 KB RAM and 128 KB FLASH memories. The MicaZ is a 2.4 GHz mote module used for enabling low-power wireless sensor networks. TinyNode584 is another ultra-low power wireless sensor node, which uses Xemics XE1205 for wireless communication and TI MSP430 microcontroller (10 K RAM, 48 K Flash) as a processing unit [14]. The physical dimension of this board is 30×40 mm^2.

The Crossbow's Mica2 sensor node works at 868 and 916 MHz ISM frequencies. Like MicaZ it uses a separate microcontroller — ATmega128L (4 K SRAM, 128 K Flash, 8 MHz). The total power consumption of this board at 3.3 supply voltage, including the radio and microcontroller, is 148. 5 mW ($= (27 + 10 + 8)*3.3$). During sleep mode, the power consumption is at μW levels.

The Mica2DOT mote is a third-generation mote module wireless platform designed for smart sensors similar to the Mica2. Its quarter-sized (25 mm) form factor makes it suitable for sensor network applications requiring low size. Mica2DOT is one of the smallest sensor nodes available in commercial domain (Fig. 6.6). The power consumption of this node is similar to Mica2 (MPR400). Mica motes run on TinyOS (TOS) distributed software operating system v1.0 [15]. They can operate on the frequencies such as 868/916 MHz, 433 MHz, or 315 MHz multichannel radio transceiver. The Mica2DOT is Mica2 compatible and uses the same microcontroller.

Another small node for sensors is T-nodes working in the 868 MHz ISM frequency band. The board uses a separate chip for

Table 6.4. Various hardware sensor nodes configurations

Model	Company	Frequency	Data rate	Trans. power (dBm)	Physical dimension	Power consumption†	
						Tx	Rx
Mica2 (MPR400)	Crossbow[1]	868/916 MHz	38.4 Kbps	−24 to +5	58 × 32 × 7 18 g (board)	27 mA @ 3.3 V	10 mA @ 3.3 V
MicaZ	Crossbow[1]	2400 MHz to 2483.5 MHz (IEEE 802.15.4)	250 Kbps	−24 to 0	58 × 32 × 7 18 g (board)	17.4 mA @ 3.3 V	19.7 mA @ 3.3 V
Mica2DOT	Crossbow[1]	868/916 433 MHz	38.4 Kbps	−20to +10	25 × 6 mm^2 3 g (board)	25 mA @ 3.3 ×	8 mA @ 3.3 V
Tmote Sky node	Moteiv	2.4 GHz (IEEE 802.15.4)	250 Kbps	−25 to 0	66 × 32.6 × 7 mm	19.5 mA @ 3 V	21. 8 mA @ 3 V
T-node	SOWNet	868, 433, 915, or 315 MHz	52.2 Kbps	−20 to −5	Diameter of 23 mm	25 mA @ 3 V	13 mA @ 3 V
Sensium™	Toumaz	868/915 MHz	50 Kbps	−23 to − 7	90 × 45 × 10 mm	3 mA @ 1 V	2.7 mA @ 1 V
Wibree	Nokia	2.4 GHz (Bluetooth)	200 Kbps	−6	—	—	—

[1] http://www.xbow.com.

† Transmitter (Tx) power values are usually for the maximum transmission power level. When the transmission power level is configured to lower values, transmitters will consume a power lower than the values in this Table.

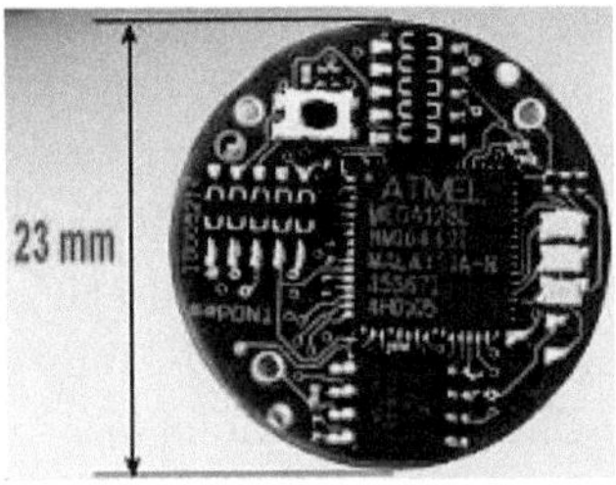

Figure 6.7. A T-node sensor node [16]. See also Color Insert.

microcontroller with 10-bit ADC, memory of 128 K flash memory and 4 Kbyte SRAM [16]. It consumes only 20 μA during sleep mode. It has a similar size to that of Mica2DOT (23 mm), shown in Fig. 6.7.

The Sensium™ sensor node uses the chip TZ1030, which uses a wireless transceiver operating at 868/915 MHz and the 8051 processor for digital processing and with 10-bit ADC. It provides 50 Kbps of data rate [17]. The low power consumption of this node is resulted from the low-supply voltage of 1 V. The node samples input signals within a dc to 250 Hz bandwidth.

Another popular node is BTnode [18], which consists of a Chipcon CC1000 radio (it is CC1010 without a microcontroller built-in) operating at 433–915 MHz, Atmel AVR microcontroller, and a Bluetooth radio (Zeevo ZV4002). In other words, a BTnode is made of a Crossbow Mica2 Mote with SRAM (256 K) and an additional Bluetooth radio. This node's power consumption is high due to the operation of two radios (~200 mW). However, it gives the advantage of connecting to another device using the Bluetooth technology. This node can be used to interface with mobile phones and the Internet using the Bluetooth link. Its microcontroller is from Atmel (ATmega 128L), having 64 + 180 Kbyte RAM, 128 Kbyte FLASH ROM, and 4 Kbyte EEPROM.

The Conexant is another node based on Bluetooth wireless technology. It uses CX72303 RF transceiver and CX81400 baseband controller to establish a Bluetooth communication link [19]. The CX81400 integrates an RISC processor ARM7TDMI having 192 KB ROM and 192 KB RAM.

Among the available Bluetooth modules Wibree is probably the most suitable node for a lowpower sensor network application.

The device has been designed by Nokia to have similar power consumption and battery life as ZigBee. This device similar to ZigBee applications, because of its small form factor, can be used in health care, fitness, security, wrist watches and home entertainment industries. It operates on coin cell batteries. Wibree operates in 2.4 GHz band with a data rate up to 1 Mbps with a 10 m distance similar to any other Bluetooth technologies.

In addition to these commercially available sensor nodes, there are also nodes that have been developed specifically for medical monitoring. In reference [20], a very compact system has been developed for a wearable ECG sensor node that can communicate wirelessly with a base station. The node uses Nordics nRF24E1, a 2.4 GHz RF transceiver together with an embedded 8051-compatible microcontroller (DW8051) with 512 byte ROM, 4 K RAM, and 9-bit ADC. The board has the size of 26 (L) $\times$ 15 (W) $\times$ 7 (H) mm^3.

BSN nodes developed by Imperial College [21] are TinyOS-based sensor nodes The node also uses CC2420 from TI and MSP430 microcontroller families with 2 KB RAM, 60 KB flash, and 12-bit ADCs. The node has a size of 26 mm. The power consumption of these nodes is similar to the power consumption of CC2420 transceiver given in Table 6.3. The sensor board in reference [22] uses the Nordic nRF905 transceiver however operating at 915 MHz ISM with 0 dBm and 10 Kbps. This sensor node includes 3-axis accelerometer SCA3000 and has active power of 7.8 mW and 8.2 mW for transmitter and receiver respectively. Its dimension is as small as a one dollar coin.

Another sensor node [23] developed for a medical sensor network uses the medical bands MICS and WMTS for applications in the hospital environment. The board uses AMIS and CC1010 transceivers for the radio system and PIC16F785 microcontroller having 10-bit, 52 kHz sampling clock , and 368 + 256 RAM. The board has a dimension of 30 mm $\times$ 80 mm.

The power consumption given Table 6.4 defines the life time of a sensor node. For example, a continuously active sensor node using the CC24XX family (this chip family is used for ZigBee and also for Bluetooth) can last about seven hours when using a battery with a capacity of 300 mAh, assuming the transmitter and receiver working with the same amount of time (the nodes are always active either there is transmission (the transmitter is active) or the receiving (the receiver is active)). For monitoring patients in emergency and disaster events, sensors measuring critical vital signals are required to be active (like ECG signal) as these vital signals are crucial for the patient's life. However, there may be some sensors information needed to be monitored once a while. The life time of such a sensor will be similar to current cellular phones we use in our daily life. It is important to note that the power consumption values given in Table 6.4 is for the maximum transmitter power. When the distance between the control unit and the sensor nodes is short, which is likely the case for the targeted medical applications, a lower transmission power can be configured in the transceiver of the sensor nodes to increase the battery life.

6.2.3 *Design of Implanted Sensors Nodes for WBAN*

Electrocardiogram (ECG) and temperature recording have been used for more than 50 years in medical diagnosis to understand biological activities [24] The early implantable devices are constructed with simple electronic structures to make the devices small enough so that they can be inserted in the body. A simple transmitter connected to a sensor is used to send the signal from inside the body to external devices for tracking certain organs' physiological parameters. Recent implantable systems have focused on more complex microelectronics systems with the use of the advanced integrated circuit technology. Some milestones of implantable devices are given below:

- **1957** — First wearable **pacemaker** (in 1958 fully implant)
- **1984** — **The Australian cochlear implant** approved by FDA
- **2000** — First clinical trial **wireless endoscope** (electronic pill)
- Study of **Bionic eye** since 1990

In addition to these projects, there is a vast area of different implantable systems, being designed by designers, such as cardioverter defibrillators, micoelectrode arrays for detecting and stimulating brain functions, and electronic pills for drug delivery systems. Moreover, there are implantable biosensors developments to conduct animal studies.

Electronics for implantable sensors are designed taking into account very specific requirements for a given wireless telemetry application. Wireless technologies such as ZigBee, WLANs, and Bluetooth (IEEE 802.15.1) developed in the commercial domain cannot be used directly in medical implants because of the following reasons: (1) they have been optimized for general use; (2) the device size exceeds the required size limitation of the current implant technology, (3) their frequency bands are very crowded, and (4) they do not meet the safety-related radiation. The existing advanced wireless systems operate at 2.4 GHz ISM band and suffer from strong interference from each other when they are located in the same environment [25]. Thus an implant system should have a different transmission band for an interference-free wireless link as the transmitted information could be related to patients' critical conditions.

In implantable telemetry systems, to optimize the power and data telemetry links independently, the dual-band (i.e., two-inductive links) has been used [26]. The power and data signals are carried on two wireless links. For implantable systems like the cochlear implant and bionic eye, one wireless data link (i.e., forward data link) is used to transmit data from the external unit to inside the skin to stimulate the nerves in order to help the patient restore hearing and sight. These devices also require the transmitting of status information regarding the operation of the device, which is done based on a third link-back telemetry. There are also systems such as pacemakers, electronic pill, and biosensors that send physiological data such as ECG, pressure, or video images from inside the body to the external unit and then to the monitoring station in order to track a patient's condition. The latter implantable systems may or may not require the forward data link depending on whether data or a control signal is required from the external unit to control the implantable system.

Figure 6.8 shows the basic building blocks of implant systems. The power signal is received and then regulated to provide supply for the wireless data links, the sensors, and the signal

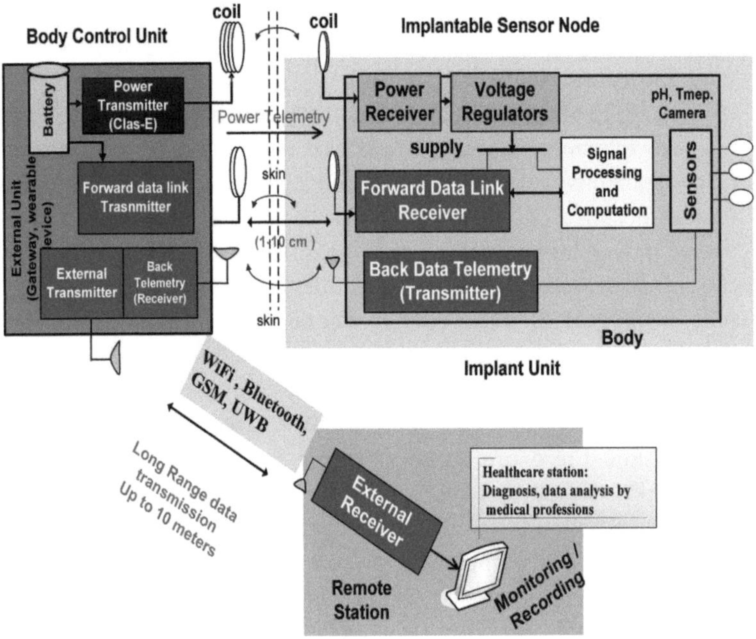

Figure 6.8. An implanted sensor node with the external unit and wireless data links. See also Color Insert.

acquisition unit, which processes and amplifies the senor data. The external unit may contain another wireless link that will control the communication between the implant devices and the remote monitoring station. The patient will wear the external unit, which will be on the surface of the body close to the implanted device. A remote wireless receiver will be used to receive data from the external unit for monitoring and analyzing the data. As the required specifications are not restricted for this link, commercial wireless devices can be used to undertake this duty. This additional wireless link will provide patient free in a room in the hospital environment.

Implantable systems have been constructed using inductive links between the external and implanted units with a distance of a few centimeters. The main reason behind the short distance and the inductive link is the need of wireless powering. Inductive links are mainly used to power implants to eliminate the use of batteries. The wireless module is designed with a very simple communication scheme such as binary ASK (amplitude-shift keying), FSK (frequency-shift keying), and PSK (phase-shift keying) modulators and demodulators [26].

The 13.56 MHz ISM band is usually the most common frequency for such telemetry systems and is also used for RFID applications. A low transmission frequency usually less than 20 MHz is utilized mainly because of the simplicity in the design and to avoid the use of power-hungry blocks such as mixer, oscillators to maintain the miniature size of the implant. For example in reference [27], PSK is used at 20 MHz carrier frequency for the wireless data link for a retina prosthesis. ASK is used for a bionic eye system in reference [28] with 1–10 MHz frequencies and for a recent cochlear implant in reference [29] at 2.5 MHz transmission frequency. In addition, there are some attempts for general implants and neural prostheses systems employing FSK [30, 31] operating at 480 kHz and 5/10 MHz frequencies, respectively. The most suitable frequency for such implantable system requiring low frequency forward data transmission would be the ISM band 13.56 MHz as given in Table 6.2. Otherwise, designers would have to obtain permissions from communication authorities (e.g., FCC) to use a frequency rather than ISM. In addition to the wireless power link, utilizing a low frequency transmission has been attractive to eliminate the absorption of more energy [32].

There are some issues with the communication schemes of implantable systems for forward data telemetry and back data telemetry. For example, when many similar types of implantable systems are used in the same environment or for the same patient, the simple communication systems mentioned above will face the problem of interference and lack the multiuser (i.e., multi-access) capability. For a frequency less than 20 MHz, you are only allowed to use a few kHz bandwidths, which limit the use of the multi-access (also known as multiuser) communication techniques to allow several implantable devices work in the same environment. For example, allowable bandwidth for 13.56 MHz ISM (the industrial, scientific, and medical) band is regulated with 14 kHz. Thus a more advanced wireless technology will be required in the future to accommodate better radio links for medical implants enabling safe and reliable communication links [33].

The above issue has become more evident in practice when more advanced medical implants such as wireless implantable cardioverter defibrillators and electronic pills (i.e., wireless endoscopes) started to appear in clinical environments. These modern devices need a much longer range for the wireless telemetry in addition to a higher bandwidth. Moving to higher operation frequencies was the only way to increase the range and dedicate enough spectrums for a reliable communication. Thus, in order to alleviate some of the issues mentioned above, international authorities have allocated a band at 402–405 MHz with 300 kHz channels (MICS band). With the use of this band, it is aimed to deliver high level of comfort, better mobility, and better patient care [34, 8].

The MICS band has been proposed to permit individuals and medical practitioners to utilize ultra-low power medical implant devices such as cardiac pacemakers and defibrillators, particularly without causing interference to other users of the electromagnetic radio spectrum. However, unlike the low-frequency inductive links, at high frequencies like 400 MHZ a wireless transceiver requires the use of radio-frequency (RF) blocks such as voltage-controlled oscillators (VCOs), mixer and phase-locked loops (PLLs) to down convert (or up convert for the transmitter case) the frequency signal in order to process it using the integrated circuit technology. These blocks are constructed using inductors and capacitors on chip or off chip, which increase the physical size of the wireless chip [9, 35]. Medical implants have a physical limitation for the electronics of the wireless telemetry and cannot afford to accommodate such blocks. Thus high-level of integration, which results in miniaturization and low power consumption, should be met for wireless designs at MICS band. So far only one company, Zarlink, designed a transceiver for implantable applications. Zarlink's transceiver still requires some external components for its full functions [35]. Although with an MICS link a data could be transferred from or to the implant within a distance of a few meters, when the implant system requires the use of a wireless power or wirelessly charging the battery, the external unit still will need to be attached to the body close or adjacent to the implant system for an efficient power transfer as indicated in Fig. 6.8. It is still necessary to utilize the power signal at a lower frequency, which also eliminates the possible interference issue from the power signal to the data signals [27].

Medical implant telemetry can be categorized into two groups: high data rate and low data rate systems. As an example, the wireless capsule endoscope and the retina prosthesis are the recent implanted systems requiring a large amount of data delivered to or outside the body. Implantable multichannel neural recording systems and multichannel monitoring of continuous signals electroencephalogram (EEG) also necessitate a high data rate communication. As an example, scientists aim to achieve the recording of more than 100 channels in order to simultaneously record brain functions; a data rate more than 20 Mbps is required. A similar figure is also useful for a wireless endoscope implant to obtain higher resolution and detailed real-time video images. MICS has been divided into channels with 300 kHz bandwidth and thus cannot provide such data rates. Thus for these types of implantable systems requiring high data rate, UWB will be the technology to investigate [12].

6.3 WBAN Systems

The early telemedicine systems have used the WMTS for patient monitoring at hospitals. As an example, the LX-5160, which is a WMTS telemetry transmitter (608~614 MHz WMTS Band), has been developed by Fukuda Denshi USA [36]. It transmits ECG and respiration parameters to a central telemetry receiver, which can be displayed on a remote computer. This device has a dimension of 5.4 (W) × 2.2 (D) × 8.6 (H) cm and a weight of 80 gram. Transceiver-608 telemetry device developed by Mindray [10] also uses a WMTS transmitter for patient monitoring. This telemetry uses a 12-lead ECG and has a dimension of (13.0 × 7.5 × 3.1 cm). Its weight is 181.4 g. The wireless telemetry is based on channelized WMTS frequency meaning uses different frequency for transmit and receive signals. For example, 608–614 MHz (transmit) WMTS channelized 1395–1400 MHz and 1427–1429.5 MHz (receive) WMTS are used. Welch Allyn is another company developing patient monitoring system for vital signs monitoring for hospital and sportive applications [37]. They have developed micropaq® Wearable Monitor that displays heart rate, one or two ECG leads, motion-tolerant SpO_2 and pulse bar. The device can also connect wirelessly to a central monitoring station. This device has a large dimension similar to a mobile phone.

Comparing to the wireless module discussed before, these devices are bulky and are not appropriate sensor node when considering a WBAN application. Thus new telemedicine systems are currently being developed using new small size sensor nodes described earlier.

Recently Corventis, Inc. has developed a small sensor node called PiiX™ for wireless cardiovascular solutions [38]. It uses a wireless gateway device named zLink™ to communicate with the monitoring center. The device has some limitations. For example, PiiX is intended for single patient use and cannot be used for patients with implantable devices. The unique features of this device are leadless and water-resistant. It eliminates the cumbersome leads and wires, and can still operate during showering or sleeping. Piix sensor is quite small and can easily be attached to the body like a plaster. Figure 6.9 shows a picture of PiiX sensor node from Corventis and micropaq® Wearable Monitor from WelchAllyn.

A UK-based company Toumaz [17, 39] is developing a digital plaster where the sensor node can be attached to the human

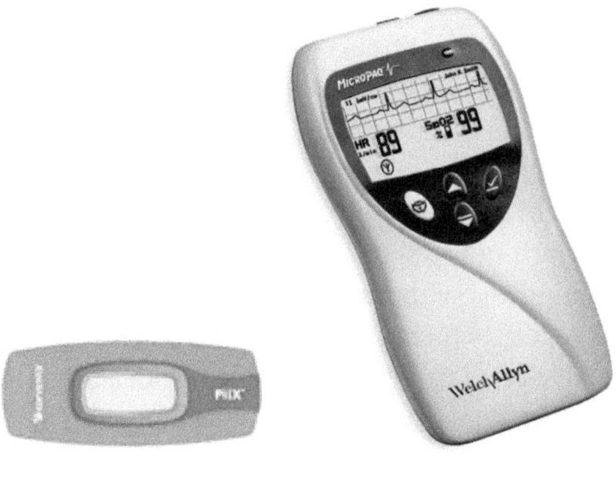

Figure 6.9. Figure PiiX sensor node from Corventis [38] and micropaq® Wearable Monitor from WelchAllyn [37]. See also Color Insert.

body with a plaster. The wireless sensor node is based on ECG and temperature data. The data received from the plaster can be displayed on a PDA device. It can also be transferred to a monitoring station via WLAN link using the PDA as a gateway device. The digital plaster can communicate with PDA up to 3 meter distance. The plaster is based on the chip TZ1030 which uses the 8051 processor for digital processing a wireless transceiver.

The Fitbit has developed a wireless sensor node that a tracks motion of a human body. The Fitbit sensor contains a 3D motion sensor that tracks the motion in three dimensions and converts to useful information such as calories burned, steps taken, distance traveled, sleep quality, etc. It can be worn on the waist, in pocket or on undergarments. Nike and adidas have produced similar body-worn devices Nike+iPod and miCoach Pacer, respectively, for sport activities [40].

Here in we will discuss some of ongoing projects for WBAN applications. These projects have mainly used the sensor nodes provided in Table 6.4. Some have designed their own sensor nodes by using the wireless transceivers in Table 6.3 together with one of famous microcontroller — ATMega128L, MSP430, and/or 88051. There have been a number of different developments for WBAN systems in the literature. The goal here is to compare the existing systems in terms of their suitability for a large scale implementation for WBAN applications. We will discuss whether the available systems demonstrate a multi-patient or a single-patient monitoring. A comparison of some of current implemented WBAN systems is given in Table 6.5.

The project in reference [41] presents the AID-N triage system for disaster applications. It uses the MicaZ or TmoteSky platform from Crossbow Technology (2.4 GHz Radios (802.15.4) with CC2420 transceiver chip) in the sensor node for data collection. The sensor node communicates with another ZigBee device that is attached to a personal server, which is a laptop and PDA. Software programs have been developed to gather vital signs from a disaster scene. This system requires WLAN connections with an IEEE 802.11 link for transmission of patients' data to a remote server in a medical centre. If there is no network connection, the data will be stored in a device or can be monitored on a PDA in an emergency vehicle until patients are taken to a hospital.

Table 6.5. WBAN-based telemedicine systems

Existing systems	Wireless device	Sensors	Comments
CodeBlue [42]	802.15.4 (2.4 GHz)	Pulse oximeter, EKG, motion-activity	It uses MicaZ/T-mote sensor nodes.
WiiSARD [43]	802.11	Pulse oximeter	It uses PDA device as a sensor in order to use Wi-Fi link to connect the Internet.
(Gao, 2007) [41]	ZigBee (2.4 GHz)	SpO_2 , ECG, BP	It uses mesh networking for multi-patient, and WLAN for remote destinations.
(Park, 2006)	nRF24E1 radio (2.4 GHz)	ECG	The ECG sensor communicates to a computer wirelessly or a base station with the Wi-Fi link. No multi-patient scenario.
(Espina, 2008) [44]	IEEE 802.15.4 (2.4 GHz)	ECG, PPG, blood pressure	Single-user data is presented.
(Chen, 2009) [1]	IEEE 802.15.4 (ZigBee)	8–Channel EEG	Single patient. Internet data transmission is provided. No wireless gateways.
(Zhang, 2009) [2]	Bluetooth (2.4 GHz)	ECG	Although it mentions it is done for two persons ECG, one continuous ECG monitoring data is shown. The control device is directly connected to a home computer via USB (no wireless gateways). No data transmission through the Internet.
(Anliker, 2004) [45]	GSM	Blood pressure, SpO_2, 1-lead ECG	Multiple sensors have been integrated in one hand-held device.

(Contd.)

Table 6.5. (*Continued*)

Existing systems	Wireless device	Sensors	Comments
(Yuce, 2008) [23]	MICS, WMTS	Temperature, pulse rate, single lead ECG/EEG	he design targets monitoring of physiological ameters for multi-patient. It uses multi-hoping technique. Long-range data transmission through the Internet. Several wireless gateways to increase mobility.
TZ1038 Sensium [17,39]	868/915 MHz	Single lead ECG, temperature	Data received and displayed on a PDA device. Also transferred to a monitoring station via WLAN link using the PDA as a gateway device.
(Jovanov, 2005) [1]	ZigBee (2.4 GHz)	Activity sensor	The control device connects with a computer for data display directly or wirelessly using a Wi-Fi link. It is a multi-hopping system with multi-sensors; however, one patient data has been implemented. No Internet transmission has been indicated.
BSN Node [21]	IEEE 802.15.4 (2.4 GHz)	3-lead ECG, 2-lead ECG, and SpO_2 sensors	Multichannel sensor node. Sensor signals can be gathered, displayed, and analysed on a PDA. The PDA works as a wireless gateway for remote destinations.

(Wang , 2009)	nRF905 (915 MHz)	3-D acceleration	The sensor nodes communicate directly to a base-station board that includes an aMSP430F149 microprocessor and an nRF905. Data from single body is tested.
(Yuce, 2009) [12]	UWB	Multichannel EEG/ECG	The design targets monitoring of multichannel continuous physiological ameters for implantable and wearable systems. No wireless gateways.
Human++ [48]	nRF24E1 radio (2.4 GHz)	Single-lead ECG, EEG	An ECG node is designed using flexible polyimide material. No wireless gateways.
HeatlhGear [46]	Bluetooth (2.4 GHz)	Oximeter	The sensor communicates with a cell phone using the Bluetooth link. No wireless gateways. Individual use.

The CodeBlue project uses sensor nodes based on the popular MicaZ and T-mote designs [42] for a multi-patient monitoring system. The WiiSARD team [43] developed a prototype system for pulse oximeter sensor. It operates with the Wi-Fi link to enable a medical response through the Internet. It uses a PDA device as a senor which has a large size which is not attractive for a medical sensor node.

In reference [44], Espina et al. presented an IEEE 802.15.4-based wearable telemedicine system that monitors continuous cuff-less blood pressure and ECG signals. The sensor data was measured on a single user. Date is displayed from the sensor to a PDA or a wristwatch device. Anliker and his colleagues [45] designed a wrist-worn device to measure multiple physiological parameters from one person. Sensors like blood pressure, SpO_2, one lead ECG all have been integrated in one handheld device. The data is then transmitted using a GSM data link to a computer similar to Fig. 6.1. The device acts the same as a regular cellular phone since it operates with the GSM network. However, measurements more than one patient at the same time have not been provided. HealthGear project uses a cell phone as a central processing unit that receives signals from body sensors [46]. It uses the Bluetooth technology to connect a Bluetooth-enabled cell phone. The sensor is a Nonin's flex oximeter used to measure a single user's blood oxygen level (SpO_2). The Berkeley Tricorder sensor node is another Bluetooth-based WBAN system developed at Berkley [47] for ambulatory health monitoring systems. The system utilizes a multichannel wireless sensor node, including parameters such as ECG, EMG, blood oxygenation, respiration, and motion all on one board. The data are received and stored on a PC via a Bluetooth link.

Another system in reference [20] designs very small, wearable (probably one of the smallest custom-made sensor nodes available at the present time) ECG sensor that communicates wirelessly with a base station connected to a computer. The system works similar to a multi-hopping network by using the Wi-Fi link. Although a multi-patient data monitoring is not presented, this system can be used for measurements and monitoring of vital signs from multiple patients.

A wireless patient monitoring system based on an ECG patch has been developed for cardiac monitoring by IMEC Belgium. IMEC's Human++ program develops wearable wireless devices [48]. The current ECG node electronic is assembled on flexible polyimide substrate that can easily be integrated in textile.

There are also ongoing projects to develop sensor platforms operating at UWB (3–10 GHz) [12], MICS, and WMTS bands [49]. A WBAN system for multi-patient medical monitoring is presented in reference [23]. It uses a multi-hoping technique through wireless gateways to transfer data from sensors on several human bodies to remote locations. It uses the Internet to allow access at any location. In reference [12], a low-power WBAN system has been developed using UWB technology for implantable multichannel neural recording for brain–computer interface applications and 8-channel ECG for patient monitoring. The ECG data is monitored at a remote station.

Table 6.5 shows the current progress in WBAN applications. Real-time and simultaneous monitoring is required from multi-sensors on multi-bodies for future WBAN systems. System should take the advantage of the Internet for data transmission to remote locations, which will provide an access to a person's data anywhere in a city. Future WBAN systems should also develop small-size sensor boards to meet the size and power constraints. Most of the current systems use the commercially available ZigBee (IEEE 802.15.4) devices and boards, which are designed for many other applications and, thus, are not entirely wearable.

For longer range communications, the wireless nodes should be able to communicate to remote stations through a wireless, portable gateway (it could be a watch, PALM, iPhone, mobile phone, etc.). The wireless transmission technique between the gateway device and remote monitoring stations could be done using one of wireless protocols such as ZigBee, Bluetooth, or Wi-Fi for using the Internet.

The implanted device will most likely use the MICS band or a frequency around 400 MHz. The wearable nodes in a WBAN system will use an ISM band for low data rate communication and UWB when a high data rate transmission is required.

6.4 A WBAN-Based Multi-Patient Monitoring System

A WBAN system that has been designed for health care applications is presented in this section. The system is based on a multi-hopping network technique that can be used in medical environments for remote monitoring of physiological parameters from multiple patient bodies [52]. The system is depicted in Fig. 6.10. The data is transferred to remote stations through the local area network or the Internet already available in medical centers.

An example of sensor node design for medical monitoring is depicted in Fig. 6.11. The analog front-end of the sensor node for EEC, EEG, and EMG uses the circuit given in Fig. 6.3. The pulse rate sensor consists of an infrared emitter (SFH487-880nm) and a phototransistor (SFH309FA). Light is shone through the tissues, and the variation in blood volume alters the amount of light falling on the detector. When the heart beats, a pressure wave moves out along the arteries at a few meters per second. This pressure wave can be felt at the wrist, but it also causes an increase in the blood volume, which can be detected by a plethysmograph. Human heartbeat ranges between 50 bpm and 200 bpm. These figures indicate that the frequency range of human heartbeat is between 0.83 Hz and 3.33 Hz: A high-pass filter and a low-pass filter are used to form a band-pass

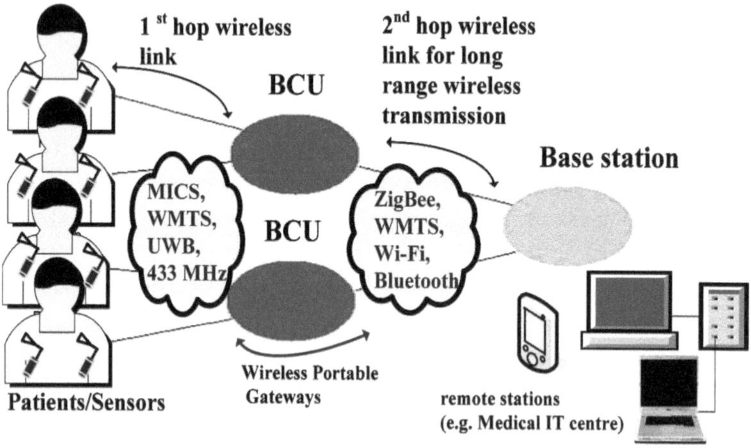

Figure 6.10. Multi-hopping WBAN-based multi-patient monitoring.

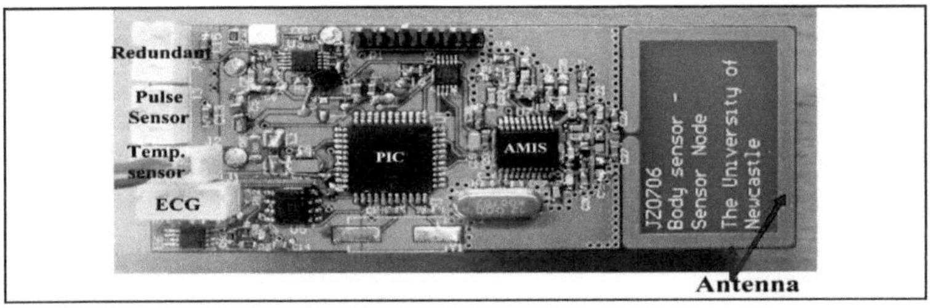

Figure 6.11. A 4-channel sensor node. See also Color Insert.

filter to remove unwanted environment signals. The optical part of the pulse sensor consumes quite large power [49]. It is also possible to detect the pulse rate from the ECG signal as explained later in Fig. 6.16. The temperature sensor uses the IC LM 35 for temperature measurement. The IC produces an analog voltage with respect to temperature. The ADC in the microcontroller samples the voltage and converts it to a digital/data for RF wireless transmission.

The primary function of BCUs is to collect data from sensor nodes via the first wireless link and forward these data to a remote PC (i.e., control station) for further analysis. Two types of BCU devices are needed in order to provide a complete WBAN transmission coverage in a medical center. One BCU is designed to be connected to a computer (Fig. 6.12a) via the USB port, while the other BCU is used to function as an intermediate device (Fig. 6.12b) that presents a second wireless link for a longer range wireless sensor network. The latter case is more suitable for large medical centers and functions as a portable gateway device. Although both BCUs can be used for multi-patient monitoring, the first BCU type (BCU-1) can also be useful for private usage at home or in a room of a hospital for single-patient monitoring.

The sensor nodes/BCU hardware requires a microcontroller and a wireless transceiver to coordinate all activities. The BCU-1/sensor nodes consist of a microcontroller PIC16F87 and a wireless transceiver (AMIS52100 IC and CC1000). Both AMIS and CC1000 can generate MICS band frequencies. Especially when CC100 is used, the operation frequency can be configured to 433 MHz ISM, WMTS bands as well as MICS.

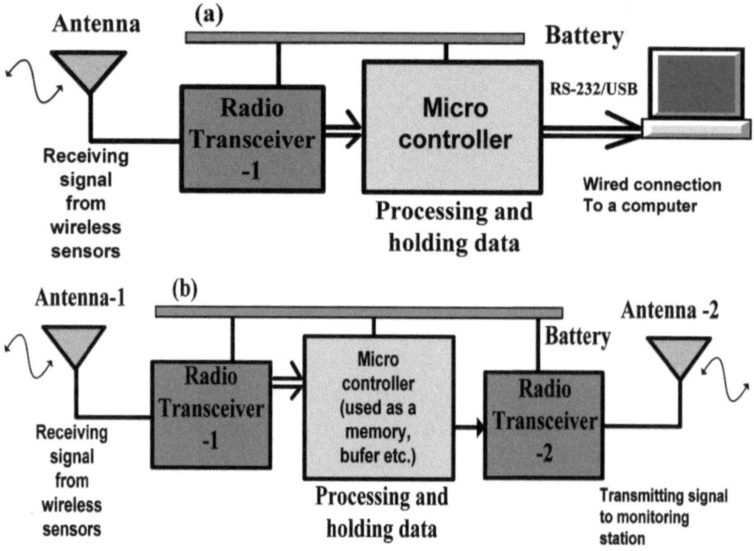

Figure 6.12. Block diagram of BCUs: (a) wired BCU, (b) portable BCU. See also Color Insert.

In addition to these chips, we use another transceiver, the CC1010 chip from TI (this chip contains CC1000 and a microcontroller built in), on the intermediate BCU (BCU-2) board to develop a wireless transmission and networking between BCUs and the remote base station. The CC1010 and CC1000 transceiver chips have the capability to transmit anywhere within 300 and 1000 MHz (It was tuned to WMTS band for the second wireless link in our prototype system). The AMIS IC has a data rate capability of 19 Kbps, while CC1010 provides 76 Kbps. The BCU-2 device in a WBAN can also be composed of standards such as ZigBee and the 802.11 Wi-Fi standards to accommodate and interface with different wireless platforms and to connect to the Internet for remote monitoring. A summary of devices used in BCUs and sensor node boards is given in Table 6.6. The hardware implementations of BCU-1 and BCU-2 are given in Figs. 6.13 and 6.14.

The modulation technique used in CC10xx chips is FSK with a frequency step of 64 kHz, which is acquired by loading the crystal. FSK has better interference rejection and causes less spectrum splatter normally seen in ASK systems. The picture in Fig. 6.15

Table 6.6. Summary of devices used in sensor nodes and BCUs

Device features	AMIS (52100)	CC1000	CC1010*	PIC16F887
Size (mm)	7.5 × 7.8	9.7 × 6.4	12.9 × 12.9	17.53 × 17.53
Modulation	ASK/OOK	FSK/OOK	FSK	—
Sensitivity	−117 dBm	−109 dBm	−107 dBm	—
Power: Transmitter	TX: 25 mA	Tx: 26.7 mA	Tx: 26.6 mA	< 0.6 mA (active)
Receiver	Rx: 7.5 mA	Rx: 7.4 mA	Rx: 9.1 mA	
Data Rate	19 Kbps	76.8 Kbps	76.9 Kbps	—
Sniffing**	500 nA	200 nA	~ 0 mA	—
Memory (RAM)	—	—	2048 + 128	368 + 256
Additional Features	—	—	10-bit, 22.7 kHz sampling frequency	10-bit 52 kHz sampling frequency

*Note: CC1010 has a microcontroller built-in.
** Sniff mode enables the receiver to wake up or operates at times to "sniff" received RF signals and then return to "sleep" or "wait" mode if a signal is not detected.

Figure 6.13. A wired BCU (BCU-1). See also Color Insert.

Figure 6.14. Intermediate body control unit (BCU-2). This device is shared by more than one patient and is portable. It contains dual wireless transceiver to support two directional wireless links. See also Color Insert.

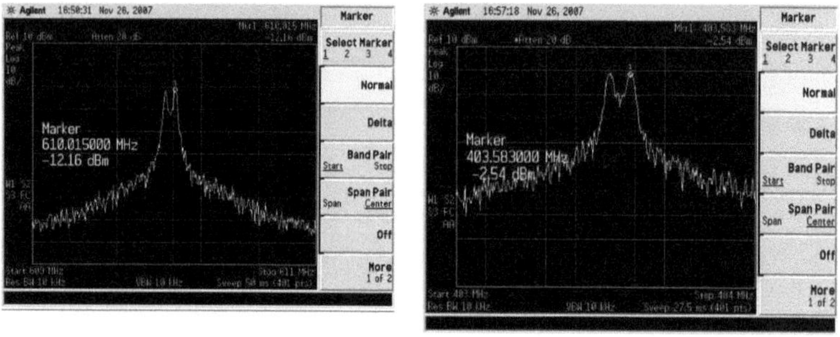

Figure 6.15. Spectrums generated for MICS and WMTS frequencies with Chipcon CC1010.

demonstrates the FSK modulation for MICS and WMTS bands. What is necessary to realize from this picture is that the bandwidth of the signal is well within the bandwidth of the WMTS band (608–615 MHz).

In order to provide communication from sensors to a personal computer, a wired serial port interface, such as RS-232 standard, should be used to connect the base station to monitor the received body signals. The interface system here uses RS-232 serial interface and USB connections.

Figure 6.16 is an ECG signal obtained from our set up. The ECG signal is transmitted from a sensor node to the computer. Each sensor node representing only one patient can only have one ECG. In order to eliminate the DC noise (50 Hz/60 Hz interference), a recursive filter has been software implemented to obtain an accurate ECG signal (see appendix for notch filter implementation). As shown in Fig. 6.10, by clicking on 50 Hz filter, the recursive notch filter operates on the received ECG signal.

High computation programs like MATLAB can further be used to analyze data automatically and to warn medical professionals when the value of a critical data goes outside the safe margin. A warning signal could be easily generated by the program at the PC to warn health professionals to track the sensor node generating a warning signal. This feature will strengthen the reliability and safety in implementation, which would be useful for patients' lives.

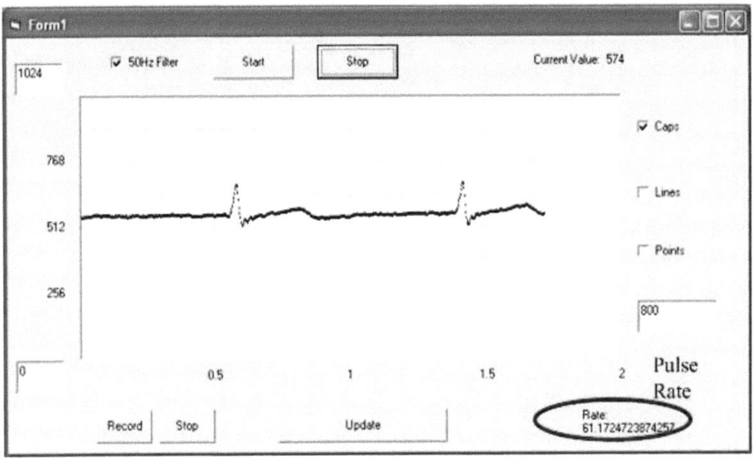

Figure 6.16. A wireless ECG monitoring showing pulse rate as well.

A medical data from a single human body can be monitored within 300 ms for a distance of 1 m when the WBAN system is implemented in a four–patient scenario. When the active patient is doing all sort of activities, e.g., sitting standing, turning around the other patients, the average delay is 1 s up to 5 m and 2 s at the distance of 10 m. These values show that the WBAN system exhibits a correct, timely, and reliable communication performance up to 10 m with a time performance less than 2 s for a multi-patient scenario [52].

6.4.1 *Software Programs and Monitoring*

In order to monitor data in body area network applications, several computer programs should be developed. The necessary software programs have been identified in Fig. 6.17 for this specific application. A software called GATEWAY should be developed at the monitoring PC to control the communication with the BCU to get readings from sensors and then forward them through another network/Internet to an application on a remote PC (at a remote location). While performing this task, the GATEWAY will also verify the data integrity and schedules retransmission, if required. Another software program is developed at the remote PC (called BSN),

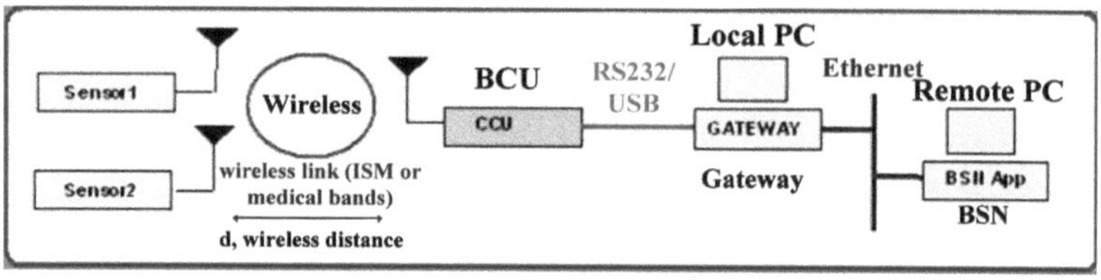

Figure 6.17. Software programs for the proposed WBAN application. See also Color Insert.

which gets readings from GATEWAY via the network/Internet. These readings are stored in the remote PC for analysis. A graphical user interface (GUI) at the local PC as well as at remote PCs should be designed to display the human body data. Both the data received from the BCU and the data sent to the BSN can also been shown by the GUI in text or graphical formats. In case of medical application, the physiological signals of patients can be accessed by medical staff anywhere in the medical center as long as their computers are connected to the local area network in the building.

The BSN application is designed to collect and store readings automatically so that no person is required to be stationed at the application. It gets sensor information and readings from the GATEAWAY and will ask retransmission if error is detected. It undertakes the administration of patients' particulars such as assigning new sensor ID to patients, segregating sensor readings from different patients, and storing them into the data base. The GUI at the monitoring PCs allows medical personnel to enter patients' information. It can also display live monitoring graphs on the screen (Fig. 6.18). Every sensor device has a unique sensor ID and must be registered under a patient name before they are used. In the event that an unregistered sensor node is used, all its readings received will be discarded by the BSN application

Since all sensor nodes of a body are communicating with the same BCU, the data is prefixed with an identifier, which is used to identify the source of the data. As mentioned earlier, to reduce collisions further between data sent, a firmware (medium access control (MAC) protocol) is written to control data transmissions. The communication between the sensors and the BCU is bi-directional as to support a multiuser (i.e., multi-patient) communications.

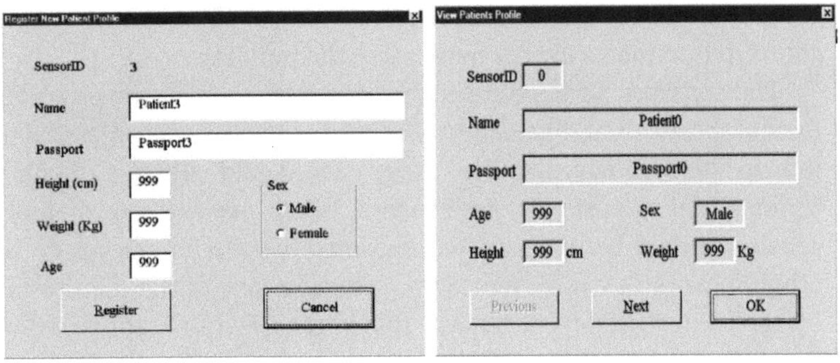

Figure 6.18. Example of a graphical menu registering and viewing patients.

The GATEAWAY menu shows a GUI at the local PC, which is designed to configure the source and destination socket and port numbers for data transfer on the Ethernet (See Fig. 6.19). This GUI ensures portability if the BSN application or the GATEWAY need to be relocated, it can be configured to work without additional changes in the codes. Both the data received from the BCU and the data sent to the BSN can also been shown here in text format. The physiological signals of patients can be accessed by medical

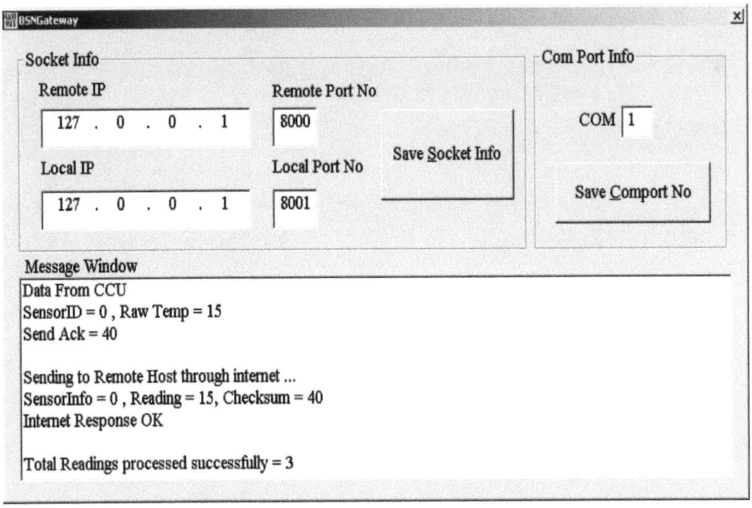

Figure 6.19. The GATEWAY menu.

staff anywhere in the medical center as long as their computers are connected to the local area network in the building.

The software packages used for monitoring and storing data (in the database) should be user friendly so that medical professionals utilize them efficiently. The programs should allow a medical professional to easily set up sensors. When needed, one or more sensors should be removed without affecting the operating of the other sensors in the system. Software programs at the monitoring PC can be designed in such a way that when a sensor stops working, the data from the remaining sensors should still be received and monitored. Before a new sensor is deployed for a particular person, a new sensor ID (i.e., user ID) should be assigned at the monitoring PC so that the MAC protocol used for the data communication will enable a communication from that sensor. The MAC protocols mentioned above can integrate an error-checking mechanism to allow the monitoring of correct data only and discard when a data is faulty.

A database server is developed to maintain data integrity, which is necessary for big medical centers. Monitoring of ECG for an individual patient is shown in Fig. 6.20. In this GUI, more detailed patient's particulars can be seen from the database. Clicking on

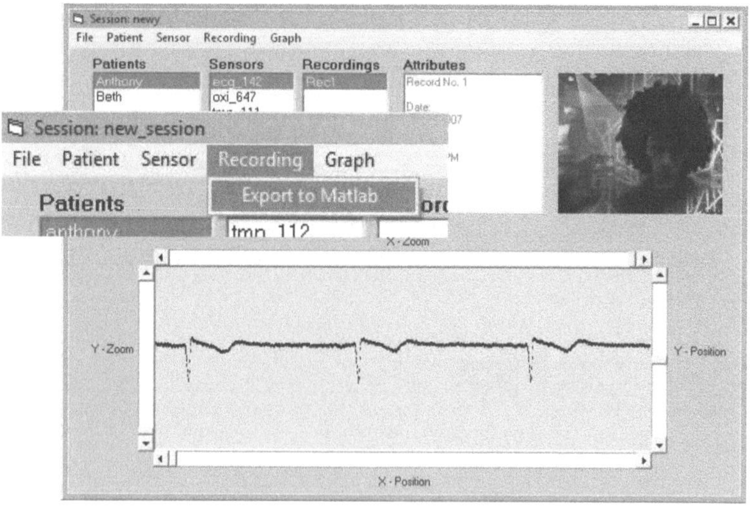

Figure 6.20. An example of ECG Monitoring. See also Color Insert.

a patient shows the sensors attached to them and their personal information/picture. Clicking on a sensor displays the sensor's information (e.g., interval). It also brings up how many recordings are available. Single clicking a record shows when the record was made. Double clicking displays the record on the graph. In the following figure, it is arranged for an ECG monitoring. It is very handy to have a data file compatible with a computation software program like MATLAB for signal analysis as shown in the window.

6.5 Conclusion

We have discussed implementation issues and presented details of techniques for design of wireless sensors in WBAN applications. Portable and wireless gateway nodes are used to connect the sensor nodes to the local area network or the Internet already available for long-range access points. Wireless standards such as ZigBee (IEEE 802.15.4) and Bluetooth (IEEE 802.15.1) are popular low-power technologies for communication between sensors and the control device. New wireless devices specifically targeting at WBAN application can be designed based on MICS, WMTS, 433 MHz ISM as well as UWB bands. Miniaturization of the sensor node electronics, especially the sizes of the microcontroller, the wireless chip, the battery, and low-power consumption are current hardware related issues for small sensor nodes. TI's CC2420 and CC1010 transceivers and Nordic's transceivers like nRF24E1 are current popular choices for wireless module in a sensor node. The successful implementation of a WBAN system should operate and coexists with other network devices and should provide wearable, wireless (no wire connections), easy to remove, and attach sensor nodes, leading to increased mobility of patients and flexibility.

Any future projects based on WBAN should take under consideration the co-existence of other network devices operating in similar frequency bands. The majority of WBAN systems cover only wearable sensor nodes. There is very limited information on integration of wearable and implantable systems together. Implanted or wearable sensor nodes should be able to work together in a WBAN system without any data collisions.

Acknowledgements

I would like to thank Anthony Bott and Ng Peng Choong for their help in developing software programs and boards.

Appendix

Digital Notch Filter Designs

A digital notch filter can be designed to filter out the 50/60 Hz noise for continuous physiological parameters. We used the website in [53] as a reference. Two parameters should be defined: f, the center frequency, and BW, the bandwidth. Finding the notch filter coefficients:

$$\begin{aligned}
R &= 1 - 3 \times \mathrm{BW} \\
\mathrm{BW} &= 0.033 = 13.2\ \mathrm{Hz} \\
\therefore \quad R &= 0.901 \\
f &= \frac{50}{400} = 0.125 \\
K &= \frac{1 - 2 \times 0.901 \times \cos(2\pi f) + 0.901^2}{2 - 2 \times \cos(2\pi f)} \\
\therefore \quad K &= 0.917731353 \\
a_0 &= K, a_1 = -2K\cos(2\pi f) = -1.297868126, a_2 = K \\
b_1 &= 2R\cos(2\pi f) = 1.27420642, b_2 = -R^2 = -0.811801
\end{aligned}$$

These are used in the equation

$$\begin{aligned}
y[n] = a_0x[n] + a_1x[n-1] + a_2x[n-2] \\
+ \cdots + b_1y[n-1] + b_2[n-2]
\end{aligned}$$

Note that BW and f are fractions of the sampling frequency. All calculations are done in radians. The bandwidth 13.2 Hz, which could be wide for the application, can easily be decreased to a nominal value.

The system requires that three previous inputs and the last two outputs are stored. Whenever a new value is calculated, the array is shifted so that $x[n-1] = x[n]$, etc. It is possible to increase the number of coefficients but at the expense of CPU resources and the increased risk of instability. Decreasing the bandwidth of the filter also has limits and will cause problems if it becomes too small.

References

1. Jovanov, E., Milenkovic, A., Otto, C., and de Groen, P. C. (March 2005) A Wireless Body Area Network of Intelligent Motion Sensors for Computer-Assisted Physical Rehabilitation, *Journal of NeuroEngineering and Rehabilitation*, **2(6)**, doi:10.1186/1743-0003-2-6.
2. http://focus.ti.com/lit/ds/symlink/ina321.pdf, 2010.
3. http://www.analog.com/static/imported-files/data_sheets/AD620.pdf.
4. E. Company-Bosch and E. Hartmann , ECG Front-End Design is Simplified with Micro-Converter® " Analog Dialogue 37-11, November (2003).
5. P. A. Pour, T. Gulrez, O. AlZoubi, G. Gargiulo, and R. A. Calvo (15–18 December 2008) Brain-Computer Interface: Next Generation Thought Controlled Distributed Video Game Development Platform, *IEEE Symposium on Computational Intelligence and Games*, pp. 251–257.
6. S. Hanna (2009) Regulations and Standards for Wireless Medical Applications, *ISMICT* 2009.
7. K. Finkenzeller (April 2003), *RFID Handbook*, 2nd edn, John Wiley and Sons, New York, USA.
8. FCC Rules and Regulations (January 2003) MICS Band Plan, *Table of Frequency Allocations*, Part 95.
9. A. Tekin, M. R. Yuce, and W. Liu (2008) Integrated VCOs for Medical Implant Transceivers, *VLSI Design*, 2008, Article ID 912536, 10 pages, doi:10.1155/2008/912536.
10. http://www.na.mindray.com/pdf/panorama_WMTS_specifications_us.pdf.
11. http://www.fda.gov/MedicalDevices/Safety/MedSunMedicalProductSafetyNetwork/ucm127778.htm, 2010.
12. M. R. Yuce, H. C. Keong, M. Chae (October 2009) Wideband Communication for Implantable and Wearable Systems, *IEEE Transactions on Microwave Theory and Techniques*, **57(2)**, pp. 2597–2604.

13. Moteiv Corporation (2010) http://sentilla.com/files/pdf/eol/tmote-sky-datasheet.pdf.
14. http://www.tinynode.com. Ref: Tinynode.
15. http://webs.cs.berkeley.edu/tos/.
16. http://www.snm.ethz.ch/Projects/T-Nodes.
17. A. Wong, G. Kathiresan, C. Chan, O. Eljamaly, O. Omeni, D. McDonagh, A. Burdett, and C. Toumazou (2008) A 1 V Wireless Transceiver for an Ultra-Low-Power SoC for Biotelemetry Applications, *IEEE Journal of Solid-State Circuits*, **43**, pp 1511–1521.
18. http://www.btnode.ethz.ch/Documentation/BTnodeRev3Sensor Guide.
19. http://www.palowireless.com/database/conexant/101685b.pdf.
20. C. Park, P. H. Chou, Y. Bai, R. Matthews, and A. Hibbs (2006) An Ultra-Wearable, Wireless, Low Power ECG Monitoring System, in *Proceedings of IEEE BioCAS*, pp. 241–244.
21. G. Z. Yang (2006) *Body Sensor Networks*, Springer-Verlag London.
22. B. Wang, et al. (11–13 June 2009) A Body Sensor Networks Development Platform for Pervasive Healthcare, *in Proceedings of the 3rd International Conference on Bioinformatics and Biomedical Engineering*, pp. 1–4.
23. M. R. Yuce, and C. K. Ho (August 2008) Implementation of Body Area Networks Based on MICS/WMTS Medical Bands for Healthcare Systems, *IEEE Engineering in Medicine and Biology Society Conference (IEEE EMBC08)*, pp. 3417–3421.
24. R. S Mackay, B Jacobson (1961) Radio Telemetering From Within the Human Body, *Science*, **134**, pp. 1196–1202
25. S. Y. Shin, H. S. Park, and W. H. Kwon (August 2007) Mutual Interference Analysis of IEEE 802.15.4 and IEEE 802.11b, *Computer Networks*, **51**, pp. 3338–3353.
26. W. Liu, et al. (September–October 2005) Implantable Biomimetic Microelectronic Systems Design, *IEEE Engineering in Medicine and Biology Magazine*, **24**, pp. 66.
27. M. Zhou, M. R. Yuce, and W. Liu (September 2008) A Non-Coherent DPSK Data Receiver With Interference Cancellation for Dual-Band Transcutaneous Telemetries, *IEEE Journal of Solid-State Circuits*, **43**, pp. 2203–2012.
28. W. Liu, et al. (October 2001) A Neuro-Stimulus Chip With Telemetry Unit for Retinal Prosthetic Device, *IEEE Journal of Solid-State Circuits*, **35**, pp. 1487–1497.

29. S. K. An, et al. (June 2007) Design for a Simplified Cochlear Implant System, *IEEE Transactions on Biomedical Engineering*, part 1, 54, pp. 973.
30. P. R. Troyk, and G. A. DeMichele (September 2003) Inductively-Coupled Power and Data Link for Neural Prostheses Using a Class-E Oscillator and FSK Modulation, in *Proceedings of the IEEE International Conference on Engineering in Medicine and Biology Society*, pp. 3376–3379.
31. M. Ghovanloo, and K. Najafi (December 2004) A Wideband Frequency Shift Keying Wireless Link for Biomedical Implants, *IEEE Transactions on Circuits Systems II*, **51**, pp. 2374–2383,.
32. M. A. Stuchly, A. Krazewski, S. S. Stuchly, and A. M. Smith (1982) Dielectric Properties of Animal Tissues in Vivo at Radio and Microwave Frequency: Comparison Between Species, *Physics in Medicine and Biology*, **27**, pp. 927–936.
33. D. Halperin, et al. (March 2008) Security and Privacy for Implantable Medical Devices, *IEEE Pervasive Computing*, **7(1)**, pp. 30–39.
34. Australian Communications Authority, Radio Frequency Planning Group (October 2003) Planning for Medical Implant Communications Systems and Related Devices, *Proposals Paper*, http://www.acma.gov.au/.
35. P. D. Bradley (December 2006) An Ultra-Low Power, High-Performance Medical Implant Communication System (MICS) Transceiver for Implantable Devices, in *Proceedings of the IEEE Biomedical Circuits and Systems Conference (BioCAS 2006)*, pp. 158–161, also at http://www.zarlink.com/zarlink/hs/4889.htm.
36. http://www.fukuda.com/fukuda_usa/lx-5160.html, 2010.
37. http://www.welchallyn.com, Micropaq® Monitor.
38. http://www.corventis.com/AP/nuvant.asp, 2010.
39. A. Burdett (2006) Low Power Device Communication, *IET Seminar on Biocompatible Materials and Devices*, http://www.toumaz.com/.
40. www.micoach.com.
41. T. Gao, et al. (September 2007) The Advanced Health and Disaster Aid Network: A Light-Weight Wireless Medical System for Triage, *IEEE Transactions on Biomedical Circuits and Systems*, **1(3)**, pp. 203–2007.
42. V. Shnayder, B. Chen, K. Lorincz, T. R. F. F. Jones, and M Welsh (2005) Sensor Network for Medical Care, *Technical Report TR-08-05*, Division of Engineering and Applied Sciences, Harvard University.
43. https://wiisard.org.

44. J. Espina, T. Falck, J. Muehlsteff, Y. Jin, M. A. Adán, and X. Aubert (June 2008) Wearable Body Sensor Network Towards Continuous Cuff-less Blood Pressure Monitoring, in *Proceedings of the 5th International Workshop on Wearable and Implantable Body Sensor Networks*, pp. 28–32.
45. U. Anliker, et al. (2004) AMON: A Wearable Multiparameter Medical Monitoring and Alert System, *IEEE Transactions on Information Technology in Biomedicine*, **8**, pp. 415–427.
46. N. Oliver, and F. Flores-Mangas (2006) HelathGear: A Real-Time Wearable System for Monitoring and Analysing Physiological Signals, in *Proceedings of the IEEE International Workshop on Wearable and Implantable Body Sensor Networks (BSN 2006)*, pp. 1–5.
47. R. Naima, and J. Canny (June 2009) The Berkeley Tricorder: Ambulatory Health Monitoring, in *Proceedings of the 2009 6th International Workshop on Wearable and Implantable Body Sensor Networks*, pp. 53–58.
48. J. Penders, at al. (December 2008) Human++: From Technology to Emerging Health Monitoring Concepts, in *Proceedings of the 5th International Summer School and Symposium on Medical Devices and Biosensors (ISSS-MDBS 2008)*, pp. 94–98.
49. M. R. Yuce, et al. (December 2007) Wireless Body Sensor Network Using Medical Implant Band, *Journal of Medical Systems*, **31**, pp. 467–474.
50. H. Chen, W. Wu, and J. Lee (March 2009) A WBAN-Based Real-Time Electroencephalogram Monitoring System: Design and Implementation, *Journal of Medical Systems.*, **34**, pp. 303–311.
51. Y. Zhang, and H. Xiao (November 2009) Bluetooth-Based Sensor Networks for Remotely Monitoring the Physiological Signals of a Patient, *IEEE Transactions on Information Technology in Biomedicine*, **13(6)**, pp. 1040–1048.
52. 52. M. R. Yuce (July 2010), "Implementation of Wireless Body Area Networks for Healthcare Systems," *Sensors & Actuators: A. Physical*, **162**, pp. 116-129.
53. http://www.dspguide.com/ch19/3.htm.

Chapter 7

Wireless Body Area Network Implementations for Ambulatory Health Monitoring

Reza Naima[a] and John Canny[b]

[a] *Department of Bioengineering, University of California, Berkeley, 306 Stanley Hall #1762, Berkeley, CA 94720-1762*

[b] *Electrical Engineering and Computer Sciences, University of California, Berkeley, 387 Soda Hall, MC1776,Berkeley, CA 94720-1776*

reza@reza.net; jfc@cs.berkeley.edu

The design and implementation of an ambulatory health monitoring device is a process of matching the medical need to available technology. The process starts with an understanding of the requirements, which leads to the design of sensors, filters, amplifiers, etc. There are a number of additional decisions to be made that range from the battery chemistry to the wireless technology that must also match the underlying requirements. Finally, some existing implementations will be reviewed.

7.1 Design Process

The first step in developing a device is to understand the medical need in detail. It is not sufficient to simply have an understanding

Wireless Body Area Networks: Technology, Implementation, and Applications
Edited by Mehmet R. Yuce and Jamil Y. Khan

ISBN 978-981-4316-71-2 (Hardcover), ISBN 978-981-4241-57-1 (eBook)
www.panstanford.com

of the physiology, pathology, etc., and to build a device accordingly. From the physician's perspective, the device might be too complex to use, the data might be too verbose to analyze, or the physician might not trust the data compared to the existing standards. From the patients' perspective, the device might be too cumbersome or irritating to wear, the battery might die too quickly, the device might require too much maintenance, or they might not see sufficient value in the device to comply with the doctor's instructions. There are many other stakeholders that must also be considered through the design process: nurses or other caregivers, insurance companies or other sources of reimbursement. There is a great book that discusses the design process for medical devices in detail [1], and is highly recommended.

Once the high-level application, set of sensing modalities, data storage, and wireless telemetry mechanism are determined, the requirements of the individual components need to be flushed out. It isn't sufficient to "record ECG"; rather, the application will result in requirements that need to be satisfied by the design. For example, many ECG circuits will bandwidth limit the signal to 40 Hz, and thus only require a sampling rate around 80 Hz (based on Nyquist–Shannon sampling theorem). However, if the data is to be used to measure heart-rate variability, then a sampling rate of 1 kHz is required to have the required resolution. It is easy to miss this step and to start designing a circuit based on a number of assumptions which might not be valid — only to find out months later and have to spend considerably more time and effort trying to fix the design or work around the flaws. This is a lesson the author has learned the hard way — spend the time understanding the application and the requirements, and the interdependence of requirements between the various components.

To help understand the design process, we will give examples involving an ambulatory health monitor developed at the Berkeley Institute of Design, named the Tricorder (Fig. 7.1). The original goal for the Tricorder was a small and highly integrated device, which records a number of parameters that can be used to assess the wellness-state of the user. The data was intended to be used by health care providers to better monitor the health state of the user, to help in the treatment of any chronic condition, and to provide

Figure 7.1. The Berkeley Tricorder. See also Color Insert.

advanced notice of any possible health problems. Based on our application, we developed the following high-level design goals:

- A means of data storage for at least 24 h and a mechanism for offloading said data.
- Remote telemetry for real-time data viewing.
- A comfortable means of wearing the device.
- As many relevant sensing modalities as possible in a single highly integrated form factor.

Based on these design goals, we decided on Bluetooth for the wireless telemetry, microSD card for the data storage, and a number of sensing modalities, detailed in Table 7.1. Examples involving the Tricorder will be presented in the various sections of this chapter.

7.2 Existing WBAN Implementations

There are a large number of existing WBAN (wireless body area network) implementations for an array of applications. Any attempt at categorizing implementations is artificial, and the distinctions

Table 7.1. Device parameter summary

Modality	Sampling rate	Bits/sample	Comment
ECG	256 Hz	12	2-Stage HPF; RFI filer; single channel
EMG	256 Hz	12	1-Stage HPF; RFI filter
Respiration (bioimpedance)	256 Hz	12 Phase; 12 Mag.	4-Electrode configuration; 350 μA current at 50 kHz
Acceleration (3-Axis)	256 Hz	8/Axis	
Blood oxygenation	256 Hz	15	Reflective forehead sensor; 2-stage amp with DC-offset subtraction

between different deployments can be very different from the perspective of two different users. We will attempt to group together various common characteristics between some of the different implementations used in the noninvasive health/medical space based on my experiences and research, and present it as a starting point.

Other sections in the chapter are dedicated to the implementation of sensing modalities at a low level — this section will focus on the differences at a higher level: the hardware design paradigm, the firmware implementation, and mechanism by which data is handled.

7.2.1 *Hardware Paradigms*

There is a big tradeoff between implementations that are designed and built from the ground up, and designs which use OEM (original equipment manufacturer) components to accelerate development. Fully custom designs will generally take longer to develop, test, and debug — but they will provide the smallest form factor, lowest power consumption, and best fit for your application. These implementations might have higher up front costs associated with all the required development tools, but will cost lower to produce in volume.

Examples of highly integrated circuits include a non-contact ECG electrode with an integrated amplifier [2]. In this case, the electrode form factor requires a highly integrated device, which is not possible with OEM options. The same is true with another group that has developed a button-sized wireless pulse oximeter [3]. Even greater levels of integration can be achieved by fabricating an application-specific integrated circuit (ASIC) containing all the required components for a WBAN system on a single chip [4], or just a portion [5].

A majority of WBAN devices utilize at least one OEM component — ranging from small PCBs that implement the wireless link to fully integrated sensing devices such as a pulse oximeter, which provides the blood oxygen saturation percentage data over a serial stream [6]. The use of OEM components can greatly accelerate the design cycle; however, each module often implements redundant components such as the power stage and microcontroller's which consume more power, take up more space, might not provide an ideal form factor, and have higher costs.

The most common OEM modules are ZigBee-based motes. Motes[1] are a class of devices that are conform to the 802.15.4 communications standard and allow for a data transfer and relay between motes in a mesh configuration. Motes can include an ADC (analog-to-digital converter) and serial ports to interface with custom sensing boards, such as the MICAz mote from Crossbow. The Telos moteiv also includes temperature, humidity, and light intensity sensors. Their popularity is based on their low levels of power consumption, a well-established networking interface, and the ability to relay data between motes.

There are many mote-based implementations, such as the group at the University of Alabama, which has a built a custom ECG module to interface with a Telos Mote [7, 8].

Another group has proposed a system that utilizes a number of OEM modules, including a Nonin ipod pulse oximeter, a belt-based respiration rate sensor, and MICA based motes [6]. A joint effort between Stanford University and NASA has resulted in the Lifeguard [9], a platform that utilizes a number of OEM components such as the Nonin Xpod pulse oximeter and the Accutracker II blood pressure monitor.

7.2.2 *Firmware*

There are two main camps in terms of how the firmware operates — either as a single application, or as an operating system that

[1]Dr. Kris Pister, inventor of motes, defines a mote as a single dust particle

runs the application. As with the hardware example, writing a stand-alone application provides the greatest flexibility in how to access the hardware and utilize its resources, whereas implementing an operating system, such as TinyOS[2], can help accelerate the development cycle and provide a large set of built-in tools but introduces overhead into the application and imposes some limits in terms of how the application is structured.

TinyOS is the primary operating system running in most of the motes to handle network communications, and is used by a number of projects. The CodeBlue group at Harvard University have developed a number of sensors that use TinyOS motes for data acquisition and telemetry [10] and evaluated their use for triage [11].

A group at the Imperial College London has utilized a TinyOS-based sensor module developed at the University of California, Berkeley to accelerate the development cycle of a number of biosensors. These include an ECG module and a pulse oximeter module, each of which can relay data directly to a PDA [12].

7.2.3 *The Data*

Once data is collected, it can be stored, processed, transmitted, or any combination thereof. For data storage, the predominant mechanism is by the use of a secure digital card, especially the microSD form factor, which is 15 mm × 11 mm × 1 mm. These devices require standard Serial Peripheral Interface (SPI), present on most microcontrollers. Their use is further promoted by the availability of numerous royalty-free interface libraries and the availability of multi-gigabyte cards. This is especially important for devices such as the Berkeley Tricorder, which can generate up to 823 MB of data in 24 h [13]. It should be noted that during a write cycle, SD cards use considerable bursts of current, which can introduce noise to your system. For the Berkeley Tricorder, we found this noise was coupled onto the LED current, which produced noise from the pulse oximeter signal (visible in Fig. 7.15). A redesigned current source is expected to correct this.

[2]http://www.tinyos.net/

Processing of data can involve digital filtration to remove signal noise, or calculating the percentage oxygen saturation from the output of the pulse oximeter photodiode and lookup tables. Processing of data consumes limited CPU cycles and increases the power consumption of the microcontroller. Hence, as much data processing as possible should be performed on more powerful devices (i.e., a cell phone or a personal computer).

In terms of telemetry, there are a large number of possible implementations that depend on: the desired telemetry range, requirements of data relaying, the operating frequency, and the requried bandwidth. As a general rule of thumb, the technology that meets all the requiremets while minimizing the power consumption and PCB real-estate will provide the best option.

7.3 Signal Acquisition

The number of possible sensing modalities is very large — a discussion of all the possible modalities is beyond the scope of this chapter. Instead, we will discuss some common strategies used in interfacing analog and digital components, with examples in the modalities used in the medical space.

The strategy for any sensing modality is to start with a characterization of the signal of interest in the context of how that data is to be used. The parameters of interest include the frequency bandwidth, the required dynamic range, and source impedance. It is always critical to start with signal source and design the rest of the system from that point; otherwise you might find yourself in the position of being forced to redesigning portions of your circuit at the expense of extra time and cost.

In addition to designing the circuit based on the signal of interest, it must also be designed to be immune to noise coupled from the power supply, from radio-frequency interference (RFI), or from other components on the circuit board while adhering to various safety requirements.

A similar strategy exists for sensing modalities that do not require any analog interface, such as an accelerometer that communicates through a digital interface. In this situation, it's important to understand data's application and to design around those requirements. For example, an accelerometer that is used for fall detection might require a high sampling rate, but not many bits of resolution, whereas an accelerometer used to measure cardiac output (i.e., balistocardiograph) would require significantly higher resolution but a lower sampling rate.

7.3.1 *Frequency Bandwidth of Interest*

An understanding of the frequency bandwidth ($f_{\min} - f_{\max}$) of interest (FOI) for any particular sensing modality is perhaps the single most important characteristic to understand. For analog-based signals, it will determine how the front-end interface is designed, will influence the choice of the amplifier, and determine the sampling requirements for the ADC. The FOIs are best determined by doing research on the modality of interest, especially for any biologically originating signals as most of them have been well characterized.

The FOI is often derived by an examination of the power spectrum of the modality of interest. For example, consider the power spectrum of an ECG signal sampled at 256 Hz (Fig. 7.2). By 40 Hz, we can see a 15 dB drop in the power of the signal, which means that the importance of the 40 Hz frequency is $10^{\frac{-15}{10}} =$

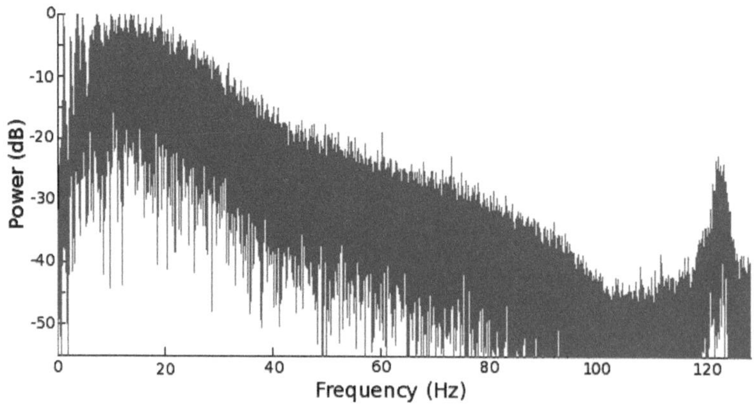

Figure 7.2. Power spectrum of ECG signal. The peak at 120 Hz is a harmonic of the 60 Hz power-line interference. See also Color Insert.

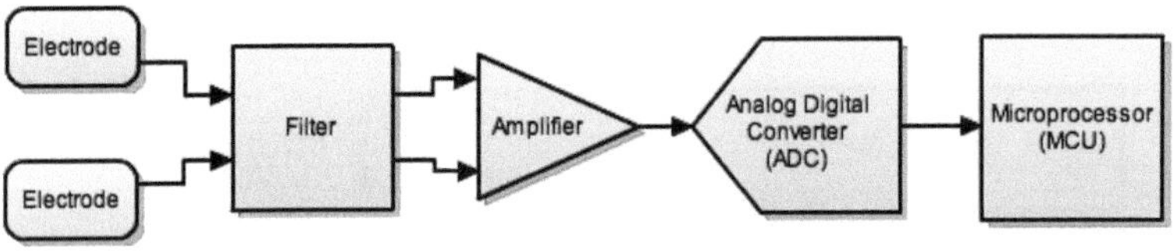

Figure 7.3. Biopotential signal transduction pathway

3.1% that of the dominant frequency. The importance continues to drop down to −40 dB, or $10^{\frac{-40}{10}} = 0.01\%$. Depending on the application, this data can help determine what range of frequencies are of interest.

7.3.2 *Measuring Surface Biopotentials*

There are a number of sensing modalities that are based on detecting and amplifying biopotentials from the skin. These include the electromyogram (EMG) for detecting the activity of muscle fibers, the electrocardiogram (ECG) for detecting the electrical activity of the heart, the electroencephalogram (EEG) for detecting the electrical activity of the brain, the electroretinogram (ERG) for measuring the activity associated with eye movement, among others. All these modalities operate under the same basic mechanism (Fig. 7.3).

7.3.2.1 The electrode

The first site of signal transduction occurs at the skin-electrode interface, where ionic potentials are converted into electrical potentials through reduction/oxidation reactions that follow the basic form

$$C \rightleftharpoons C^{n+} + ne^{-} \tag{7.1}$$

$$A^{m-} \rightleftharpoons A + me^{-} \tag{7.2}$$

where C and A represent the cation and the anion, respectively, and m and n represent their respective valencies. These reactions are reversible for most electrode/electrolyte combinations, but not all. As a result, it is wise to not be too creative with electrode

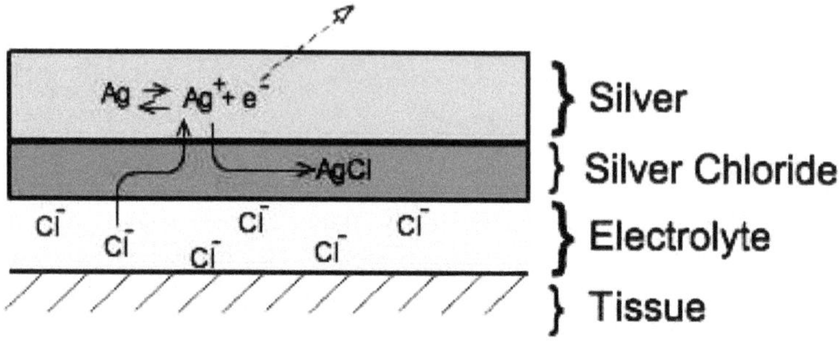

Figure 7.4. Silver-silver chloride electrode redox reaction.

choices. There are many electrode/electrolyte combinations that have various advantages and disadvantages, but we will focus on the most commonly used one: the silver-silver chloride (Ag-AgCl) electrode. Among other benefits, the electrode chemistry will help minimize the effects of motion artifacts [14].

The Ag-AgCl electrode consists of a layer of silver surrounded by a very thin layer of silver chloride. A gel containing a high concentration of Cl^- ions and saturated with AgCl is often used in conjunction with the Ag-AgCl electrodes (Fig. 7.4).

The interface of the Ag-AgCl electrode, the electrolyte, and the skin form a equilibrium, which is governed by the following equations:

$$\mathrm{Ag} \rightleftharpoons \mathrm{Ag}^+ + \mathrm{e}^- \tag{7.3}$$

$$\mathrm{Ag}^+ + \mathrm{Cl}^- \rightleftharpoons \mathrm{AgCl} \tag{7.4}$$

Metallic sliver will give up electrons and spontaneously combine with free chloride ions to form silver chloride, which due to its limited solubility, will precipitate out of solution adding to the silver chloride layer. These reactions occur constantly and reversibly, but the concentrations of the various constituents remain the same averaged over time resulting in no net current. However, if the potential of the tissue at the location of the electrode changes, then the reactions will be driven in one direction or another causing a net flow of electrons into or out of the electrode, which can be measured.

It is also possible to measure the electrical potential differences using capacitative electrodes that do not have a galvanic connection [15]. These electrodes can be fabricated with a thin layer of nonconducting dielectric material and offer benefits, including being able to sense potential differences though clothes, but suffer from a very high sensitivity to external noise and an attenuation in the low-frequency components of the signal.

7.3.2.2 Filtering

As mentioned in Section 7.3.1, different biological processes exhibit electrical activity at different frequencies, and should be filtered accordingly. Beyond filtering for unwanted noise, a low-pass filter is always required to avoid aliasing, while a high-pass filter is required to bias the signal. Furthermore, it is always a wise idea to add an RFI filter [16] to prevent output offset errors due to RF rectification inside the amplifier.

7.3.2.3 Amplifier

Picking an optimal operational amplifier (OpAmp) for the application is a critical step in the design in a biopotential measurement device. Care must be taken that the amplifier does not contribute additional noise to the system, that its input impedance is sufficiently greater than the source impedance of the signal to prevent distorting the signal, and that it can provide sufficient gain at the frequencies of interest. A number of other issues exist, such as verifying that the amplifier is able to operate at the supply voltage of the PCB or that the quiescent current consumption is sufficiently low, but these issues are more obvious and do not require as much attention.

As most biosignals are based on the difference in potential from two different sites, the use of an instrumentation amplifier (InAmp) is highly recommended. An InAmp is designed to provide balanced, high input-impedance (typically 10^9 Ω) inputs with low bias currents, and a low impedance output with low DC-offset error. An InAmp can be constructed with two or three OpAmps; however, using a monolithic single-package InAmp has the advantage of

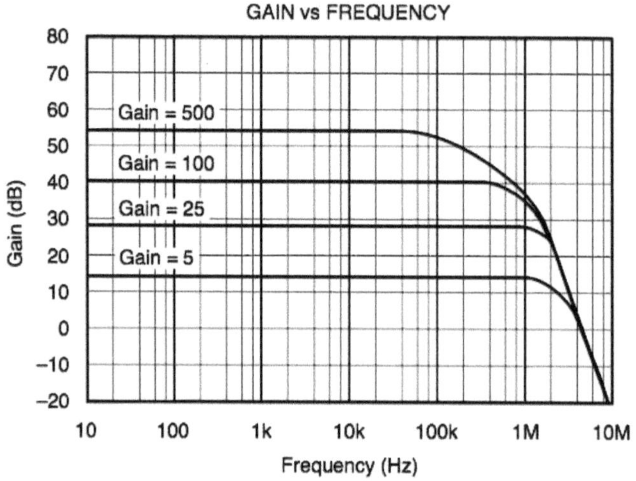

Figure 7.5. Texas Instruments INA331 Instrumentation Amplifier Gain-Frequency Bode Plot.

having highly matching passive components that are often trimmed to provide lower errors and very high common-mode rejection. Additionally, a monolithic InAmp will use less PCB space to help reduce the WBAN size. There are a number of free online references that help explain OpAmps [17] and InAmps [18].

The first step in picking the right amplifier is to see if it's able to provide a flat gain response over the range of frequencies of interest (Section 7.3.1) by examining the Bode plot, or the gain-frequency response of the amplifier on a log-log plot (Fig. 7.5).

Although most passive components can generate noise, referred to as thermal noise, the noise from active components (i.e., OpAmps/InAmps, transistors, etc.) usually dominates the signal. There are a number of noise sources in OpAmps and InAmps, but the main ones to be concerned with are input bias currents and input-referenced voltage noise. For biopotentials, the source impedance is often very high (upward of 10^5 Ω), where the bias-current noise will dominate. Only by understanding the characteristics of the signal requiring amplification can you determine what amplifier requirements exist for your implementation.

7.3.2.4 Analog Digital Converter and Microcontroller

The number of low power microcontrollers (MCU) is ever increasing with certain applications allowing a device to remain operational for years with a single battery. Most contemporary MCUs include ADC and digital to analog converter (DAC), which greatly simplify the design; however, do not feel restricted to using the provided peripherals if they do not satisfy your design requirements.

For the design of the Berkeley Tricorder, we required an MCU that provided us with multiple fast, high-precision ADC inputs for data acquisition as well as a DAC for our SpO_2 implementation. We needed multiple serial (UART, SPI, I^2C) interfaces, a DSP for digital filtering, and low power consumption for extended battery life. We found the Texas Instruments MSP430 MCU a perfect fit for our needs. Our current design utilizes the MSP430F2618, which is fast (16 Mhz), and has large RAM (8 kb). The built-in ADC is capable of 200 k samples per second with a 12-bit resolution and is multiplexed across eight channels.

7.3.3 *Electrocardiograph*

The electrocardiograph is a device that measures the electrical activity of the heart. The heart can be thought of as four distinct pumps, which are synchronized to beat in a specific order. Blood from the body first fills the right atrium, which is essentially a low-pressure pump that fills the right ventricle. The right ventricle is a higher pressure pump that pumps blood into the lungs. When blood returns from the lungs, it similarly fills the left atrium, which fills the left ventricle whose job is to pump blood at high pressure throughout the body.

The timing of these pumps is coordinated by means of the heart tissue itself. A region known as the sino-atrial node, known as the pacemake of the heart, initiates an electrical depolarization that causes muscle tissue nearby to start contracting. A healthy heart is able to direct the wave of electrical depolarization in such a way to cause a highly orchestrated series of contractions resulting in a synchronized pumping action.

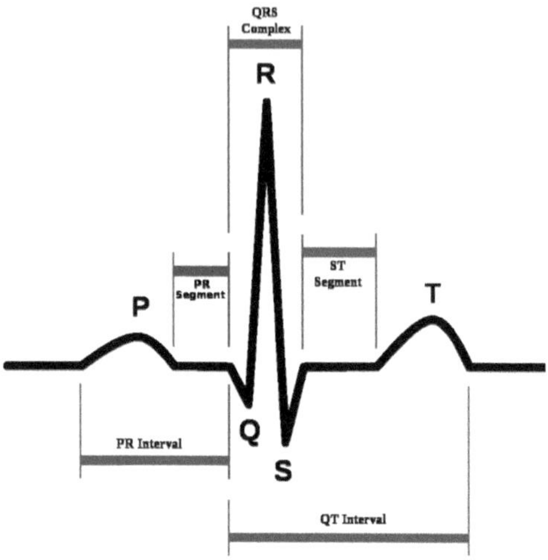

Figure 7.6. ECG waveform. See also Color Insert.

When the heart initiates a contraction, a potential difference forms between different regions of the heart and follows the wave of depolarization as it progresses though the heart. The potential difference can be treated like a dipole that moves through a volume conductor, generating potential differences on the surface of the body. This potential difference is what the ECG detects, and the placement of the electrodes provides different views of the heart's electrical activity (Fig. 7.6).

A damaged heart, as is the case after a heart attack, can result in problems with the propagation of the wave of electrical depolarization resulting in a number of chronic conditions. By utilizing an ECG to detect how the electrical activity of a heart has been altered, a cardiologist can determine how the heart has been damaged.

There are many additional uses for an ECG. The simplest use is to measure the heart rate, which is performed by measuring the main left-ventricular contraction, which appears as the strong positive deflection known the R-peak.

ECGs are described by either the number of channels or the number of electrodes. A channel refers to the number of different differential pairs, whereas the number of electrodes simply refers to the number of electrodes. The same electrode can be used as a reference for multiple differential pairs, but when people speak about 5- or 12-electrode ECGs, standard known configurations are used. At a minimum, two electrodes are required to generate an ECG trace, but 3-electrode, 5-electrode, and 12-electrode configurations can provide additional views of the heart's electrical activity to aid in monitoring and diagnosis. In this section, we will detail the implementation of a single-channel ECG, which can be extrapolated out to configurations requiring additional channels.

The range of frequencies of interest for an ECG depends on the application. Often, the distinction is made between monitoring-ECG and diagnostic-ECG. Monitoring-ECG is used for routine examination of the ECG waveform, and is more heavily filtered to remove noise artifacts with a pass-band of 0.5 Hz to 40 Hz. Diagnostic-ECG is filtered less to provide a more detailed waveform for diagnosing cardiac problems, and typically ranges from 0.05–150 Hz. You can see the relative contributions at the various frequencies in the power spectrum in Fig. 7.2. The peak at 120 Hz results from the first harmonic of the 60 Hz power-line noise; the 60 Hz peak is not noticeable due to the high signal-to-noise ratio at that frequency.

The heart is a large muscle capable of generating a strong signal on the surface of the skin, in the range of several millivolts in amplitude, which makes it a fairly easy signal to detect. For a two-electrode configuration, placing an electrode on either side of the chest will work; however, it will be useful to understand the logic behind some standard ECG configurations in determining where to place the electrodes.

The first electrode configuration proposed was by Einthoven (Fig. 7.7), and uses two channels and three electrodes. The electrodes are placed on the left arm, right arm, and the left leg, and forms a triangle with the heart in the center. The differential voltage between any two electrode locations defines a view from the lead defined by electrodes, denoted as I, II, and III. Given the symmetry of the triangle, only two differential pairs are required to be measured, and the third lead can be determined from the sum or difference of the two differential measurements [19]. For example, if both positive electrodes are placed on the right foot (V_F), and the negative lead from one channel is placed on the left arm (V_L) and the

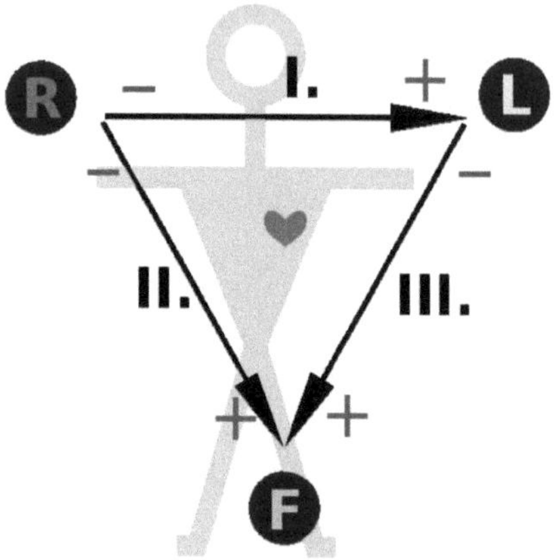

Figure 7.7. Einthoven ECG lead locations (from wikiCommons). See also Color Insert.

other is placed on the right arm (V_R), then you have the following relationships:

$$V_{\mathrm{II}} = V_F - V_R \tag{7.5}$$

$$V_{\mathrm{III}} = V_F - V_L \tag{7.6}$$

It can then be shown that the difference of equation 7.6 from 7.5 will yield V_I.

$$V_{\mathrm{II}} - V_{\mathrm{III}} = (V_F - V_R) - (V_F - V_L)$$

$$V_{\mathrm{II}} - V_{\mathrm{III}} = V_L - V_R = V_{\mathrm{I}} \tag{7.7}$$

There are a number of other electrode configurations worth considering [20–22], but it's important to note that differential surface potentials can be arithmetically combined to form different electrical views as long as more than one channel is being used.

If the electrodes are close to any other large muscles, such as the pectorals, then they can also pick up the electrical activity associated with the muscle activity. It is possible to build a device known as an electromyogram (EMG) to measure the activity of arbitrary muscles, which will be discussed in Section 7.3.4.

After deciding where the electrode, are to be placed, leads must be attached connecting the electrodes to the amplification circuitry. The leads should be kept as close to one another, and ideally be twisted together to prevent picking up electromagnetic noise differentially. The instrumentation amplifier can do a very good job of rejecting common-mode noise, which appears on both wires; however, the further the wires are from each other, the greater the probability that noise can be coupled onto just one wire, which will show up as a differential signal and amplified by the InAmp.

7.3.3.1 Berkeley Tricorder

The Berkeley Tricorder was designed to have a monitoring-ECG rather than a diagnostic-ECG as our use-case required any detected abnormalities to be escalated to a cardiologist for a workup. The design utilizes a second-order high-pass filter biased at the mid-supply voltage followed by radio-frequency interference filter and an instrumentation amplifier with a gain of 1000. A second-order high pass filter was required to minimize the DC bias drift resulting from motion artifacts while maintaining a low -3 dB corner frequency (0.79 Hz). A low-pass filter was not required as the instrumentation amplifier has a natural roll-off starting at 200 Hz at a gain of 1000. Although this frequency is higher than required, it simplified the design to use the built-in low-pass filter than to add an additional one.

The output of the instrumentation amplifier is biased to mid-rail $\left(\frac{3.3\text{ V}}{2} = 1.65\text{ V}\right)$, and the ADC has a positive/negative reference voltage of 3.3V/0V, respectively. Thus, the output of the amplifier can swing ± 1.65 V before saturating. For a gain of 1000, this translates to an input voltage of $\left(\frac{\pm 1.65\text{ V}}{1000} = \pm 1.65\text{ mV}\right)$ from the skin surface potential.

7.3.4 *Electromyogram*

The focus of this section will be on surface EMG measurements. Although transcutaneous ("needle") EMG electrodes are useful for a number of applications [23], it is not likely that a WBAN implementation will have this requirement and the safety concerns associated with placing electrodes under the skin is beyond the scope of this chapter.

A lot of the same methodologies apply to EMG as it does to ECG, after all, the ECG is a specialized case where the electrical activity of the myocardium is being measured. As with the ECG, the EMG measures surface potentials generated by muscle fibers during contraction. There are a large number of muscles that can be measured, all with different applications ranging from kinesiology studies [25] of motion to stress detection [26].

Unlike ECG, the EMG signal has higher frequency components as noted in its power spectrum (Fig. 7.8) and thus require high sampling rates. It is typically accepted that the EMG signal has a bandwidth of 500 Hz, thus minimum a sampling rate of 1 kHz is typically used due the Nyquist sampling theorem. Given the shape of the power spectrum, a bandpass filter between 5–10 Hz and 500 Hz is typical. The band-pass filter is created by placing a low-pass filter of 500 Hz in series with a high-pass filter at 5–10 Hz. However, if only the envelope of the signal is required, rather than the individual peaks, the signal can be rectified, low-pass filtered at 50 Hz, and sampled at 100 Hz [27] (Fig. 7.9). EMG surface potentials can range from 0–6 mV peak to peak, depending on the muscle being

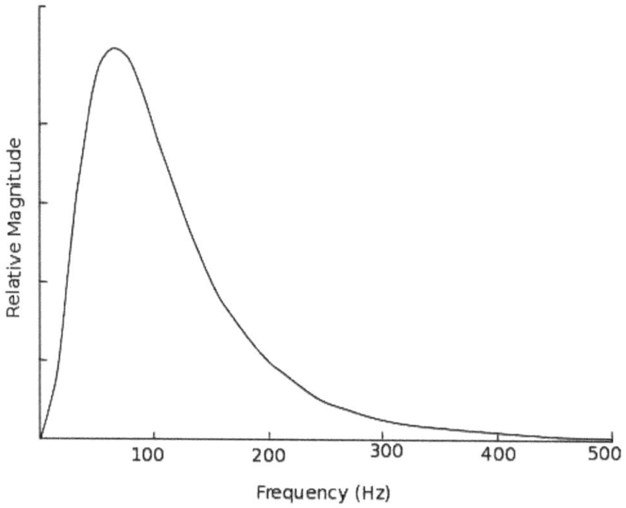

Figure 7.8. EMG power spectrum (redrawn from [24]).

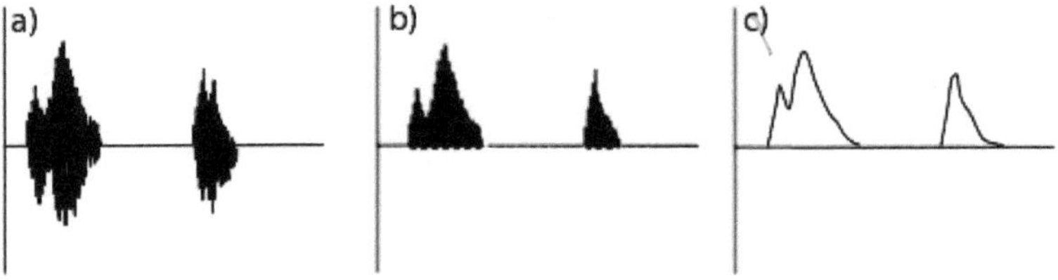

Figure 7.9. (a) Raw EMG signal, (b) rectified EMG signal, (c) low pass filtered.

investigated. Hence, the amplifier should provide sufficient gain to utilize the full dynamic range of the ADC without saturating.

For a more detailed examination of the requirements associated with EMG measurements, the reader is directed to reference [24].

7.3.4.1 The Berkeley Tricorder

Our initial interest in an EMG sensor was to measure back tension by measuring muscle activity as a proxy for stress. For this, we did not require a very fast input stage and reused the bulk of the ECG design. Given that the bandwidth of the EMG signal is in the range of 20–500 Hz, we could be more aggressive with the HPF. A single stage with a a −3 dB of 7.9 Hz was chosen and found to be effective. The amplifier we chose demonstrates a gain roll-off starting at 300 Hz, and allows us to sample the signal at a slower rate. This is sufficient to measure a general level of muscle activity. However, for a more diagnostically relevant EMG implementation, we are considering migrating to a faster amplifier and higher sample rates.

7.3.5 *Pulse Oximetry (SpO_2)*

Blood oxygenation is a critical parameter that can help diagnose conditions of pulmonary distress, hypovolemia [28], vascular perfusion, among other. Blood oxygenation has a secondary benefit in that it can be used to compute pulse transit time (PTT), or the amount of time between the ejection of blood from the heart to the pulse arrival time at a peripheral location. It has been shown that there is a direct correlation between PTT and arterial blood pressure [29].

Pulse oximetry is a method by which the variation in the absorption of light by tissue is used to determine the percentage saturation of blood hemoglobin by oxygen. The modern technique was invented by Takuo Aoyagi, and works based on the following two principles. The first involves the macromolecule responsible for oxygen transport, hemoglobin. The iron containing macromolecule found inside blood cells binds to oxygen molecules and helps transport them at much higher concentrations than possible with freely dissolved oxygen in the blood. Deoxygenated hemoglobin (Hb) absorbs red light more than oxygenated hemoglobin (HbO_2), giving HbO_2 a brighter red color. The opposite is true for infrared light, where HbO_2 absorbs infrared light more than Hb (Fig. 7.10). This means that by measuring the ratio of absorption of red versus infrared light, the percentage oxygenation of the hemoglobin can be determined in samples of blood. However, there are many other types of tissues found in the body that also absorb varying levels of red and infrared light, so the simple technique cannot be applied to blood in the body.

The second principle, which was the key to Takuo's discovery, was that it is possible to measure the variation in amplitude of light being absorbed due to the pulsatile nature of blood. As the

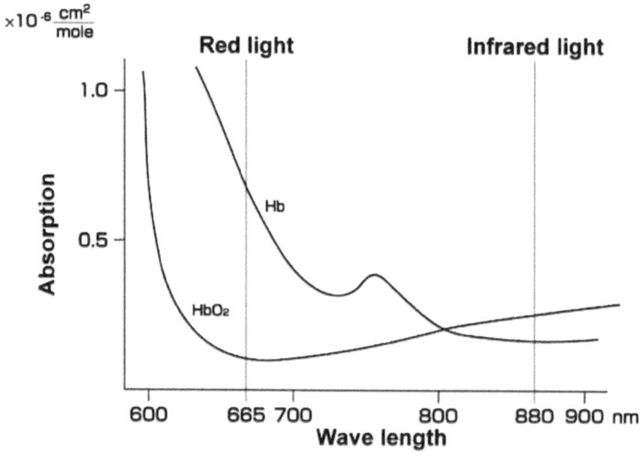

Figure 7.10. Hemoglobin light absorption curves.

heart beats, a pressure wave travels through the arteries to the capillaries, which change in volume as they fill and empty with blood. By examining the ratio of absorption for just the pulsatile portion of the signal, you are able to compare red and infrared light absorption by the portion of blood that is changing, allowing for a direct measurement of a component of blood rather than that of any other tissue components.

The implementation of this technique is simple: measure the ratio of the pulsatile portion of red to infrared light as it passes and is absorbed by tissue, multiplied by a calibration value. Both these steps are performed by shining alternating red and infrared light from closely mounted (ideally on the same die) LEDs through tissue (i.e., finger, ear) and measure intensity of transmitted light by a PIN photodiode. A PIN photodiode is sensitive light detector. The resulting waveform will resemble Fig. 7.11. Next, in software, the amplitude of the waveform is measured for both light frequencies (AC_{RED}, AC_{IR}) as well as the DC bias of the waveforms (DC_{RED}, DC_{IR}). The ratio $\frac{AC_{RED}}{AC_{IR}}$ needs to be corrected for variations in light intensity from the differed LEDs, the difference in absorption of the different frequencies of light by the tissue, as well as the varying sensitivity of the photodiode to the different frequencies of light. The ratio of DC biases $\left(\frac{DC_{IR}}{DC_{RED}}\right)$ takes into account all of these variables,

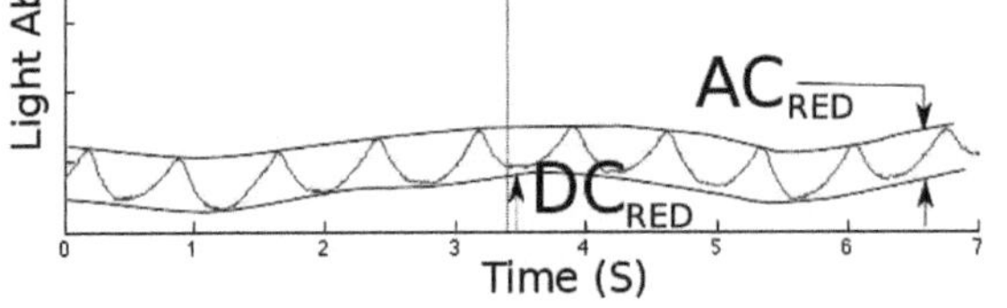

Figure 7.11. Light absorption through tissue by infrared (top curve) and red (bottom curve) light.

and the product of the $\frac{AC_{RED}}{AC_{IR}}$ and the correction factor $\frac{DC_{IR}}{DC_{RED}}$ will result in a ratio $\left(R = \frac{DC_{IR}}{DC_{RED}}\frac{AC_{RED}}{AC_{IR}}\right)$ which directly corresponds to the percentage saturation of hemoglobin by oxygen (SpO_2).

Note that most literature defines R as $\frac{\log_{10}\left(\frac{AC_{RED}}{DC_{RED}}\right)}{\log_{10}\left(\frac{AC_{IR}}{DC_{IR}}\right)}$. However, as the ratio R is an arbitrary number used in conjunction with an empirically derived mapping, there is no need to take the logarithms. This simplified explanation of the process does not explain the underlying theory; for more information, the reader is directed to reference [30].

As contemporary amplifiers are very fast compared to the rate of the heart beat, it is possible to only have the LEDs on for a very short period of time, take a measurement, then shut off the LEDs to conserve power for some period of time. An additional light intensity measurement should be taken while the LEDs are off in order to subtract out any background light from both the infrared and red light signals. The duration of the LEDs on time to the off time is referred to as the duty cycle, and the lower the duty cycle, the less power the circuit consumes, which is an important consideration for a battery-powered device.

Although a theoretical correlation between R and the percentage SpO_2 exists, sufficient deviation exists to necessitate use of an empirically derived mapping of R and percentage SpO_2. The empirical mapping is made by simultaneous measurements of R through pulse oximetry and blood oxygen though samples drawn from a subject while the subject is being deprived oxygen. Based on these measurements, a conversion factor that maps R values to SpO_2 percentages can be determined, which is valid for the device and LED/photodiode configuration. A new empirically derived mapping will have to be made if the circuit or any of the components change. Also, the mapping is only valid down to approximately 80% blood saturation as depriving oxygen to the test subjects below that point is dangerous. This is the case with all commercially available monitors. However, given that there are number of commercially available and calibrated pulse oximeters, it is sufficient to use one of these devices to derive an R-SpO_2 mapping, rather than drawing and analyzing blood from test subjects.

The description thus far has been for a transmissive pulse oximeter that works by measuring the levels of light absorbed by blood as light passes through tissue. There exists another type referred to as a reflective pulse oximeter that works by measuring the levels of light absorbed by blood as the light travels through tissue and backscatters back to the same side of the tissue where the LEDs are. The most prominent placement for such a sensor is the forehead, but other locations have also been evaluated in reference [31].

There are a number of tradeoffs that should be considered early in the design process:

- The SNR is proportional to the volume of blood in the tissue being measured, thus; tissues that suffer from poor blood flow or vasoconstriction due to cold temperatures will result in a poor signal.
- The greater the SNR, the further the LEDs and photodiode are from each other, as the light travels through and is affected by more tissue, but additional LED current is required (and LEDs that can handle higher currents) which decreases battery life.
- The faster the amplifier and ADC circuitry, the shorter the duty cycle of the LEDs, which results in lower power consumption at the expense of additional noise due to higher cutoff frequency for the low-pass filters

There are many different SpO_2 circuit implemetnations, such as this one by Texas Instruments [32]. However, the basic building blocks consist of noise-free current source to drive either a RED or IR LED, a PIN photodiode to pick up the light transmitted, a transimpedance amplifier to convert the current from the photodiode to a voltage, and a DC subtraction and amplification stage to remove the DC bias and further amplify the signal.

7.3.5.1 The Berkeley Tricorder

In order to derive blood oxygen saturation, determine pulse arrival time, and keep costs low, we needed to record the SpO_2 waveform rather than to use an OEM device, which only computes the oxygen

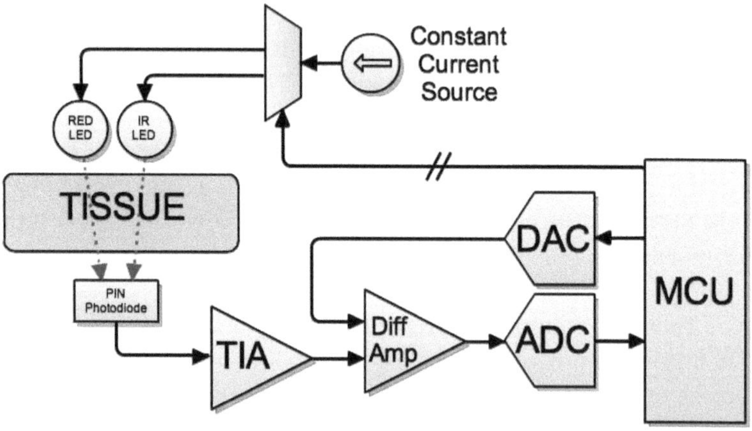

Figure 7.12. Block diagram of pulse oximeter.

saturation percentage. Being able to visually inspect the SpO_2 waveform is also important to the physician as a figure of merit for the SpO_2 calculation. Poor perfusion or poor SpO_2 placement can lead to erroneous oxygenation.

Our design (Fig. 7.12) called for a circuit capable of illuminating one of the two LEDs with a constant current and measuring the amount of light picked up from a nearby PIN photodiode. We decided to do our initial development and testing using a commercial reflective SpO_2 sensor manufactured by Nellcor (Max-Fast, Nellcor/Tyco, Pleasanton, CA). Nellcor produces a series of disposable and reusable sensors, which utilize a DB-9 connector with identical pinouts allowing for the ability to quickly test different devices and configurations.

The output stage of our device consists of a constant-current H-Bridge LED driver. The LEDs on the Nellcor sensor have their anode and cathodes connected. By driving one side high and the other low, one can choose which LED to illuminate. It is critical to drive the LEDs at a constant current to avoid coupling supply noise onto the received signal.

The input stage consists of a transimpedance amplifier (TIA) connected to a PIN photodiode in the sensor followed by a differential amplifier used to subtract out the dc-offset of the signal and add additional gain. The offset subtraction signal is generated from the MSP430's on-board DAC. A feedback loop keeps the SpO_2 signal from saturating by adjusting the DAC output. The DAC value is also stored and used for the blood oxygenation computation.

The system also measures background levels by repeating the measurements with both LEDs are turned off. These values are also used in the blood oxygenation calculation. All the data is then stored and/or transmitted for further processing to extract the relevant features.

7.3.6 *Respiration*

Monitoring respiration is important in a large number of cases. It's critical to monitor respiration in patients with cardiopulmonary issues such as congestive heart failure and patients who are on medication that suppresses breathing. It's also useful in diagnosing a number of conditions such as systemic inflammatory response syndrome (SIRS).

7.3.6.1 The Berkeley Tricorder

Our original attempt at respiration rate detection relied on acoustic pickup with a microphone, but we found bioimpedance to be a much more reliable indicator of respiration rate. Bioimpedance also allows us to measure relative breathing tidal volumes and to detect coughs. Bioimpedance can also be used for a number of other diagnostic measurements beyond respiration such as measuring cardiac output [33] and body fat [34], although we have not investigated these applications.

A high frequency sinusoidal signal (50 kHz) is used so that the effects of the skin impedance are minimized through capacitative coupling. We chose a 350 μA current at 50 kHz in order to provide us with a good signal, while being significantly lower than the maximum safe current as defined by IEC60601-1-2005. Using a pair of electrodes connected to a high impedance differential amplifier, we can accurately measure the voltage difference between the electrodes, which is directly related to the impedance between those points.

We originally planned on building a fully analog impedance measurement system; however, we were able to significantly reduce the system complexity by utilizing Analog Device's AD5933 impedance analysis IC. The AD5933 has previously been shown effective in bioimpedance measurements [35]. Our output stage consisted of a voltage-to-current circuit followed by a 3.3 nF DC blocking capacitor. The input stage utilizes the ECG electrodes, passes the signal through an RFI filter to a programmable differential amplifier which feeds the amplified signal back to the AD5933. Although the AD5933 should be very effective at filtering out any high-frequency noise, we felt it would be prudent to add an RFI filter to the input stage — given its importance in ECG measurements. The IC then computes the real and imaginary components of the signal, which we convert to a magnitude and phase. The magnitude of the impedance is proportional to the chest volume, and breathing can easily be seen as a variation of the impedance (Fig. 7.15, bottom trace). From this signal, we can determine respiration rate and approximate tidal volume.

7.3.7 *Accelerometry*

Most health monitors incorporate an accelerometer. With this, one can determine body orientation, activity levels, and perform fall detection. With additional processing, it is possible to deduce what activity the subject is engaged in [36, 37–39]. Furthermore, it is very easy to implement — hence its prevalence.

7.3.7.1 The Berkeley Tricorder

Our first implementation utilized a 3-axis analog accelerometer, which was connected to the MSP430 by means of three analog input lines. Although this approach worked well, it used too many ADC channels of the MSP430. We later switched to the LIS302DL manufactured by STmicroelectronics. Not only does it provide a digital serial interface, but it also has built-in hardware fall detection. Although initially designed to detect the fall of a portable electronic device, the IC can be used to detect if the patient falls down without taxing the microcontroller.

7.4 Wireless Interface

The number of wireless options is ever changing and evolving with no clear dominant player at this point. One must consider all the requirements based on the end application to determine the right technology. Issues to consider include transmission range, data rate, interoperability, and operating frequency. It is also insufficient to simply present a table with the comparison of the various features as many of the values will differ based on the manufacturer and operating environment.

We will discuss the requirements for the Berkeley Tricorder and the rational behind the choice of its wireless technology, followed by a discussion on some of the issues that are important to consider in making a decision for the wireless data link. Additionally, Table 7.2 at the end of this section provides a general summary for a number of wireless technologies. The complied data should not be considered concrete as there are many different implementations of the various technologies that it should only be considered a starting point.

7.4.1 *The Berkeley Tricorder*

Telemetry is an important requirement for a medical device as it provides a means for the diagnostic data to be viewed in real-time, or to be transmitted to caregiver or a more powerful system to perform computational analysis of the data. We chose Bluetooth to satisfy our telemetry needs for a number of reasons.

Table 7.2. Comparison of wireless technology standards

	ZigBee	Bluetooth	ANT	Wi-Fi	Wi-Fi
IEEE Standard	802.15.4	802.15.1	Propriatary	802.11b	802.11g
Frequency	868 Mhz/915 Mhz/ 2.4 Ghz	2.4 Ghz	2.4 Ghz	2.4 Ghz	2.4 Ghz
Max Rate Over Air	20/40/250 Kbps	1–24 Mbps	1 Mbps	11 Mbps	54 Mbps
Max Achievable Rate	10/20/125 Kbps	–	20 Kbps	6.3 Mbps	31.4 Mbps
Range (meters)	10–75 (1500 for Pro)	1–100	10	10–100	10–100
Current (mA)	1–50	1–35	1–22	100–350	100–350

- High level of penetration in consumer devices such as cell phones and laptops.
- Standardized profiles (serial, audio, object transfer, dial-up networking, etc) requiring no additional development effort.
- High-speed wireless (v1.2 Bluetooth is rated at 1 Mbit/s).
- Lower power consumption per bit than competing devices.
- Single IC implementation.
- Multiple built-in interfaces peripherals to help simplify device design such as an ADC/DAC/USB interface

Rather than using a monolithic serial to Bluetooth module, we have incorporated the CSR plug'n'go BlueCore3 chipset directly. Not only does this give us full access to many peripheral interfaces, but it also significantly reduces the device cost. The BlueCore3 IC provides audio input/output, a USB interface, 16 general-purpose I/O lines, which can be used for I^2C or SPI communication, ADC, DAC, and a serial interface. The firmware can be configured to provide many different profiles including object exchange (OBEX) to facilitate bulk data transfer, and more importantly, dial-up networking (DUN) to allow telemetry without the need for any custom phone software. Our current implementation utilizes Bluetooth as a serial device and interfaces with the MSP430 at up to 1 Mbps.

7.4.2 *Power Consumption*

Of primary importance to many WBAN projects is the power consumption of the wireless data link, which in most cases is the dominant power consumer for the device. As we stated earlier, it's almost impossible to put together a table that's capable of comparing, apples to apples, the power consumption of the various radio technologies. This is because it's impossible to separate the radio technology from an implementation of that technology. Consider a radio transceiver that generates the RF signal. There are many transceivers to choose from which all have different power consumption characteristics, even if they operate at the same frequency, have identical output power, and modulation. For example, both Texas Instruments CC2500 IC and Nordic

Semiconductor's nRF2401A IC operate at 2.4 Ghz and can support Gaussian Frequency-Shift Keying (GFSK) modulation, which is used for Bluetooth. However, the nRF2401A datasheet specifies 13 mA at 0 dBm (1 mW) output power while the CC2500 specifies 21.2 mA. Even this comparison is unfair as there are many other differences between the feature set of the components, but it demonstrate the difficulty in comparing power consumption between technologies.

Newer implementations of a protocol can vary the power consumption of a device — depending on the feature set. For example, Bluetooth version 4.0 defines Bluetooth low energy technology, which trades off bandwidth for a decrease in power consumption (50% to 99% reduction based on use case) while doubling the maximum range.

The duty cycle of the radio will have a huge impact in regards to power consumption. If your wireless technology allows for a maximum bandwidth of 1 Mbps, but your application only requires a 10 Kbps stream to be transmitted, then you should be able to shut down the radio 99% of the time to save on power consumption. This is just a very rough approximation, and it's never so so clear cut. There is additional time required for the radio to exit low power, it needs to power on at regular intervals to see if some other transmitter is sending it data, it needs to deal with the protocol overhead, and if there is a problem with the data link, then the radio will need to retransmit data.

7.4.3 *Data Range and Transmit Power*

The range of a wireless data link is dependent on the power of the transmitter, the quality and orientation of the antennas (TX & RX), the frequency of the transmitter, and the surrounding material. Let's first discuss how the power from the radio is sent to the antenna, through space, and to the receiver and the various points of power loss along the path.

The transmit power plays a very large role in the range of a wireless link, and how much of this signal makes it to the receiver will determine the range of the link. RF power is measured in dBm which decibel value in reference to 1 mW of transmit power, and is defined as

$$P = \frac{10^{(x/10)}}{1000} \tag{7.8}$$

where P is the transmit power in watts and x is the dBm. Note that 0 dBm is equal to 1 mW, and the power doubles for every 3 dBm, or is halved for every −3 dBm. The RF transceiver will generate this power, and it will be sent to an antenna; however, impedance mismatches between the transmitter, the PCB trace(s) (or coax cable(s)), and the antenna can reduce the amount of transmitted power.

The type of antenna and its radiation pattern will also play a role in how efficiently the power is transmitted. The simplest antenna is formed on the PCB by a specially designed trace pattern. These are the simplest and cheapest solutions, but not necessarily the best. The use of a chip antenna is a good compromise with generally improved performance while utilizing minimal board space at a low cost.

The radiation pattern of the antenna determines how the antenna can focus energy in a particular direction. The ability of an antenna to direct its energy is measured in dBi, or the decibels of radiated power for a particular radiation geometry as compared to an isotropic antenna, which radiates equally in all directions. Antennas exist that can focus the transmission energy giving significant gain in one orientation, but at the expense of being significantly weaker in other orientations. If the body-worn RF device is to be consistently oriented in reference to a remote receiver, then the transmit range can be extended by using a directional antenna with a high dBi. However, for most applications, the relative antenna orientations cannot be guaranteed and thus an omnidirectional antenna will be preferred to maximize transmit power independent of the subject's orientation.

Once the RF power hits the antenna and is transmitted in free space, the RF energy degrades with the square of the distance between the transmitter and receiver. Furthermore, the ability for the receiving antenna to pick up a signal also degrades with the square of the frequency. The loss attributed to these two factors determines the free-space path loss (FSPL) and is defined as the ratio of transmit power (P_t) to receive power (P_r). For an isotropic antenna,

$$P_t = 4\pi d^2 S \tag{7.9}$$

and

$$P_r = \frac{S\lambda^2}{4\pi} \tag{7.10}$$

where S is power per unit area, λ is the wavelength, and d is the distance between the transmitter and receiver. Considering the ratio of Eqs. 7.9 and 7.10

$$FSPL = \frac{P_t}{P_r} = \frac{4\pi Sd^2}{S\lambda^2/4\pi} = 16\pi^2\frac{d^2}{\lambda^2} = \left(4\pi\frac{d}{\lambda}\right)^2$$

Substituting in $\lambda = c/f$, where f is the frequency and c is the speed of light,

$$FSPL = \left(\frac{4\pi}{c}df\right)^2 \tag{7.11}$$

Expressed in dB,

$$FSPL(dB) = 10\log\left(\frac{4\pi}{c}df\right)^2 \tag{7.12}$$

For a 2.4 Ghz transmitter with a receiver 10 m away, the free space path loss will be in the order of $10\log((4\pi/3\times10^8)(10\times2.4\times10^9)) \approx 30$ dB.

However, the losses will be greatly increased if there is no clear line of sight between the transmitter and receiver. This loss is dependent on both the frequency as well as the blocking material, such as the building walls or the human body. Consider 2.4 Ghz radiation — it is used by microwave ovens to heat water in foods. Likewise, given that the human body consists mostly of water, the energy of a 2.4 Ghz RF link (about 1,000,000 weaker than a microwave) is absorbed and heavily attenuated by the human body. This is a huge problem with many of the technologies that operate at 2.4 Ghz, such as certain ZigBee and Wi-Fi implementations, and Bluetooth. If the body-worn device is on the front of the subject's chest, and the receiver unit is located behind the subject, severe signal degradation will occur.

7.4.4 *Data Rate*

The maximum data rate is determined by the operating frequency and the modulation technique of the technology in question. This (often sited) theoretic upper limit does not include the protocol overhead and assumes a high quality RF link. One should consider the maximum required data rate for the application at hand, determine the location of any bottlenecks in the system, and verify that the chosen technology can satisfy that data link given real values, not the ones found in datasheets.

For example, consider the case of two ZigBee-based sensors, each sampling with a resolution of 12 bits at 4,000 samples per second (sps) with the data being received by a PDA mounted with a ZigBee receiver. If every sample is to be sent as a 16-bit word, then each sensor will generate 64,000 bits per second of data (bps). This means that the receiver must be able to handle 128 Kbps. ZigBee is rated to handle 250 Kbps according to the spec sheet, but this doesn't include all the protocol overheads or any retransmissions due to loss. A good rule of thumb for ZigBee is that you should expect about half specified transmission rate with a good data link, or 125 Kbps. In this case, a reduction in the sampling rate or some other technique to lower the data rate would be required — or using a different technology that can handle the higher data rates.

7.4.5 *Human Safety*

As discussed in Section 7.4.3, the human body is capable of absorbing RF energy and converting it to heat. Although a radio transmitter is orders of magnitude less powerful than a microwave oven, it is important to consider how the energy can affect the body or implants such as pacemakers, and so as to minimize any potential health hazards. There are a number of different regulatory bodies that have established safety guidelines for the use of RF transmitters that should be referenced. In the United States, the FCC guidelines were derived from the recommendations of the National Council on Radiation Protection and Measurements (NCRP) and the Institute of Electrical and Electronics Engineers (IEEE). Many European countries use the guidelines established by the

International Commission on Non-Ionizing Radiation Protection (ICNIRP). The NCRP, IEEE, and ICNIRP guidelines specify different permissible energy levels based on the frequency and on whether part of the body or the whole body is exposed.

The FCC's guidelines can be found in OET bulletins 56 [3] and 65[4], and are highly recommended readings. As far as RF emissions are concerned, the FCC differentiates between whole-body exposure from partial-body exposure from a smaller device, such as a cellphone or WBAN device. The FCC limits for partial-body exposure, defined as a radiating source operating within 20 cm of the user, are based on the specific absorption rate (SAR) of a particular frequency of energy per unit mass of tissue. For the general public, 1.6 W/kg averaged over 1 g of tissue are permissible, though less restrictive limits are defined by the regulatory agencies in other countries. For example, the CENELEC-imposed limit in the United Kingdom is 2 W/kg averaged over 10 g of tissue[5].

The shielding in pacemakers and other implantable devices should prevent most interference from a WBAN device; however, contemporary implantable devices are designed for bidirectional RF communication and thus a risk interfering with the operation of the device exists. As a result of a study performed in 1997 [38], the FCC recommends that the transmitter should be at least 6 inches away from the pacemaker.

7.4.6 *Security*

For most WBAN-type applications, especially if health-related information is being gathered, then data security is a must. Depending on the country the device is to operate in, various government regulations apply, such as the Health Insurance Portability and Accountability Act (HIPAA) in the United States. Violations of these requirements can result in incarceration, so care must be taken in the implementation of the authentication of peers and encryption of data to be transmitted.

---

[3]http://www.fcc.gov/oet/info/documents/bulletins/#56
[4]http://www.fcc.gov/oet/info/documents/bulletins/#65
[5]http://webstore.iec.ch/webstore/webstore.nsf/artnum/033746

Many of the wireless technologies support some level of encryption, but as security expert Bruce Schneier often states, it's not the encryption that fails, but the implementation of the encryption. Consider the WEP encryption used by many Wi-Fi routers. The encryption used is RC4, which is still secure. However, the implementation of the encryption is faulty, and as a result a WEP password can be determined within minutes using standard tools. In the case of WEP, this is because the encryption vectors are reused often if there is sufficient traffic allowing for cryptographic attacks. This is a fundamental flaw in WEP, and similar flaws might exist in the implementation of the security of your wireless technology — or in your particular implementation of a technology.

If you are required to use a particular technology and you need to comply with regulations such as HIPPA, but the security provided by your particular technology is insufficient, there is no reason you can't implement your own additional layer of authentication or security. Bruce Schneier's book *Applied Cryptography* is a great resource in understanding the fundamentals of encryption and authentication if you plan on implementing your own. Some encryption schemes are simple to implement and can be done using spare clock cycles in a microprocessor — others might require more power computational power on a complex programmable logic device (CPLD) or digital signal processor (DSP). However, care must be taken if a multi-IC solution is used where a secret key might be transmitted from one IC to another in an unencrypted fashion.

7.5 Batteries

In choosing what type of battery to use to power a WBAN device, one must consider the battery size, capacity/power density, nominal operating voltage, internal resistance, and safety. The right battery for a WBAN implementation is the result of weighing all the options relative to the design requirements and making an informed decision.

Size is important in such that it helps determine the minimum physical dimensions of the WBAN device — though this often at the expense of capacity. The smallest battery options include coin-cells, which can be as small as 4.8 mm in diameter and 1.2 mm in height, and lithium-polymer ion (LiPo) batteries, which can be manufactured to any specified size. Coin-cell batteries are convenient that they can be mounted directly to the PCB; however, it also means that the end user needs to have access to the PCB to change the battery. And although rechargeable coin-cell batteries are available, they suffer from higher level of self-discharge and lower charge densities than other rechargeable technologies, making them less practical.

Capacity is measured in ampere hours (Ah), or milliampere hours (mAh) for smaller batteries. The runtime can be determined by examining the discharge graph for the battery chemistry you are interested in. If you are using a linear regulator, then you must maintain the battery voltage above the sum of the operating voltage of your device and the dropout voltage from the linear regulator. Consider the case of a lithium-ion battery in Fig. 7.13. If your microcontroller requires 3.3 V to operate at, and your linear regulator has a 150 mV dropout voltage, then the minimum battery voltage required is 3.45 V. The point at which the discharge curve intersects with the minimum required voltage will indicate your

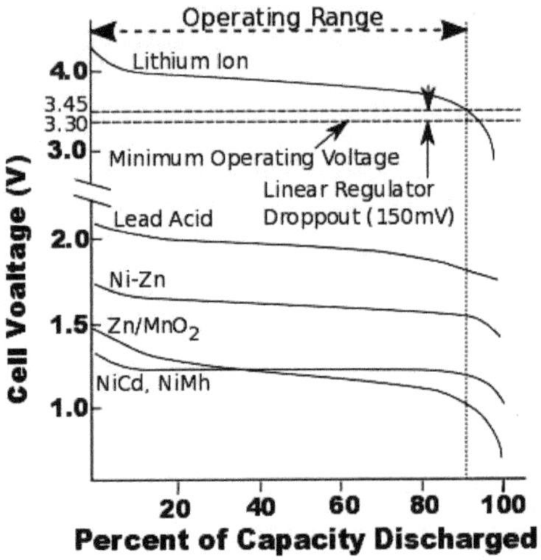

Figure 7.13. Discharge graph for various battery chemistries (adapted with permission from mpoweruk.com).

runtime. For a single-cell lithium-ion battery, you can operate down to 90% of your battery's capacity, which means 10% of the capacity is wasted. By lowering the required voltage, you can decrease this wasted capacity.

If your device uses a switching mode power supply (SMPS), you will be limited by whichever voltage is higher: the minimum safe operating voltage of the battery or the minimum operating voltage of the switching mode supply. However, an SMPS will introduce noise onto your supply lines, and possibly onto any analog signals in your device, unless great care is taken in its implementation.

The nominal voltage is the characteristic operating voltage of the battery, and if the required voltage is greater than the nominal voltage of a single cell, then putting additional cells in series will increase the voltage. For example, a LiPo battery has a nominal voltage of 3.7 V. Two batteries in series will produce a voltage of 7.4 V.

The discharge rate of batteries often normalized and presented as the discharge rate in ampere hours divided by the capacity of the battery in ampere hours and is presented as fractions of C. $1C$ represents fully discharging a battery in one hour, $2C$ in half an hour, and $0.5C$ in 2 hours. The discharge curve of a battery will be depressed as the battery ages, at higher discharge currents, and at lower operating temperatures. For example, in the case of Fig. 7.14, operating a LiPo battery at 0°C will decrease its capacity by 17% versus at 25°C.

The internal resistance of a battery limits the maximum available current that the WBAN device will consume. Although the WBAN device might be designed with a low average current requirement, various components on the device can consume high current in bursts, especially the wireless transmitter. For example, if your wireless transceiver is a GSM cell module, then you can expect bursts of power consumption in the range of 2 amps while your WBAN device averages less than 100 mA of current. If the battery has a high internal resistance, then the voltage provided by the battery can dip below the required levels, causing the digital components on your device to latch up, and/or result in erroneous data on the analog side. The use of large-capacitance low effective serial resistance (ESR) capacitors can help buffer the power usage and minimize these effects.

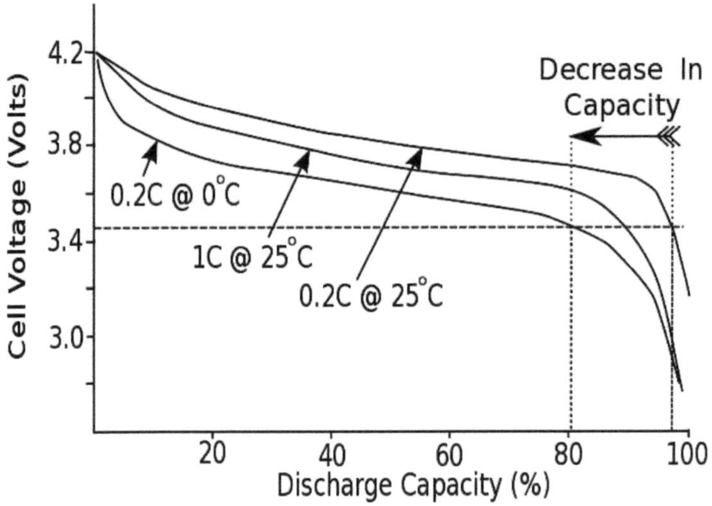

Figure 7.14. Discharge graph for lithium-ion at various temperatures and currents.

Of the various battery options, LiPo batteries have very low internal resistance and can provide many amps of current if required making them an ideal choice. However, it is important to provide a low-resistance path to the ICs that will be consuming the bursts of power or else there will be little benefit in using a battery with low internal resistance. This is often accomplished by the use of dedicated supply and ground planes on the PCB. This technique will also help minimize potential supply voltage differences in the circuit, which can result in unexpected behaviour in analog circuits. Another advantage to LiPo batteries is that they can be manufactured in to fit any required shape, though the size is directly proportional to the amount of charge the battery can store.

If the end user is expected to replace the batteries often, then the use of a standard battery should be considered. In the United States, they include sizes denoted as AAA, AA, or 9-Volt. These batteries are easily found in both rechargeable and nonrechargeable flavors. The main advantage in using these types of batteries is their availability.

In terms of safety, it is important to design the WBAN device to prevent the battery, or the cables attached to the battery, from contacting an unintentional part of the device. For example, if the

WBAN device includes ECG electrodes, and the battery accidentally contacts the electrode leads, then sufficient current can flow through the heart to cause fibrillation and kill the user. Furthermore, some battery chemistries, such as LiPo, are flammable/explosive if the enclosure is punctured and can result in sever burns. Thus, it is important to provide an adequate enclosure for the battery to guarantee that the user or some other component cannot damage the battery.

7.6 Final Thoughts and the Berkeley Tricorder

The Berkeley Tricorder, currently in its 7th version, has been used in a number of studies. These include two to evaluate its ability to record ambulatory data, one on evaluating activities of daily living, and to evaluate sports training. A sample of the data acquired from the ambulatory study can be found in Fig. 7.15. The EMG signal was derived from electrodes placed over the latissimus dorsi muscle on the back. A reflective SpO_2 sensor manufactured by Nellcor (Max-Fast, Nellcor/Tyco, Pleasanton, CA) was placed on the forehead as it minimized motion artifacts. The system performed well with

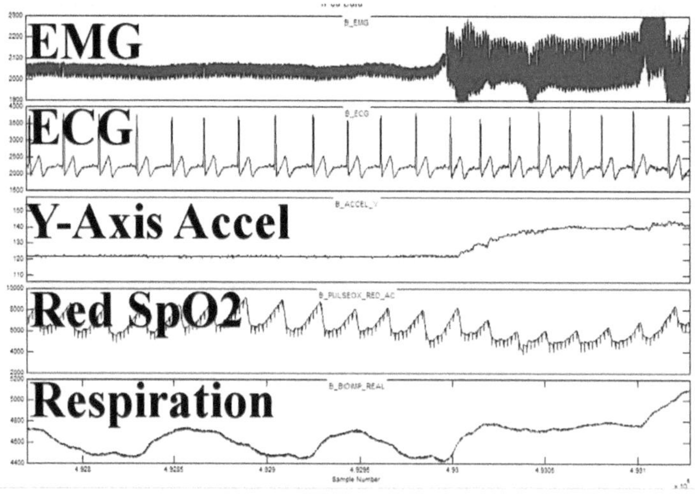

Figure 7.15. Sample data from the Berkeley Tricorder. See also Color Insert.

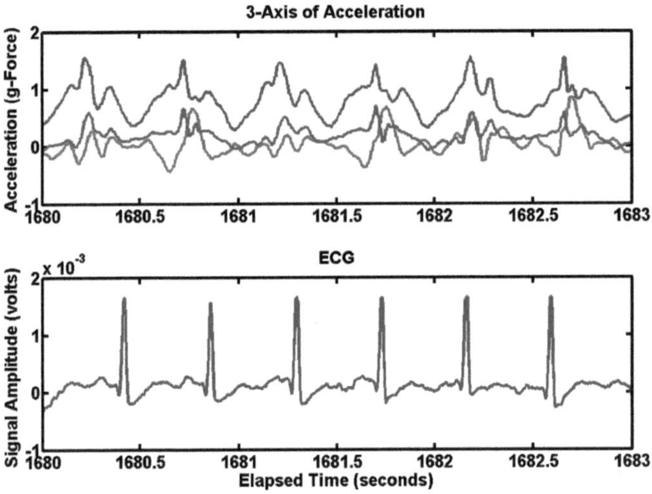

Figure 7.16. ECG signal (lower plot) captured while subject in motion (upper plot). See also Color Insert.

minimal motion artifacts. Fig. 7.16 shows the ECG signal while the subject was walking.

Many lessons have been learned through the process of designing the Berkely Tricorder, and this chaper has tried to convey some of those learned lessons and expereiences. There is no cookbook methodology of desinging a WBAN device; rather, the need to understand the problem and formulate a solution while understanding the various trade-offs. This same theme has been reitterated in many sections of this chapter and is perhaps the single most important lesson this author has learned.

References

1. S. Zenios, J. Makower, P. Yock, T. J. Brinton, U. N. Kumar, L. Denend, and T. M. Krummel (January 2009) *Biodesign: The Process of Innovating Medical Technologies*, New York, Cambridge University Press.
2. Y. Chi, S. Deiss, and G. Cauwenberghs (2009) Non-contact low power EEG/ECG electrode for high density wearable biopotential sensor networks, in *Proceedings of the 2009 Sixth International Workshop on Wearable and Implantable Body Sensor Networks*, **00**, pp. 246–250.

3. Y. Mendelson, R. J. Duckworth, and G. Comtois (July 2006) A wearable reflectance pulse oximeter for remote physiological monitoring in *Proceedings of the International Conference of IEEE Engineering in Medicine and Biology Society 2006*, **1**, pp. 912–915.
4. T. H. Teo, G. K. Lim, D. S. David, K. H. Tan, P. K. Gopalakrishnan, and R. Singh (2007) Ultralow power sensor node for wireless health monitoring system in *Proceedings of ISCAS 2007*, pp. 2363–2366.
5. L. Wang, J. Y. Zhang, J. Huang, L. K. Chen, L. Yu, Y. T. Zhang, and G. Z. Yang (April 2009) A field programmable analogue ASIC towards low power processing-on-node BSNs, in *Proceedings of the Sixth International Workshop on Wearable and Implantable Body Sensor Networks*, pp. 1–4 doi: 10.1109/P3644.54.
6. E. Teaw, G. Hou, M. Gouzman, K. W. Tang, A. Kesluk, M. Kane, and J. Farrell (August 2005) A wireless health monitoring system, in *Proceedings of the 2005 IEEE International Conference on Information Acquisition*, pp. 247–252.
7. C. Otto, A. Milenkovic, C. Sanders, and E. Jovanov (January 2006) System architecture of a wireless body area sensor network for ubiquitous health monitoring, *Journal of Mobile Multimedia* **1(4)**, pp. 307–326.
8. E. Jovanov, A. Milenkovic, C. Otto, and P. C. de Groen (March 2005) A wireless body area network of intelligent motion sensors for computer assisted physical rehabilitation, pp. 1–10.
9. K. Montgomery, C. Mundt, G. Thonier, A. Tellier, U. U. Barker, R. Ricks, L. Giovangrandi, P. Davies, Y. Cagle, J. Swain, J. Hines, and G. Kovacs (April 2004) Lifeguard — A personal physiological monitor for extreme environments, pp. 1–4.
10. D. Malan, T. Fulford-Jones, M. Welsh, and S. Moulton (2004) CodeBlue: An ad hoc sensor network infrastructure for emergency medical care, in *International Workshop on Wearable and Implantable Body Sensor Networks*, **5** pp.
11. T. Gao, D. Greenspan, and M. Welsh (December 2005) Improving patient monitoring and tracking in emergency response, pp. 1–6.
12. B. P. Lo, S. Thiemjarus, R. King, and G. -Z. Yang (May 2009) Body sensor network — A wireless sensor platform for pervasive healthcare monitoring, pp. 77–80.
13. R. Naima and J. Canny (April 2009) The Berkeley tricorder: Ambulatory health monitoring, *2009 Sixth International Workshop on Wearable and Implantable Body Sensor Networks*, pp. 53–58 doi: 10.1109/P3644.57.

14. R. WEBSTER, Medical instrumentation sm 3e, *lavoisier.fr* (January 1997) URL http:// www.lavoisier.fr/notice/frLWO6A2LARRWLKO.html.
15. M. Oehler, V. Ling, K. Melhorn, and M. Schilling (2008) A multichannel portable ECG system with capacitive sensors, *Physiological Measurement*, **29**, pp. 783–793.
16. C. Kitchin, L. Counts, and M. Gerstenhaber Reducing RFI rectification errors in in-amp circuits, Analog Devices Inc, Application Note (AN-671).
17. B. Carter and T. Brown (2001) *Handbook of Operational Amplifier Applications*, Texas Instruments Application Report, *SBOA092A*.
18. C. Kitchin and L. Counts (January 2000) A designer's guide to instrumentation amplifiers, *school.mech.uwa.edu.au* URL http://school.mech.uwa.edu.au/mechatronics/El-resources/Kitchin-Op-Amplifiers%2520Guide.pdf.
19. F. Wilson, A. Macleod, and P. Barker (December 1931) The potential variations produced by the heart beat at the apices of Einthoven's triangle, *American Heart Journal*, **7(2)**, pp. 207–211, URL http://linkinghub.elsevier.com/retrieve/pii/S0002870331904110.
20. E. Goldberger (1942) A simple, indifferent, electrocardiographic electrode of zero potential and a technique of obtaining augmented, unipolar, extremity leads, *American Heart Journal*, **23(4)**, pp. 483–492.
21. H. Burger and J. V. Milaan (1948) Heart-vector and leads, *British Heart Journal*, **10(4)**, pp. 229.
22. J. Malmivuo, R. Plonsey, and J. Cameron (January 1995) Bioelectromagnetism: Principles and applications of bioelectric and biomagnetic fields, *Informa Pharma Science* URL http://www.informapharmascience.com/doi/abs/10.3109/03091909609009004.
23. G. Rau, E. Schulte, and C. Disselhorst-Klug (2004) From cell to movement: To what answers does EMG really contribute? *Journal of Electromyography and Kinesiology*, **14(5)**, pp. 611–617.
24. B. Gerdle, S. Karlsson, S. Day, and M. Djupsjobacka (1999) Acquisition, processing and analysis of the surface electromyogram, in *Modern Techniques in Neuroscience* (ed. U. Windhorst and H. Johansson), Berlin, Springer Verlag, pp. 705–755.
25. J. Ahlgren (1967) Kinesiology of the mandible an EMG study, *Acta Odontologica*, **25(6)**, pp. 593–612.
26. U. Lundberg, R. Kadefors, B. Melin, G. Palmerud, P. Hassmen, M. Engstrom, and I. E. Dohns (1994) Psychophysiological stress and EMG

activity of the trapezius muscle, *International Journal of Behavioral Medicine*, **1(4)**, pp. 354–370.

27. R. Merletti (1999) Standards for reporting EMG data, *Journal of Electromyography and Kinesiology*, **9(1)**, pp. 3–4.
28. M. Shamir and C. Weissman (January 2003) Plethysmographic waveform variation as an indicator to hypovolemia, *Anesthesia and Analgesia* URL http://www.anesthesia-analgesia.org/content/97/2/602.2.full.
29. L. Geddes, M. Voelz, and C. Babbs (January 2007) Pulse transit time as an indicator of arterial blood pressure, *Psychophysiology* DOI: 10.1111/j.1469-8986.1981.tb01545.x, URL http://www3.interscience.wiley.com/journal/119574157/abstract.
30. T. Rusch, R. Sankar, and J. Scharf (1996) Signal processing methods for pulse oximetry, *Computers in Biology and Medicine*, **26(2)**, pp. 143–159.
31. A. Tobola and C. Douniama (June 2007) Evaluation of alternative derivation areas for plethysmography and pulse oximetry, in *Proceedings of SENSATION Second International Conference, Monitoring Sleep and Sleepiness with New Sensors Within Medical and Industrial Applications* pp. 1–2.
32. V. Markandey (June 2009) Pulse oximeter implementation on the tms320vc5505 DSP medical development kit (MDK), Texas Instruments Application Report SPRAB37 pp. 1–28 URL http://focus.ti.com/lit/an/sprab37/sprab37.pdf.
33. G. Cotter, Y. Moshkovitz, E. Kaluski, A. Cohen, and H. Miller (January 2004) Accurate, noninvasive continuous monitoring of cardiac output by whole-body electrical bioimpedance, *Chest* URL http://www.ncbi.nlm.nih.gov/pubmed/15078756.
34. R. Sung, P. Lau, C. Yu, and P. Lam (January 2001) Measurement of body fat using leg to leg bioimpedance, *Archives of Disease in Childhood* URL http://www.fetalneonatal.com/content/85/3/263.full.
35. F. Seoane, J. Ferreira, and J. Sanch'ez (January 2008) An analog front-end enables electrical impedance spectroscopy system on-chip for biomedical applications, *Physiological Measurement* URL http://iopscience.iop.org/0967-3334/29/6/S23/pdf/0967-3334_29_6_S23.pdf.
36. U. Maurer, A. Smailagic, D. Siewiorek, and M. Deisher (2006) Activity recognition and monitoring using multiple sensors on different body positions, in *Proceedings of the International Workshop on Wearable and Implantable Body Sensor Networks (BSN'06)* URL http://www.ece.cmu.edu/~bfrench/ewatch_files/papers/ewatch%2520papers/01612909.pdf.

37. S. Patterson, D. Krantz, L. Montgomery, P. Deuster, S. Hedges, and L. Nebel (2007) Automated physical activity monitoring: validation and comparison with physiological and self-report measures, *Psychophysiology*, **30(3)**, pp. 296–305.

38. D. Hayes, P. Wang, and D. Reynolds (January 1997) Interference with cardiac pacemakers by cellular telephones, *The New England Journal of Medicine* URL http://content.nejm.org/cgi/content/abstract/336/21/1473.

39. M. J. Mathie, B. G. Celler, N. H. Lovell, and A. C. F. Coster (2004) Classification of basic daily movements using a triaxial accelerometer, *Medical and Biological Engineering and Computing*, 42 (5), 679–687, DOI: 10.1007/BF0234755,URL: http://www.springerlink.com/content/0140-0118/42/5/.

Chapter 8

Ambulatory Recording of Biopotential Signals: Constraints and Challenges for Analog Design

Refet Firat Yazicioglu,[a] Sunyoung Kim,[a] Tom Torfs,[a] Julien Penders,[b] Buxi Singh Dilpreet,[b] Inaki Romero,[b] and Chris Van Hoof[a, b]

[a] *IMEC, Kapeldreef 75, Leuven, 3001, Belgium*
[b] *IMEC-nl/Holst Centre, High Tech Campus 48, 5656 AE Eindhoven, The Netherlands*
firat@imec.be

The monitoring of biopotential signals such as EEG, ECG, and EMG is a common procedure in modern clinical practice. The instruments that can monitor these signals are traditionally realized by combining precision building blocks with powerful DSP modules. The growing interest toward the improvement of patients' quality of life and the use of biopotential signals in nonmedical applications such as entertainment, sports, and brain-computer interfaces requires the implementation of miniaturized and wireless biopotential acquisition systems with ultra-low power dissipation. This has dramatically changed the way of developing instruments for the extraction of biopotential signals, placing stringent constraints on the design of analog front-end circuits that can be used in

Wireless Body Area Networks: Technology, Implementation, and Applications
Edited by Mehmet R. Yuce and Jamil Y. Khan

ISBN 978-981-4316-71-2 (Hardcover), ISBN 978-981-4241-57-1 (eBook)
www.panstanford.com

ambulatory biopotential monitoring applications. In addition, the ambulatory monitoring of patients has introduced new challenges that can jeopardize the signal integrity.

This chapter will focus on the design of analog-integrated circuits that can be used in ambulatory biopotential monitoring applications. Several constraints in terms of power dissipation and signal-to-noise ratio will be addressed. Later, an important challenge in ambulatory biomedical signal monitoring, namely motion artifacts, will be introduced and different strategies to tackle this problem will be explained.

8.1 Introduction: The Need for Portable Medical Electronics Systems

The need for portable medical electronics stems from the fact that the current health care routine is very much centralized by heavily focusing on the hospitalization of patients. This requires that people adapt to this data-centric approach, which is indeed very efficient for curative care delivery but, on the other hand, very expensive. Keeping this in mind, demographical changes indicate that the continuously aging population is leading to significant rise in chronic diseases, resulting in ever-increasing health care costs as shown in Fig. 8.1 [1]. Such increase in costs not only creates a significant burden on the economical system, but also reduces the efficiency of health care delivery.

Therefore, there is a strong interest to introduce the parts of health care cycle into the daily routine of people. It is expected that this will significantly reduce hospitalization costs and also help in increasing the efficiency of health care delivery. The initial interest is to introduce the practice of signal monitoring and analysis into people's daily routine through ambulatory and continuous monitoring systems with miniature size and wearable form factor. This can significantly reduce the hospitalization requirements of people and assist health care professionals in data analysis, significantly reducing the time and the cost of long-term patient monitoring. In the long run, such monitoring systems will evolve to

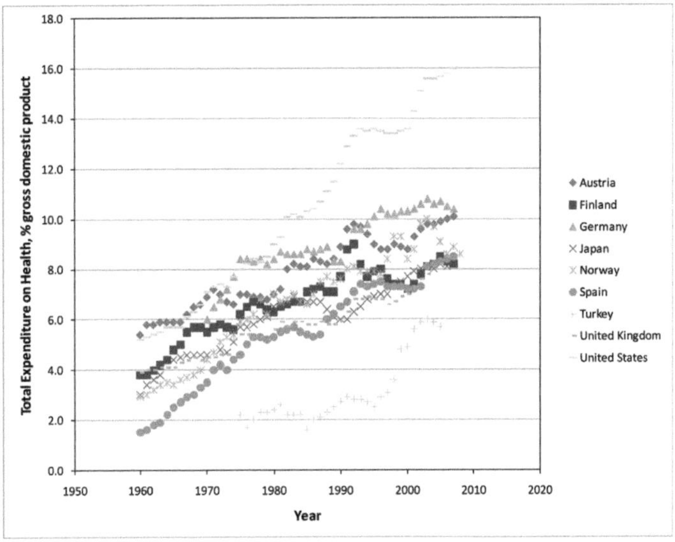

Figure 8.1. Change of total expenditure on health (percentage of gross domestic product) over years. See also Color Insert.

smarter systems capable of performing reliable diagnostics based on acquired medical signals, and even perform partial treatment and/or suppression of the disorder through drug delivery and/or electrical/visual stimulation systems.

The key for the realization of such a vision is a technology platform that can enable the collection of data from patients and communicate these to medical professionals in a reliable manner. The e-Health project, "The use, in the health sector, of digital data — transmitted, stored, and retrieved electronically — in support of health care, both at the local site and at a distance." as defined by WHO, can be the platform for the support and the diagnostics of patients. E-health is claiming to offer the potential to reduce medical cost, enable personalized health care, deliver remote health services, and increase the delivery efficiency in real-time. Therefore, the gathering of fast, reliable, and continuous medical information from patients lies in the center of the e-Health project further addressing the necessity for the realization of miniature and smart systems for medical signal monitoring.

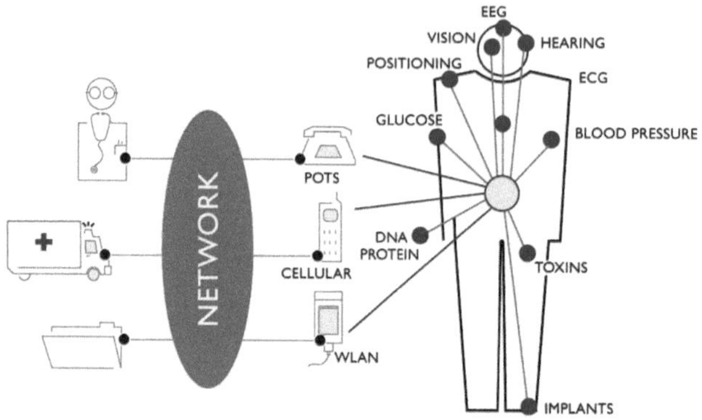

Figure 8.2. Technology vision for future medical monitoring and diagnostics systems [2]. People will be carrying their personalized sensor nodes, collecting medical information from the patient and their surroundings, analyze it, and communicate the results to medical professionals with wireless communication.

This requirement is being addressed by body area networks (BAN) [2], which consist of smart and miniaturized sensor nodes collecting information from patients and its environment, processing this information, and wirelessly communicating the results to medical professionals. The main challenge behind the realization of these sensor nodes is the fact that the available power budget for performing these functionalities is strictly limited due to the small form factor of these sensor nodes. Hence, research focuses on the realization of power-efficient implementation for each and every building block of this smart system.

Therefore, this chapter will focus on the analog front-end part of these miniature sensor nodes and how to implement low-power analog front-end circuits, which not only extract biomedical signals with high signal quality, but also provide a method to monitor signal integrity problems in ambulatory monitoring systems.

The outline of the chapter is as follows: Section 8.2 will present the basics of biopotential signal acquisition. Section 8.3 will address constrains and challenges in analog circuit design for the monitoring of biopotential signals in portable/ambulatory biopotential recording systems. Section 8.4 will describe how to design power efficient instrumentation amplifiers extracting biopotential signals in a power-efficient manner. Last but not the least, Section 8.5 will describe signal integrity problems in ambulatory measurement systems.

8.2 Basics of Biopotential Signal Acquisition

Figure 8.3 shows the characteristics of biopotential signals as presented in reference [3]. These signals can be grouped in two categories. The signals in the first group are electrocardiogram (ECG) and electromyogram (EMG) signals, which are due to muscular activity, i.e., activity due to the cardiac muscles and the skeletal muscles, respectively. On the other hand, the signals in the second group are due to neural activity within the brain, and the naming convention changes according to the measurement invasiveness and the focus of the measurement. The action potential (AP) measurements are invasive measurements and refer to the signals' single neurons. Similarly, local field potential (LFP) measurements are also invasive, but the signal of interest is the average activity of group of neurons. It should be noted that since these two measurements are invasive, large biopotential signals can be picked up. On the other hand, as the measurements move toward noninvasive methods such as electrocorticogram (ECoG),

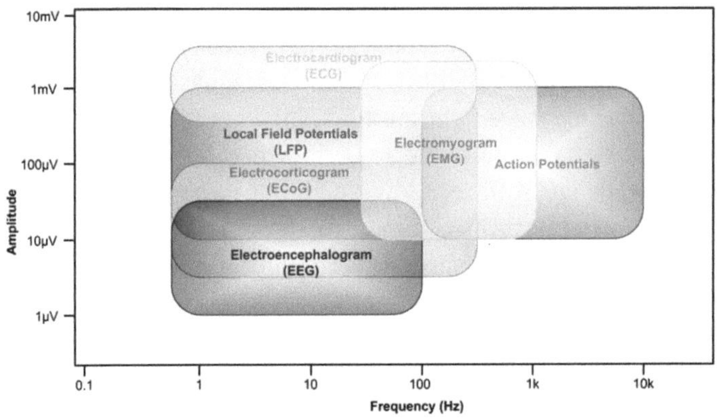

Figure 8.3. Frequency and amplitude characteristics of biopotential signals — most commonly monitored signals in medical practice [3]. See also Color Insert.

where the signals are measured on the surface of the cortex, and electroencephalogram (EEG), where the signals are measured above the skull, the signal amplitudes decrease significantly, requiring higher performance analog building blocks for signal acquisition.

The acquisition of biopotential signals requires two main elements: 1) biopotential electrodes and 2) an instrumentation amplifier. The purpose of the electrodes is to convert the ionic current within the body into electronic current so that the instrumentation amplifier can process this signal for amplification [3]. The measurement of biopotential signals can be grouped into three categories according to the electrode configurations as:

- Type I: Bipolar measurement with third electrode biasing
- Type II: Bipolar measurement with lead biasing
- Type III: Unipolar measurement

Type I measurement configuration is intended for precision measurements, where the common-mode signals need to be rejected from the measurement. However, this measurement needs the use of a third electrode for setting the DC potential of the patient. On the other hand, Type II enables the bipolar measurement of signals without using the third biasing electrode. This is generally the preferred measurement method for applications requiring the use of minimum number of electrodes, where the common-mode signals need to be rejected. The third, and the last measurement method, is the unipolar measurement method. This method is generally used, where the signal amplitudes are sufficiently large, or the common-mode signals are sufficient low, such that it is not required, by the application, to reject the common-mode signals.

Similar to the measurement configuration, the design of an instrumentation amplifier also depends on the application. For applications where the signal levels are very small, together with Type I measurement a high-performance instrumentation amplifier is required. On the other hand, an instrumentation amplifier with lower performance specifications may be sufficient for measurements with larger signals enabling the implementation of a lower power instrumentation amplifier.

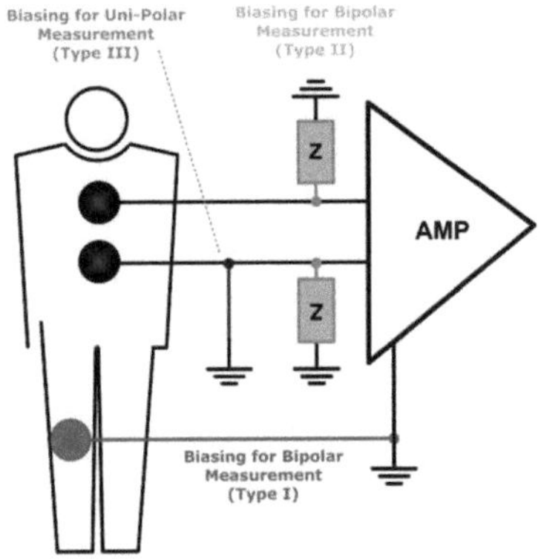

Figure 8.4. Different measurement and biasing methods for biopotential signal acquisition. See also Color Insert.

8.3 Constrains and Challanges

The previous section has given a brief introduction to the measurement of biopotential signals. Of course, such an ideal model indicates that the implementation of an analog front-end can be a straightforward process and that simple amplifier architectures can be used to achieve very low-power analog front-end circuits. However, this is unfortunately not the case for biopotential measurements. There are significant aggressors and error sources that create challenges and constrains for instrumentation amplifier design. This section will summarize these constrains and challenges, which will enable the analog designer to select the best amplifier topology to be used in different applications of biopotential measurements.

The nonidealities in biopotential measurements can be due to:

i. Electrostatic Interference
ii. Biopotential Electrodes
iii. Instrumentation Amplifier

These three main groups of nonidealities are shown in Fig. 8.5 for a Type I measurement. The first nonideality comes from the interference to human body appearing due to the electrostatic coupling to surrounding mains, which results in a displacement current flowing through the human body [4] leading to the

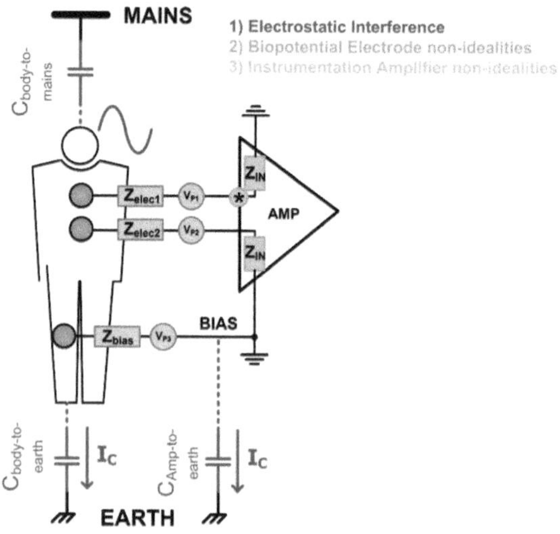

Figure 8.5. Measurement of biopotential signals (Type I measurement) introducing the nonidealities from different sources. See also Color Insert.

generation of an AC potential over the body. This potential appears as a common-mode signal to the instrumentation amplifier and requires sufficiently large common-mode rejection ratio (CMRR) so that we can reject the large common-mode signals and extract the weak biopotential signals.

The second nonideality is due to the electrodes. An electrode sitting on a skin, or on a tissue, results in complex impedance and a polarization voltage between the signal source and the input of the instrumentation amplifier [3]. The latter leads to a net DC input to the instrumentation amplifier, which can be rejected by introducing AC coupling prior to amplification. On the other hand, the prior, combined with the finite input impedance of the instrumentation amplifier, leads to the conversion of the common-mode signals into differential mode according to:

$$\frac{V_{\mathrm{DM}}}{V_{\mathrm{CM}}} = \frac{\Delta Z_{\mathrm{ELEC}}}{Z_{\mathrm{IN}}} \tag{8.1}$$

where V_{DM} is the differential mode signal at the input of the amplifier, V_{CM} is the common-mode signal on the human body, ΔZ_{ELEC} is the mismatch of the impedances between two recording electrodes, and Z_{IN} is the input impedance of the instrumentation amplifier. Therefore, it can be seen that even if we can implement an instrumentation amplifier with infinitely large CMRR, the finite input impedance and the mismatch of electrode impedance lead to the conversion of the common-mode signal into differential mode. This means that the affective CMRR is significantly reduced by the finite input impedance of the amplifier.

Last but not the least, the nonidealities from the instrumentation amplifier can affect the accuracy of the biopotential recordings. Two major nonidealities with significant impact on the quality of recordings are the noise and the finite input impedance of the instrumentation amplifier. The latter, unless sufficiently large, may load the electrodes, as well as, leading to the conversion of common-mode signals into differential mode. The prior, on the other hand, significantly reduces the signal-to-noise ratio of biopotential recordings due to the fact that biopotential signals have very low frequency behavior [5].

As a result of this discussion, we can divide the main performance specifications of instrumentation amplifiers into four groups:

i. DC Headroom
ii. Common-Mode Rejection Ratio
iii. Input Impedance
iv. Noise Performance

It should be noted that the quantitative specifications for these requirements will be defined by the application, and the choice of the instrumentation amplifier should be able to meet these specifications. A very good example on how application defines the instrumentation amplifier specifications can be seen from different brain activity measurement methods as shown in Fig. 8.6.

During EEG measurements, the electrodes are placed above the skull. This leads to biopotential signals with very weak amplitudes, which make them very sensitive to various aggressors. On the other hand, due to the less strict material choice, the electrodes are more stable leading to more relaxed DC headroom requirements. This means that instrumentation amplifiers addressing EEG applications should realize very high performance, i.e., high CMRR, very high input impedance, and very low noise, whereas, the DC headroom specifications are not very strict.

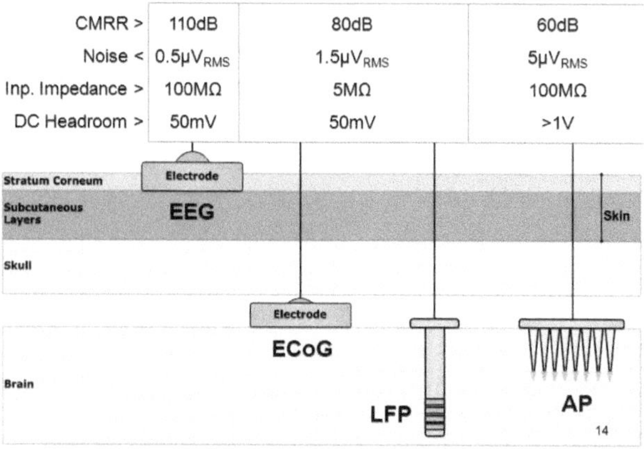

Figure 8.6. Different measurement methods of brain activity showing how the application defines the instrumentation amplifier specifications.

On the other hand, if we have a look at more invasive measurements, for instance ECoG measurements, the signal amplitudes are larger, relaxing noise, CMRR, and input impedance specifications. This means that circuits targeting these applications may be more aggressive on low-power consumption rather than the noninvasive measurements of biopotential signals.

Finally, if we have a look at more invasive measurements such as AP measurements, the signal amplitudes even further increase, which means that the noise and CMRR requirements can be more relaxed. On the other hand, the size of the electrodes is significantly reduced, which leads to a very large electrode impedance. In addition to that, very strict biocompatibility requirements limit the choice of electrode material. Hence, even larger polarization voltages can be seen while recording biopotential signals from such electrodes. Hence, instrumentation amplifiers implementing very large input impedance and incorporating large DC headroom are required.

Therefore, the conclusion is that the design of instrumentation amplifiers for different biopotential recording applications targets different specifications. The next section will review these different instrumentation amplifier architectures and why they suit very well for specific applications.

8.4 Design of Instrumentation Amplifiers for Biopotential Recordings

The most common instrumentation amplifiers that are being used for biopotential recordings can be grouped into two main categories:

i. Uncompensated Instrumentation Amplifiers
ii. Compensated Instrumentation Amplifiers

The name compensation represents the use of circuit techniques that reduce the effects of flicker noise in CMOS transistors and mismatch related errors. Therefore, the amplifiers utilizing compensation are being used in applications requiring very high performance instrumentation amplifiers. On the other hand, the uncompensated amplifiers focus on aggressively improving the power efficiency of the instrumentation amplifiers, which is one of the most important criterions in implantable applications. Therefore, this section will describe the most commonly used instrumentation amplifiers for the extraction of biopotential signals. Their key advantages and disadvantages will be reviewed.

8.4.1 *Uncompensated Instrumentation Amplifiers*

There are two main types of instrumentation amplifiers that do not use any compensation for the amplifier nonidealities. These instrumentation amplifiers differ in the sense that the first type uses the ratio of resistors to define the gain of the instrumentation amplifier, or namely resistive instrumentation amplifiers, whereas the second type uses the ratio of capacitors, or so called capacitive instrumentation amplifiers.

The typical architecture of the resistive instrumentation amplifiers is shown in Fig. 8.7 [6]. The instrumentation amplifier consists of an input transconductance stage and an output transimpedance stage. The input stage converts the differential input voltage into a current over the resistor R_1. This current is converted into a voltage at the transimpedance stage. Hence, the voltage gain of the architecture is defined by:

$$A_V = \frac{R_2}{R_1} \tag{8.2}$$

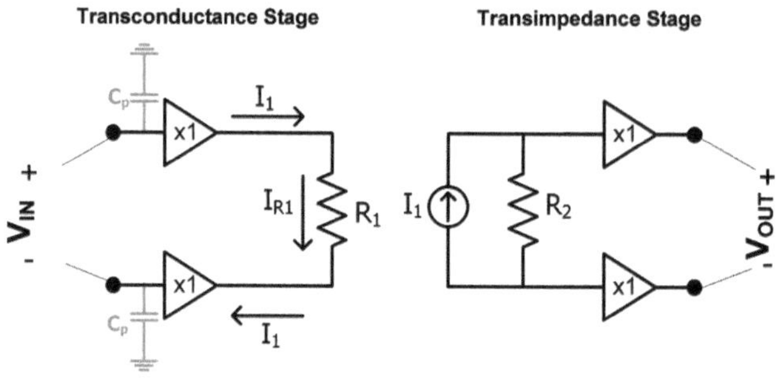

Figure 8.7. Simplified architecture of the resistive instrumentation amplifier topology [6] that does not make use of any compensation technique for improving CMRR and reducing noise.

If we analyze the important characteristics of this instrumentation amplifier topology, the critical performance specifications can be summarized as:

i. **DC Headroom:** The lack of AC coupling significantly reduces the DC headroom of this instrumentation amplifier. Additional circuitry is required for the introduction of AC coupling.
ii. **CMRR:** It is defined by components mismatches. Especially, the gain mismatch of the input buffers will lead to the conversion of the common-mode voltage into differential current over the input resistor R_1. As a result, the CMRR of this amplifier topology will be in the moderate-low range.
iii. **Input Impedance:** One of the key advantages of this architecture is its input impedance, which is solely defined by the parasitic capacitances at the inputs of the transconductance stage. This simply means that the input impedance of this architecture is the maximum possible value for a given instrumentation amplifier topology.
iv. **Noise Performance:** The input-referred noise of this instrumentation amplifier architecture is defined by:

$$\overline{v_{\text{IA}}^2} = 2 \times \overline{v_{\text{BUF}}^2} + \overline{v_{R_1}^2} + \frac{\overline{v_{R_2}^2}}{A_{\text{V}}^2} \tag{8.3}$$

This shows that the noise of the instrumentation amplifier is defined by the noise of the input buffers together with the resistors defining the transconductance of the input stage. It

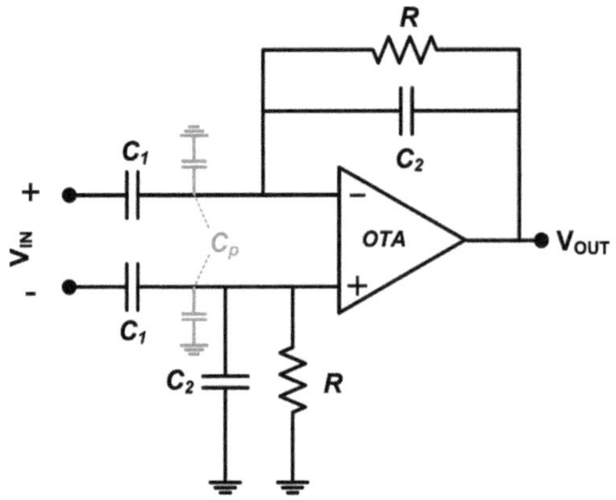

Figure 8.8. Simplified architecture of the capacitive instrumentation amplifier topology [8].

should be noted that the noise of the buffers will be dominated by flicker noise, significantly reducing the signal-to-noise ratio of the measurements at low frequencies, i.e., the range of biopotential signals.

As an alternative to the resistive instrumentation amplifier, the architecture of the capacitive instrumentation amplifier is shown in Fig. 8.8 [7, 8]. It is similar to a differential resistive feedback amplifier, but resistors are replaced with capacitors, hence the gain is defined as:

$$A_V = \frac{C_1}{C_2} \tag{8.4}$$

In such a configuration, the DC voltages at high impedance nodes are defined by the resistors. Note that these resistors are implemented using transistors operating in weak inversion, implementing very large values on-chip by using a very small area [8]. The critical performance specifications can be summarized as:

i. **DC Headroom:** This architecture is inherently AC coupled. Supply level DC voltages can be rejected by this instrumentation amplifier.
ii. **CMRR:** It is defined by the components mismatches. Hence, the expected CMRR is also in the moderate-low range.

iii. **Input Impedance:** The input impedance is defined by the input capacitances C_1 and can be given as:

$$Z_{IN} = \frac{1}{jwC_1} \quad (8.5)$$

Hence, the input impedance of this architecture is also very large considering the fact that biopotential signals have very low frequency behavior.

iv. **Noise Performance:** The total input referred noise of the instrumentation amplifier is mainly dominated by the noise of the core operational transconductance amplifier (OTA) as represented in Eq. 8.6. The only nonideal effect that elevates the instrumentation amplifier noise above the core OTA noise level is the presence of the parasitic capacitance C_p. The value of this parasitic capacitance appears at the numerator of the noise equation. Hence, the capacitors C_1 and C_2 should be sized properly to minimize the effect of C_p. In addition, the core OTA also suffers from flicker noise. This significantly increases the noise of the OTA for low-frequency applications.

$$\overline{v_{IA}^2} = \left(\frac{C_1 + C_2 + C_p}{c_1}\right)^2 \times \overline{v_{OTA}^2} \quad (8.6)$$

As a conclusion, if we compare the two uncompensated instrumentation amplifier topologies, both of the architectures suffer from large 1/f noise and limited CMRR, but the capacitive one presents a clear advantage of having maximal DC headroom, which is an important requirement for the extraction of AP and LFP signals.

On the other hand, neither of the instrumentation amplifiers has sufficient performance for the recording of weaker biopotential signals, where the 1/f noise and limited CMRR are important error sources. Hence, compensated instrumentation amplifiers are used for applications requiring very low noise levels.

8.4.2 *Compensated Instrumentation Amplifiers*

A well-known technique that reduces flicker noise and improves the CMRR of amplifier is called chopper modulation (Fig. 8.9) [9]. This technique works on the principle that the input signal can be modulated to a frequency location, where flicker noise and

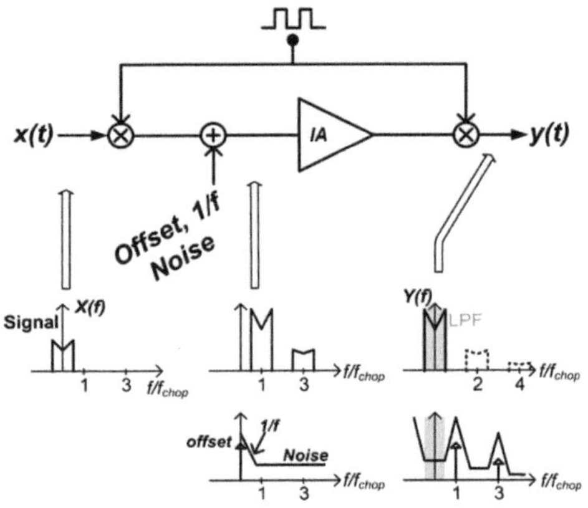

Figure 8.9. Operating principle of chopper modulation. Signal aggressors are modulated to out of the signal band.

other DC errors are not present, passed through the amplifier and demodulated back. In such an implementation, square-wave modulation is used, where the modulators are implemented using cross-coupled switches toggling at alternating clock phases. These modulators only modulate the differential input signal, but they are transparent to common-mode signals. Therefore, not only the flicker noise of CMOS transistors can be pushed out of the signal band, but also the CMRR can be increased.

The only problem that prevents the use of chopper-modulated instrumentation amplifiers for biopotential measurements is the fact that this technique inherently leads to the implementation of DC-coupled instrumentation amplifiers. A simple solution to this problem can be the use of off-chip passives implementing high-pass filter characteristics. However, this can lead to the use of large number of external components, especially important for multichannel EEG recording applications with high channel count.

As an alternative to off-chip passives, a recently proposed technique utilizes a DC servo to implement AC coupling (Fig. 8.10) [10]. In this architecture, the input of the DC servo input is connected to the output of the amplifier after the demodulator. It checks the

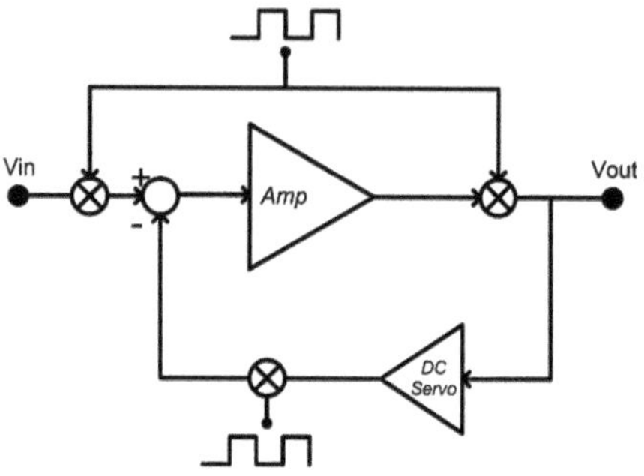

Figure 8.10. Implementation of an AC-coupled chopper-modulated instrumentation amplifier. The DC servo loop regulates the input of the core amplifier to reject the differential input impedance to chopper-modulated instrumentation amplifiers [10].

DC level of the output signal and regulates the input voltage of the amplifier to reject the input DC voltage.

Similar to uncompensated instrumentation amplifiers, there are two different instrumentation amplifier architectures, which use chopper modulation. The first architecture is called resistive compensated instrumentation amplifiers. Figure 8.11 shows the actual implementation of an AC-coupled chopper modulated instrumentation amplifier relying on resistors as the gain elements [10, 11].

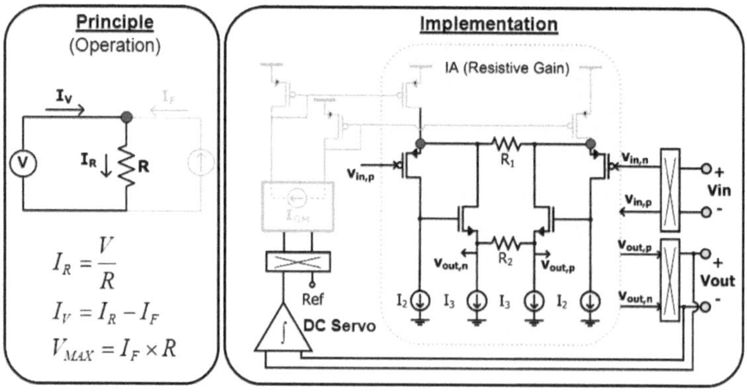

Figure 8.11. The complete architecture of a compensated instrumentation amplifier using a DC servo loop for implementing high-pass filtering characteristics [10, 11]. See also Color Insert.

The core of the instrumentation amplifier is an uncompensated resistive instrumentation amplifier, whose simplified architecture is shown in Fig. 8.7. The modulators and demodulators are located at the input and output of the resistive instrumentation amplifier. The input of the DC servo is sensing the DC level of the output and subtracts it from the input of the amplifier. This subtraction takes places at the terminals of the resistor R_1, which is actually based on the principle shown in the inset figure. This principle simply indicates that applying a voltage over a resistor leads to a current I_R, flowing through R. However, the source of this current will depend on the presence of other current sources forcing current trough R. For instance, if a current source forces a current I_F through resistor R, the voltage source only needs to supply I_R-I_F through the resistor R. This is the exact operating principle of DC servo loop presented in Fig. 8.11. The input transconductance only supplies the rest of the current due to I_{GM}. DC servo ensures that I_{GM} equals to the input DC voltage divided by R, so that our transimpedance will not receive this current to convert it into voltage.

The critical performance specifications can be summarized as follows:

i. **DC Headroom:** The DC servo loop introduces AC coupling to the chopper-modulated instrumentation amplifier. However, the maximum current that can be supplied from the DC servo limits the DC headroom to:

$$\text{DC Headroom} = I_{\text{GM,max}} \times R_1 \tag{8.7}$$

ii. **CMRR:** Chopper modulation eliminates the errors related to component mismatches. Therefore, the CMRR of instrumentation amplifiers can reach more than 120 dB.

iii. **Input Impedance:** The input impedance of this architecture is slightly reduced compared with the input impedance of an uncompensated resistive instrumentation amplifier due to the up modulation of the input signal.

iv. **Noise Performance:** The main advantage of chopper-modulated architecture is the removal of flicker noise. Hence, this architecture can implement very low noise at low frequencies.

The second compensated instrumentation amplifier architecture uses capacitive instrumentation amplifier as the core and introduces

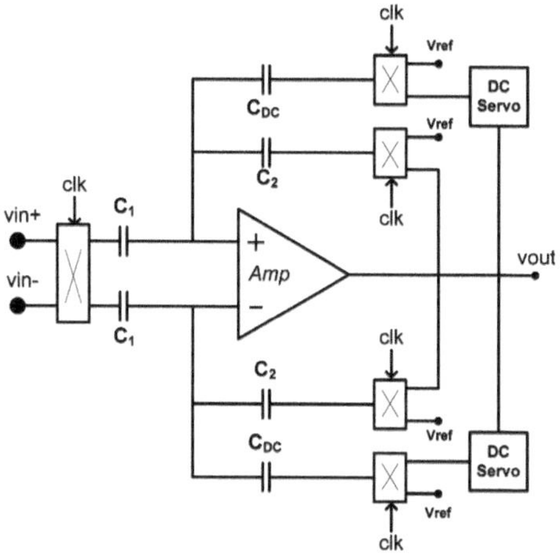

Figure 8.12. The complete architecture of a compensated capacitive instrumentation amplifier using a DC servo loop for implementing high-pass filtering characteristics [12].

chopper modulation. Figure 8.12 shows the architecture of the instrumentation amplifier [12]. Similar to the resistive instrumentation amplifier, a DC servo loop is used for the introduction of AC coupling. Different from the resistive instrumentation amplifier, the DC servo is implemented through capacitive coupling. Again, the current through the input capacitor C_1 can be seen as the addition of DC and AC input signals. In order to prevent its amplification, the current due to the DC input is subtracted through the DC servo path from the amplification signal path leading to AC coupling characteristics.

The basic specifications of this instrumentation amplifier can be summarized as:

i. **DC Headroom:** The DC servo introduces AC coupling. The DC headroom is again defined by the limitations of the DC servo as:

$$\text{DC Headroom} = \frac{C_{\text{DC}}}{C_1} \times \frac{V_{\text{DD}}}{2} \tag{8.8}$$

ii. **CMRR:** Chopper modulation increases the CMRR of the architecture.

iii. **Input Impedance:** One of the main drawbacks of this architecture is the input impedance. The modulation of the input signal reduces the equivalent input impedance. This is due to the fact that input impedance is defined by the capacitance C_1. Therefore, the input impedance of the architecture can be given as:

$$Z_{\text{IN}} = \frac{1}{j(w_{\text{in}}c_1 \pm n \times c_1)} \tag{8.9}$$

where w_{in} is the frequency of the input signal and n is defined by the harmonics of the square-wave modulation as:

$$n = \frac{4}{\pi}\sum_{k-1,3...}^{\propto} \frac{1}{k} \times k \times w_{\text{clk}} = \frac{4}{\pi}\sum_{k=1,3...}^{\propto} w_{\text{clk}} \tag{8.10}$$

The first scaling factor within the summation formula comes from the odd harmonics of a square wave, and the second scaling factor comes from the increasing frequency of the odd harmonics. These two factors cancel each other leading to a pure summation of the clock frequency as the scaling factor of C_1. Note that the effective range of C_1 is defined by the resistance of the chopping switches and the capacitance of the instrumentation amplifier.

iv. **Noise Performance:** Chopper modulation eliminates the flicker noise of the core amplifier. This means that the noise equation of this architecture can be written as Eq. 8.6, with the key improvement that the core amplifier noise is only defined by the thermal noise.

8.4.3 *Summary and Comparison of Instrumentation Amplifier Topologies*

In this section, we have presented four different instrumentation amplifier architectures, which are commonly used in biopotential measurement applications. These architectures are categorized according to their gain elements as resistive or capacitive and according to the use of compensation techniques.

It should be noted that the correct amplifier choice for a given application should combine the lowest possible power with sufficiently good performance. In addition to the discussion on amplifier specifications, Fig. 8.13 compares the power efficiencies of the instrumentation amplifiers presented in this section. It is interesting to see that the best efficiency is achieved by

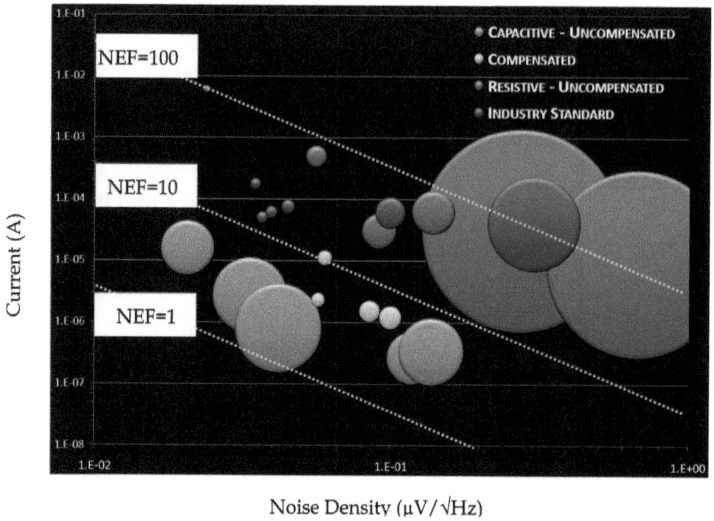

Figure 8.13. Comparison of different instrumentation amplifiers in terms of their power efficiency. Y-axis indicates the current consumption of the instrumentation amplifier and x-axis indicates the noise density of the instrumentation amplifier. Dashed lines indicate the constant contours of noise efficiency factor (NEF) lines [8]. The diameter of a circle indicates the total integrated noise of the instrumentation amplifier. See also Color Insert.

capacitive uncompensated amplifiers. This is due to the fact that this instrumentation amplifier topology uses only a single active circuit, which is the core OTA. Unfortunately, the implementations using the capacitive uncompensated instrumentation amplifier topology achieve relatively large total integrated noise. On the other hand, due to the presence of the servo loop, compensated amplifiers have slightly worse efficiency, but they have very much lower total integrated noise compared with capacitive uncompensated instrumentation amplifiers. Finally, uncompensated resistive and industry standard three opamp [13–15] instrumentation amplifiers have simply much worse power efficiencies preventing their use in ambulatory biopotential measurement applications.

It can be concluded from this discussion that capacitive uncompensated instrumentation amplifiers can be the preferred choice for applications that can live with relatively large circuit noise level.

Table 8.1. Comparison of different instrumentation amplifier architectures

	Figure 8.7 Resistive uncompensated	Figure 8.8 Capacitive uncompensated	Figure 8.11 Resistive compensated	Figure 8.12 Capacitive compensated
DC Headroom	Limited	Maximum	Limited	Limited
CMRR	Low to Moderate	Low to Moderate	High	High
Input Impedances	High	High	High	Moderate
Noise Performance	Large 1/f noise	Large 1/f noise	Low Noise	Low Noise

Last but not the least, Table 8.1 summarizes the properties of instrumentation amplifiers presented in this section and presents the comparison of their key characteristics. Not only due to their bad power efficiency but also due to their lower performance specifications, resistive uncompensated instrumentation amplifiers are not the preferred choice for portable biopotential monitoring systems. On other hand, the rest of other three instrumentation amplifier topologies are very attractive in terms of their power efficiencies as well as their performance specifications.

In addition to having the best power efficiency, uncompensated capacitive amplifiers have very large input impedance and maximal DC headroom. If we refer to Fig. 8.6, these characteristics match very well with the requirements of AP and LFP measurements.

The compensated version of this amplifier, i.e., compensated capacitive instrumentation amplifiers, additionally removes flicker noise and improves the CMRR of the instrumentation amplifier. On the other hand, the equivalent input impedance is reduced due to the capacitive nature of the input impedance. Hence, this architecture is generally used for ECoG applications where lower noise and higher CMRR performance are expected from the instrumentation amplifier, but lower input impedance values can be allowed by the application due to the use of invasive electrodes with large surface area.

Finally, resistive compensated instrumentation amplifiers have very high input impedance, low noise, and high CMRR. Therefore, their characteristics match very well with noninvasive biopotential measurements, where a high performance instrumentation amplifier is required.

8.5 Signal Integrity Problems in Ambulatory Measurements

In addition to the noise sources such as circuit noise, interference to measurement setup, and irrelevant biological signals that correlate the biopotential measurements, another important noise source in ambulatory measurements is motion artifact signals. These potentials occur in the electrode cables, in the skin, and at the electrode/electrolyte interface. While artifacts coming from cables can be reduced by appropriate electrode cables, artifacts from the skin and electrode/electrolyte interface are difficult to reduce by design.

The main source of the motion artifacts, as the name implies, is the motion present in ambulatory measurements. During ambulatory measurements, the subject/patient is continuously moving. The relative movement of electrode and tissue leads to significant voltage fluctuations called motion artifact signal. It is assumed that the source of the motion artifact signal is the fluctuations in the polarization generated at the electrode-tissue interface and within the tissue [16].

These motion artifact signals may have large amplitudes, and also they may have similar frequency spectrum as biopotential signals. Hence, amplitude-based or frequency-based algorithms, or the algorithms combining both methods, can significantly suffer from these motion artifacts reducing the reliability of biopotential measurements in ambulatory conditions. Therefore, there is a strong need for the methods and circuits that can enable the differentiation between the medically irrelevant motion artifact signals and the biopotential signals.

8.5.1 *Methods Focusing on Motion Artifact Reduction in Biopotential Recordings*

There have been several proposals in the literature trying to tackle the motion artifact problem in ambulatory biopotential recordings. Time synchronous averaging or median is one of them [16]. This is a classic technique widely used for cleaning of signals like ECG. However, these methods require the use of several ECG beats in order to obtain one single clean beat, and, therefore, beat-to-beat variations are removed.

Another technique is the use of fixed filters [17]. This technique could give the information about beat-to-beat variations. However, it does not adapt to the time-variant properties of noise and requires some previous knowledge of the noise. Furthermore, the design of linear phase band-pass filters is critical in preventing distortion of the ECG. Wavelets have also interesting filtering properties [18]. Wavelet de-noising based on a decomposition of the signal using wavelets that are highly adapted to the ECG gives a higher resolution in lower frequencies and, therefore, could lead to better removal of noise. However, wavelet processing is computationally heavy for ultra-low power ambulatory biopotential recording systems.

An alternative technique called adaptive filtering was introduced in the 60's as a new tool for ECG de-noising [19]. Since then it has become more and more popular, and several approaches have been implemented. For noise removal, the input of an additional reference signal is required, which should be highly correlated with the noise. Several reference signals have been investigated with successful results such as motion measured by accelerometers [20], skin-electrode impedance [21], and skin stretch measured by optical sensors [22].

Finally, independent component analysis (ICA), which contains beat-to-beat information and requires no reference signal, aims to reconstruct the ECG and motion artifact signals [23]. This method is, however, also computationally expensive and requires more than one ECG lead. Table 8.2 gives a summary of existing methods that are trying to reduce the effect of motion artifact signals on biopotential recording applications.

Among these methods, adaptive filtering is gaining more and more attention since it focuses on removing motion artifacts from biopotential signals without affecting ECG signal parameters. Also another advantage of adaptive filtering is the fact that it is computationally less heavy compared with ICA analysis.

8.5.2 *Readout Circuits for Adaptive Filtering*

Compared with other methods, adaptive filtering presents several advantages. First of all it can be implemented using a single-lead

Table 8.2. Summary of methods found in the literature for artifact reduction in biopotential recordings (ECG)

	Advantages	Disadvantages
Time Synchronous Averaging	Low computational complexity	Nonstationary information removed, require fiducial point demarcation.
Filter Banks	Beat-to-beat information	No adaptation to time-varying noise
Adaptive Filtering	Adapts to time-varying noise, no previous knowledge of noise required	Choice of reference signal is critical, implying complexity in reliable noise measurement
Independent Component Analysis	No reference signal required	Multiple leads required. Computationally heavy.

biopotential measurement (unlike ICA analysis), it does not require the history of the signal (unlike time synchronous averaging), and it can successfully eliminate noise from biopotential recordings even though the noise source looks exactly similar to the biopotential signal of interest (unlike filter banks). Therefore, there has been a specific attention on adaptive filtering over the last years. However, a critical requirement for adaptive filters is the extraction of a signal from the biopotential measurement system that is correlated with the motion artifact but uncorrelated with ECG signals.

A possible approach to tackle this problem is to collect data from other sensors that have maximum correlation with motion artifact signals and minimal correlation with biopotential signals. Several approaches has been tried in the literature, such as using accelerometers to sense the subjects' movement, electrode bend sensors to detect the motion of electrodes, skin stretch sensors to sense the movement of skin, and electrode impedance measurement circuits to correlate electrode impedance measurement with the motion artifact [22, 24–27].

Among these possibilities, the use of electrode-tissue impedance as the reference signal input to adaptive filters is gaining even more attention, since it does not require the use of an additional sensor and also since the characteristics of the electrode-tissue interface is actually responsible for motion artifact signals. Therefore, different circuit topologies that focus on the simultaneous measurement of biopotential signals with the electrode-tissue impedance have

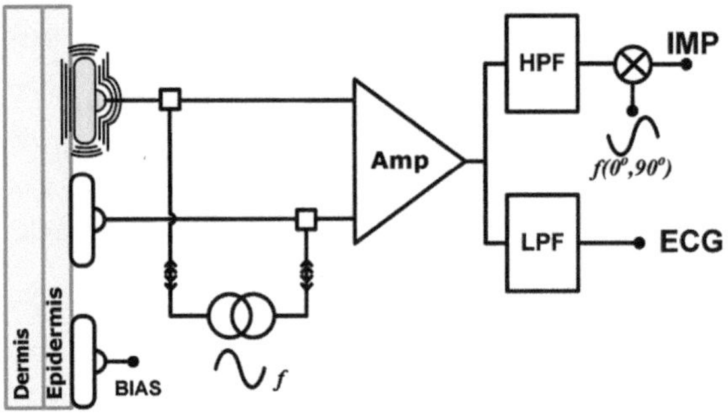

Figure 8.14. A circuit topology proposed by [26] to monitor biopotential signals simultaneously with the electrode-tissue impedance.

been proposed. Figure 8.14 shows the most common circuit topology for the measurement of biopotential signals with the electrode-tissue impedance [26]. A sinusoidal stimulation current is injected differentially to the measurement electrodes. The AC current frequency is selected such that it is out of the frequency band of biopotential signals. Hence, the low-pass filtering of the instrumentation amplifier output can extract biopotential signals, whereas the demodulation of the output with the AC current frequency can extract the electrode-tissue interface impedance. Although this technique can be very effective for extracting the required information, it requires the use of sinusoidal current generation blocks, which can be very power consuming in CMOS technology.

An alternative measurement method focuses on using the bias electrode for extracting the electrode-tissue impedance simultaneously with biopotential signals [27], Fig. 8.15. Instead of using a fixed bias voltage for the human body, an AC bias voltage is used. Under ideal conditions, this bias voltage appears as a common-mode signal to the instrumentation amplifier. However, as stated in Eq. 8.1, the finite input impedance of the instrumentation amplifier combined with the mismatch of the electrode impedances converts the common-mode signal into differential mode. Hence, the electrode-tissue impedance can be extracted using the formula:

$$V_{\mathrm{IMP}} = \frac{|\delta Z_{\mathrm{ELEC}}|}{Z_{\mathrm{IN}}} \times V_{\mathrm{BIAS}} \tag{8.11}$$

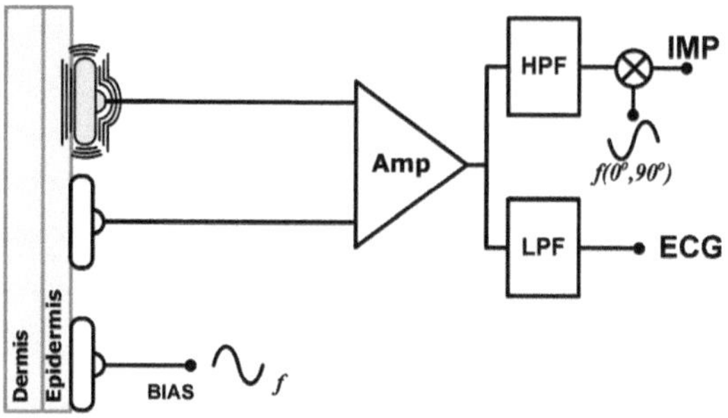

Figure 8.15. Another circuit topology proposed by [27] to monitor biopotential signals simultaneously with the electrode-tissue impedance. The method modulates the bias voltage of the patient to extract the electrode-tissue impedance.

However, it can be seen from the measurement method that this principle relies on the finite input impedance of the instrumentation amplifier, which simply contradicts with the requirements from biopotential measurement systems, i.e., as large as possible input impedance.

Last but not the least, a recent circuit topology that focuses on the measurement of electrode-tissue impedance with low-power dissipation has been proposed by [28], Fig. 8.16. This circuit uses square wave current stimulation rather than sinusoidal current stimulation. This way an integrated circuit with much lower power

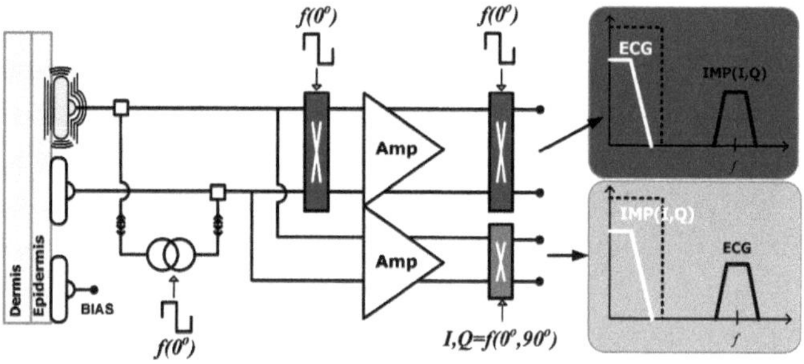

Figure 8.16. The implementation of electrode-tissue impedance measurement circuit operating simultaneously with the biopotential measurement circuit [28]. Instead of sinusoidal current sources and analog modulators/demodulators, chopper-modulation technique has been employed. See also Color Insert.

dissipation can be realized. This is due to the fact that the development of sinusoidal currents and their demodulation only relies on the use of chopper modulators leading to a significantly lower power dissipation compared with the methods using sinusoidal currents and analog modulators/demodulators.

8.6 Conclusion

As the cost of health care is increasing, the remote monitoring of people is gaining more and more attention. This leads to a strong urge for portable medical signal monitoring systems that can be integrated into people's daily lifestyle. It is expected from these systems that they can collect data, analyze these, and communicate the results to medical professionals.

One of the most important building blocks for this system is the realization of the analog front-end circuit that can consume minimal power and can still meet the challenging requirements of medical monitoring systems. In this chapter, we have reviewed the most commonly used instrumentation amplifier topologies for different medical signal acquisition. It can be seen from this review that there is not a single instrumentation amplifier topology ideal for different biomedical signal acquisition applications; rather, it is more important to consider the requirements of the application and select the instrumentation amplifier topology accordingly.

In addition to instrumentation amplifier design, an important challenge in ambulatory medical signal measurement systems is motion artifacts. These signals, though do not have any medical relevance, appear together with biomedical signals and reduce the reliability of signal analysis software. Several methods that address this problem have been presented in this chapter. The advantages and disadvantages of these different techniques have been reviewed. In particular, adaptive filtering approach has been favored due to its advantages over other methods. Last but not the least, the implementation of analog circuits that can generate the reference signal of an adaptive filter is described, and low-power techniques for such circuits have been presented.

References

1. OECD health data 2009.
2. B. Gyselinckx, R. Vullers, C. Hoof, J. Ryckaert, R. Yazicioglu, P. Fiorini, and V. Leonov (October 2006) Human++: Emerging Technology for Body Area Networks, in *2006 IFIP International Conference on Very Large Scale Integration*, pp. 175–180.
3. J. G. Webster (1992) *Medical Instrumentation: Application and Design*, 2nd edn, Houghton Mifflin, Boston (Mass.).
4. J. C. Huhta, and J. G. Webster (March 1973) 60-Hz Interference in Electrocardiography, *IEEE Transactions on Biomedical Engineering*, **20(2)**, pp. 91–101.
5. B. Razavi (2001) *Design of Analog CMOS Integrated Circuits*, McGraw-Hill.
6. C. Toumazou, et al. (1989) Novel Current-Mode Instrumentation Amplifier, *IEEE Electronics Letters*, **25(3)**, pp. 228–230.
7. R. Olsson, A. Gulari, and K. Wise (20–22 March 2003) A Fully Integrated Band-Pass Amplifier for Extracellular Neural Recording, in *Proceedings of the 1st International IEEE EMBS Conference on Neural Engineering*, pp. 165–168.
8. R. Harrison, and C. Charles (June 2003) A Low-Power Low-Noise CMOS Amplifier for Neural Recording Applications, *IEEE Journal of Solid-State Circuits*, **38(6)**, pp. 958–965.
9. C. Enz, E. Vittoz, and F. Krummenacher (June 1987) A CMOS Chopper Amplifier, *IEEE Journal of Solid-State Circuits*, **22(3)**, pp. 335–342.
10. R. F. Yazicioglu, P. Merken, R. Puers, and C. Van Hoof (6–9 February 2006) A 60 μW 60 nV/$\sqrt{}$Hz Readout Front-End for Portable Biopotential Acquisition Systems, *IEEE International Solid-State Circuits Conference Digest of Technical Papers*, pp. 109–118.
11. R. F. Yazicioglu, P. Merken, R. Puers, and C. Van Hoof (3–7 February 2008) A 200 μW Eight-Channel Acquisition ASIC for Ambulatory EEG Systems, *IEEE International Solid-State Circuits Conference Digest of Technical Papers*, pp. 164–165.
12. T. Denison, K. Consoer, A. Kelly, A. Hachenburg, and W. Santa (2007) 2.2 μW 97nV/$\sqrt{}$Hz, Chopper Stabilized Instrumentation Amplifier for EEG Detection in Chronic Implants, *IEEE International Solid-State Circuits Conference Digest of Technical Papers*, pp. 162–163.

13. M. A. Smither, D. R. Pugh, and L. M. Woolard (September 1977) CMRR Analysis of the 3-op-amp Instrumentation Amplifier, *IET Electronics Letters,* **13**, pp. 594.
14. A. C. Metting van Rijn, A. Peper, and C. A. Grimbergen (1990) High-Quality Recording of Bioelectric Events; Part 1: Interference Reduction, Theory and Practice, *Medical and Biological Engineering and Computing*, **28**, pp. 389–397.
15. M. Burke, and D. Gleeson (February 2000) A Micropower Dry-Electrode ECG Preamplifier, *IEEE Transactions on Biomedical Engineering*, **47(2)**, pp. 155–162.
16. H. Tam, and J. G. Webster (1977) Minimizing Electrode Motion Artifact by Skin Abrasion, *IEEE Journal on Biomedical Engineering*, **24**, pp. 134–139.
17. V. Afonso, W. Tompkins, T. Nguyen, K. Michler, and S. Luo (1996) Comparing Stress ECG Enhancement Algorithms, *IEEE Engineering in Medicine and Biology Magazine*, **15**, pp. 37–44.
18. P. Augustyniak (2007) Separating Cardiac And Muscular ECG Components Using Adaptive Modeling in Time-Frequency Domain, in *Proceedings of the WACBE World Congress on Bioengineering*.
19. B. Widrow, J. Glover, J. R. J. McCool, J. Kaunitz, C. Williams, R. Hearn, J. Zeidler, J. Eugene Dong, and R. Goodlin (1975) Adaptive Noise Cancelling: Principles and Applications, *IEEE Journal Proceedings*, **63**, pp. 1692–1716.
20. D. Tong, K. Bartels, and K. Honeyager (2002) Adaptive Reduction of Motion Artifact in the Electrocardiogram, in *Proceedings of the 2nd Joint EMBS/BMES Conference*, **2**, pp. 1403–1404.
21. P. H. Devlin, R. G. Mark, and J. W. Ketchum (1984) Detecting Electrode Motion Noise in ECG Signals by Monitoring Electrode Impedance, in *Proceedings of Computers in Cardiology* pp. 51–56, Los Angeles.
22. P. Hamilton, M. Curley, R. Aimi, and C. Sae-Hau (2000) Comparison of Methods for Adaptive Removal of Motion Artifact, *Computers in Cardiology*, **27**, pp. 383–386.
23. M. Milanesi, N. Martini, N. Vanello, V. Positano, M. Santarelli, R. Paradiso, D. De Rossi, and L. Landini (2006) Multichannel Techniques for Motion Artifacts Removal From Electrocardiographic Signals, in *Proceedings of the 28th Annual International Conference of the IEEE Engineering in Medicine and Biology Society*, pp. 3391–3394.
24. P. T. Gibbs, L. B. Wood, and H. H. Asada (2005) Active Motion Artifact Cancellation for Wearable Health Monitoring Sensors Using Collocated

MEMS Accelerometers, in *Proceedings of the Society of Photo-Optical Instrumentation Engineers.*

25. P. S. Hamilton, and M. G. Curley (1997) Adaptive Removal of Motion Artifact, in *Proceedings of the 19th Annual International Conference of the IEEE Engineering in Medicine and Biology Society*, pp. 297–299.
26. J. Ottenbacher, M. Kirst, L. Jatobá, U. Großmann, and W. Stork (2007) An Approach to Reliable Motion Artifact Detection for Mobile Long-Term ECG Monitoring Systems Using Dry Electrodes, in *Proceedings of the IV Latin American Congress on Biomedical Engineering.*
27. T. Degen, and T. Loeliger (August 2007) An Improved Method to Continuously Monitor the Electrode-Skin Impedance During Bioelectric Measurements, in proceedings of 29th Annual International Conference of the IEEE *Engineering in Medicine and Biology Society*, pp. 6294–6297.
28. R. F. Yazicioglu, P. Merken, R. Puers, and C. Van Hoof (February 2010) A 30 μW Analog Signal Processor ASIC for Biomedical Signal Monitoring, in *Proceedings of IEEE International Solid-State Circuits Conference Digest of Techical Papers*, pp. 124–125.

Chapter 9

Network and Medium Access Control Protocol Design for Wireless Body Area Networks

Jamil Y. Khan
School of Electrical Engineering and Computer Science,
The University of Newcastle, Callaghan,
NSW, 2308 Australia
Jamil.Khan@newcastle.edu.au

Wireless body area network (WBAN) is considered a special-purpose sensor network designed to collect and transmit physiological data within a short distance. Physiological sensors could be either implanted or placed external to a human body. A WBAN will generally operate either as a stand-alone personal area network (PAN) or as an interconnected network in telemedicine applications. The network configuration of a WBAN will depend on medical applications. In a WBAN design, the MAC protocol plays a very important role, which determines the QoS (Quality of Service) performance, energy efficiency, and reliability of a network. This chapter introduces the WBAN architecture, MAC protocols, and network design techniques for medical applications. The chapter initially introduces different network topologies followed by fundamental description of MAC protocols and its classifications,

Wireless Body Area Networks: Technology, Implementation, and Applications
Edited by Mehmet R. Yuce and Jamil Y. Khan

ISBN 978-981-4316-71-2 (Hardcover), ISBN 978-981-4241-57-1 (eBook)
www.panstanford.com

and then reviews several WBAN-specific MAC protocols. Following the MAC protocol discussions, the chapter introduces energy management techniques and patient monitoring design techniques. Some simulation results of a WBAN are presented to analyze the performance of a CSMA/CA MAC-based WBAN networks.

9.1 Introduction

The WBAN is becoming a key component of future telemedicine-based health care systems [1, 11]. Applications of WBAN in future can bring a number of key benefits to the health care system. Major benefits include collection of medical data of patients from their home or workplaces, collection of transient medical data in a cost-effective manner, patient's mobility, reduction of health care cost, improved patient care based on the availability of timely information, etc. On some occasions, collection of transient data for different diagnostic purposes could be difficult. In many cases, patients need to be kept under observation in hospitals or clinics for an extended period to detect certain medical conditions, which is quite resource intensive. Applications of WBAN could reduce the cost of patient observation and treatments because health care professionals can access those data from anywhere anytime.

A WBAN is a special-purpose wireless sensor network with specific QoS requirements [21]. Key requirements of a WBAN are low packet transmission delay, low to medium data transmission rate, near zero packet loss, minimum delay jitter, and low energy usage. Performance of a WBAN system will depend on a number of factors, which include the hardware design, applications characteristics, and network design.

In a patient monitoring system, data transmission reliability and latency is extremely important. The reliability and latency of a WBAN will largely depend on the design of physical (PHY) and medium access control (MAC) layers. For optimum network efficiency and data transmission reliability, the MAC layer needs to be designed to suit specific needs of specialized applications. Considering the importance of WBAN applications, an IEEE standard group (IEEE802.15.6 working group) is working on the development of a body area network architecture [22]. The new standard will define the PHY and MAC layer management issues, which could be used to develop a low cost, ultra-low power and highly reliable wireless network. Reliability of a medical network can be defined in terms of its main QoS parameters such as delay profile, delay jitter, and information loss rate. Performance and reliability of a WBAN will depend on the PHY and MAC layer procedures as well as on the network topology used. Following sections examine network architecture and MAC protocol design issues for a WBAN-based patient monitoring system. Before moving into the discussion of design issues, first we examine the network and MAC protocol design requirements of a WBAN. Listed below are some of the main requirements of a WBAN.

- A WBAN should be able to support a range of medical applications, which includes acquiring data samples from periodic and nonperiodic sources and transmitting information to a service node within a maximum fixed delay without any loss of critical information [2]. Also, the system should be able to support exchange of nonmedical and control information with various remotely controlled appliances.
- A WBAN should be able to operate in a power-constrained environment where power sources such as battery should operate for a reasonably longer period of time. Power savings for implantable nodes are more critical than other nodes. Power savings can be achieved by combining PHY and MAC layer procedures [3].
- A WBAN should be self-healing, secure, and reliable.
- A WBAN should support data rates between few tens of kbs to several Mbps to host a range of applications such as images and video clips.
- A WBAN should support QoS management features to offer priority services. Particularly when a critical patient is monitored, the system must guarantee the delivery of critical information to a service node. For medical data, the main QoS features will be the transmission delay and the packet loss.

- A WBAN may incorporate a narrow transmission band (400 MHz or 2.4 GHz) together with the UWB (ultra-wide band) technology to cover different environments.
- A WBAN should operate and coexists with other network devices operating in similar frequency bands. Also, a WBAN should be able to operate in a heterogeneous networking environment, where networks of different standards may cooperate with each other to acquire information from different sensors.

In this chapter, we concentrate on the network design techniques, mainly concentrating on the MAC and energy management issues. The MAC protocol implements a set of rules, which are used to coordinate packet transmissions in a shared transmission channel environment. Numerous MAC protocols have been designed to support applications in different types of networks. In this chapter, we review several MAC protocols that could be used for WBAN applications. This chapter also introduces an interconnected WBAN architecture, which can be used to support remote patient monitoring applications. Rest of the chapter is organized as follows. Section 9.2 introduces the network topology and configurations used in wireless sensor networks. Section 9.3 introduces the basic properties and classifications of MAC protocol. Section 9.4 discusses scheduled and polling MAC protocols and their design features. Section 9.5 discusses the random access protocols and their design features. Section 9.6 presents a hybrid MAC protocol design proposed by the IEEE 802.15.4 standard. Section 9.7 discusses the energy management issues in a WBAN. Section 9.8 presents the design of several patient monitoring networks. Section 9.9 presents some simulation results of a ZigBee-based patient monitoring system. Conclusions are presented in Section 9.10.

9.2 Network Topologies and Configurations

A communication network can be designed using a number of different topologies that define the geometrical position of communication nodes [12]. A network topology is generally selected

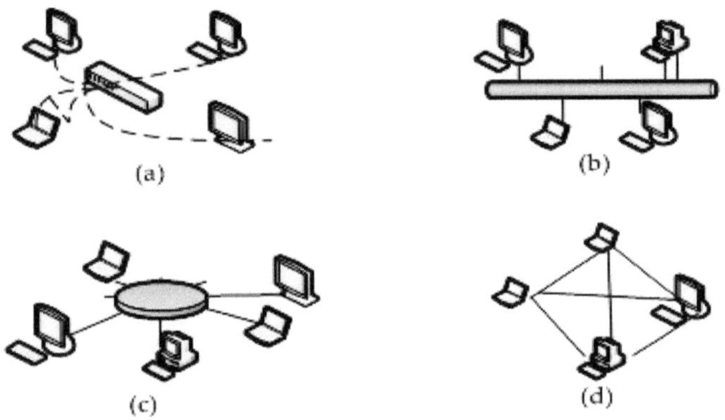

Figure 9.1. Network topologies: (a) centralized star topology; (b) distributed bus topology; (c) ring topology, and (d) fully connected mesh topology.

based on application scenarios. There are five standard topologies generally used: *star, bus, tree, ring, and mesh*. Figure 9.1 shows commonly used four network topologies. The star topology consists of a centralized controller where all nodes communicate directly with the central controller to exchange information. Information between nodes is also exchanged via the central controller. The star configuration is suitable for WBAN applications where a central controller can collect and aggregate all physiological data from different sensors. The bus structure supports communication among various nodes in a decentralized manner. The topology has been developed to support operation of random access protocols with minimum protocol control overhead. Bus structure could be used for wired body area network applications. The ring topology is used to transmit packets in a contention free mode using a token, which is a form of distributed polling network. In the ring topology, nodes are connected in a serial closed-loop fashion where a control packet, known as the *token*, schedules the transmission of packets. A ring network may not be suitable in a WBAN due to longer transmission delay and maintenance of token overhead. A ring network is generally more suitable for high data rate networks with high volume of traffic. The mesh topology is used to provide full connectivity between all nodes. The topology provides many redundant transmission links, thus offering higher reliability in a large network. The mesh network topology is quite often used in large sensor networks to provide full connectivity. For a WBAN application, the mesh topology is not very useful because each body area network will generally consists of limited number of nodes. However, a mesh topology could be used in a large hospital network for multi-patient monitoring systems.

The star topology in a wireless sensor network is organized as a cluster tree topology where the star network controller becomes the cluster head. A cluster-based network forms a PAN; in case of a wireless clustered network, it is referred to as the WPAN (Wireless PAN) where the cluster head becomes the PAN coordinator [14]. All data communication in a cluster network goes through the cluster head or in case of a PAN/WPAN, through the PAN coordinator. The PAN coordinator is also responsible for resource allocation, synchronization, allocation of a PAN ID. Each node in a cluster has a unique cluster ID, which is controlled by the cluster head. The cluster head/PAN coordinator is a more computationally powerful data processing device, which collects information from all its sensor nodes. In WPAN, nodes are generally classified into two types: full functional device (FFD) and reduced functional device (RFD). The FFD can operate in three modes serving as PAN coordinator, a coordinator, or as a network node. The RFD can only act as a simple node, which forward information to a coordinator or to a PAN coordinator. RFDs can exchange information only through the PAN coordinator. To operate a cluster-based WPAN, it is necessary to activate a star configuration. In a cluster-based WPAN, a FFD is activated first, which may establish its own network and become the PAN coordinator. All devices in a PAN are identified by the PAN identifier. Once the PAN identifier is chosen, the PAN coordinator allows other nodes to join the network. WPAN network devices generally do not implement the full seven-layer protocol stack; rather, it implements only the lower two layers of the protocol stack, i.e., the physical layer (PHY) and the data link layer (DLL).

9.3 Basics of Medium Access Control Protocols

A sensor node is designed using the standard OSI (open system interconnection) model to exchange data using communication links.

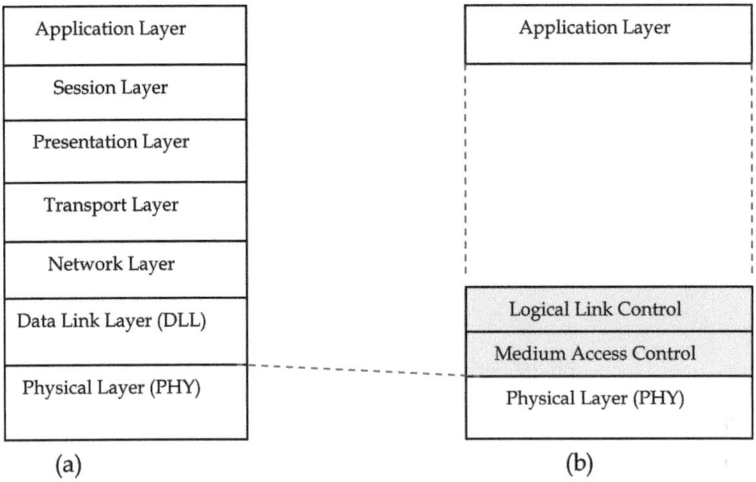

Figure 9.2. Protocol stack (a) standard OSI model, (b) three-layer WPAN model.

The basic seven-layer OSI model and the corresponding WPAN protocol stacks are shown in Fig. 9.2. A WPAN device will only implement two and half or three layers of the OSI model, which includes PHY, mostly a sublayer of the DLL, and an application layer [14]. The physical layer is responsible for the transmission and reception of physical signals using the hardware resources. The PHY layer is also responsible for conditioning of transmitted and received signals suitable for the transmission channel. A WPAN will generally only use the MAC sublayer of the DLL. The MAC protocols are used in multiuser communication networks to transmit packets in a contention-free mode. The MAC protocol allow users to transmit packets in an interference free mode while maintaining high network throughput and low delay, thus supporting QoS of different types of traffic. MAC protocols can also control transmission power of a node by optimizing packet transmission and scheduling processes. MAC protocols have been evolving over the last few decades to support different type of networks and applications. MAC protocols can be classified into two broad categories: fixed assignment- and dynamic assignment-based protocols as shown in Fig. 9.3. Fixed assignment-based protocols use the basic properties

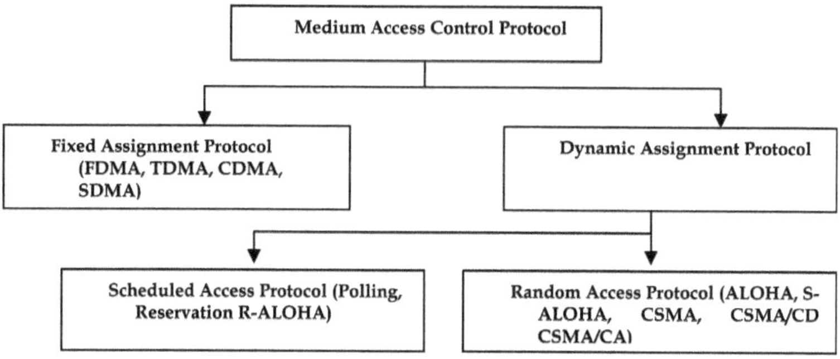

Figure 9.3. MAC protocol classifications.

(time, frequency, and code) of a transmission channel to divide transmission resources and use the allocated resources to transmit their information in an interference-free mode. Fixed allocation-based MAC protocols generally do not alter allocated transmission resources during the length of a data session. Fixed assignment protocols can also be viewed as scheduled protocols. Some of the commonly used fixed resource assignment-based protocols are: FDMA (frequency-division multiple access), TDMA (time-division multiple access), CDMA (code-division multiple access), and SDMA (space-division multiple access) [13, 14]. For WBAN applications, some of the systems use the TDMA technique to accommodate multiple sensor nodes.

Dynamic assignment-based protocols are generally known as ON demand-based protocols where transmission resources are allocated to transmitters only when an actual transmission takes place. ON demand protocols are more suitable for bursty non-periodic traffic sources. The dynamic access protocols are again divided into two subclasses: scheduled access protocol and random access protocol. A scheduled access protocol allocates transmission resources in a dynamic manner, but it coordinates transmission either in a centralized or in a distributed manner to avoid any interference or collisions. The random access protocols are also ON demand-based transmission system where transmitters may use some channel state information to transmit a packet. The channel sensing mechanism is used to reduce the probability of collisions. A random access protocol generally requires less control signaling compared to the scheduled access protocols. Transmission of control signaling has implications on a network. Large control overheads could reduce the information transmission capacity of a WBAN; also control signaling could increase the power requirements of a node or a network.

One of the key issues of a WBAN MAC protocol design is power management. Power requirements of a WBAN will be jointly determined by PHY and MAC layers. The MAC protocol can also control the physical layer of a node to control the sleep cycle of a node, which can affect the energy requirements of a node. The MAC protocol can determine sleep, wakeup, and listening periods depending on the application. Most of the medical applications will generate low duty cycle data. The MAC protocol can synchronize various application duty cycle data and intelligently control the sleep cycle of a transceiver [13, 4]. Further discussion on MAC energy management is provided in Section 9.7.

QoS requirements of a WBAN protocol are simple and straightforward. Most of the nodes will transmit information in a periodic manner using fixed size packets. It is necessary for a WBAN that packets can be transmitted in a contention-free mode with high guarantee of delivery and low packet transmission delay. For WBAN applications, most cases packet retransmission needs to be avoided because generally the sensor nodes will have little buffer and processing capabilities. Also, retransmission of packets will reduce the life of the WBAN energy source. Some of the specialized WBAN nodes connected with a tiny camera may require higher transmission bandwidth to transmit images. A WBAN can prioritize its transmission by assigning different priority value to different traffic sources. QoS requirements of WBAN applications have not been strictly defined yet. With the further development of the IEEE 802.15.6 standard, it is expected that QoS profiles for medical data will be developed for future applications.

9.3.1 *WBAN Traffic Characteristics*

Traffic sources in a WBAN are mostly medical sensors which generate data depending on the properties of physiological signals. Table 9.1 lists some of the common diagnostic medial signal

Table 9.1. Physiological signal characteristics

Parameter	Value	Frequency (Hz)	Data Rate (bits/sec)[1]
ECG Signal (single lead)	0.5–4 mV	0.01–250	4,000
Respiratory rate	2–50 breaths	0.1–10	160
Blood pressure	10–400 mm Hg	0–50	800
EEG	3–300 μV	0.5–60	960
Body temperature	32–40°C	0–0.1	1.6
EMG (Electromyogram)	10 μV–15 mV	10–500	8,000
GSR (Galvanic Skin Reflex)	30 μV–3 mV	0.03–20	320

[1] Assuming an 8 bit A/D converter is used.

properties. Data rate presented in the fourth column is calculated based on the Nyquist frequency and an 8 bit/sample A/D converter. The table shows that signal data rate and interarrival time vary quite significantly. All the traffic generated by various sensors is periodic in nature because of constant sampling rate. For patient monitoring applications, it is also possible that sensors could transmit data on demand basis, i.e., a sensor transmit data only when a sensor is being queried by a medical device. In addition to following medical data, a WBAN may also support image or video transmission from either an implanted or an on-body camera. Data rate for imaging devices will depend on application and image/video resolution and frame size. A WBAN also needs to cater for irregular events such as irregular heart beat when a patient may enter into a critical stage. WBAN can be used in an ambulatory environment where a critical patient could be monitored over a communication network. Hence, it is apparent from this discussion that a WBAN generally needs to cater for periodic data, but on occasions it may be necessary to cater for nonperiodic bursty data. Besides these signals, a WBAN may need to transmit/receive control signals from specialized medical devices. In future for specialized applications, a WBAN will probably consist of sensor nodes as well as diagnostic medical appliances.

9.4 Scheduled Protocols

Packet transmissions in a WBAN will be controlled by the MAC protocol. A scheduled MAC protocol will allocate transmission resources to all nodes in an orderly manner using either a fixed or a dynamic channel assignment technique. In a WBAN transmission resource is the radio channel. In communication networks many other scheduled protocols are used in various systems; however, for WBAN applications we focus on TDMA and polling protocols because they are most likely to be used for medical applications. For certain applications, fixed assignment protocols can be modified into adaptive assignment protocol such as ATDMA (advanced TDMA) protocol.

9.4.1 *TDMA Protocol*

Time-division multiple access (TDMA) is one of the simplest fixed assignment protocols [14]. TDMA protocol allows multiple users to share a transmission channel by dividing the transmission time among users. The TDMA protocol operates using a frame and slot structure where a frame consists of a numbers of time slots with a fixed frame length. The time slots are allocated to each transceiver that can transmit or receive information using their allocated time slots. The TDMA frame is synchronous in nature, which repeats at a constant rate. For a WBAN, TDMA links can be configured either in a full duplex or in a half duplex manner. Duplex links could be frequency-division duplex (FDD) or a time-division duplex (TDD). The implementation of FDD links is more costly because each node needs to use a duplex filter, whereas a TDD link can be easily implemented using a software scheduler on a node. For WBAN applications, most of the time traffic will be transmitted in one direction, i.e., sensor to the coordinator; hence, either a full duplex or a TDD link will be suitable. A TDMA-based WBAN can be designed using both star and bus network topology. TDMA, being a synchronous packet transmission system, can handle periodic traffic more efficiently than nonperiodic traffic. A TDMA link can provide fixed packet transmission delay, which is one of the key requirements of a WBAN. Another benefit of the protocol is the use of minimum number of overhead bits to support information transmission due to the fixed channel allocation technique. The main disadvantages of a TDMA-based system are scalability and efficient support of nonperiodic traffic sources. The energy efficiency of a TDMA-based system can be enhanced by synchronizing the transmission frame/slots with the sleep cycle of a node [5]. Energy efficiency issues are discussed in Section 9.7.

9.4.2 *Polling Protocol*

Polling is one of the dynamic scheduled access mechanisms that can be used to support packet transmission in any size network [6]. A polling network is a centralized network architecture where a central network node controls the transmission of packets, avoiding any collisions of packets. Unlike the TDMA protocol, polling protocols generally do not require any fixed allocation of resources but require higher level of signaling to transmit information. Figure 9.4 shows a typical wireless polling network and its polling sequence operating in star topology. This figure shows that all nodes in the network are polled by the central controller in a round-robin manner to receive data from these nodes. The controller sends a poll message to a specific node, which in reply can send either a data packet or a NACK (negative acknowledgement) specifying that it has no data to send. To minimize packet transmission, the controller could include the acknowledgement of the pervious transmission with the current poll message. Since a poll message is a broadcast message, all nodes in the network can receive the message. However,

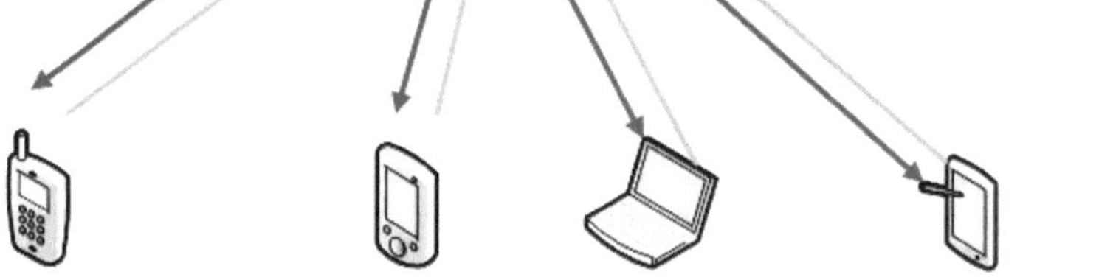

Figure 9.4. A wireless polling network showing the polling sequence. See also Color Insert.

the controller specifies in the poll message which node should reply. A polling system is more flexible than a TDMA-based system due to direct interaction between the controller and a node. In this case, each node can be polled based on its transmission requirements. For example, Table 9.1 shows that an ECG sample is produced after every 2 ms whereas an EMG sample is produced every 0.2 ms. If we assume 10 signal samples are transmitted in a packet, then the ECG node needs to be polled after every 20 ms whereas the EMG node needs to be polled every 2 ms. This flexibility is not available in a TDMA-based system. Also, in polling networks, nodes can transmit different size data packets depending on the application. These flexibilities are achieved at the expense of extra signaling and the processing capability of the central controller. Polling sequence and rate can be tailored for different applications.

Packet transmission delay in a polling network will depend on the polling cycle time [12]. As shown in Fig. 9.4, nodes are polled in a sequential manner where a node can transmit a packet after it has been polled. So the packet transmission delay will be the sum of packet queuing time and the packet transmission time. In a WBAN, propagation delay is extremely small, which can be neglected. In a network if the controller is polling all the nodes at a constant rate, then the polling cycle will be the sum of polling time and packet transmission time. For example, if the poll and acknowledgement packet transmission time is 1 ms and the data packet transmission time is 2.5 ms, then each node in Fig. 9.4 will be polled after every 14 ms assuming that all nodes always have data to transmit when polled. In this case, if the first node generates data in the middle of a polling cycle, then it must wait 8 ms before a poll message is received and then it will take another 2.5 ms to transmit the data packet to the controller, incurring a 10.5 ms delay for the transmission. For a WBAN application, if the nodes are polled by matching their data generation rate, then the polling cycle length will vary due to the number of active nodes in each polling cycle. A polling system can be tailored to cater for data with different priorities and data generation rates. A high priority node can be polled more frequently compared to lower priority nodes. Polling networks are easily scalable, particularly when the controller finds a new node it can include the new node in its polling list or delete a node from the list if a node does not respond to subsequent polls. The energy requirements of a polling network can be variable depending on the system design. Polling networks can be implemented using star, bus, and ring topologies. For WBAN application, a star topology-based polling network will be most suitable because the bus and ring topologies introduce higher polling and message transmission delay.

9.5 Random Access Protocols

Random access protocols are generally used in distributed networks without the presence of any central coordinator [12, 13, 14]. Nodes using random access protocols coordinate the transmission of packets to minimize collisions. Nodes using this class of protocols will use the transmission channels only when they have data to transmit. Random access protocols are more complex than scheduled access protocols; hence, careful approach is required to design the protocol and the network. Random access protocols are in use in local area networks for many years. This class of protocol is also becoming popular for wireless sensor network applications. The IEEE 802.15.x standard defines a number of random access protocols for a number of short-range wireless networks [7]. Bluetooth was the first short-range low-to-medium data rate MAC protocol developed for WPAN applications. Later the Bluetooth protocol was modified to develop the IEEE 802.1.5.1 standard. Subsequently, a low-power low data rate standard, the IEEE 802.15.4 MAC protocol, was developed for the wireless sensor network applications. In the IEEE 802.15.4 standard, which consists of PHY and MAC layers, functionalities have been merged with application layer protocols to develop the ZigBee standard. The IEEE 802.15.x family of protocols is becoming one of the major wireless sensor networking standards. The IEEE 8021.5.6 protocol is currently under development to support WBAN applications. It is most likely that the new WBAN standard will be based on the CSMA/CA (carrier sense multiple access/collision avoidance) standard [22].

One of the first random access protocols used in computer networks is the ALOHA protocol. Since the development of the

ALOHA protocol, many other random access protocols have been developed by further enhancing the features of the protocol. A more efficient random access protocol known as the CSMA protocol was developed from the ALOHA/S-ALOHA (slotted ALOHA) protocols [12]. Using the CSMA protocol, each transmitting node checks the status of the transmission channel to decide whether to transmit a packet or not. The philosophy of the CSMA protocol is *listen before transmit*. The CSMA protocol can operate either in persistent or in a nonpersistent mode. In the persistent mode, a node constantly monitors the status of the transmission channel and starts transmitting as soon as the channel becomes free. Using the nonpersistent mode, a node constantly monitors the status of a transmission channel. When the channel becomes free, the node then can either immediately transmit or backoff for a certain period of time to check the channel again. The decision whether to transmit a packet or to defer a packet transmission depends on the transmission probability value generated by a random number generator at the MAC layer. The nonpersistent protocol has been developed to minimize the probability of collisions by randomizing packet transmission times. The persistent protocol may offer a low transmission delay at the expense of higher level of packet collisions, whereas the nonpersistent CSMA protocol offers network stability at the expense of higher average packet transmission delays. A compromised version of the CSMA protocol is quite widely used and is known as the p-persistent protocol, where the probability of a transmission or a deferment can be adjusted by selecting appropriate value of p, which is the packet transmission probability [15]. In a CSMA/CA network, an ACK packet is used to inform the transmitter about the status of a transmission. A receiver transmits an ACK packet when a packet is successfully received. A transmitter may not receive an ACK packet if either the transmitted packet is incorrectly received due to channel errors or packet transmission is unsuccessful due to collisions. A transmitter can successfully transmit a packet if all other transmitter remains silent for a period of maximum propagation delay, t_p, of the network. After the propagation delay, all nodes of a network become aware of the transmission and will not attempt to transmit any packet while the current transmission continues.

Channel sensing and *p* persistent techniques are combined to improve the performance of a network by trying to avoid collisions. CSMA/CA-based protocols also use another collision avoidance technique particularly to reduce the *hidden* user problem [13]. A wireless terminal is considered to be a *hidden* terminal if the terminal is unable to receive transmitted signal(s) from its neighboring terminals because that particular terminal could in a RF (radio frequency) shadow region compared to its neighboring terminals. In such case, a hidden terminal could increase the packet-collision level in a network. A *hidden* terminal may be unaware of other transmission or transmission attempts and inadvertently interfere with other transmissions. The CSMA/CA protocol uses the RTS/CTS (ready to send/clear to send) signaling to reduce the *hidden* terminal problem. However, for a WBAN application, the *hidden* terminal will not be a problem due to the size of the network and the proximity of terminals.

Random access protocols can offer several advantages for WBAN applications. The first advantage is the support of periodic and nonperiodic traffic with equal efficiencies. Using a random access protocol, a transmitter can initiate a packet transmission on its own only when a packet is generated at the node. Unlike the scheduled access protocol, the transmitter does not need to wait for a time slot or a poll message to initiate a transmission. In a WBAN when periodic and nonperiodic traffic are mixed, the problem of collision reduces due to randomization of packet generation rate. Second, the main advantage of the random access protocol is the energy consumption [16]. A node using the random access protocol can stay in the sleep state most of the time, waking up only when transmitting a packet. Another key advantage of a random access protocol is the scalability. The networking size can easily be increased or decreased without the need for any software or hardware upgrade. The main disadvantage of the random access protocol is variable packet delay due to collisions at the higher traffic load. However, this issue is not a problem for WBAN applications if the volume of traffic is kept relatively lower compared to the network transmission capacity. Random access protocols have been widely accepted for sensor network applications. Industrial standards such as ZigBee, Bluetooth, etc are quite widely used in sensor network applications.

Both ZigBee and Bluetooth standards are based on the CSMA/CA-based MAC standards, which can be successfully used in WBAN applications.

9.6 Hybrid MAC Protocol

Previous sections discussed scheduling and random access protocols, which are used in different wireless sensor networks. For sensor network applications, the IEEE 802.15.4 standard defines a hybrid packet transmission structure, which allows transmitters to use both contention and guaranteed transmission modes [14, 17, 18]. The hybrid transmission structure has been developed to efficiently support periodic, intermittent, and repetitive low latency data. The IEEE 802.15.4 standard supports the hybrid transmission mode using the super frame structure as shown in Fig. 9.5. The super frame structure is controlled by the PAN coordinator. The coordinator periodically sends beacons that identify the PAN. The PAN is identified by a 16 bit PAN identifier. The beacon frame also contains information, a list of outstanding frames, and other system-related parameters. The time interval between beacon signals is constant, and users can select a value which is a multiple of 15.38 ms. The maximum beacon interval time could go up to 252 ms. Two consecutive beacon signals form a *super frame* separated by 16 equally sized time slots as shown in Fig. 9.5. The super frame is subdivided into active and inactive periods. All nodes, including the coordinator, can turn off their transceivers and go into a sleep

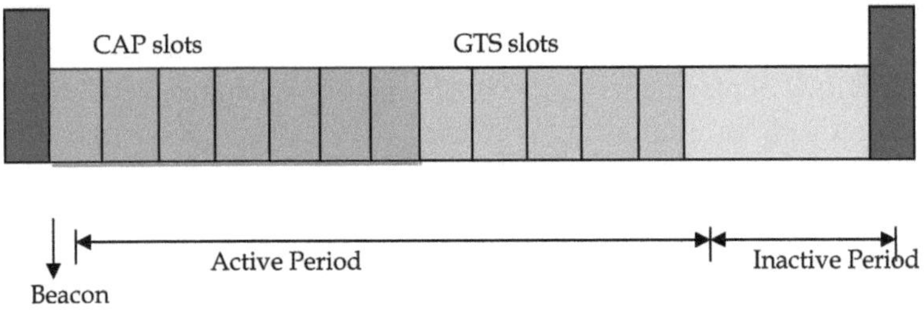

Figure 9.5. IEEE 802.15.4 MAC super frame structure. See also Color Insert.

state during the inactive period of a super frame. The active period consists of 16 timeslots, where each timeslot is divided into two groups. One set of slots forms the *contention access period* (CAP) and the rest of the slots become part of the *guaranteed time slots* (GTS). The length of the active and inactive periods is configurable, hence allowing system design flexibility to accommodate different types of applications. The CAP slots can be accessed using the standard CSMA/CA procedure, whereas the GTS slots are allocated by the coordinator when a request is made by a node. The flag in the request packet indicates whether the slot will be used to transmit or receive. Upon receiving a slot request, the coordinator will allocate a slot when appropriate resources are available. GTS slots are organized in the same way as the TDMA slots. When GTS slots are used for transmission, a node can schedule its sleep cycle and wake up just before the time slot starts and send the packet. On the other hand, when a CAP slot is used, the node must wake up in advance and perform carrier sensing or other collision avoidance procedures before it can initiate a transmission.

The super frame structure based on the hybrid MAC protocols could be well suited for WBAN applications where medical data with different QoS requirements will be transmitted. Sensors of lower priority data could be transmitted using CAP slots, whereas high priority data could be transmitted using GTS slots. The super frame structure is configurable where the active and inactive periods can also be adjusted.

9.7 Energy Management in a WBAN

Energy management of nodes is one of the crucial requirements of WBAN deployments [19]. WBAN's could be deployed for patient monitoring continuously for several days. In such deployments, it is important that nodes are designed properly so that the battery lasts for the duration of the deployment. Energy requirements of a node will depend of a number of factors, which includes the PHY layer and hardware design, MAC layer-controlled sleep cycle, MAC packet scheduling, and the application requirements. Source of energy of WBAN nodes will be a battery whose life cycle will depend on the way a node will drain the energy. There are a number of battery phenomena that could affect both charge and discharge outcomes. Two important phenomena are the rate-capacity effect and the recovery effect. The rate-capacity effect of a battery is the idea that drawing a large, continuous current from a battery will lead to it depleting faster than expected. In an ideal battery, it is assumed that the voltage stays constant for the entire life of the battery, and then drops to zero. In the real world, however, voltage is not constant over the entire life of the battery; instead it drops during the discharge. This drop in voltage during the discharge varies in severity from battery to battery. In all cases though, it leads to a perceived drop in battery capacity. Figure 9.6 shows the typical rate capacity characteristics of a battery for a continuous and an intermittent load situation [6]. An intermittent transmission is a typically ON/OFF scenario when a transmitter sends data for a brief period and then moves into an inactive state. When a transmitter moves to an inactive state, the battery load will be low, allowing the battery to recover. Figure 9.6 shows that an intermittent application could prolong the life of a battery due to the recovery effect. For a continuous discharge, the slope remains fairly constant as expected. For an intermittent discharge, however, the battery is able to recover some of the lost charge, resulting in a piecewise-continuous discharge slope. This results in the battery having a longer lifetime.

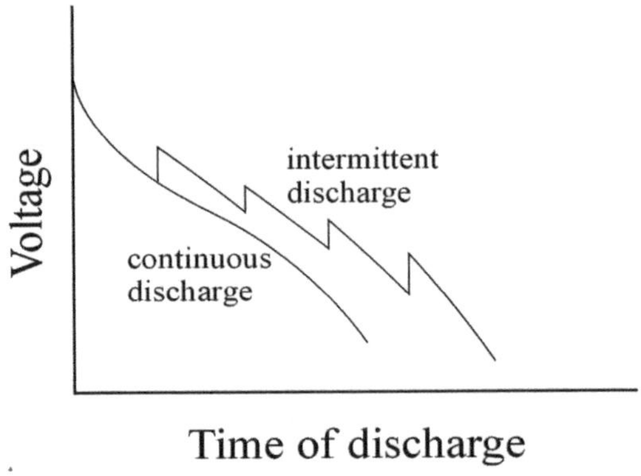

Figure 9.6. Battery charging and discharging phenomena.

As shown in Fig. 9.6, the battery lifetime of WBAN node can be sufficiently extended by appropriately designing the packet transmission schedule in a network. It is possible to optimize the power requirements of a sensor node by using the following techniques.

- Lower data transmission frequency by the use of data aggregation technique.
- Reduction of proportional frame overhead by selecting appropriate payload size.
- Appropriate use of power management algorithms to coordinate the sleep cycle.
- Placement of nodes in a WBAN to avoid poor transmission channel conditions.

To achieve significant advantages, each of the above techniques needs to be carefully designed and integrated with the MAC protocol. Lower the data transmission frequency can reduce the power requirements of a sensor node. Average current consumed by a sensor node is described by Eq. 9.1. The equation clearly shows that transmission duration and average transmitter current requirement will increase with more frequent transmission of data. In addition to power requirement for data transmission, there are other power requirements. For example, for a CSMA/CA-based network for every packet transmission the node needs to listen to the channel when the node acts as a receiver. On the basis of channel sensing result, it either transmits or backoff for a certain duration. It is necessary for reduced power consumption that during the backoff period, the transmitter goes to the standby mode. The transmitter could be switched ON by a timer for further channel sensing and transmission decision-making. This intermittent ON/OFF process of the transmitter will help to prolong the battery. Data aggregation technique will lower frequency of data transmission as well as reduce the frame overhead [10]. For example, EEG data samples are generated at a rate of 60 Hz, i.e., samples are generated after every 16.66 ms. If we use 8 or 12 bit A/D converter and transmit each sample individually, then the payload size will be very small. Instead of a single sample if a number of EEG samples are combined and transmitted every second, then the payload size becomes

480 bits or 720 bits depending on the chosen A/D converter word size. In this case the transmitter will use far more less power to transmit without introducing unnecessary delays. Similarly the data aggregation technique can be applied for a multichannel sensor networks as described in Chapter 5. In case of a multichannel WBAN, samples from different channels will be aggregated and encapsulated in a single packet. The use of data aggregation technique has another advantage for a random access network. The data aggregation technique will reduce the number of packet transmission attempts for a CSMA/CA network, thus reducing the packet collision probability and increasing the QoS of a WBAN. In a random access network such as ZigBee or Bluetooth network, transmission of fewer large packets will reduce the collision level as well as the transmission power requirements.

$$I_{avg} = T_{txon} \times I_{txon} + T_{rxon} \times I_{rxon} + (1 - T_{txon} - T_{rxon}) \times I_{stb} \quad (9.1)$$

T_{txon} is the transmission duration including packet transmission and other time necessary to support the transmission.
T_{rxon} is the packet reception time.
I_{txon} is the average current consumed by the transmitter.
I_{rxon} is the average current consumed by the receiver.
I_{stb} is the current consumed by the transceiver module in the standby mode.

The coordination of sleep cycle is important for the node power management. On the transmitter side, sleep cycle can be easily managed because the transmitter blocks can be turned ON or OFF by the MAC layer depending on a node's transmission requirements. The transmitter circuit will only be turned ON when the MAC layer receives packet from its higher layer. Switching the receiver block between ON/OFF modes is a critical issue. A node needs to know when it is expecting a packet or a control packet. For a TDMA or a poll-based system, the receiver can easily work out the packet reception time. In this case, the receiver can be left in the standby or sleep mode until the packet reception time. However, using the CSMA/CA protocol the packet reception time is not fixed until the transmitter and receiver work out the packet transmission schedule. In case of WBAN applications, sensor nodes will be mostly transmitting information packets. Sensor nodes will only receive acknowledgement packets or occasional control packets. On the other hand, the coordinator or the PAN coordinator will receive information and control packets from different sensor nodes; hence, it may be necessary for the coordinator the keep the receiver circuit ON for the duration of its operation. A receiver can save more power when receiving data from periodic sources. However, a WBAN node should be able to receive data from different type of sources, i.e., periodic and nonperiodic sources.

Power transmission requirement of a node is also influenced by transmission channel conditions. Although a WBAN will operate within a small area, it is possible that the receive signal will be corrupted when transmission loss increases due to the placement of a node or a posture of a body. Most of the modern transmitters will transmit at a minimum power level, which will allow the receiver to correctly receive the transmitted information. For a multi-patient monitoring system, it is important that mutual interference should be minimum; hence, all transmitters must transmit at their optimum power level. An optimum power level is defined as the minimum power level at which a receiver can receive packets with zero bit error rate (BER). The transmitter power level requirement will increase with increasing transmission losses. A transmitter can adjust its transmission power level based on the feedback from the receiver, which informs the link condition. A receiver can send a feedback using a positive ACK when a packet is received correctly or a NACK (negative acknowledgement) in case a packet is received incorrectly. For a WBAN, transmission of ACK or NACK could introduce additional traffic load. It may be possible that ACK or NACK mechanism can be used only for critical data leaving other nodes to transmit data at a predetermined power level. In this case, a receiver may occasionally send explicit instruction to the transmitter to change its power level to maintain zero BER level.

To examine the effect of sensor node power management issues, we use the CC2420 chip parameters for sample calculations. The receiver module of the chip operates on a fixed power level, which consumes 18.8 mA current. The transmitter power requirement depends on the transmitted RF (radio frequency) signal power. Transmitter current values are listed in Table 9.2 for different

Table 9.2. CC2420 IC current consumptions

RF signal Power Level (dBm)	Transmitter Current (mA)
−25	8.5
−15	9.9
−10	11
−5	14
0	17.4

transmitted RF signal powers. The table shows that the chip can operate at six different power levels. Strongest RF signal will consume an average current of 17.4 mA, whereas the weakest RF signal will consume 8.4 mA of current. The standby current of the IC is 1 μA. The required RF signal power level will be determined by the transmission loss value. The transmission loss value can be calculated using the Eq. 9.2. For WBAN applications, additional losses can add with the transmission loss. Major additional loss component value is the tissue absorption particularly for implanted nodes. The received signal strength should be at least equal to or greater than the receiver sensitivity. The receiver sensitivity of the CC2420 IC is -95 dBm. If a transmitter is generating a RF signal at −25 dBm, then the total transmission loss should be lower than 70 dBm.

$$PL\,(d)\ \mathrm{dB} = PL\,(d_0) + 10\gamma \log_{10}\left(\frac{d}{d_0}\right) + X_\sigma, \qquad (9.2)$$

where $PL(d_0)$ is the path loss at the reference distance, $d_0$is the reference distance, and X_σ is the shadowing variance.

Let's now analyze the power requirements of a CSMA/CA-based transmitter. The CSMA/CA protocol uses a series of procedures before a node can transmit a packet. Figure 9.7 shows the packet transmission sequence of the CSMA/CA protocol. When the transmitter senses a channel and initially it finds busy, then it waits until the channel is idle for the DIFS (distributed intra-frame spacing) duration. After the DIFS waits the transmitter enters

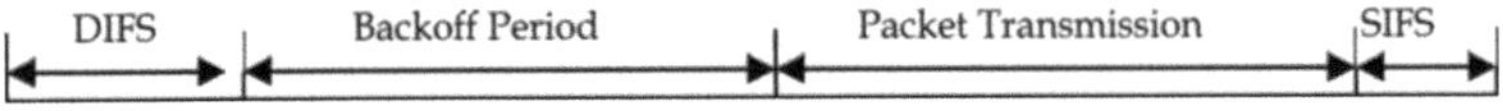

Figure 9.7. Packet transmission sequence of the CSMA/CA protocol.

into a backoff period. Following the backoff, the transmitter could transmit in a contention mode, i.e., transmit a packet with a risk of collision. After the transmission of a packet, the terminal will wait for a period known as SIFS (short IFS) [12]. After the SIFS, the transmitter is expected to receive an ACK packet. Typical CSMA/CA timing parameters are as follows. The DIFS length is 50 μs, minimum back period is 140 μs, and the SIFS value is 10 μs. The packet transmission time will depend on the packet size. The IEEE 802.15.4 CSMA/CA packet will consist of PHY layer header of 6 byte, MAC layer header 5 byte assuming a 2 byte address field and a 2 byte FCS (frame check sequence). Total transmission time of a packet can be calculated using Eq. 9.3 neglecting the propagation delay. For a 40 byte/320 bit payload, the packet length will be 51 byte, and the packet transmission time of the CC2420 transmitter will be 1.63 ms. Table 9.3 lists current consumption values of different CSMA/CA parameters for different RF signal levels. Values listed Table 9.3 is calculated based on the assumptions that a packet can be transmitted in the first attempt; hence, these are best case values. For example, when the channel is sensed and found to be busy, the transmitter will backoff and resense the channel after the backoff period. A node senses the channel after each backoff period until it finds the channel free for a DIFS period when it could initiate a packet transmission. In this case power transmission requirements of packet can significantly increase. If we assume that a WBAN node is using two 900 mAH AA batteries, then a node can transmit 30,706 packets at 0 dBm RF signal level. The same node can transmit 60,851 packets at −25 dBm RF signal level effectively doubling the lifetime of a battery for the best transmission case condition. In a

Table 9.3. Current consumption values for different CSMA/CA parameters

RF Signal (dBm)	DIFS (μA)	SIFS (nA)	Backoff (nA)	Packet transmission (μA)	Total (μA)
0	0.94	0.01	0.14	28.36	29.31
−5	0.94	0.01	0.14	22.82	23.76
−10	0.94	0.01	0.14	17.93	18.87
−15	0.94	0.01	0.14	16.13	17.07
−25	0.94	0.01	0.14	13.85	14.79

WBAN node, if one packet is transmitted per second, then the battery can potentially last up to 17 h for the −25 dBm RF signal level. Packet retransmissions can affect the life of a battery. However, in the best transmission case, the battery can last slightly longer than the above value due to the battery recharging phenomena as shown in Fig. 9.6.

$$T_{\text{packet}} = T_{\text{DIFS}} + T_{\text{BO}} + \frac{L}{R} + T_{\text{SIFS}} \tag{9.3}$$

where T_{DIFS} is the DIFS period, T_{BO} is the backoff period, L is the packet length in bits, R is the transmission rate in bits/s, and T_{SIFS} is the short interframe spacing.

Discussions in this section clearly show that the battery lifetime of WBAN will depend on selected MAC parameters and traffic characteristics. Data aggregation and packet scheduling technique will significantly influence the power requirements of a node. For a polling network node, transmission power requirements will depend on the design of polling message and cycle.

9.8 Patient Monitoring Network Design

In this section, we will introduce the general rules and techniques of WBAN design. As mentioned earlier, WBAN is a special-purpose, low-to-medium rate sensor network with a typical operating diameter of about 1 m. The number of nodes in a typical WBAN is expected to be around 2 to 5 with each node connected to a number of sensors. So the data rate requirements per node will not be significantly high unless a node is connected to a camera or supporting multichannel ECG data. Main QoS requirements of a WBAN are low latency, no or very minimum packet loss, and low power transmission to avoid any tissue heating or damage. Two main networking requirements of a WBAN are low power consumption and the network should be a scalable one. The size of WBAN could range from a single-body network to an interconnected multi-body network operating as a multi-hop network as shown in Fig. 9.8. To design a WBAN, the following design issues need to be considered.

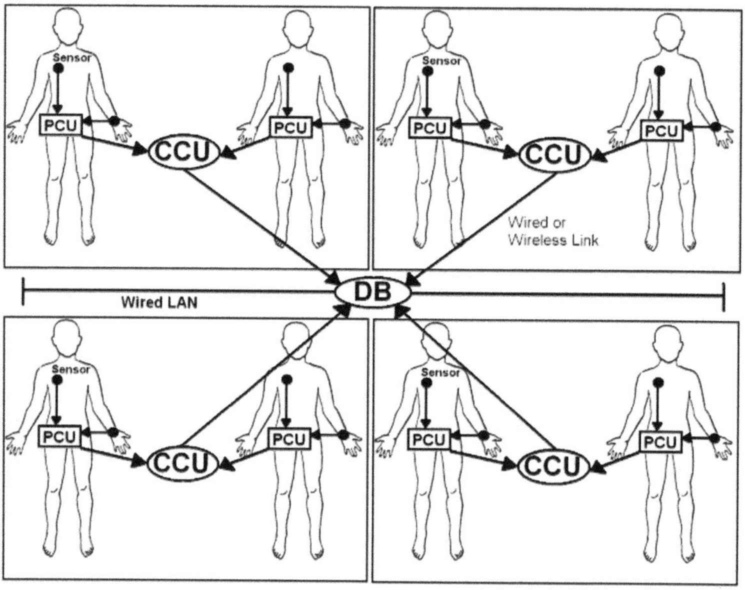

Figure 9.8. A WBAN-based multi-patient monitoring system. Each patient's body forms a PAN and consists of sensor nodes and the PCU.

9.8.1 *Transmission Capacity Requirements*

It is necessary to develop the data profile of a network. Most of the data will be generated from physiological sensors. It is important to examine all physiological signal characteristics and calculate the minimum and maximum transmission data requirements of each signal. Data must be characterized in terms of their interarrival rate, which is related to the sampling rate and the data burst size. The designer should also consider the dynamic range of signals and select the sampling rate and A/D converters appropriately. Next, the aggregate data volume should also be considered. The aggregate data rate is important for the selection of the networking standard and the MAC protocol. The total aggregate data rate of a network should not exceed 85–90% capacity of a scheduled access network. In case of a random access network, the aggregate data rate should be below 75% of the maximum capacity of the network. Performance of a network could be significantly affected by the aggregate data rate of a network. It is also important to work out the peak-to-average data transmission rate of a network. In a WBAN, most of the time data rate will be steady, but monitoring a critical patient may occasionally generate peak traffic. If the peak-to-average traffic rate is high, then occasional data loss can happen due to congestions or buffer overflow.

9.8.2 *PHY and MAC Layer Parameter Selection*

It is important to select PHY and MAC parameters appropriately to satisfy the requirements of a WBAN [8]. Although PHY layer parameter selections are difficult because in most cases PHY parameters are decided by the hardware design. Important PHY layer parameters include the modulation and coding techniques that can influence the performance of a sensor node. The modulation and coding parameters determine the receiver sensitivity and effective packet error rate (PER) based on the channel BER. Selection of MAC parameters is very crucial in determining performance of a network. Sections 9.4–9.7 have discussed various features of various MAC protocols. MAC protocol selection will influence the power consumption, QoS values, and scalability issues. Scalability could be an important issue for the WBAN deployment. A WBAN should offer flexibility to health care professionals so that they either increase or decrease the number of nodes on a body without any major change in software or hardware. It will be useful if the MAC protocol can adaptively accommodate varying number of sensor nodes. The TDMA protocol will not be very scalable unless software modifications are done on the network controller side. Using a polling network, it is possible to add or remove sensor nodes without major changes [11]. From the scalability point of view, the CSMA/CA protocol offers the best advantage. In case of CSMA/CA protocol, since there are no central coordination necessary, any new sensor node will read the broadcast message of its PAN coordinator and will register with PAN coordinator for data exchange within a WBAN.

Most of the sensor networking standards such as code blue, Wibree, Mica2, etc. have selected the CSMA/CA MAC protocol to develop medical applications [21]. As mentioned above, a CSMA/

CA-based system is adaptive in nature, requires minimum control signaling, and supports required QoS for a WBAN. Also, the CSMA/CA protocol can be optimized for different application scenario by appropriate selecting transmission and priority parameters. The IEEE 802.15.4-based hybrid protocol could be a good choice for future systems where low priority nodes can transmit packet using the contention mode, whereas the GTS mode could be used for high priority nodes. Hybrid protocol will also offer excellent scalability feature.

9.8.3 *Network Configuration*

The network configuration of a WBAN depends on the deployment scenario. A WBAN can be deployed either in a stand-alone mode or in an interconnected mode [20]. A stand-alone mode could be used for a single-patient monitoring system, whereas an interconnected mode could be used for either monitoring multiple patients or monitoring a patient remotely. A stand-alone mode will simply form a PAN with a CCU (central coordination unit) as the PAN coordinator. In this mode, the CCU will collect all sensor data and store them for further processing [9]. The protocol architecture will be very simple where application layer can directly communicate with the MAC layer for transmitting and receiving information. In this mode, we don't need to consider mobility because it is most likely that all nodes will be attached to a single human body and the CCU will also be located either on the body or close to the body in a fixed location.

Multi-patient or remote monitoring system will require a complex networking configuration where we need to construct a multi-hop network as shown in Fig. 9.8. In this case, each body will form a single PAN and all PANs need to be connected to a PCU (patient coordinator unit — a wearable device that can be carried, wearable by the patient). Each CCU will act as a router for their PAN and transmit all its sensor data to the PCU. The PCU could be connected to a hospital LAN (local area network) or to the Internet. All CCU's can transmit their packets in the broadcast mode with PCU address in the MAC address field. The multi-hop network needs to introduce some form of packet scheduler to minimize the contention if the CSMA/CA protocol is used. In the multi-hop networking scenario, all sensor nodes, CCUs, and PCU will transmit packets in the contention mode; hence, it is necessary that some form of power control and packet scheduling need to be introduced. For example, sensor nodes in each PAN should transmit their packets with a minimum power sufficient for their CCU to receive packets with zero BER. If sensor nodes transmit at a higher power level than necessary, then it may be possible packet transmissions in other PANs will be affected. It is also necessary that CCUs should introduce some form of pseudo synchronized packet scheduling so that all of them should not try to send their packets at the same time. Each CCU will individually adjust its transmitter's power based on its PCU link characteristics.

A multi-patient or a stand-alone patient monitoring system could be converted to a remote patient monitoring system by introducing the TCP/IP (transmission control protocol/Internet protocol) stacks either at the CCU or at the PCU. The TCP/IP protocol stack will allow these nodes to communicate with external networks. In this case no modifications will be necessary for the sensor nodes because they still will be sending their data within their PAN.

Some deployment scenario of WBANs may require the network to support mobility. For example, mobility support is necessary when a patient is monitored while they are carrying out their day-to-day activities either at home or at workplace. Mobility can introduce two additional problems: the variation of link characteristics and routing data to the appropriate node. If a mobile patient is supported within an indoor area, then it is most likely an access point will be used within the home or work area, which can connect it with the CCU to exchange data. In that case, we might use a Wi-Fi access point to connect with the CCU. It is preferable to use a Wi-Fi access point because it can support a larger cell size — up to 300 m link distance and relatively cheaper to implement. In this case, we need to develop a CCU with dual protocol stack. If we assume a ZigBee-based PAN and a Wi-Fi-based access point, then CCU will act as a router and it will have vertically split protocol stack supporting the ZigBee protocol on the PAN side and Wi-Fi on the CCU to access point link. Basically in this case, the CCU will be built using dual radio IEEE 802.15.4 and IEEE 802.11. If a WBAN requires mobility at a higher speed for sports training applications, then wireless standards such as GSM or UMTS or WiMAX should be used because Wi-Fi can only support mobility up to 5 km/h.

9.9 Performance Analysis of a WBAN

In this section, we will analyze the performance of a ZigBee-based patient monitoring system. Figure 9.8 shows a typical WBAN network architecture for mult-patient monitoring applications. The data from each sensor are transmitted to a PCU. The PCU aggregates sensor data and transmit them to a CCU located at a short distance. The PCU can gather data from both implant and external nodes. The CCU acts as an intermediate network device that forward the collected medical data to a patient database (DB) where remote monitoring devices can retrieve patient's data for health care professionals. Each patient body forms a PAN where a PCU acts as the PAN coordinator. In this WBAN application, the CCU acts as the PAN coordinator for all PCUs, which can be considered a second-tier network. The network operates in a multi-hop fashion where each link transmits data using the CSMA/CA protocol. Relay nodes such as the PCU and the CCU can be configured in two ways. One of the configurations is that these nodes simply act as forwarding node, i.e., it receives a packet and simply forwards the packet to the next link. Other configuration is that these nodes receive multiple packets from sensor nodes and then encapsulates multiple packets and forwards a larger packet. The former approach can be implemented with nodes with low-processing capability, whereas the latter approach reduces the number of transmission access on the forward link, thus reducing the probability of collisions for a ZigBee-based system.

In this section, we present the performance of the above WBAN for multi-patient monitoring applications using an OPNET simulation model [23]. Using the simulation model, initially we analyze the performance of the data aggregation technique. To analyze the performance of data aggregation technique, we use a single PAN model. Figure 9.9 shows the packet loss and packet generation rate of the simulated WBAN for different payload sizes. Result shows that with the increasing payload size, the throughput

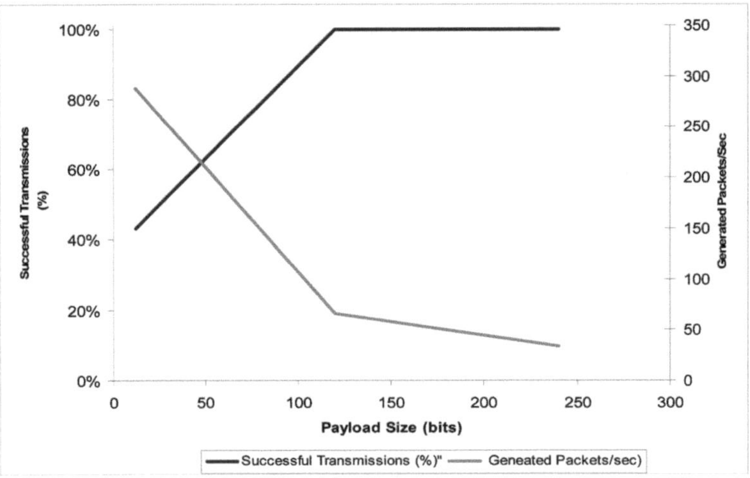

Figure 9.9. WBAN packet loss and packet generation rate for different data aggregation rate/payload size. See also Color Insert.

of WBAN increases due to fewer transmission attempts. The figure shows that with single sample/payload (12 bits), the network only managed to successfully transmit 120 packets, offering a success rate of 40%. When the average payload size is increased to 120 bits/packet, then the number of transmission attempts is dropped to 50 packets/second, offering 100% throughput in the network. The improvement in the efficiency can be achieved due to lower contention level in the network, which resulted fewer packet losses due to the packet dropping threshold. In this simulation, we used five retransmission attempts as the packet dropping threshold, which means if a packet is unable to successfully transmit a packet in five successive attempts, then the packet is dropped from the transmission queue. The delay profile of the WBAN for different payload sizes is also investigated and presented in Fig. 9.10. The plot shows that the average packet delay decreases with the increasing payload size. Both results (Figs. 9.9 and 9.10) show the same trend.

Next we investigate the performance of an interconnected WBAN to analyze the performance of a multi-patient monitoring system. In the proposed scenario, it is assumed that all patients' sensor

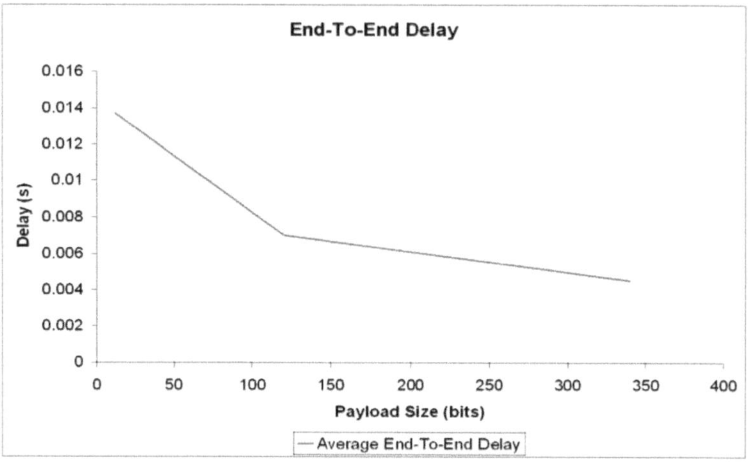

Figure 9.10. WBAN delay profile for different payload sizes.

nodes transmit their data via the PCU and CCU to the database. In this model, PCUs and CCUs are simply acting as a relay node, and forwarding those data to the DB. In order to minimize the number of packets, we use the data aggregation technique at the source node. In this simulation, we connect all sensors on a body using two sensor nodes transmitting data to the PCU. One of the sensor nodes (sensor 1) aggregates ECG, body temperature, and blood pH data and transmits these data after every 145 ms generating 7 packets/second. The other sensor node (sensor 2) aggregates data from the blood flow, blood pressure, and respiratory rate sensors, and transmits these data after every 485 ms generating 3 packets/second. This aggregated packet structure allows 73 data samples with a payload size of 880 bits/packet. Using the aggregation technique, each PAN generates only 10 packets/second. The effective data rate requirement of sensor 1 and sensor 2 becomes 6.05 kbps and 1.92 kbps, respectively.

The delay profile for a multi-patient networking scenario is depicted in Fig. 9.11. The plot shows the average end-to-end delay, which is the total delay over 3 hops (sensor→PCU, PCU→CCU, and CCU→DB). It is observed that the delay increases with increasing number of patients. In this case, the main reason for the increase

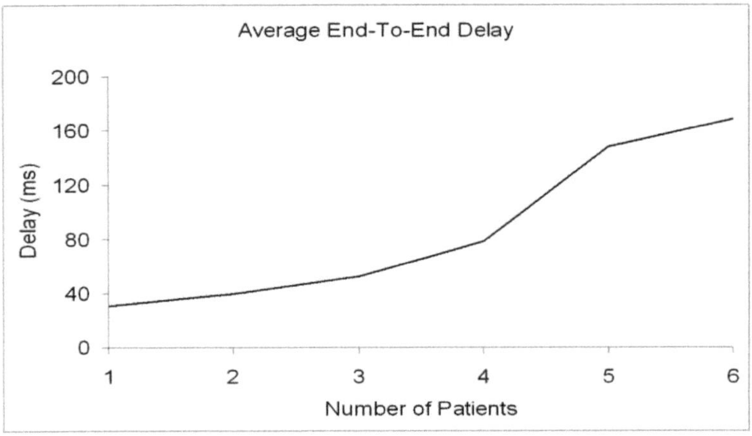

Figure 9.11. Delay profile of the multi-patient monitoring system.

in delay is caused by the contention or interference from different PANs. For an operating ZigBee network, sensor nodes on each patient's body forms a working PAN with the PCU as the coordinator. In a multi-hop networking, collisions can happen either within a PAN or between PANs when simultaneous transmission goes on in different PANs. Also, transmission from the PCU to CCU or from the CCU to DB transmissions can be interfered by the transmissions from other PANs if the IEEE 802.15.4 MAC is used on all the links. In our simulation, all the wireless links used the CSMA/CA protocol. The delay plot shows that the delay profile for up to six patients is quite acceptable where the average packet delay is about 170 ms. A monitoring system will be able to transmit large number of samples within this 170 ms due to the aggregation technique used.

9.10 Conclusions

This chapter presented WBAN network design techniques for various application scenarios. WBAN is currently considered a developing technology, which needs to be standardized for clinical applications. There are many networking techniques that need to be appropriately developed to support medical applications. Since the MAC protocol is one of the key components of a WBAN design, it is necessary to select appropriate protocol and configure them to achieve optimum performance. This chapter presented theoretical discussions on MAC protocol and network design techniques, which will be useful for medical, IT professionals for future medical application development. Simulation results presented in Section 9.9 can act as a guideline for WBAN design engineers. Section 9.9 also demonstrates the importance of network simulation for the development of an efficient WBAN architecture.

Acknowledgement

Author wishes to acknowledge contributions of Mr. Garrick Bulger and Mr. Benjamin Harding. Mr. Bulger graduated with B.E. Electrical Engineering from the University of Newcastle in 2007. Mr. Harding graduated with B.E. Computer Engineering from the University of Newcastle in 2009.

References

Journals

1. Soomro, A., and Cavalcanti, D. (2007) Opportunities and Challenges in Using WPAN and WLAN Technologies in Medical Environments, *IEEE Communications Magazine*, **45(2)**, pp. 114–122.
2. Golmie, N., Cypher, D., and Rebala, O. (2005) Performance analysis for Low Rate Wireless Technologies for Medical Applications, *Computer Communications*, **28**, pp. 1266–1275.
3. Omeni, O., Wong, A. C. W., Burdett, A. J., and Tomazou, C. (2008) Energy Efficient Medium Access Protocol for Wireless Body Area Networks, *IEEE Transactions on Biomedical Circuits and Systems*, **2(4)**, pp. 251–258.
4. Demirkol, I., Ersoy, C., and Boazici, F. A. (2006) MAC Protocols for Wireless Sensor Networks: A Survey, *IEEE Communications Magazine*, **44(4)**, pp. 115–121.
5. Su, H., and Zhang, Xi. (2009) Battery Dynamics Driven TDMA MAC Protocols for Wireless Body Area Monitoring Networks in Healthcare

Applications, *IEEE Journal on Selected Areas in Communications*, **27(4)**, pp. 424–434.

6. Sharon, O., and Altman, E. (2001) An Efficient Polling MAC for Wireless LANs, *IEEE/ACM Transactions on Networking*, **9(4)**, pp. 439–451.
7. Gutierrez, J. A., Naeve, M., Callaway Ed., Bourgeois, M., Mitter, V., and Heile, B. (2001), IEEE802.15.4: A Developing Standard for Low Power Low Cost Wireless Personal Area Network, *IEEE Network*, **15(5)**, pp. 12–19.
8. Ullah, S., Higgins, H., Islam, S. M. R., Khan, P., and Kwak, K. S. (2009) On PHY and MAC Performance in Body Sensor Networks, *EURASIP Journal on Wireless Communications and Networking*, 2009, pp. 1–7.
9. Yuce, M. R., Ng, P. C., and Khan, J. Y. (2008) Monitoring of Physiological Parameters From Multiple Patients Using Wireless Sensor Network, *Journal of Medical Systems*, **32(5)**, pp. 433–441.
10. Baek, S. J., Veciana, G. de, and Su, X. (2004) Minimizing Energy Consumption in Large Scale Sensor Networks Through Distributed Data Compression and Hierarchical Aggregation, *IEEE Journal on Selected Areas of Communications*, **22(6)**, pp. 1130–1140.

Book

11. Khan, J. Y., and Yuce, M. R. (2010) Wireless Body Area Network (WBAN) for Medical Applications, in *New Developments in Biomedical Engineering*, (ed. Campolo, D.), In-Teh, Croatia, pp. 591–627.
12. Leon-Garcia, A., and Widjaja, I. (2004) *Communication Networks: Fundamental Concepts and Key Architecture*, McGraw Hill, New York.
13. Karl, H., and Willig, A. (2005) *Protocols and Architecture for Wireless Sensor Networks*, John Wiley, England.
14. Ganz, A., Ganz, Z., and Wongthavarawat, K. (2004) IEEE802.15, in *Multimedia Wireless Networks: Technologies, Standards and QoS*, Prentice Hall, New Jersey, pp. 165–185.

Proceedings

15. Chan-Soo , H., Kibeom, S., and Cioffi, J. M. (2006) Opportunistic *p*-persistent CSMA in Wireless Networks, in *Proceedings of the IEEE International Communications Conference*, pp. 183–188.
16. Timmons, N. F., and Scanlon, W. G. (2004) Analysis of the Performance of IEEE 802.15.4 for Medical Sensor Body Area Networking, in *Proceedings of the IEEE SENCON*, pp. 16–24.

17. Zhang, Y., and Dolman, G. (2009) A New Priority-Guaranteed MAC Protocol for Emerging Body Area Networks, in *Proceedings of the IEEE International Conference on Wireless and Mobile Communications,* pp. 140–145.
18. Li, C., Wang, L., Li, J., Zhen, B., Huan-Bang, Li., and Kohno, R. (2009) Scalable and Robust Medium Access Control in Wireless Body Area Networks, in *Proceedings of the 20th IEEE International Symposium on Personal, Indoor and Mobile Radio Communications 2009,* pp. 2127–2131.
19. Khan, J. Y., Yuce, M. R., and Harding, B. (2010) Battery Life Cycle and Transmission Power Profile Analysis of a Wireless Body Area Network With Implanted Nodes, in *Proceedings of the 4th International Symposium on Medical Information and Communication Technology.*
20. Khan, J. Y., Yuce, M. R., and Karami, F. (2008) Performance Evaluation of a Wireless Body Area Sensor Network for Remote Patient Monitoring, in *Proceedings of the 30th Annual International IEEE Engineering in Medicine and Biology Society Conference.*

Technical Report

21. Shnayder, V., Chen, B.-R., Lorincz, K., Fulford-Jones, T. R. F., and Welsh, M. (2005) Sensor Networks for Medical Care, *Technical Report TR-08-05,* Division of Engineering and Applied Sciences, Harvard University.

Website

22. http://www.ieee802.org/15/pub/TG6.html.
23. http://www.opnet.com.

Chapter 10

Power Management in Body Area Networks for Health Care Applications

Vijay Sivaraman,[a] Ashay Dhamdhere,[b] and Alison Burdett[b]

[a]*School of Electrical Engineering and Telecommunications, The University of New South Wales, Sydney, NSW 2052, Australia.*

[b]*Toumaz Technology Limited, 115 Milton Park, Abingdon, OX14 4RZ, United Kingdom.*

{vijay, ashay}@unsw.edu.au; alison.burdett@toumaz.com

For an increasing number of people living with chronic medical conditions, wearable wireless sensor devices can provide nonintrusive yet continuous physiological monitoring, allowing effective clinical management without compromising quality of life. Truly non-intrusive sensor devices will have light weight and small form factor, placing fundamental constraints on the available energy, which necessitates very careful energy management at all layers. This chapter investigates the opportunities and challenges in the use of dynamic radio transmit power control for prolonging the lifetime of such energy-constrained, body-wearable sensor devices.

We first present extensive empirical evidence that the wireless link quality can change rapidly in body area networks (BAN), and a fixed transmit power results in either wasted energy (when the link is good) or low reliability (when the link is bad). We quantify

Wireless Body Area Networks: Technology, Implementation, and Applications
Edited by Mehmet R. Yuce and Jamil Y. Khan

ISBN 978-981-4316-71-2 (Hardcover), ISBN 978-981-4241-57-1 (eBook)
www.panstanford.com

the potential gains of dynamic power control in body-worn devices by benchmarking off-line the energy savings achievable for a given level of reliability. We then propose a class of schemes feasible for practical implementation that adapt transmit power in real time based on feedback information from the receiver. We profile their performance against the off-line benchmark and provide guidelines on how the parameters can be tuned to achieve the desired trade-off between energy savings and reliability within the chosen operating environment. Finally, we implement and profile our scheme on a MicaZ mote-based platform, and also report preliminary results from the ultra-low-power integrated health care monitoring platform we are developing at Toumaz Technology. Our work sets the stage for a holistic approach to power management, incorporating innovations across all layers of BAN design.

10.1 Introduction

Lifestyle changes combined with an aging population and poor diet are contributing to an ever-increasing number of people living with chronic medical conditions requiring ongoing clinical management. This has resulted in a heavy burden on health care systems that are primarily geared toward treating acute conditions. Wireless sensor network technologies have the potential to offer large-scale and cost-effective solutions to this problem. Outfitting patients with tiny, wearable, vital-signs sensors would allow continuous monitoring by caregivers in hospitals and aged-care facilities, and long-term monitoring by individuals in their own homes.

To successfully deploy BANs that can perform long-term and continuous health care monitoring, it is critical that the wearable devices be small and lightweight, lest they would be too intrusive on patient's lifestyle. This places fundamental limitations on the battery energy available to the device over its lifetime. Typical prototype devices in use today, such as MicaZ motes [1] used in Harvard's CodeBlue [2] project, operate on a pair of AA batteries that provide a few watt-hours (a few tens of kilo-joules) of energy. Truly wearable health monitoring devices are emerging that have orders of magnitude lower battery capacity — at Toumaz Technology we

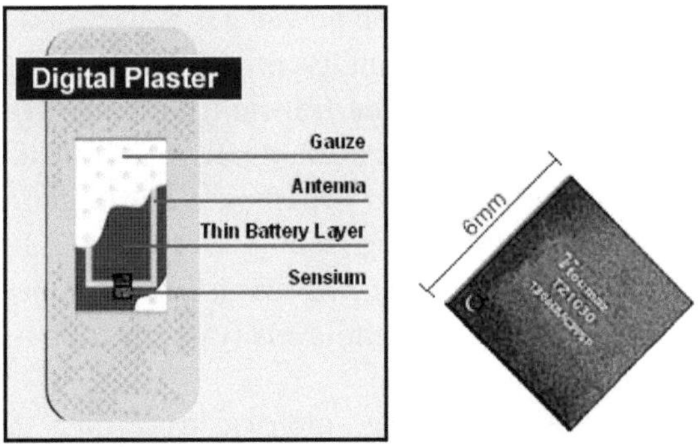

Figure 10.1. Toumaz Sensium™ digital Plaster. See also Color Insert.

are building a new generation of single-chip low-cost disposable "digital plasters" (shown in Fig. 10.1) that provide nonintrusive ultra-low power monitoring of ECG, temperature, blood glucose, and oxygen levels. Our Sensium™ chip operates on a flexible paper-thin printed battery [3] with a capacity of around 20 mW-h (approximately 70 J). Such stringent energy constraints necessitate very careful energy management.

Communication is the most energy-consuming operation that a sensor node performs [4] and can be optimised at multiple layers of the communication stack. At the physical layer, we at Toumaz have innovated an ultra-low-power radio [5] suited to BANs: our radio provides a proprietary 50 Kbps wireless link over a distance of 2–10 m and consumes 2.7 mW at a transmit strength of −7 dBm (compare this to the CC2420 radio [6] in MicaZ motes that consumes 22.5 mW for −7 dBm output). At the data-link layer, energy can be saved by intelligent medium access control (MAC) protocols that duty-cycle the radio, i.e., by turning the radio off whenever packet transmission or receipt is not expected. Several such MAC protocols have been developed in the literature (see [7] for a survey). The B-MAC [8] protocol included in the TinyOS distribution provides versatility to the application in controlling the duty-cycling of the radio, while at Toumaz we have developed our proprietary MAC protocol [9] suited to BANs. However, these MAC protocols only control *when* the radio is switched on, they do not determine the *output power* of the radio when it is on. The focus of this work is to study the impact of transmission power (for any given transmission schedule) in trading off reliability of the time-varying wireless link for energy efficiency at the transmitting node. We note that the ability to control the transmission power is available on most platforms: the CC2420 radio in Crossbow's MicaZ motes provides 32 transmission levels (ranging from −25 dBm to 0 dBm output) selectable at run-time by configuring a register, while our Sensium^{TM} platform similarly supports eight levels (ranging from −23 dBm to −7 dBm output).

First, we present extensive empirical trace data that profiles the temporal fluctuations in the body area wireless channel. The *large* variations show fixed transmit power to be suboptimal: when link quality is poor, low transmit levels result in reduced reliability, whereas when link quality is good, high transmit levels waste energy. Furthermore, the *rapid* variations render existing schemes (discussed in detail in Section 10.2), that adapt transmit power over long time scales (hours and days), inappropriate for use in BANs. Using trace data we compute off-line the "optimal" power control scheme, i.e., one that minimizes energy usage subject to a given lower bound on reliability. Though infeasible to realize in practise, the optimal gives insight into the potential benefits and fundamental limitations of adaptive power control in BANs, and also provides a benchmark against which practical schemes can be compared.

Second, we develop a class of practical on-line schemes that dynamically adapt transmission power based on receiver feedback. These schemes are easy to implement and can be tuned for desired trade-off between energy savings and communication reliability. We show conservative, balanced, and aggressive adaptations of our scheme that progressively achieve higher energy savings (14–30%) in exchange for higher packet losses (up to 10%). We also provide guidelines on identifying algorithm parameter settings appropriate to application requirements and operating conditions.

Third, we present a real-time implementation of our power control scheme on a MicaZ mote-based platform, demonstrating that energy savings are achievable even with imperfect feedback information. We also present preliminary observations of the efficacy of our scheme in the ultra-low power platform for continuous health care monitoring being developed at Toumaz Technology. Our work shows adaptive transmit power control as a low-cost way of extending the battery-life of severely energy-constrained body wearable devices and opens the doors to further optimizations customised for specific deployment scenarios.

The rest of this chapter is organised as follows: Section 10.2 briefly discusses prior work in the area of adaptive power control. Section 10.3 presents empirical observations on channel variability in BANs and motivates dynamic power control. Optimal off-line power control is explored in Section 10.4, while practical on-line schemes are proposed and analyzed in Section 10.5. Section 10.6 describes our implementation and experiments on the MicaZ mote and Toumaz Sensium^{TM} platforms, while conclusions and directions for future work are presented in Section 10.7.

10.2 Related Work

Transmit power control has been studied extensively in the literature in several contexts with different objectives. A large number of works, e.g., references [10–15], consider the IEEE 802.11 environment for wireless ad hoc networks and propose adjusting transmit power or transmit rate for data packets based on several factors such as wireless channel conditions (probed via RTS/CTS messages), payload length, etc. In spite of the useful insights from these works, there are significant differences between the IEEE 802.11 WLAN (wireless local area network) environment and a BAN for health care monitoring such as: (a) WLAN devices send sporadic data while BAN devices typically send data periodically, (b) packet sizes are much smaller in BANs than in WLANs, and (c) BAN devices operate on or very near the human body, which makes radio propagation (and hence channel conditions) in BANs markedly different from WLANs. These important differences merit study of power control in the specific context of BANs, which to the best of our knowledge has not been undertaken to date.

A large body of work in power control has also targeted multi-hop networks with the objective of enhancing throughput [16], increasing connectivity [17, 18] or reliability [19], and reducing delays [20]. Joint routing, scheduling, and power control schemes have also been proposed [21, 22]. Our work considers a single-hop BAN where such issues do not arise. Moreover, the environment presented by a BAN is much more dynamic than the psuedo-static scenarios considered by these works.

More relevant to this work are existing power control schemes that target energy savings in wireless sensor networks. The study in refference [23] proposes two algorithms that adapt transmission power by exchange of information among nodes and based on signal attenuation, while reference [24] proposes a linear prediction model for estimating the optimal transmission power based on measured link quality. However, these studies have targeted static deployments (such as for environmental or structural monitoring applications) wherein variability in wireless link quality has been shown empirically [25, 26] to be slow. In contrast, this work considers wearable mobile devices for which the wireless link quality can change significantly and rapidly since it is very susceptible to position and orientation of the human body [27]. To the best of our knowledge, adaptive power control for body-wearable devices has not been explored by other researchers before.

Two recent papers make interesting observations that are complementary to our work: the authors in reference [28] investigate the number of different power levels that can be effectively leveraged by power control algorithms in an indoor WLAN environment. They show that, due to multipath and fading effects, there is significant overlap between the RSSI distributions for nearby power levels, making them practically indistinguishable at the receiver, and claim that as few as four power levels may suffice to make power control attractive. We leverage this fact when we shift from the MicaZ (32 power levels) to the Sensium^{TM} (8 power levels). Lastly, the study in reference [29] points out that power control by itself does not lead to significant energy savings unless complemented by a good MAC protocol that has low duty-cycling of the radio. We concur with this observation and note that the MAC protocol used in the Sensium^{TM} platform has very low duty cycle.

10.3 The Case for Transmit Power Control in Body Area Networks

We begin by empirically profiling the temporal variations in the quality of the wireless link between a body-worn device and a fixed base-station, as a patient wearing the device performs various activities. The patient was played by the first author. Our experiments in this section use the MicaZ motes from Crossbow Technologies [1], while some preliminary results that use the Toumaz Sensium^{TM} platform are presented in Section 10.6. In each experiment the device was strapped around the patient's chest, simulating continuous monitoring of heartbeat and ECG. We also conducted several experiments in which the device was strapped around the patient's arm (for monitoring blood pH and glucose); the results were qualitatively similar and are omitted here due to lack of space. The experiments were conducted indoors in an office space containing 10 cubicles. The base-station was placed close to one side of the room at an elevation (atop a shelf) to provide better line-of-sight coverage across the office space.

The MicaZ mote operates in the 2.4 GHz frequency band and can support a 250 Kbps data rate. It supports 32 RF output power levels, controllable at run-time via a register; the output power (in dBm) and corresponding energy consumption rate (in mW) for various levels are shown in Table 10.1. Since our goal is to save energy at the body-worn device (which typically has lower energy resources than the base-station), our experiments involve emitting packets periodically from the body-worn device at various power levels,

Table 10.1. Characteristics of the MicaZ CC2420 radio

Transmit level	Output (dBm)	Power (mW)
31	0	31.3
27	−1	29.7
23	−3	27.4
19	−5	25.0
15	−7	22.5
11	−10	20.2
7	−15	17.9
3	−25	15.3

and measuring the link quality at the receiver (base-station). Two metrics of link quality are available on the mote platform: received signal strength indicator (RSSI), which is computed internally in the radio by averaging the signal power over eight symbol periods of the incoming packet, and link quality indicator (LQI) metric, which measures the chip error rate for the first eight symbols of the incoming packet. For a static scenario, previous research [24, 30, 31] has observed LQI readings to suffer from early saturation and be relatively less stable. We conducted experiments to verify this in a BAN scenario. Figure 10.2a,b show the RSSI and LQI as a function of transmit power level for various scenarios of patient distance and orientation relative to the base-station. While the RSSI seems to show a smooth increase with transmit power, the LQI saturates early on for several scenarios and also seems to exhibit a large variance (shown as error-bars), which makes the readings less reliable. Figure 10.2c,d show the RSSI and LQI as a function of

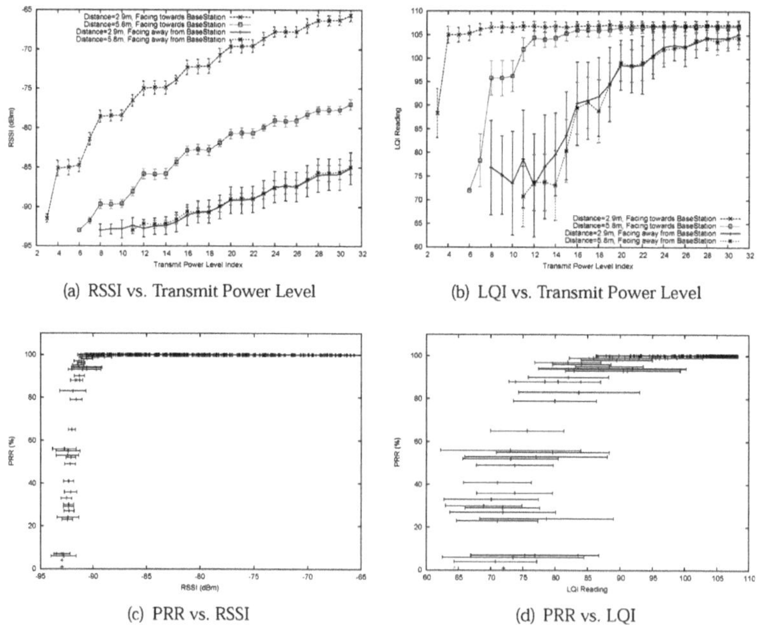

(a) RSSI vs. Transmit Power Level

(b) LQI vs. Transmit Power Level

(c) PRR vs. RSSI

(d) PRR vs. LQI

Figure 10.2. Comparison of RSSI and LQI as indicators of channel quality in a BAN.

the packet receive ratio (PRR) observed across several experiments. While the RSSI seems to provide a good estimate of packet loss rates (e.g., RSSI of −90 dBm or larger always corresponds to PRR of 95% or more), the LQI is seen to be a much weaker indicator of PRR due to its high variance. Our subsequent study, therefore, uses RSSI as an indicator of channel quality.

We profile, at different radio transmit levels, the changes in link quality with time as patients perform their routine activities involving resting, moving, turning, etc. To compare link quality at different power levels, we should ideally take *simultaneous* measurements at all power levels, which is infeasible. As an approximation, we make the body-worn device transmit every packet multiple times in quick succession at 16 different transmit levels 31, 29, 27, ... 1. The receiver (base-station) can thus record, more-or-less simultaneously, the signal strength corresponding to each transmit level. Measurements for three scenarios are described next.

10.3.1 *Normal Walk*

This scenario has the patient walking back and forth in the room for a few minutes at a normal walking pace; the patient stays between 1 and 8 m from the base-station at all times. The body device, strapped onto the patient's chest, generates a packet every second (which is not entirely unrealistic for a heartbeat/ECG monitor) and transmits it at 16 different output levels. The RSSI is recorded at the base-station for each packet at each transmit level, and plotted in Fig. 10.3a against time for four of the transmit levels. For any fixed transmit power, the received signal strength fluctuates widely: at fixed maximum transmit output (level 31 at 0 dBm), the signal strength at the receiver changes from −64 dBm (at 34 s) to −94 dBm (at 86 s — a change of 30 dBm under a minute. There are nevertheless some discernible trends: for example, in the interval 30–50 s, the received signal is consistently above −72 dBm (at the maximum transmit level) due to the clear line-of-sight presented by the patient walking toward the base-station, while the subsequent interval 50–70 s exhibits RSSI below −75 dBm (again at the maximum transmit level) due to the patient turning and

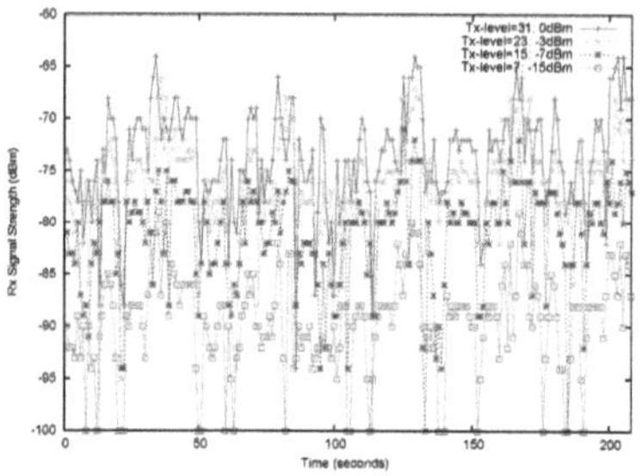

(a) Normal Walk

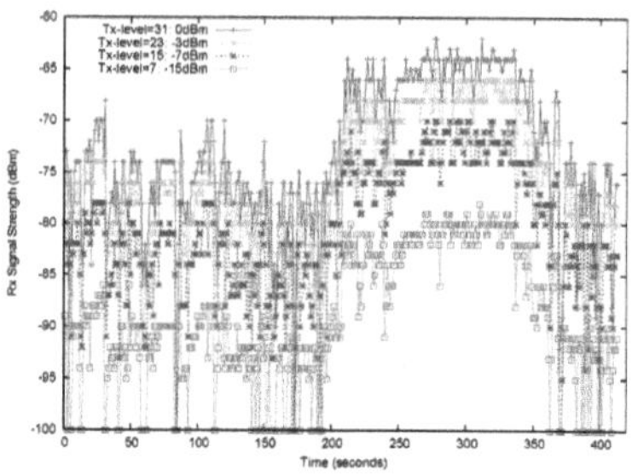

(b) Slow Walk

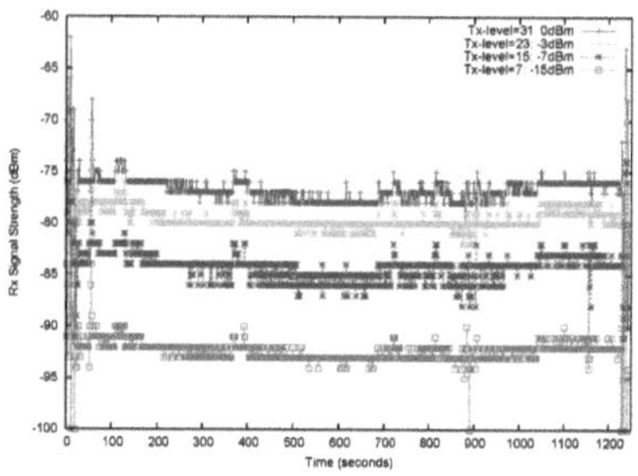

(c) Resting

Figure 10.3. RSSI vs. time for various patient scenarios. See also Color Insert.

blocking the line-of-sight with his body. The question of whether these patterns present opportunities for energy savings by adapting transmit power will be tackled in Section 10.4.

10.3.2 *Slow Walk*

In this scenario, we consider a slow-moving person (such as an elderly or a handicapped person with restricted mobility) who takes an exaggeratedly long time (over 6 min) to walk a distance of 3 m and back. As before, packets are transmitted every second at several power levels, and the received signal strength at the base-station is recorded and depicted in Fig. 10.3b. The trend in the plot is very evident: the RSSI is fairly low for the first half, when the patient's body blocks the line-of-sight between the body-worn device and the base-station, and then rises to a perceptibly higher value in the second part of the experiment when the patient is walking facing the base-station (barring the last few seconds when the patient turns again). This scenario depicts the shortcomings of fixed transmit power: a low transmit level would result in weak signals (and packet loss) during the first half, while a high transmit level would unnecessarily waste energy in the latter half. Such a scenario, therefore, presents opportunities for saving energy by reducing transmit power adaptively when the good channel persists.

10.3.3 *Resting*

In this scenario, the patient sits down to rest for approximately 20 min on a chair at a distance of about 6 m from the base-station. Fig. 10.3c plots the RSSI over the entire period, at several transmit levels. The wireless link is found to be fairly stable when the patient is at rest (in spite of a few other people moving around at several points in the experiment). This is in some sense an "ideal" environment with tremendous potential for energy savings, particularly with patients who are resting for a major part of the day. These energy savings would be unattainable if the transmit level were fixed, since a fixed setting would have to cater to the worst-case scenario of a poor channel.

Having gained an understanding of the wireless channel under various patient activity scenarios, the next section quantifies the potential benefits of adaptive transmit power control.

10.4 Optimal Off-Line Transmit Power Control

To quantify the potential benefits of adaptive transmit power control, we compute what the "optimal" transmission level might be for each of the scenarios considered before. We define the "optimal" as the lowest required transmit power level (as a function of time) to achieve a minimum target RSSI. Based on our studies in Section 10.3 relating packet loss with RSSI for the MicaZ motes in a BAN setting, we choose a conservative target RSSI of −85 dBm. The computation of the optimal transmission level defined thus is done off-line, i.e., using the traces shown in the previous section. For each scenario, at each time instant, we know the RSSI for each transmit power level, and we can, therefore, identify the lowest transmit power at which the signal strength at the receiver is no lower than the threshold of −85 dBm (if all received signal strengths are below the lower threshold we set the transmit level to be the maximum). We note that such a scheme is not implementable in practice, since it would require the transmitter to have instantaneous knowledge of the RSSI at the receiver for each choice of transmit power level, which is infeasible given that the channel varies with time.

The optimal transmit levels, and their associated RSSI values, for each of the three scenarios are depicted as a function of time in Fig. 10.4. Subplot (a) shows, for the normal walk scenario, that the optimal changes rapidly to track the rapid fluctuations in channel quality, thereby maintaining a fairly stable RSSI as shown in subplot (b): for example, in the time interval 50–75 sec, the optimal transmit level fluctuates multiple times between a high of 29 and a low of 9. Based on the energy draw for each transmit power level (shown in Table 10.1), we can compute the energy savings of optimal power control to be around 34% as compared to using the maximum transmit power. However, as the rapid fluctuations in the optimal level indicates, a practical scheme is unlikely to be able to predict the current optimal transmit level based on prior channel quality.

The optimal transmit power for a slow walk in Fig. 10.4c shows high sensitivity to body orientation, even when the motion is very slow. The rapid changes during the first 200 s arise from minor variations in the patient's body orientation while blocking the line-of-sight between the body-worn device and the base-station (indeed a few packets are lost even at the highest transmit power). But when the patient turns (at approximately 200 s), there is a clear line-of-sight, and the wireless link is relatively stable permitting the optimal transmit power to remain low for a considerable length of time (more than 2 min). This indicates that if the body orientation is favorable, periods of slow activity could be capitalized by a transmit control scheme to save energy without compromising realiability.

When the patient is resting, the link is fairly stable and the optimal transmit power level is near-constant as shown in Fig. 10.4e, which permits an energy savings of over 38% compared to maximum transmit power. It would seem the quiescent wireless channel in this case gives ample opportunity for practical schemes to reduce transmit power without sacrificing reliability. The design of such schemes is discussed next.

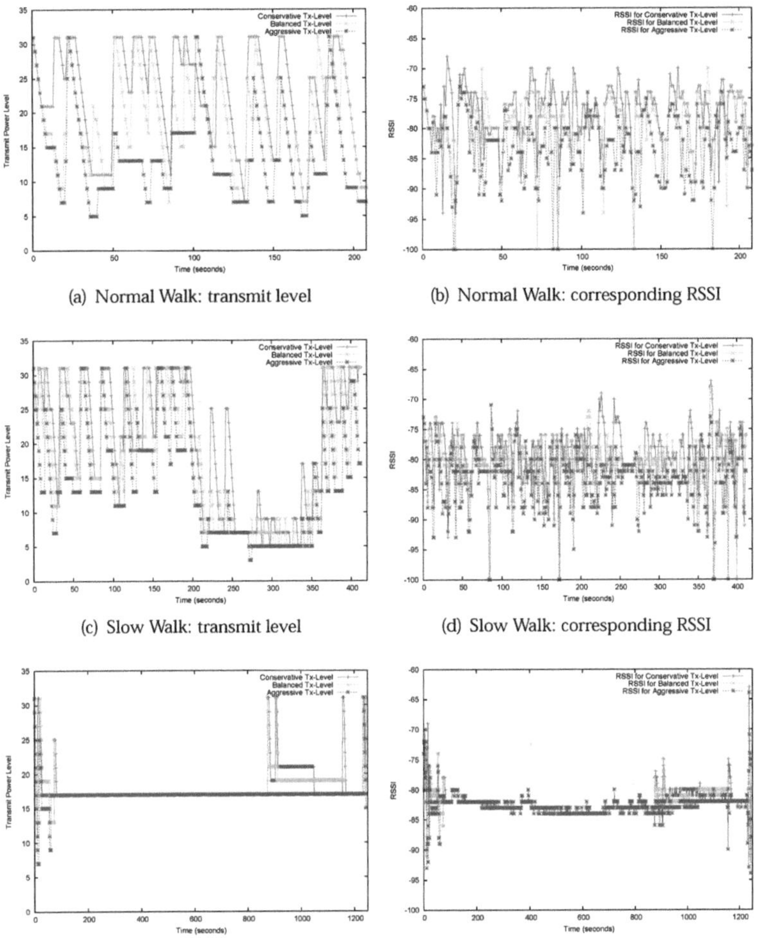

(a) Normal Walk: transmit level

(b) Normal Walk: corresponding RSSI

(c) Slow Walk: transmit level

(d) Slow Walk: corresponding RSSI

(e) Resting: transmit level

(f) Resting: corresponding RSSI

Figure 10.4. Optimal transmit power and associated RSSI for a normal walk, slow walk, and resting position. See also Color Insert.

10.5 Practical On-Line Transmit Power Control

The optimal transmit power control scheme above was performed off-line and required the sender to have *a priori* knowledge of the link quality at the receiver, which is infeasible in reality. This section develops practical algorithms that are then benchmarked against the optimal.

A *predictive* approach to designing on-line schemes for power control would require a wireless channel model for BANs. Modeling the propagation of electromagnetic waves around the human body, e.g., via "creeping waves" [32], is fairly complex as it needs to account for the permittivity and conductivity of the different layers of bone and tissue in the human anatomy. Additionally, the model has to contend with changes in orientation of the human body, mobility of the patient, and other spatio-temporal aspects (such as room layout, people in the vicinity, etc.). We believe such predictive models are too complex to implement on energy-constrained wearable devices and do not pursue them in our work.

We focus instead on *reactive* schemes that adjust transmit power based on feedback from the receiver (inspired by the way TCP adjusts its transmission rate in reaction to congestion in the Internet). Specifically, the base-station measures the RSSI for each received packet and feeds it back to the body-worn device in the acknowledgment packet. In this section, we assume that the feedback information is perfect (i.e., acknowledgment packets are never lost); this assumption will be relaxed in Section 10.6.1.

10.5.1 *A Simple and Flexible Class of Schemes*

At its core, any reactive power control scheme must ramp up transmit power when the channel quality deteriorates (so as to avoid packet loss) and decrease transmit power when the channel quality improves (in order to save energy). We propose a general class of schemes that allow these operations to be tuned via appropriate parameter settings.

Algorithm 10.1 depicts our class of schemes and is characterized by four parameters: α_u, α_d, T_L, and T_H. The scheme maintains a running average $\bar{R}$ of the RSSI, computed by exponentially weighted averaging (steps 2 and 4) of each newly obtained sample. The averaging weight α_u for a sample representing an improving channel can, in general, be different from the weight α_d used for a sample representing a deteriorating channel; this gives flexibility to a scheme in reacting differently to a perceived increase or decrease

Algorithm 10.1 A class of power control schemes

Require: R {RSSI from the current sample}
Require: $\bar{R}$ {Average RSSI}
1: **if** $R \leq \bar{R}$ **then**
2:     $\bar{R} \leftarrow \alpha_d R + (1 - \alpha_d)\bar{R}$
3: **else** $\{R > \bar{R}\}$
4:     $\bar{R} \leftarrow \alpha_u R + (1 - \alpha_u)\bar{R}$
5: **end if**
6: **if** $\bar{R} < T_L$ **then**
7:     Double the transmit power
8: **else if** $\bar{R} > T_H$ **then**
9:     Reduce the transmit power by a constant
10: **else** $\{T_L \leq \bar{R} \leq T_H\}$
11:     No action is required
12: **end if**

in channel quality, thereby placing different emphasis on energy savings versus packet loss.

The scheme increases or reduces transmit power by comparing the running RSSI average $\bar{R}$ to lower and upper thresholds T_L and T_H. Based on our previous analysis of RSSI and packet loss (Fig. 10.2c), we believe a lower threshold $T_L = -85$ dbm is appropriate for the MicaZ platform. If $\bar{R}$ falls below this threshold (step 6), the transmit power is immediately doubled (step 7) to avoid imminent packet loss in the deteriorating channel. If, on the other hand, the average RSSI exceeds an upper threshold T_H (step 8), the transmit power is reduced by a small amount (step 9). Our experimental traces indicate that $T_H = -80$ dBm is appropriate for the MicaZ platform; lower values make the target RSSI range $[T_L, T_H]$ too narrow, while larger values lead to higher transmit power (and hence higher energy usage) than required. Readers may note the similarity of our scheme to TCP in that the transmit power increase is multiplicative while the decrease is additive, and thus has a clear bias toward reacting faster to a deteriorating channel than an improving one.

The class of schemes outlined in algorithm 10.1 is fairly easy to implement in platforms with very limited CPU and memory resources. It is also very flexible, and the parameters α_u and α_d can, in particular, be tuned for appropriate trade-off between energy efficiency and reliability. The performance of the schemes for specific parameter settings is discussed next.

10.5.2 *Example Adaptations of the General Scheme*

A *conservative* approach to energy savings might be warranted in applications where data loss is critical. A relatively high value $\alpha_d = 0.8$ could be used so that a low RSSI value triggers a rapid ramp-up in transmit power, while keeping $\alpha_u = 0.2$ low so that transmit power is reduced cautiously when good channel conditions prevail. Applications in which energy is at a high premium and data loss is not as critical may adopt an *aggressive* strategy with a high $\alpha_u = 0.8$ that reacts quickly to improvements in channel conditions while setting $\alpha_d = 0.2$ so that a transient bad channel is ignored. A *balanced* approach that equally values energy savings and packet loss may place equal emphasis on whether the channel is getting better or worse by setting $\alpha_u = \alpha_d = \alpha = 0.8$.

We tested the efficacy of the conservative, aggressive, and balanced schemes as described above on the trace data for the three scenarios described earlier. The transmitter is assumed to know the RSSI for each transmitted packet via feedback from the receiver; it performs the exponential averaging using the parameters listed for each scheme, and chooses an appropriate transmit power level for the subsequent packet transmission. Figure 10.5 shows the transmit power level and the corresponding RSSI for each of the three schemes for each scenario. The average power draw per packet, as well as packet loss rates, for each of the schemes under the different scenarios is summarized in Table 10.2.

For normal walk, Fig. 10.5a shows that all schemes exhibit considerable fluctuation in their transmit power, since the channel varies quite rapidly. The conservative scheme yields good link reliability (only 1.4% loss) but has high energy usage (30.58% above optimal), while the aggressive scheme yields good energy savings (5.73% within optimal) by sacrificing link quality (9.6% packet loss). As one might expect, the energy usage and packet loss

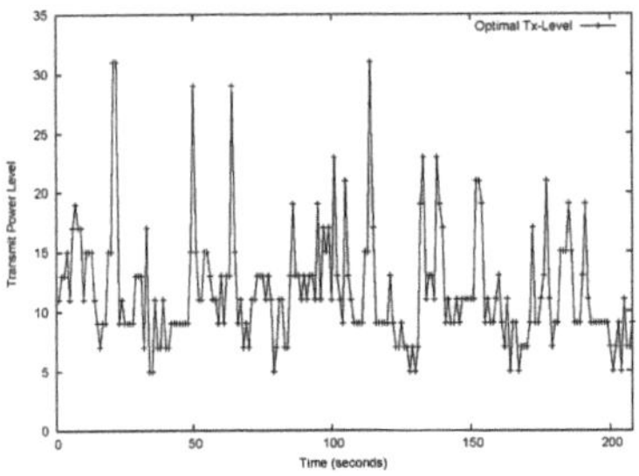

(a) Normal Walk: Optimal transmit level

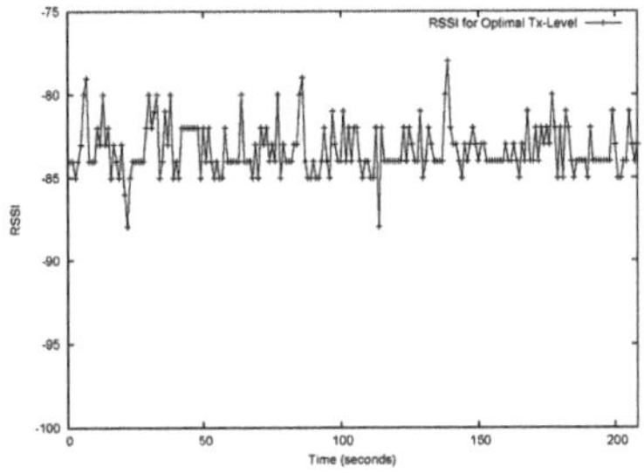

(b) Normal Walk: RSSI for optimal transmit level

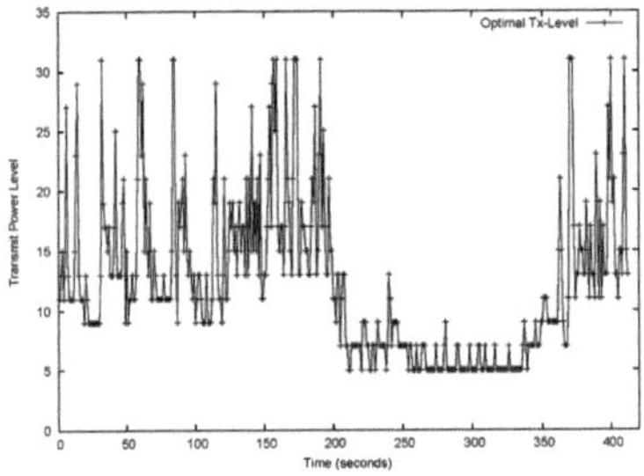

(c) Slow Walk: Optimal transmit level

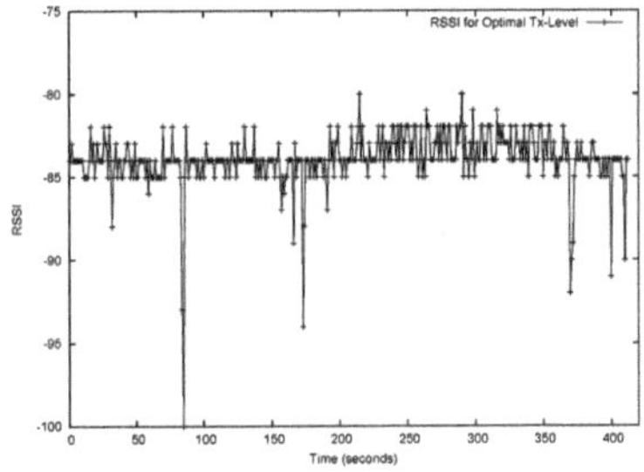

(d) Slow Walk: RSSI for optimal transmit level

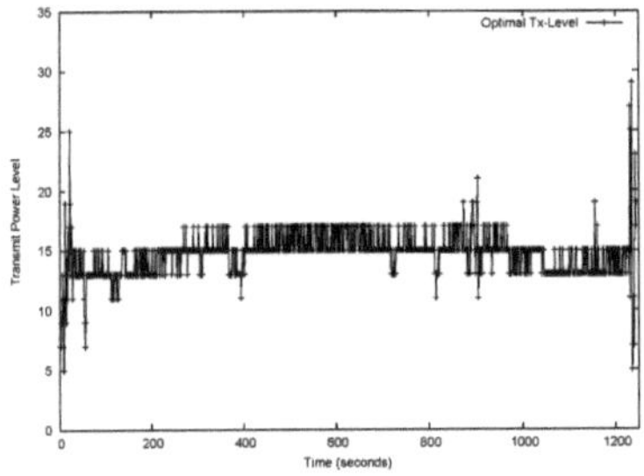

(e) Resting: Optimal transmit level

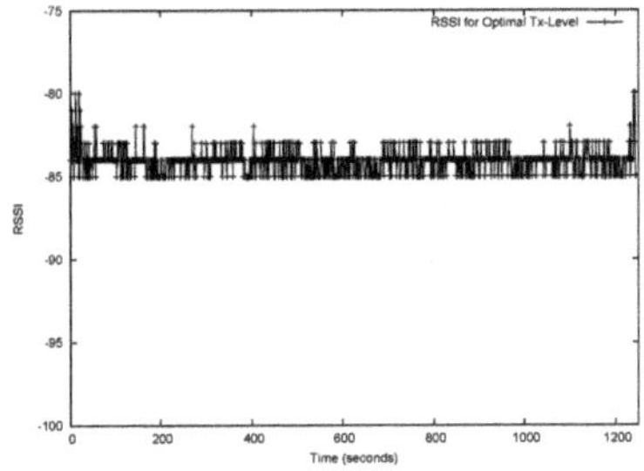

(f) Resting: RSSI for optimal transmit level

Figure 10.5. Transmit power and RSSI under conservative, balanced, and aggressive schemes for various scenarios.

under the balanced scheme are between those of the conservative and aggressive.

For the slow walk scenario, Fig. 10.5 shows the aggressive scheme to be fairly stable in the interval 210–350 s when the patient is walking slowly facing the base-station, whereas the conservative scheme shows some fluctuations in that region. Again,

Table 10.2. Power draw averaged across packets, and loss rate, for various power control schemes

	Normal Walk		Slow Walk		Resting	
Scheme	power (mW)	loss (%)	power (mW)	loss (%)	power (mW)	loss (%)
Maximum	31.32	0	31.32	1.2	31.32	0
Optimal	20.60	0	20.74	1.2	22.35	0
Conservative	26.90	1.4	25.52	2.18	24.36	0.08
Balanced	25.15	3.85	24.11	3.85	24.22	0.16
Aggressive	21.78	9.6	21.96	7.28	23.74	0.32

the conservative scheme uses 16.21% more energy than the aggressive scheme, but has a packet loss rate much lower than the aggressive scheme, showing that energy and reliability can be traded-off in our schemes by choosing parameters appropriately. As before, the balanced scheme lies between the other two in both its metrics.

The transmit level under the three schemes for the resting scenario, shown in Fig. 10.5e, is fairly stable, barring some rogue glitches in the link quality (e.g., at 905 s) that the conservative scheme over reacts to. The energy savings, as well as the packet loss ratios, are comparable under all three schemes, indicating that under quiescent channel conditions the parameter settings do not influence the algorithm performance significantly.

10.5.3 *Tuning the Parameters*

In this section, we undertake a more detailed study of the impact of the algorithm parameters α_u and α_d on the performance of our class of schemes. We conducted several experiments in which the patient is fairly active, since the parameters have a larger impact on energy and loss under such scenarios. Figure 10.6 plots the energy consumption (left column) and packet loss (right column) of our scheme for various parameters settings for a representative scenario. Figures 10.6a,b show that for any fixed α_d, the energy consumption falls monotonically with α_u: this is because a larger α_u makes the scheme react faster to an improving channel,

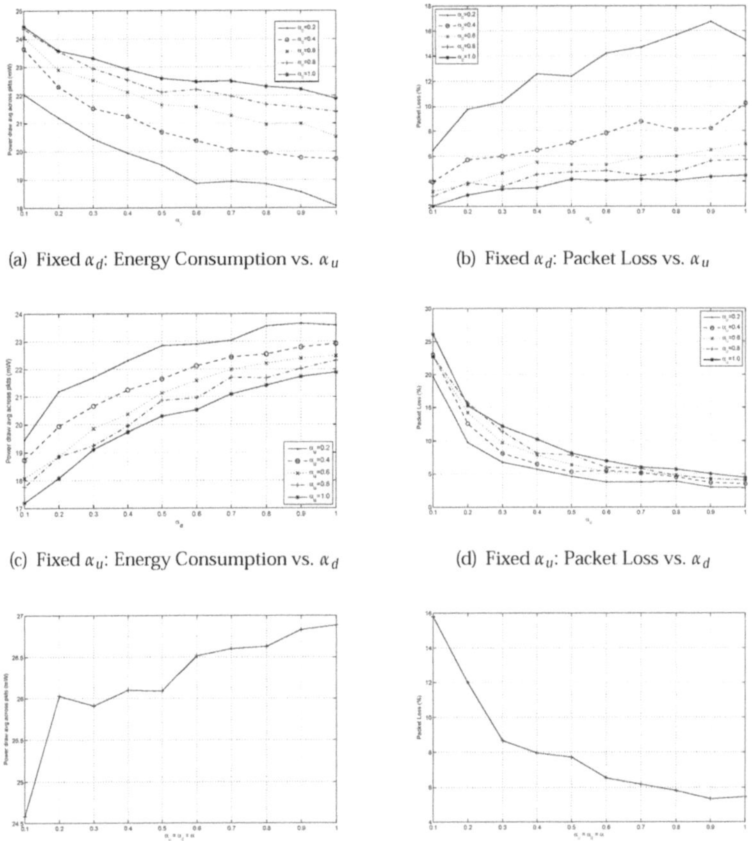

(a) Fixed α_d: Energy Consumption vs. α_u

(b) Fixed α_d: Packet Loss vs. α_u

(c) Fixed α_u: Energy Consumption vs. α_d

(d) Fixed α_u: Packet Loss vs. α_d

(e) $\alpha_u = \alpha_d = \alpha$: Energy Consumption vs. α

(f) $\alpha_u = \alpha_d = \alpha$: Packet Loss vs. α

Figure 10.6. Energy consumption and packet loss for various parameter settings.

thereby reducing transmit power to save energy. When α_u is held constant and α_d increases, the scheme reacts more quickly to a degrading channel, and the packet loss rate diminishes (Fig. 10.6d) at the expense of increased energy consumption (Fig. 10.6c). The plots show clearly that increasing α_u pulls the algorithm in the direction favoring energy savings, while increasing α_d pulls it toward reliability — these opposite trends justifying our decision to devise a scheme with two separate α values. For a balanced scheme that uses

$\alpha_u = \alpha_d = \alpha$, Fig. 10.6e,f show that a scheme that is very reactive to both good and bad channel conditions (i.e., high α) improves loss performance as expected, but incurs a penalty in terms of slightly increased energy usage.

An interesting question concerns the optimization of the parameters for a given operating scenario. For the trace considered in this section, we found that ($\alpha_u = 1.0, \alpha_d = 0.3$) yields maximum energy savings when the loss budget is high at 15%, implying that the ramp-down in transmit power should be immediate while a ramp-up happens slowly over several poor RSSI samples. When the loss margin is stringent at 5%, energy savings are maximized for ($\alpha_u = 0.8, \alpha_d = 1.0$), i.e., the scheme should react immediately to a degrading channel and relatively slowly to an improving channel. Though the optimal settings obtained here do not transfer directly to other scenarios, they provide insight which could conceivably be used by a system to adjust the parameters at run-time based on application requirements and operating conditions.

10.6 Prototyping and Experimentation

This section reports on a real-time implementation of our power control schemes on a MicaZ mote-based testbed, as well as preliminary experiments on the Toumaz Sensium^{TM} platform.

10.6.1 *MicaZ Mote Platform*

The previous section evaluated the performance of the transmit power control schemes using trace data wherein the sender (body-worn device) is assumed to know the RSSI at the receiver (base-station) for each of the packets it has thus far transmitted. In a real operational setting, such feedback information would be contained in acknowledgment packets (from the receiver to the sender), which may also experience loss. This section undertakes a real-time implementation of our scheme on body-worn MicaZ motes to evaluate the efficacy of our power control scheme under imperfect feedback. We note that the base-station is assumed to have abundant energy reserves and does not implement power control, always using the highest power level for transmitting acknowledgments.

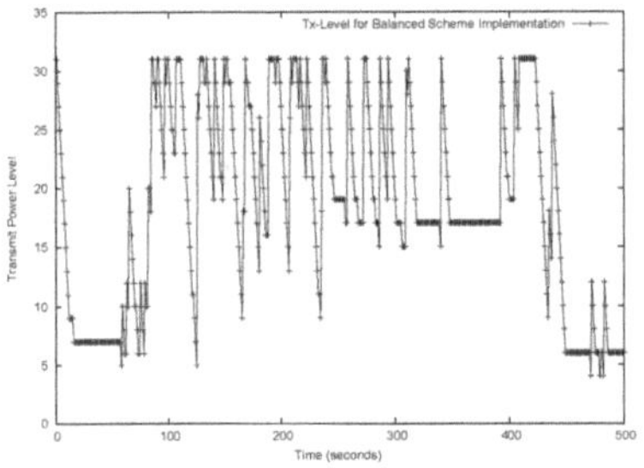

(a) Transmit level

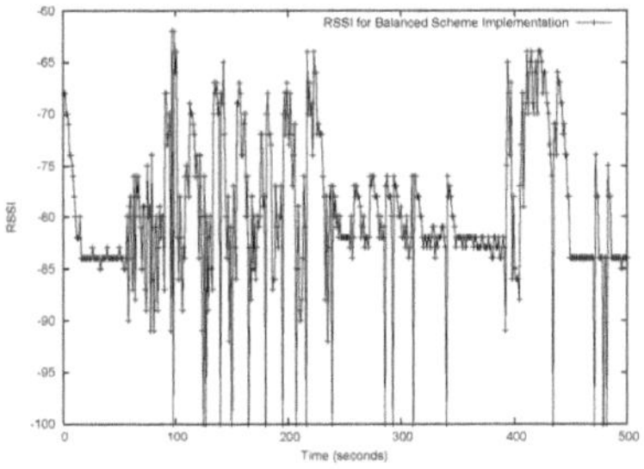

(b) RSSI

Figure 10.7. Transmit power and associated RSSI for real-time implementation of balanced scheme on the MicaZ platform.

The implementation performs exactly the steps shown in algorithm 10.1, with the additional step that if an acknowledgment is not received for a transmitted packet, the RSSI for that sample is taken to be −100 dBm (thereby signaling a bad channel to the algorithm). We present results with balanced parameter setting $\alpha_u = \alpha_d = \alpha = 0.8$ for a scenario where the patient undertakes a mix of walking and resting. Figure 10.7a shows the transmit power level under our scheme and Fig. 10.7b the corresponding RSSI. Our scheme performs quite well, yielding average energy consumption rate per packet of 20.23 mW (a 35.4% savings in energy compared to using maximum transmit power), and packet loss rate of 3.8%. Unfortunately, there is no way to benchmark this result (recall that computing the optimal requires a trace that includes the RSSI for all power levels at all times), but a careful look at the plots will show regions where the RSSI is high and yet the transmit power is not reduced, for example in the interval (410, 420) s. It was found that several acknowledgment packets were lost in this interval (the overall acknowledgment loss rate for this scenario was 11.82%), and the absence of feedback information forced our scheme to use a higher transmit power level than would have been necessary with perfect feedback. High acknowledgment loss rates arising from asymmetric link qualities can have a negative impact on the efficacy of feedback-based power control schemes and merit deeper study in future work.

10.6.2 Toumaz Sensium^{TM} Platform

A major goal of this project is to evaluate the costs and benefits of adaptive power control in the real-world continuous health monitoring platform being developed by Toumaz Technology. Unfortunately, the digital plaster (which includes the Sensium^{TM} chip, printed battery, etched antenna, and water-protective covering) is still in the packaging process; so we present here preliminary results obtained from using a Sensium^{TM} chip mounted on a development board strapped to the patient's chest, as preliminary indicators on the feasibility and benefits of adaptive transmit power control in the Sensium^{TM} platform.

The hardware-optimized design of the Sensium^{TM} presented several constraints in our experimentation: (i) the radio has only eight transmit power levels (unlike 32 in the MicaZ radio), and the energy savings are upper-bounded at 35% due to the limited transmission power range (Table 10.3a), (ii) the RSSI readout is only 3 bits (compared to 8 bits in the MicaZ), which gives only a coarse estimate of the RSSI (Table 10.3b), (iii) a sleep time of at least 1 s is imposed between successive packet transmissions (for energy efficiency), which restricts our ability to sample the channel at various transmit power levels without substantial change in patient position/orientation, (iv) the RSSI feedback from receiver to sender requires software modification of the acknowledgment packets, which introduces a 1 s lag in the feedback. These limitations not withstanding, we believe there is great value in experimenting

Table 10.3. Sensium^{TM} Radio transmit and receive characteristics (a) Transmit Characteristics (b) Receive Characteristics

Tx level	output (dBm)	power (mW)		
7	−6	2.8	7	> −35
6	−7	2.7	6	−46 to −35
5	−9	2.6	5	−52 to −46
4	−10	2.5	4	−58 to −52
3	−12	2.4	3	−64 to −58
2	−15	2.2	2	−70 to −64
1	−18	2.0	1	−76 to −70
0	−22	1.8	0	< −76

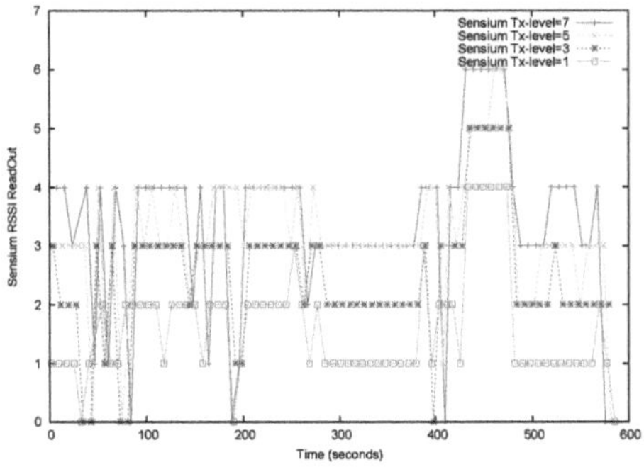

Figure 10.8. RSSI vs. time using the Sensium^{TM} platform. See also Color Insert.

with this real-world wearable device, and these limitations can be addressed in subsequent revisions of the Sensium^{TM} hardware.

As with the motes, we strap the Sensium^{TM} to the patient's chest and make it transmit a packet every second, cycling through the eight available power levels, while the RSSI is recorded at the base-station. A 10 min extract of the recorded RSSI at various transmit power levels is shown in Fig. 10.8; in this scenario, the patient is walking back and forth in the interval 30–80 s, during which period the RSSI fluctuates rapidly, while in the intervals 280–380 s and 420–470 s the patient is stationary (facing away and toward the base-station, respectively) and the RSSI is stable. This confirms that the body area wireless channel presents significant temporal fluctuations in the 862–870 MHz frequency range (much like the 2.4 GHz range of the MicaZ radio), and fixed transmit power is suboptimal.

For the above trace, we compute off-line the optimal transmission power schedule for desired RSSI level of 3 (corresponding to the range −64 to −58 dBm). Figure 10.9 shows the optimal transmit power level (at the sender) and the corresponding RSSI level (at the receiver) and clearly depicts that the transmit power level can be reduced during quiescent periods such as 420–470 s, while preserving reliability during periods when channel conditions are poor. For this scenario, optimal transmit power control uses 14.63% less energy than fixed maximum transmit power, which is a significant saving for this severely energy-constrained device.

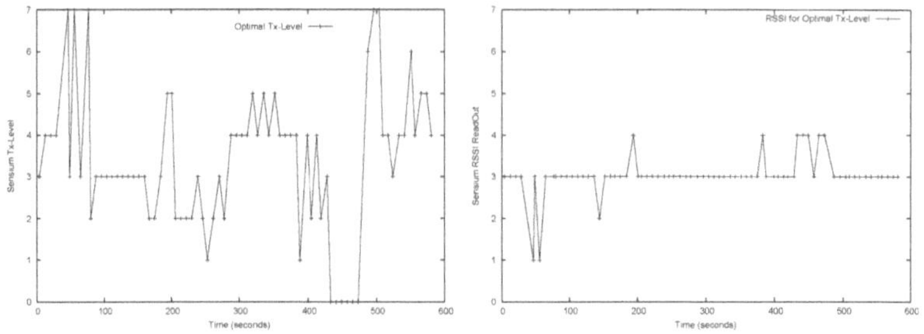

(a) Sensium: Optimal transmit level (b) Sensium: RSSI for optimal transmit level

Figure 10.9. Optimal transmit power and associated RSSI for the Sensium™.

Further results from our testing on the Sensium™ platform are reported in our subsequent work [33].

10.7 Conclusions and Future Work

This chapter outlines the potential benefits and limitations of adaptive radio transmit power control as a means of saving precious energy in body-wearable sensor devices used for medical monitoring. We experimentally profiled the radio channel quality under different scenarios of patient activity and showed that fixed transmit power either wastes energy or sacrifices reliability. We then quantified the theoretical benefits of adaptive transmit power control and showed that across different scenarios it can save nearly 35% energy without compromising reliability. We then developed a general class of practical power control schemes suitable for BANs, and showed specific instances that save 14–30% energy (as compared to using maximum transmit power) in exchange for 1–10% packet losses. We demonstrated that the adjustment of parameters allow our schemes to achieve different trade-offs between energy savings and reliability, making them suitable across diverse applications in different operating conditions. Finally, we demonstrated that a real-time implementation of our scheme on the MicaZ mote-based platform is effective even in the presence of imperfect feedback information and presented preliminary experimental results indicating the potential for saving precious energy in Toumaz's real-world platform for continuous health care monitoring.

Our work on dynamic power control in BANs, summarized in this chapter from our publications [33–35], can be extended in several ways. There is much further study required in exploring its potential for specific health monitoring environments (e.g., critical care in hospitals, aged care, athlete monitoring, etc.) which have different characteristics in terms of patient mobility, periodicity, and criticality of collected data. There is also scope for more extensive experimentation with truly wearable health monitoring devices used by real patients.

References

1. Crossbow-Technologies, Mica2 and MicaZ motes, URL http://www.xbow.com.
2. V. Schnayder et al. (2005) Sensor Networks for Medical Care, Technical Report TR-08-05, Division of Engineering and Applied Science, Harvard University.
3. Power-Paper, Power Patch Platform, URL http://www.powerpaper.com.
4. D. Culler, D. Estrin, and M. Srivastava (August 2004) Overview of Sensor Networks, *IEEE Computers*, **37(8)**, pp. 41–49.
5. A. Wong, G. Kathiresan, T. Chan, O. Eljamaly, and A. Burdett (September 2007) A 1 V Wireless Transceiver for an Ultra Low Power SoC for Biotelemetry Applications, in *ESSDERC/ESSCIRC*, Munich, Germany.
6. Chipcon, CC2420: 2.4 GHz IEEE 802.15.4 / ZigBee-ready RF Transceiver, URL http://www.chipcon.com.
7. K. Langendoen and G. Halkes (2005) Energy-Efficient Medium Access Control, *Embedded Systems Handbook*, CRC Press.
8. J. Polastre, J. Hill, and D. Culler (November 2004) Versatile Low Power Media Access for Wireless Sensor Networks, in *ACM SenSys*, Baltimore, MD, pp. 95–107.
9. O. Omeni, O. Eljamaly, and A. Burdett (August 2007) Energy Efficient Medium Access Protocol for Wireless Medical Body Area Sensor Networks, in *Proceedings of the IEEE-EMBS Symposium on Medical Devices and Biosensors*, Cambridge, UK, pp. 29–32.

10. S.-L. Wu, Y.-C. Tseng, and J.-P. Sheu (September 2000) Intelligent Medium Access for Mobile Ad Hoc Networks With Busytones and Power Control, *IEEE Journal on Selected Areas in Communications*, **18(9)**, pp. 1647–1657.
11. J.-P. Ebert, B. Stremmel, E. Wiederhold, and A. Wolisz (September 2000) An Energy-Efficient Power Control Approach for WLANs, *Journal of Communications and Network*, **2(3)**, pp. 197–206.
12. J. Monks, V. Bharghavan, and W. Hwu (April 2001) A Power Controlled Multiple Access Protocol for Wireless Packet Networks, in *IEEE Infocom*, Alaska, pp. 219–228.
13. J. Pavon and S. Choi (May 2003) Link Adaptation Strategy for IEEE 802.11 WLAN via Received Signal Strength Measurement, in *IEEE ICC*, Anchorage, AK, pp. 1108–1113.
14. E. -S. Jung and N. Vaidya (January 2005) A Power Controlled MAC Protocol for Ad Hoc Networks, *Wireless Networks*, **11(1–2)**, pp. 55–66.
15. D. Qiao, S. Choi, and K. G. Shin (October 2007) Interference Analysis and Transmit Power Control in IEEE 802.11a/h Wireless LANs, *IEEE/ACM Transactions on Networking*, **15(5)**, pp. 1007–1020.
16. T. ElBatt, S. Krishnamurthy, D. Connors, and S. Dao (June 2000) Power Management for Throughput Enhancement in Wireless Ad Hoc Networks, in *IEEE ICC*, New Orleans, LA, pp. 1506–1513.
17. R. Ramanathan and R. Hain (March 2000) Topology Control of Multihop Wireless Networks Using Transmit Power Adjustment, in *IEEE Infocom*, Tel-Aviv, Israel, pp. 404–413.
18. M. Kubisch, H. Karl, A. Wolisz, et al. (March 2003) Distributed Algorithms for Transmission Power Control in Wireless Sensor Networks, in *IEEE WCNC*, New Orleans, LA.
19. D. Son, B. Krishnamachari, and J. Heidemann (October 2004) Experimental Study of the Effects of Transmission Power Control and Blacklisting in Wireless Sensor Networks, in *IEEE SECON*, Santa Clara, CA, pp. 289–298.
20. O. Chipara, Z. He, G. Xing, Q. Chen, X. Wang, C. Lu, J. Stankovic, and T. Abdelzaher (June 2006) Real-Time Power-Aware Routing in Sensor Networks, in *IEEE IWQoS*, New Haven, CT, pp. 83–92.
21. G. Xing, C. Lu, Y. Zhang, Q. Huang, and R. Pless (May 2005) Minimum Power Configuration in Wireless Sensor Networks, in *ACM MobiHoc*, Urbana-Champaign, IL, USA, pp. 390–401.

22. I. C. Paschalidis, W. Lai, and D. Starobinski (February 2007) Asymptotically Optimal Transmission Policies for Large-Scale Low-Power Wireless Sensor Networks, *IEEE/ACM Transactions on Networking*, **15(1)**, pp. 105–118.
23. L. Correia et al. (October 2005) Transmission Power Control in MAC Protocols for Wireless Sensor Networks, in *ACM/IEEE MSWiM*, Montreal, Canada, pp. 282–289.
24. S. Lin, J. Zhang, G. Zhou, L. Gu, T. He, and J. Stankovic (November 2006) ATPC: Adaptive Transmission Power Control for Wireless Sensor Networks, in *ACM SenSys*, Boulder, CO, pp. 223–236.
25. G. Zhou, T. He, S. Krishnamurthy, and J. Stankovic (June 2004) Impact of Radio Irregularity on Wireless Sensor Networks, in *ACM MobiSys*, Boston, MA, pp. 125–138.
26. J. Zhao and R. Govindan (November 2003) Understanding Packet Delivery Performance in Dense Wireless Sensor Networks, in *ACM SenSys*, Los Angeles, CA.
27. P. S. Hall and Y. Hao (2006) *Antennas and Propagation for Body-Centric Wireless Communications*, Artech House.
28. V. Shrivastava, D. Agrawal, A. Mishra, S. Banerjee, and T. Nadeem (October 2007) Understanding the Limitations of Transmit Power Control for Indoor WLANs, in *ACM/Usenix Internet Measurement Conference (IMC)*, San Diego, CA, pp. 351–364.
29. J. Jeong, D. Culler, and J.-H. Oh (June 2007) Empirical Analysis of Transmission Power Control Algorithms for Wireless Sensor Networks, in *International Conference on Networked Sensing Systems (INSS'07)*, Kanazawa, Japan, pp. 27–34.
30. K. Srinivasan, P. Dutta, A. Tavakoli, and P. Lewis (November 2006) Understanding the Causes of Packet Delivery Success and Failure in Dense Wireless Sensor Networks, in *ACM SenSys*, Boulder, CO, pp. 419–420.
31. K. Srinivasan and P. Lewis (May 2006) RSSI Is Under Appreciated, in *Workshop on Embedded Networked Sensors (EmNets)*, Boston, MA.
32. J. Ryckaert, P. D. Doncker, R. Meys, A. de Le Hoye, and S. Donnay (April 2004) Channel Model for Wireless Communications Around Human Body, *Electronics Letters*, **40(9)**, pp. 543–544.
33. A. Dhamdhere, V. Sivaraman, and A. Burdett (December 2008) Experiments in Adaptive Power Control for Truly Wearable Biomedical Sensor Devices, in *International Workshop on Adaptation in Wireless Sensor Networks (AWSN)*, Sydney, Australia.

34. S. Xiao, A. Dhamdhere, V. Sivaraman, and A. Burdett (January 2009) Transmission Power Control in Body Area Sensor Networks for Healthcare Monitoring, *IEEE Journal on Selected Areas in Communications*, **27(1)**, pp. 37–48.

35. A. Dhamdhere, V. Sivaraman, V. Mathur, and S. Xiao (December 2008) Algorithms for Transmission Power Control in Biomedical Wireless Sensor Networks, in *IEEE Workshop on Wireless Network Algorithms (WiNA)*, Yilan, Taiwan.

Chapter 11

Channel Modeling of Narrowband Body-Centric Wireless Communication Systems

Simon L. Cotton and William G. Scanlon
ECIT Institute, Queen's University Belfast,
Queen's Island, Belfast, United Kingdom
simon.cotton@qub.ac.uk; w.scanlon@qub.ac.uk

The modeling of wireless channels in body-centric communication systems is of paramount importance when designing antennas, protocols, and wireless transceiver hardware. In this chapter, we briefly introduce the concept of *on*-body, *off*-body, and *body*-to-*body* communication channels. We then focus specifically on the modeling of narrowband on-body communication channels, which are an integral part of today's wireless body area networks (WBANs). As well as identifying the factors responsible for shaping fading characteristics in narrowband WBAN systems, we discuss a number of the models commonly used to describe the first- and second-order characteristics of amplitude distribution. The Akaike Information Criterion is introduced as a method of ranking competing channel models, and a worked example is provided to step the reader through the modeling process, as well as providing information on how to simulate the channels. The chapter then concludes with

Wireless Body Area Networks: Technology, Implementation, and Applications
Edited by Mehmet R. Yuce and Jamil Y. Khan

ISBN 978-981-4316-71-2 (Hardcover), ISBN 978-981-4241-57-1 (eBook)
www.panstanford.com

comments on the validity of the analysis presented, especially in the case of non-isotropic signal reception.

11.1 Introduction to Body-Centric Communications

Recent commercial interest in body-centric communications, wireless systems that are designed to operate in close proximity to the human body, has meant that antennas and propagation have become important focal points for research. Study in these areas will help to develop new and innovative solutions and systems geared toward applications in the consumer, medical, and military sectors. For example, requirements for specialist mobile health care have helped to fuel a revolution in body sensor network (BSN) technology, where networks of miniaturized wearable sensors with wireless functionality are attached to the human body and configured to monitor life vital signs such as body temperature, electrocardiography, and motor activity. Other recent popular uses of WBAN technology include data exchange between body-worn cellular devices and wireless headsets, and recreational sports monitoring.

At present, there are two general areas of body-centric communication channel research, broadly categorized as on-body and off-body channels. In on-body channels, such as those found in WBANs, internetworked devices positioned on the human body will typically use over-the-body surface propagation channels to communicate as shown in Fig. 11.1. At ultra-high frequency (UHF) and microwave frequencies, these wireless links are often formed using on-body surface waves and signal components reflected, diffracted, and scattered from other body parts. It should be noted that some links may also utilize free-space propagation (e.g., waist to wrist, etc.) as well as environmental multipath propagation whereby signal transmissions from an on-body node are returned toward the body from the local surroundings. Depending on network topology, some WBAN applications may employ a single body-worn master node for central processing of data and for onward transmission of this information across other wireless networks (e.g., Wi-Fi, Bluetooth, and cellular) using off-body communications

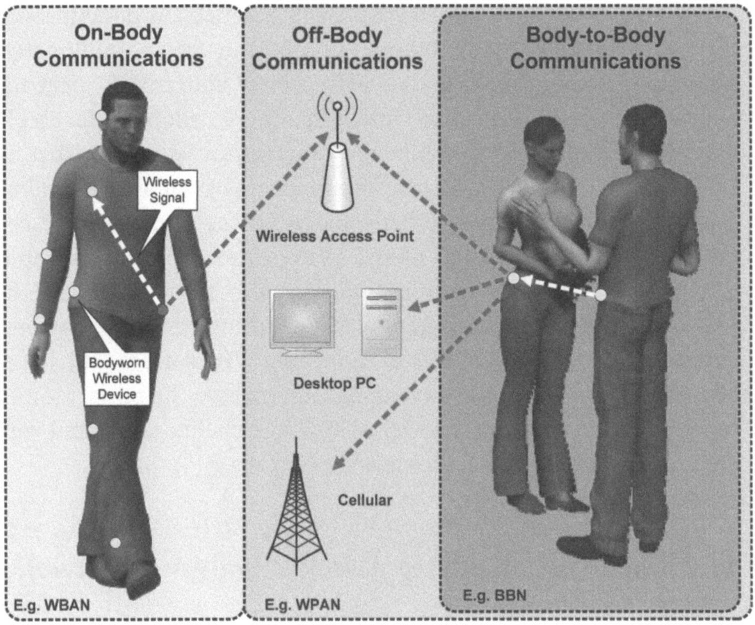

Figure 11.1. Overview of on-body, off-body, and body-to-body communications. See also Color Insert.

(Fig. 11.1). An emerging type of body-centric communications, which also utilizes off-body communications, is body-to-body networking. Body-to-body channels, such as those found in body-to-body networks (BBNs), occur when a wireless device situated on one person is communicating with a wireless device situated on another person.

Much of the current research focus for WBAN applications is centered on the unlicensed industrial, scientific, and medical (ISM) bands at 868 MHz, 915 MHz, and 2.45 GHz. In this chapter, we will focus on narrowband on-body channels at these frequencies. The interested reader is directed to the following references and those contained therein for channel studies on ultra-wideband (UWB) WBANs [1], the use of diversity in WBANs [2–4], and multiple-input multiple-output WBAN systems [5]. Those wishing to learn more about off-body and body-to-body channel characteristics are directed to the following references for narrowband off-body

[6], UWB off-body [7], narrowband and diversity in body-to-body [8] and wideband body-to-body [9]. In the future, millimeter-wave frequencies, such as the world-wide unlicensed spectrum allocations at 60 GHz, will undoubtedly be adopted for high-bandwidth applications. While the propagation characteristics in these bands are difficult with severe blocking by almost all building materials, furniture, and other objects, these same effects can be used to the advantage of the system designer. For example, the limited propagation at millimeter-wave frequencies can be used to mitigate against potential signal-to-interference problems associated with high densities of co-located WBAN users [10]. Other advantages of millimeter-wave operation include the opportunities for device miniaturization and the capacity benefits associated with short-range spatial spectrum reuse.

11.2 Channel Modeling for Wireless Body Area Networks

Wearable wireless devices for WBANs are required to be compact, lightweight, robust, unobtrusive to the user, and usually feature antennas mounted conformal to or in extremely close proximity to the body surface. These requirements must be met while still maintaining a high level of performance, reliability, and efficiency. Clearly, this will introduce design challenges at all layers in the protocol stack, but physical layer characteristics are known to be the limiting factor in the performance of body-centric wireless communication systems.

A key discriminator between body-centric systems and traditional mobile communications is the unique time- and space-varying propagation characteristics encountered. This is in part due to the human body itself, which is an extremely complex operating environment. Body-worn wireless devices are prone to antenna-body interaction effects, which include near-field coupling, radiation pattern distortion, and shifts in antenna impedance, which may degrade the efficiency of the body-worn system and reduce signal reliability [11]. Because the body acts as its own frame of reference, body-centric communication channels are susceptible to time-varying, received signal characteristics, which may range from quasi-periodic to completely stochastic depending on the person's current physical state. The electromagnetic problem is made even more difficult to analyze as signal reception in body-centric systems may range from isotropic to strongly nonisotropic [12], depending on the degree of body shadowing, attenuation of the on-body creeping wave [13] (if present), and the nature of the local surroundings. Within cluttered multipath environments, inhomogeneous, irregularly spaced dielectric structures, including nearby pedestrians, also cause the signal to undergo reflection, diffraction, scattering, and absorption in varying degrees, creating a spatially distributed electromagnetic field.

To further complicate the analysis of WBAN systems, the magnitude of each of these factors will also be dependent on the geometry of the on-body communication link, i.e., where nodes are positioned on the body relative to one another. This effect is demonstrated by the time series shown in Fig. 11.2, which shows the signal received by nodes positioned on the right ankle and elbow (as shown in Fig. 11.1) while the user was performing walking movements in an anechoic environment. While it may be possible

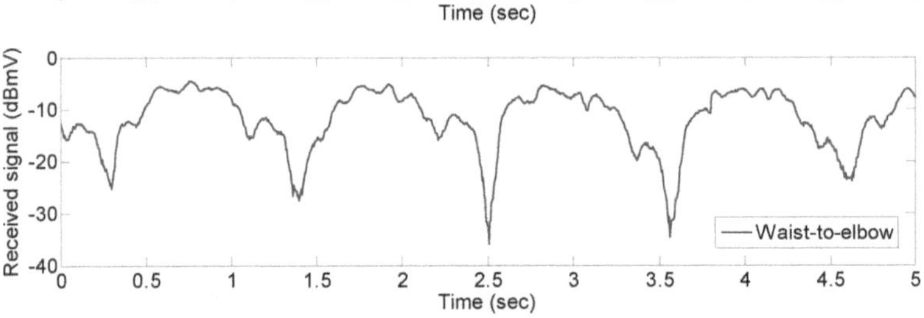

Figure 11.2. Comparison of waist-to-ankle and waist-to-elbow on-body channels at 2.45 GHz while the user was walking within an anechoic environment. The measurement scenario and setup are described in reference [12].

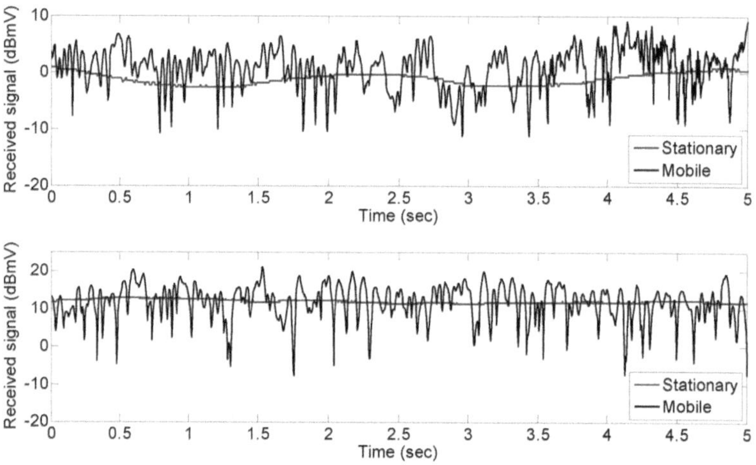

Figure 11.3. Comparison of waist-to-right head on-body channels at 2.45 GHz while the user was stationary and then walking within (a) an outdoor environment and (b) an open office area. The measurement scenario and setup are described in reference [15].

to identify a degree of channel repetition related to the user's body movements in each of the traces, it is quite clear that the distribution of the received signal in both amplitude and time is dissimilar. As discussed above, there can also be a considerable contrast in wireless channel characteristics for the same on-body link over different user states and environments. This is highlighted in Fig. 11.3 for a signal received by a right-head positioned node from a transmitter at the waist. Here the user is stationary in an outdoor environment, and the received signal is relatively constant, although a slight periodic variation due to respiration can be observed. When the user moves into an office area environment, the channel is again relatively stationary. However, the mean signal level becomes elevated in this case by around 10 dB. For these scenarios (i.e., when the person is relatively stationary), it has been proposed that the lognormal distribution may be used to describe the path gain [14]. Another important effect of environmental multipath for on-body channels is the alteration of the channel's autocorrelation function. As shown in reference [15], this effect is particularly prominent for scenarios when the user is mobile. For anechoic environments, many links may display identifiable quasi-periodic trends. However, within rich multipath environments, the signal contributions from off-body directions become important parts of the on-body channel. Signal components returned to the body act to raise the signal level (as shown in Fig. 11.3) and reduce the periodicity superimposed on the channel by the user's movements. Furthermore, as noted in reference [16], the variation in mean signal level experienced in different environments, and when the WBAN user performs a range of very different exercises, means that on-body channels will be expected to exhibit nonstationary channel statistics.

These characteristics mean that, except in the simplest of cases, it will be very difficult to develop generic analytical models that accurately model propagation characteristics in body-centric systems. Simulation, even on today's multiprocessor machines, is computationally intensive although parallelization of finite-difference time-domain (FDTD) computations [17] may help improve computation times. Given the large number of variable parameters found in body-centric communications, it is often easier to analyze these channels using statistics, where given a particular scenario, a probability level is assigned to a channel output parameter such as the received signal. To further simplify the analysis, it is also useful to investigate environments with specific user actions separately (e.g., standing in hallway, walking in open office area, running outdoors, etc.) to create a set of baseline channel characteristics for WBAN operation. In most cases, this will allow the on-body channel's statistics to be considered as wide sense stationary, that is the first and second moments do not vary with respect to time. At present, the general consensus from the literature is that there is no unified model for received signal amplitude distribution in WBANs; rather, it is accepted that particular models perform best for certain links and scenarios. A range of statistical models have been used to describe signal variation in WBANs; these include Rayleigh, Rice, Nakagami, Weibull, and lognormal, which will be discussed in detail in the next section.

11.2.1 *Statistical Distribution of the Fading Signal in WBANs*

In this chapter, we restrict ourselves to the discussion of flat fading on-body channels, often referred to as *"narrowband"* on-body channels. A wireless signal will undergo flat fading if the channel has constant gain and linear phase response over a bandwidth B_c greater than the bandwidth of the transmitted signal B_s, i.e., $B_s \ll B_c$. Signal variation in narrowband body-centric communication channels is often reported using a combination of first-order statistics, such as probability density functions (PDFs) and cumulative distribution functions (CDFs), and second-order statistics, such as level crossing rates (LCRs) and average fade duration (AFDs). Second-order statistics are of particular interest when designing mobile radio systems as they give a direct indication to the rate of change, with respect to time, of the wireless channel. Among their uses is the design of error correcting codes, optimization of interleaver size, and system throughput analysis as well as channel modeling.

11.2.1.1 Rayleigh and Rice distributions

There are a wide range of probability densities that have been used to describe the distribution of signal fading in WBANs. Among these are the popular Rayleigh and Rice fading models. Under Rayleigh fading, the received signal is viewed as the resultant of a large number of scattered signal components, each with random amplitude and uniform phase. The envelope R of a fading signal that follows a Rayleigh distribution may be expressed in terms of the in-phase, X, and quadrature, Y, components of the fading signal as

$$R = \sqrt{X^2 + Y^2} \tag{11.1}$$

where X and Y are mutually independent Gaussian processes with $E(X) = E(Y) = 0$, and $E(X^2) = E(Y^2) = \sigma^2$. The probability density function for this random variable may be written as

$$p_R(r) = \frac{r}{\sigma^2} \exp\left(-\frac{r^2}{2\sigma^2}\right), \quad r \geq 0 \tag{11.2}$$

where $E(R^2) = 2\sigma^2$ is the average envelope power. Rayleigh fading has been reported for on-body links where communications occur around the human torso at both 915 MHz and 2.45 GHz [18]. The authors explain that diffracting components that travel distances greater than 35 cm around the back of the body become significantly attenuated. Instead, the received signal is made up of reflected components, which follow a Rayleigh distribution. In body-centric channels, where a dominant signal component (e.g., strong on-body reflection or dominant creeping wave component) exists, it may be more appropriate to model the received signal distribution as following a Rice distribution. In this instance, X and Y are again assumed to be mutually independent Gaussian processes such that $E\left(X^2\right) = E\left(Y^2\right) = \sigma^2$ with nonzero means $E(X) = \mu_X$ and $E(Y) = \mu_Y$, respectively. The PDF of the Rice distribution may be written as

$$p_R(r) = \frac{r}{\sigma^2} \exp\left(-\frac{\left(r^2 + s^2\right)}{2\sigma^2}\right) I_0\left(\frac{rs}{\sigma^2}\right), \quad r \geq 0 \qquad (11.3)$$

where $s^2 = \mu_X^2 + \mu_Y^2$ is the power contained within the dominant component and $E\left(R^2\right) = s^2 + 2\sigma^2$. In the absence of a dominant component (i.e., $s = 0$), the Rice PDF becomes equivalent to the Rayleigh PDF given in Eq. 11.2. Another important figure, used to characterize the degree of fading in a wireless channel, is the Rice factor, k, which is defined as the ratio of the square of the dominant component to the scattered power, i.e., $k = s^2/2\sigma^2$. When $k \to 0$, and hence the dominant component s decreases, the fading becomes closer to Rayleigh fading, and as $k \to \infty$, the channel no longer exhibits fading. Ricean fading for on-body communications has been widely reported in the literature (868 MHz [19], 915 MHz [18], and 2.45 GHz [16, 18]). In reference [18], Ricean fading was reported for on-body link distances less than 25 cm. It was also observed that as the separation distance of the on-body link increases, the k parameter decreases, due to greater attenuation of the creeping wave component. The authors of reference [16] report that the magnitude of the k factor also changes with time. In this study, on-body channel characteristics were monitored over a wide range of everyday environments, as a test subject performed various activities such as driving a car, walking outdoors and indoors, and sitting.

11.2.1.2 Nakagami distribution

The PDF of a Nakagami fading signal envelope is given by

$$p_R(r) = \frac{2m^m r^{2m-1}}{\Gamma(m)\,\Omega^m} \exp\left(-\frac{mr^2}{\Omega}\right), \quad r \geq 0 \qquad (11.4)$$

where $m = E^2(r^2)/\text{var}(r^2)$ and $\Omega = E\left(r^2\right)$. A physical model for Nakagami fading has been proposed in reference [20], which describes the received signal as the sum of random vectors all with random amplitude and phase. It has also been shown that Nakagami random variables may be generated as the sum of m independent and identically distributed Rayleigh components with mean power Ω/m [21]. The Nakagami-m distribution is extremely versatile and holds a number of important relationships with the Rayleigh and Rice distributions. When $m = 1$, Eq. 11.4 becomes equivalent to the Rayleigh PDF given in Eq. 11.2. For $m > 1$, the Rice k factor can be approximated by the Nakagami-m distribution through the following relationship [22]:

$$k = \frac{\sqrt{m^2 - m}}{m - \sqrt{m^2 - m}}, \quad m > 1 \tag{11.5}$$

Moreover, for high values of the Nakagami-m parameter ($m \gg 1$), the Nakagami-m distribution can be used to approximate the lognormal distribution [23]. Another attractive feature of the Nakagami-m distribution is that it can be used to model fading scenarios that are worse than Rayleigh, i.e., when $m < 1$. When $m = 1/2$, the Nakagami distribution is equivalent to the semi-Gaussian distribution.

In reference [24], for WBAN channel measurements made at 2.45 GHz, it was observed that when the test subject was stationary, Nakagami-m parameters were always much larger than 1, irrespective of the operating environment. Under anechoic environmental conditions, when the user became mobile, m parameters were observed to decrease due to an increase in the diffuse contributions from the body and interruption of the on-body creeping wave component if present. In a similar fashion to Rice k factors reported in reference [18], the Nakagami-m parameter was typically observed to decrease with increasing separation distance between on-body antennas. Fading in on-body channels was also observed to increase when the user moved into a multipath environment, showing that the local surroundings are an important consideration for on-body systems. For on-body links operating at 4.5 GHz [25], Nakagami fading has also been found to occur for scenarios when the test subject performed walking movements in an anechoic chamber.

11.2.1.3 Weibull distribution

In a Weibull fading channel, the received signal is obtained as a nonlinear function of the modulus of multipath components, where the nonlinearity is expressed in terms of the power parameter α. Yacoub [26] has shown that a Weibull-distributed envelope may be obtained from two independent zero mean Gaussian random variables with identical variance such that $R^\alpha = X^2 + Y^2$. Following from this, the PDF of a Weibull-distributed envelope may be written as [26]

$$p_R(r) = \frac{\alpha r^{\alpha-1}}{\lambda} \exp\left(-\frac{r^\alpha}{\lambda}\right) \tag{11.6}$$

where $\lambda = E(R^\alpha)$. In a similar fashion to both the Rice and Nakagami distributions, the Rayleigh distribution may be obtained from the Weibull distribution by setting $\alpha = 2$ and the negative exponential distribution by setting $\alpha = 1$ [27]. Weibull fading has been reported for on-body measurements made at 2.36 GHz while the test subject was walking and running [28] and at 4.5 GHz for walking and standing up/sitting down motions [25]. In reference [25], it is suggested that the Weibull model should be used for on-body links that experience severe fading such as the shoulder and ankle, while the user was walking, and the entire trunk action, while the user was repeating standing up/sitting down motions.

11.2.1.4 Lognormal distribution

The lognormal PDF for an envelope R with mean μ and standard deviation σ may be expressed by

$$p_R(r) = \frac{1}{r\sigma\sqrt{2\pi}} \exp\left(-\frac{[\ln(r) - \mu]^2}{2\sigma}\right) \tag{11.7}$$

where $r > 0$. The lognormal distribution has traditionally been associated with shadow fading in wireless channels. It has also been reported both for body shadowing [14] and recently observed in small-scale fading [23]. Lognormal random variables can be viewed as the result of a number of multiplicative factors that become additive under logarithmic transformation. A physical interpretation of lognormal fading in on-body channels has been proposed in reference [29], where the received signal is the result of diffraction, reflection, energy absorption, and antenna losses. The occurrence of lognormal fading has generally been observed in on-body channels in which the user is stationary [14, 15, 25] but has also been reported for dynamic scenarios [24].

11.2.2 *Higher Order Statistics*

While the distribution of received signal amplitude is of interest, quite often it is necessary to model not only the first-order statistics of a wireless channel, but also the second-order characteristics to understand how the received signal is distributed in time and how often it crosses a particular threshold level.

11.2.2.1 Level crossing rate and average fade duration

The level crossing rate and average fade duration are two important second-order statistics for fading channels. They give a direct indication as to the rate of change of a wireless channel through their relationship with the maximum Doppler frequency, f_m. Both the LCR and AFD are of great importance in the design of mobile radio systems and the analysis of their performance [30]. The LCR, N_R, at a specified threshold level, R, is defined as the expected rate at which a signal crosses R in a positive direction [31] and is given by

$$N_R(r) = \int_0^\infty \dot{r}\, p_{\dot{R},R}(\dot{r}, r)\, d\dot{r} \tag{11.8}$$

where $\dot{r}$ is the time derivative and $p_{\dot{R},R}(\dot{r}, r)$ is the joint probability density of $\dot{R}$ and R.

For the case of isotropic scattering, and based on the knowledge that the Rayleigh [31], Rice [32], Nakagami [33], Weibull [34], and lognormal (after logarithmic transformation) [23] fading envelopes have Gaussian time derivatives, and their envelopes and respective slopes are independent, closed form expressions for the LCR of these fading signals have been derived. Defining $\rho = r/r\sqrt{\Omega}$, as an example, Eq. 11.9 shows the form of the LCR equation for a

Nakagami-m-distributed received signal envelope as derived in [33]

$$N_R(r) = \sqrt{2\pi} f_m \frac{m^{m-(1/2)}}{\Gamma(m)} \rho^{2m-1} \exp\left(-m\rho^2\right) \tag{11.9}$$

where f_m is the maximum Doppler frequencies and $\Gamma(\bullet)$ is the Gamma function. In a similar fashion, the average fade duration, T_R, is defined as the average length of time spent below a specified threshold level

$$T_R(r) = \frac{F_R(r)}{N_R} \tag{11.10}$$

where $F_R(r)$ is the cumulative distribution function of r. The corresponding AFD equation for a Nakagami-m-distributed received envelope may be written as [33]

$$T_R(r) = \frac{\Gamma\left(m, m\rho^2\right)}{\sqrt{2\pi} f_m m^{m-(1/2)} \rho^{2m-1} \exp\left(-m\rho^2\right)} \tag{11.11}$$

where $\Gamma(\bullet, \bullet)$ is the incomplete Gamma function. It should be noted that as signal reception in WBANs may often be nonisotropic [12], Eqs. 11.9 and 11.11 and those contained in [31], [32], [34], and [23], which were derived under the assumption of an isotropic scattered signal contribution, will only provide an approximation of the LCR and AFD in on-body channels in which the signal reception is nonisotropic.

11.3 Parameter Estimation and Model Selection

Having acquired on-body channel measurements, an important step in the data analysis procedure is the selection of the appropriate fading model to represent the data. Before this is possible, it is necessary to find parameter estimates for the candidate models. The method typically employed is that of maximum likelihood estimation (MLE).

11.3.1 *Maximum Likelihood Estimation*

MLE operates by determining the parameter estimates for a given model that maximize the probability of the sample data. For the Rayleigh distribution given in Eq. 11.2, if $r_1, r_2, \ldots, r_n$ are independent observations of the received signal, then the joint probability of the observations is

$$L(\sigma|R) \equiv p_R(R|\sigma) = \prod_{i=1}^{n} p_R(r_i|\sigma) \tag{11.12}$$

where $L(\bullet)$ represents the likelihood function, which is the probability of obtaining σ, given the received envelope R. The *posteriori* probability of a Rayleigh-distributed envelope may be expressed in the form

$$L(\sigma|R) = \prod_{i=1}^{n} (r_i)\left(\sigma^{-2n}\right) \exp\left[-\sum_{i=1}^{n}\left(\frac{r_i^2}{2\sigma^2}\right)\right] \tag{11.13}$$

The parameter σ for which the likelihood function is a maximum is found by first taking the logarithm of Eq. 11.13, which is

$$\ln[L(\sigma|R)] = \sum_{i=1}^{n} \ln(r_i) - 2n\ln(\sigma) - \sum_{i=1}^{n}\left(\frac{r_i^2}{2\sigma^2}\right) \tag{11.14}$$

Now calculation has become easier as repeated multiplication has been converted to addition. To obtain a maximum, the partial differential of Eq. 11.14 is taken with respect to σ and the result set equal to 0. With some rearrangement, the maximum likelihood estimator of σ is given by

$$\hat{\sigma} = \sqrt{\sum_{i=1}^{n}\left(\frac{r_i^2}{2n}\right)} \tag{11.15}$$

The log-likelihood reaches a maximum at the same point as the original function and the nature of the stationary point may be determined from its second derivative. As σ is a single parameter, the Hessian (matrix of partial second derivatives) is scalar. The Fisher information, $I(\sigma)$, is now

$$I(\sigma) = -E\left[\frac{2n}{\sigma^2} - \frac{3\sum_{i=1}^{n} r_i^2}{\sigma^4}\right] \tag{11.16}$$

Maximum likelihood estimators for the lognormal distribution can be found using a similar process, while estimators for the Rice [35], Nakagami-*m* [36], and Weibull [37] distributions can be found therein.

11.3.2 *Akaike Information Criterion*

The Akaike information criterion [38] (AIC) is a measure of the goodness of fit of a chosen statistical model, which takes into account the number of model parameters (i.e., complexity). Using the AIC, competing models may be ranked according to their respective scores and the model with the lowest AIC index selected as the most likely candidate. The second-order AIC, often denoted by AIC_c, which corrects for small sample size bias, may be written as [39]

$$AIC_c = -2\ln\left[L\left(\hat{\theta}|\text{data}\right)\right] + 2P + \frac{2P\left(P+1\right)}{\left(n-P-1\right)} \tag{11.17}$$

where $\ln[L(\hat{\theta}|\text{data})]$ is the maximized log-likelihood for the parameters θ, given the data set and model under test, P is the number of adjustable parameters available in the chosen model, and n is the sample size. As noted in reference [29], the AIC_c has a strong theoretical motivation as it provides an estimate of the Kullback–Leibler information loss and aims to find the model that will minimize the difference between the true and hypothesized distributions. Two other metrics associated with the AIC_c which may be used to interpret its output, are the delta Akaike and Akaike weight. The delta Akaike of a model i, denoted by Δ_i, is simply a measure of each model relative to the best model, i.e.,

$$\Delta_i = AIC_{c,i} - \min\left(AIC_c\right) \tag{11.18}$$

Interpretation of the delta Akaike is straightforward; the optimum model from the set of candidate models has a delta Akaike of 0. Burnham and Anderson [39] state that a $\Delta_i < 2$ suggests that substantial evidence for the adoption of a model exists, values between 3 and 7 indicate that the model has considerably less support, and a $\Delta_i > 10$ suggests that the model is very unlikely. The Akaike weight of a model i, denoted by w_i, represents the ratio of the delta Akaike to the whole set of R candidate models and is calculated as

$$w_i = \exp\left(-\tfrac{\Delta_i}{2}\right)\Big/\sum_{r=1}^{R}\exp\left(-\tfrac{\Delta_r}{2}\right) \tag{11.19}$$

The Akaike weights indicate the probability that a model is the best among the set of candidate models based upon the highest weight on a scale of 0 to 1. Furthermore, the Akaike weights can be compared with best model (j) to determine the evidence ratio, w_j/w_i, i.e., the degree to which one model is better than another. As a concluding remark on the AIC_c, it should be noted that its use to select the appropriate fading model should be approached with a relative degree of caution. The AIC_c can only select the best model from a set of candidate models; therefore, if the set consists of a selection of poor fitting models, the AIC_c can only select the best candidate from those models. Therefore, it is recommended that the selected model and estimated parameters are also validated visually, to ensure that the model adequately describes the data at hand.

11.3.3 *Worked Example*

11.3.3.1 Model selection

As an example of the narrowband channel modeling process, we now present a statistical analysis of the received signal at the antenna positions shown in Fig. 11.1 from a transmitter mounted on the user's left waist. In this example, the user was mobile in both an anechoic chamber and a reverberation chamber. The antennas used were 5 mm higher mode patches [40] mounted parallel to the body surface; these were designed to exploit the on-body creeping wave mode of propagation, therefore, maximizing on-body coupling. The received signal measurements were obtained using a vector network analyzer configured to record the forward path gain (S_{21}) at 2.45 GHz with a sample rate of 200 Hz. Full details of the channel measurement campaign are provided in reference [12]. For the analysis of fading characteristics, it is quite common to remove the global or local mean depending on the scenario being analyzed. In this example, we remove the global mean from the received signal and report this value separately so that the reader may generate the time correlated random variables associated with the fading distribution using the simulation methods described at the end of this section and subsequently modify the mean value of the envelope to match the system scenario being considered.

We consider as our candidate models for these scenarios the distributions described in Section 11.2.1. Quite often, the cumulative

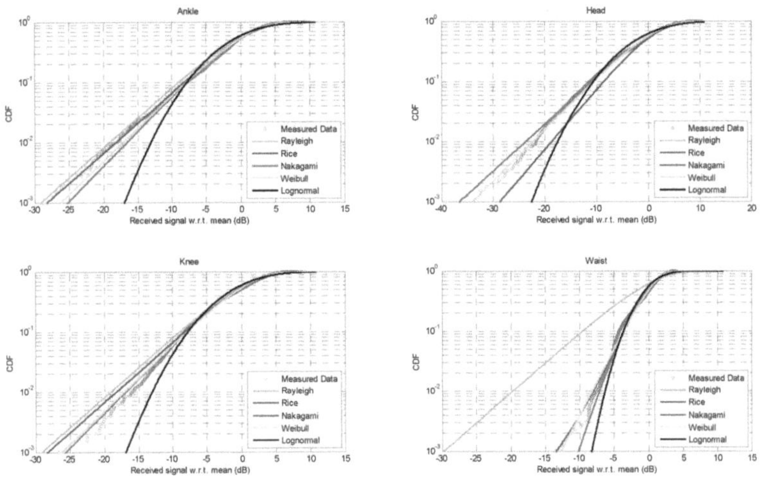

Figure 11.4. A selection of empirical CDF plots and maximum likelihood estimated theoretical plots while the user was mobile in the anechoic chamber. See also Color Insert.

distribution function (CDF) $F(x) = \int_{-\infty}^{x} p_R(r)\,dr$ of the received signal is plotted on log-log scales to expand the tail of the CDF as it is the deepest fades that cause the greatest degradation in wireless system performance. A selection of the empirical CDFs for the ankle, head, knee, and waist as the user was mobile in an anechoic chamber are shown in Fig. 11.4 alongside the theoretical plots. All parameter estimates for the competing models were obtained using maximum likelihood estimation on data sets of 4000 samples per link and are given in Table 11.1, including those not shown in Fig. 11.4. The estimated fading parameters for each of the models show significant variation between each of the on-body links. As expected, and reported elsewhere [24, 29], fading appears to increase the greater the on-body separation distance between nodes. Table 11.2 shows the Akaike weights calculated using Eq. 11.19 for all positions. We can see for this scenario, the Weibull and Rice models both account for three on-body channels and the Nakagami distribution just one. The Rayleigh model, which has only one adjustable parameter, accounted for none of the channels as did the lognormal model.

In the reverberation chamber, the high level of environmental multipath means that the fading observed in the on-body links

Table 11.1. MLE parameter estimates user mobile anechoic chamber

Position	(σ) Rayl	Rice (s, σ)	Nak-m (m, Ω)	Weibull (λ, α)	Lognormal (μ, σ)	f_m (Hz)	Mean (dBmV)
Ankle	0.78	0.74, 0.58	1.17, 1.23	1.1281, 2.20	−0.14, 0.59	5.20	−14.96
Chest	0.72	0.99, 0.15	10.76, 1.02	1.06, 7.37	−0.01, 0.16	5.82	1.21
Elbow	0.78	0.81, 0.53	1.06, 1.21	1.12, 2.26	−0.17, 0.68	3.20	−9.76
Head	0.82	0.00, 0.82	0.78, 1.35	1.12, 1.72	−0.23, 0.77	5.00	−24.40
Knee	0.79	0.73, 0.59	1.14, 1.23	1.13, 2.20	−0.14, 0.58	7.14	−9.13
Waist	0.73	0.96, 0.27	3.59, 1.07	1.10, 4.44	−0.04, 0.30	2.90	0.10
Wrist	0.83	0.00, 0.83	0.82, 1.37	1.13, 1.73	−0.21, 0.69	3.00	−10.64

Table 11.2. Akaike weights user mobile anechoic chamber

Position	Weibull	Nakagami	Rice	Rayleigh	Lognormal
Ankle	0.67	0.32	0.01	0.00	0.00
Chest	0.00	0.00	1.00	0.00	0.00
Elbow	0.00	0.00	1.00	0.00	0.00
Head	0.00	1.00	0.00	0.00	0.00
Knee	0.07	0.00	0.93	0.00	0.00
Waist	1.00	0.00	0.00	0.00	0.00
Wrist	1.00	0.00	0.00	0.00	0.00

significantly increases. Table 11.3 shows that the fading parameters for all of the distributions that contain Rayleigh as a special case (i.e., Rice, Nakagami, and Weibull) are approaching this limit. Observing the empirical CDF plots for the head and waist in particular, we see that there is very little difference in the Rice, Nakagami, and

Table 11.3. MLE parameter estimates user mobile reverberation chamber

Position	Rayl (σ)	Rice (s, σ)	Nak-m (m, Ω)	Weibull (λ, α)	Lognormal (μ, σ)	f_m (Hz)	Mean (dBmV)
Ankle	0.82	0.00, 0.82	0.87, 1.34	1.13, 1.80	−0.19, 0.67	5.90	5.16
Chest	0.80	0.52, 0.70	1.02, 1.27	1.13, 2.03	−0.16, 0.63	4.87	7.76
Elbow	0.79	.66, 0.64	1.05, 1.25	1.13, 2.09	−0.16, 0.63	5.25	8.98
Head	0.79	0.71, 0.61	1.04, 1.25	1.13, 2.10	−0.17, 0.66	5.07	7.71
Knee	0.79	0.67, 0.63	1.15, 1.24	1.13, 2.17	−0.14, 0.57	4.90	8.60
Waist	0.79	0.59, 0.68	1.01, 1.26	1.13, 2.04	−0.17, 0.65	5.44	7.71
Wrist	0.81	0.00, 0.81	0.92, 1.30	1.13, 1.89	−0.19, 0.68	7.00	7.56

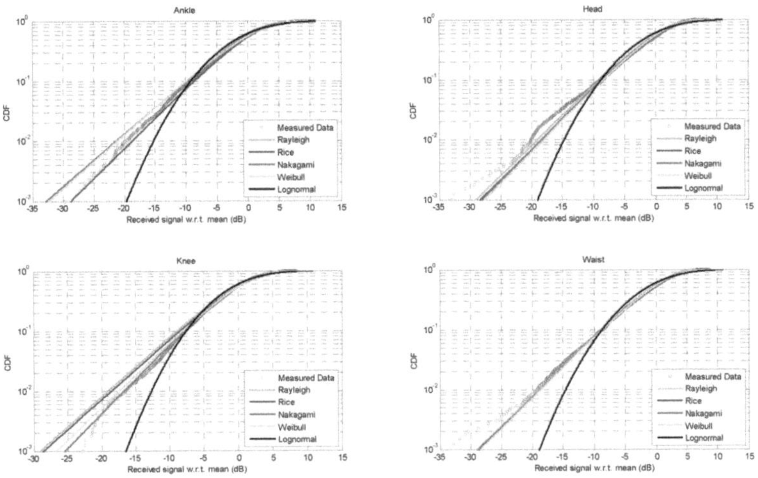

Figure 11.5. A selection of empirical CDF plots and maximum likelihood estimated theoretical plots while the user was mobile in the reverberation chamber. See also Color Insert.

Weibull plots. For this scenario, the Rice distribution was favored on four occasions, the Weibull distribution twice, and Nakagami just once. Once again, the Rayleigh and lognormal distributions were not selected. When compared with the results from the anechoic chamber, it reinforces that when a WBAN moves into an area with a considerable level of environmental multipath, the local surroundings become an important part of the overall channel model.

Table 11.4. Akaike weights user mobile reverberation chamber

Position	Weibull	Nakagami	Rice	Rayleigh	Lognormal
Ankle	1.00	0.00	0.00	0.00	0.00
Chest	0.15	0.11	0.53	0.00	0.21
Elbow	0.01	0.00	0.99	0.00	0.00
Head	0.00	0.00	1.00	0.00	0.00
Knee	0.01	0.99	0.00	0.00	0.00
Waist	0.05	0.02	0.88	0.00	0.05
Wrist	0.64	0.36	0.00	0.00	0.00

11.3.3.2 Level crossing rate

Continuing with the example and omitting the poorer fitting Rayleigh and lognormal models for clarity, Figs. 11.6 and 11.7 show the empirical LCR equations for the elbow-positioned antennas while the user was mobile in the anechoic and reverberation chambers, respectively. For both of these scenarios, the Rice distribution was chosen by the AIC_c to be the most likely candidate model for the fading distribution of the channel. The estimated maximum Doppler frequencies for these channels were obtained by minimizing the sum of the squared error between the empirical LCR and the Rice LCR model [22]. Results for the remaining channels in the anechoic and reverberation chambers were obtained using a similar method using the LCR of the AIC_c favored model and are given in Tables 11.1 and 11.3, respectively. The estimates of the maximum Doppler frequency for these scenarios are in the range 2.90 to 7.14 Hz, which suggests relatively slowly changing channels.

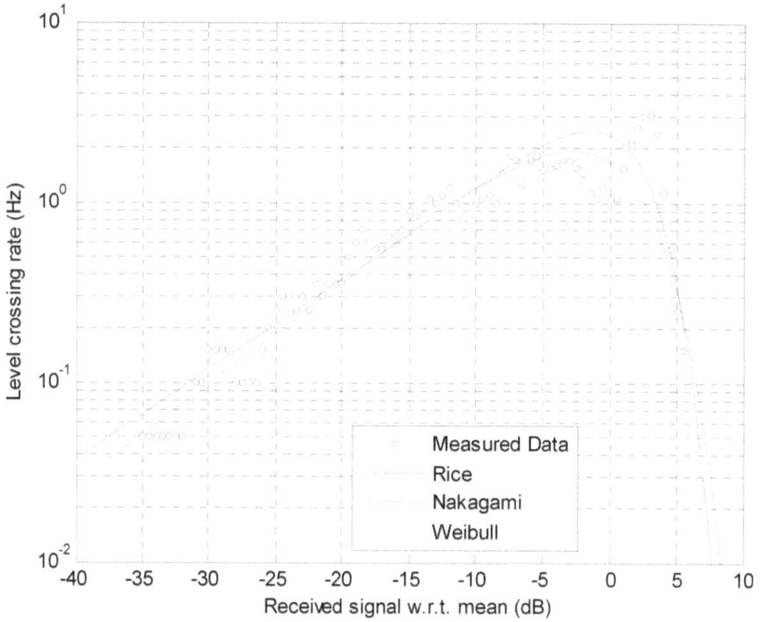

Figure 11.6. Comparison of empirical LCR with theoretical Rice, Nakagami and Weibull LCR plots for elbow-positioned antenna while the user was mobile in anechoic chamber.

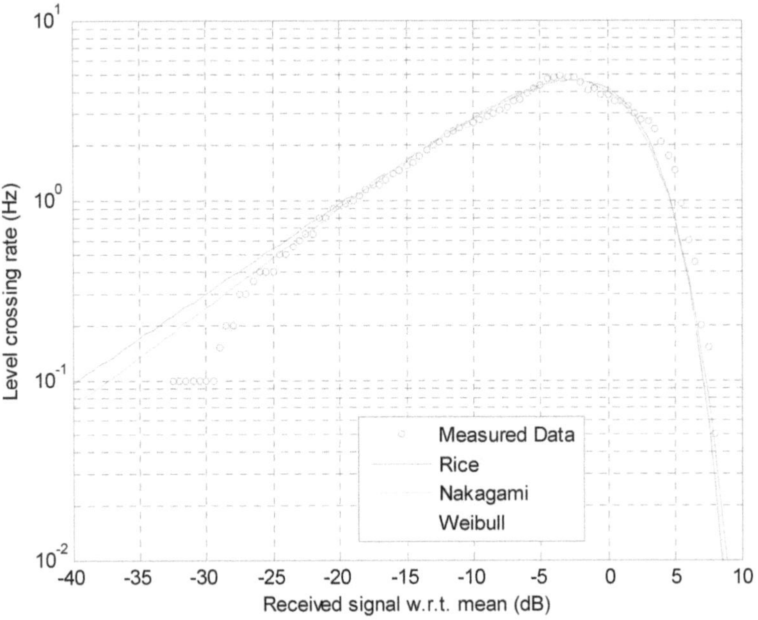

Figure 11.7. Comparison of empirical LCR with theoretical Rice, Nakagami and Weibull LCR plots for elbow-positioned antenna while the user was mobile in reverberation chamber.

11.3.3.3 Simulation of the received signal envelope

Having obtained an estimate of the maximum Doppler frequency, it becomes possible to straightforwardly simulate Rayleigh and Ricean on-body fading channels using the *rayleighchan* and *riceanchan* objects available in the communications toolbox of MATLAB. Nakagami [41][1], Weibull [42], and lognormal [23] envelopes may be generated by adopting a rank matching approach. As outlined in reference [41], where autocorrelated Nakagami random variables (RVs) are produced from time-correlated Rayleigh variates, in this method a Rayleigh envelope with the required autocorrelation properties is generated (e.g., using the *rayleighchan* object in MATLAB). Statistical rank is then exploited by rearranging the Rayleigh envelope using order statistics (i.e., smallest to largest) while keeping note of the

[1] Also describes a method for simulating Nakagami fading under nonisotropic scattering conditions.

original index of the data prior to rearrangement. At the same time, a randomly generated Nakagami, Weibull, or lognormal sequence is also sorted into an ordered vector. Now the Nakagami, Weibull, or lognormal sequence is rearranged so that its autocorrelation structure matches closely that of the original (unordered) Rayleigh reference sequence using the indexes stored earlier.

11.4 Conclusions

Due to the large number of variable parameters associated with the design of body-centric communication systems, it is necessary to make preliminary channel measurements for some well-defined scenarios, which will form the boundary conditions for normal system operations. In this chapter, we have shown how narrowband on-body channels may be modeled using some of the most commonly encountered fading distributions in wireless communications. Although there is no universal model that describes all on-body channels for every possible scenario, it is clear that probability density functions such as Rice, Nakagami, and Weibull may be used to adequately describe the channel's amplitude distribution.

One key requirement for future body-centric channel models is an understanding of the directivity of the channel (i.e., the distribution of the angle of arrival). While directional signal reception will have a negligible effect upon first-order characteristics, depending on the degree of nonisotropic signal reception, it may have an appreciable effect upon second-order statistics such as the LCR and AFD. Therefore, models for these statistics, which were derived under the assumption of isotropic signal reception, may only be used as an approximation in body-centric channels when distribution of signal arrival deviates strongly from the uniform case.

References

1. A. Sani, et al. (January 2010) Experimental characterization of UWB on-body radio channel in indoor environment considering different antennas, *IEEE Transactions on Antennas and Propagation*, **58(1)**, pp. 238–241.

2. S. L. Cotton and W. G. Scanlon (January 2009) Characterization and modeling of on-body spatial diversity within indoor environments at 868 MHz, *IEEE Transactions on Wireless Communications*, **8(1)**, pp. 176–185.
3. I. Khan and P. S. Hall (January 2009) Multiple antenna reception at 5.8 and 10 GHz for body-centric wireless communication channels, *IEEE Transactions on Antennas and Propagation*, **57(1)**, pp. 248–255.
4. A. Serra, P. Nepa, G. Manara, and P. S. Hall (2007) Diversity measurements for on-body communication systems, *IEEE Antennas Wireless Propagation Letters*, **6**, pp. 361–363.
5. I. Khan and P. S. Hall (January 2010) Experimental evaluation of MIMO capacity and correlation for narrowband body-centric wireless channels, *IEEE Transactions on Antennas and Propagation*, **58(1)**, pp. 195–202.
6. S. L. Cotton and W. G. Scanlon (2007) Characterization and modeling of the indoor radio channel at 868 MHz for a mobile body-worn wireless personal area network, *IEEE Antennas and Wireless Propagation Letters*, **6**, pp. 51–55.
7. A. A. Goulianos, T. W. C. Brown, B. G. Evans, and S. Stavrou (July 2009) Wideband power modeling and time dispersion analysis for UWB indoor off-body communications, *IEEE Transactions on Antennas and Propagation*, **57(7)**, pp. 2162–2171.
8. S. L. Cotton and W. G. Scanlon (April 2009) Channel characterization for single and multiple antenna wearable systems used for indoor body-to-body communications, *IEEE Transactions on Antennas and Propagation, Special Issue on Antennas and Propagation for Body-Centric Wireless Communications*, **57(4)**, pp. 980–990.
9. Y. Wang, I. B. Bonev, J. O. Nielsen, I. Z. Kovacs, and G. F. Pedersen (April 2009) Characterization of the indoor multiantenna body-to-body radio channel, *IEEE Transactions on Antennas and Propagation, Special Issue on Antennas and Propagation on Body-Centric Wireless Communications*, **57(4)**, pp. 972–979.
10. S. L. Cotton, W. G. Scanlon, and P. S. Hall (September 2010) A simulated study of co-channel inter-BAN interference at 2.45 GHz and 60 GHz, in *European Wireless Technology Conference*, Paris, pp. 61–64.
11. W. G. Scanlon and N. E. Evans (April 2001) Numerical analysis of body-worn UHF antenna systems, *IEE Electronics and Communication Engineering Journal*, **13(2)**, pp. 53–64.

12. S. L. Cotton, W. G. Scanlon, and G. A. Conway (2009) Autocorrelation of signal fading in wireless body area networks, in 2nd *IET Seminar on Antennas and Propagation for Body-Centric Wireless Communications*, London, pp. 1–5.
13. J. Ryckaert, P. de Doncker, R. Meys, A. de Le Hoye, and S. Donnay (April 2004) Channel model for wireless communication around human body, *Electronics Letters*, **40(9)**, pp. 543–544.
14. A. Alomainy, et al. (January 2007) Statistical analysis and performance evaluation for on-body radio propagation with microstrip patch antennas, *IEEE Transactions on Antennas and Propagation*, **55(1)**, pp. 245–248.
15. S. L. Cotton, G. A. Conway, and W. G. Scanlon (April 2009) A time-domain approach to the analysis and modeling of on-body propagation characteristics using synchronized measurements at 2.45 GHz, *IEEE Transactions on Antennas and Propagation*, **57(4)**, pp. 943–955.
16. Y. I. Nechayev, Z. H. Hu, and P. S. Hall (2009) Short-term and long-term fading of on-body transmission channels at 2.45 GHz, in *Loughborough Antennas and Propagation Conference*, Loughborough, pp. 657–660.
17. G. Z. Yang, A. Sani, S. L. Lee, Y. Zhao, and Y. Hao (2009) A simulation environment for subject-specific radio channel modeling in wireless body sensor networks, in 6th *International Workshop on Wearable and Implantable Body Sensor Networks*, pp. 23–28.
18. A. Fort, C. Desset, P. Wambacq, and L. V. Biesen (December 2007) Indoor body-area channel model for narrowband communications, *IET Microwaves, Antennas and Propagation*, **1(6)**, pp. 1197–1203.
19. S. L. Cotton and W. G. Scanlon (2006) A statistical analysis of indoor multipath fading for a narrowband wireless body area network, in 17th *IEEE International Symposium on Personal, Indoor and Mobile Radio Communications*, pp. 1–5.
20. W. R. Braun and U. Dersch (1991) A physical mobile radio channel model, *IEEE Transactions on Vehicular Technology*, **40(2)**, pp. 472–482.
21. J. C. S. S. Filho and M. D. Yacoub (December 2009) On the second-order statistics of Nakagami fading simulators, *IEEE Transactions on Communications*, **57(12)**, pp. 3543–3546.
22. G. L. Stüber (2002) *Principles of Mobile Communication*, 2nd edn, Kluwer Academic Publishers, New York.
23. S. L. Cotton and W. G. Scanlon (2007) Higher order statistics for lognormal small-scale fading in mobile radio channels, *IEEE Antennas and Wireless Propagation Letters*, **6**, pp. 540–543.

24. S. L. Cotton and W. G. Scanlon (January 2009) An experimental investigation into the influence of user state and environment on fading characteristics in wireless body area networks at 2.45 GHz, *IEEE Transactions on Wireless Communications*, **8(1)**, pp. 6–12.
25. M. Kim and J. I. Takada (2009) Statistical model for 4.5 GHz narrowband on-body propagation channel with specific actions, *IEEE Antennas and Wireless Propagation Letters*, **8**, pp. 1250–1254.
26. M. D. Yacoub, D. B. da Costa, U. S. Dias, and G. Fraidenraich (2005) Joint statistics for two correlated Weibull variates, *IEEE Antennas and Wireless Propagation Letters*, **4**, pp. 129–132.
27. M. D. Yacoub (January 2007) The α-μ Distribution: A Physical Fading Model for the Stacy Distribution, *IEEE Transactions on Vehicular Technology*, **56(1)**, pp. 27–34.
28. D. B. Smith, et al. (November 2009) Temporal correlation of dynamic on-body area radio channel, *Electronics Letters*, **45(25)**, pp. 1212–1213.
29. A. Fort, C. Desset, P. de Doncker, P. Wambacq, and L. van Biesen (June 2006) An ultra-wideband body area propagation channel model—from statistics to implementation, *IEEE Transactions on Microwave Theory and Techniques*, **54(4)**, pp. 1820–1826.
30. N. Youssef, T. Munakata, and M. Takeda (September 1996) Fade statistics in Nakagami fading environments, in *Proceedings of the IEEE 4th International Symposium on Spread Spectrum Techniques and Applications*, pp. 1244–1247.
31. W. C. Jakes (1974) *Microwave Mobile Communications*, Wiley, New York.
32. S. O. Rice (January 1948) Statistical properties of a sine wave plus random noise, *Bell System Technical Journal*, **27**, pp. 109–157.
33. M. D. Yacoub, J. E. V. Bautista, and L. Guerra de Rezende Guedes (May 1999) On higher order statistics of the Nakagami-m distribution, *IEEE Transactions on Vehicular Technology*, **48(3)**, pp. 790–794.
34. N. C. Sagias, D. A. Zogas, G. K. Karagiannidis, and G. S. Tombras (June 2004) Channel capacity and second-order statistics in Weibull fading, *IEEE Communications Letters*, **8(6)**, pp. 377–379.
35. J. Sijbers, A. J. den Dekker, P. Scheunders, and D. van Dyck (1998) Maximum likelihood estimation of Ricean distribution parameters, *IEEE Transactions on Medical Imaging*, **17**, pp. 357–361.
36. J. Cheng and N. Beaulieu (2001) Maximum likelihood-based estimation of the Nakagami-*m* parameter, *IEEE Communications Letters*, **5**, pp. 101–103.

37. A. C. Cohen (November 1965) Maximum likelihood estimation in the Weibull distribution based on complete and on censored samples, *Technometrics*, **7(4)**, pp. 579–588.
38. H. Akaike (1974) A new look at the statistical model identification, *IEEE Transactions on Automatic Control*, **19(6)**, pp. 716–723.
39. K. P. Burnham and D. R. Anderson (2002) *Model Selection and Multimodel Inference: A Practical Information-Theoretic Approach*, 2nd edn, Springer-Verlag, New York.
40. G. A. Conway and W. G. Scanlon (April 2009) Antennas for over-body surface communication at 2.45 GHz, *IEEE Transactions on Antennas and Propagation*, **57(4)**, pp. 844–855.
41. J. C. S. S. Filho, M. D. Yacoub, and G. Fraidenraich (March 2007) A simple accurate method for generating autocorrelated Nakagami-*m* envelope sequences, *IEEE Communications Letters*, **11**, pp. 231–233.
42. D. B. Smith, et al. (2009) A simulator for the dynamic on-body area propagation channel, in *Antennas and Propagation Society International Symposium*, Charleston, pp. 1–5.

Chapter 12

Antenna Design and Propagation for WBAN Applications

Tharaka Dissanayake
School of Electrical Engineering and Computer Science, University of Newcastle, Callaghan, NSW2308, Australia
tharakad@ieee.org

An essential part of wireless body area networks (WBANs), like for any wireless system, is the transducer that converts electrical signals into electromagnetic waves and vice-versa. This has been done by antennas for medium-long range communications or inductive coils for very short range communication. Usually the communication in WBAN is either short range or medium range. The on-body and implanted sensor nodes are essentially small devices that are, at most, of the size of a standard mobile phone. There are matured enough technologies to design low-profile antennas for mobile wireless devices. The transformation of those technologies into WBAN applications will be discussed. One of the main concerns of antennas in WBAN is the interaction of near field electric and magnetic fields with the human body. This phenomenon is responsible for significant changes in antenna characteristics, and researchers have found methods to mitigate these effects for on-body antennas. Some of these methods, especially the antenna

Wireless Body Area Networks: Technology, Implementation, and Applications
Edited by Mehmet R. Yuce and Jamil Y. Khan

ISBN 978-981-4316-71-2 (Hardcover), ISBN 978-981-4241-57-1 (eBook)
www.panstanford.com

shielding, will be discussed. It is even more challenging to design in-body (implanted) antennas with surrounding dissipative body tissue at high frequencies. Recent research into the efficiency enhancement of implanted antennas will be discussed in this chapter. Widely used classes of antennas in wideband and narrowband WBAN, including volume conduction antennas, will be presented as examples. The commonly used design and experimental methods of WBAN antenna evaluation will be discussed. A brief overview of antenna theory and modern simulation techniques will be incorporated, where necessary, for the readers who are not familiar with the material.

12.1 Introduction

Antennas can be considered reciprocal devices that convert currents into field and fields into current. While reciprocity of antennas does not have the same meaning for wideband operation, all characterizing parameters are defined for both transmitting and receiving antennas. For the sake of those who are not familiar with the terminology used in antenna design, let us first define several core definitions of antenna engineering. It is not by any means a complete list but the most commonly found in design literature.

12.1.1 *Antenna Gain*

Gain of the antenna is defined very much different to the gain of an amplifier. In the context of an amplifier or any other active or passive circuit component, gain relates the input currents, voltages, or power to the output currents, voltages, or power. This is exactly the case in antennas; however, the difference is that the reference is not an input parameter, but the properties of a hypothetical isotropic antenna. The concept of gain of an antenna, essentially a passive device, is defined as the radiation intensity (in watt/square meter) in a given direction $I(\theta, \phi)$compared to the intensity I_0 of an isotropic antenna driven by the same power source. Hence,

$$\text{gain}(dBi) = 10\log\left(\frac{I(\theta, \phi)}{I_0}\right) \tag{12.1}$$

Note that I_0 is not a function of elevation and azimuth angles θ and ϕ, respectively. Some engineers use dipole antenna as the reference antenna. The difference can be noted in the units of the gain. For the isotropic reference, the unit is dBi. More often than not gain is specified as a single value, which is the maximum out of all directions.

12.1.2 *Return Loss*

One of the first things an antenna designer attempts to minimize is the reflection of power at the antenna-transmission line interface. Typical feed chain consists of a source, transmission line with characteristic impedance matched to internal impedance of the source and the antenna. Matching networks may be employed to minimize power reflections at each interface. The reflection coefficient is defined as

$$\Gamma = \frac{Z_L - Z_o}{Z_L + Z_o} \tag{12.2}$$

where Z_L is the input impedance of the antenna and Z_o is the characteristic impedance of the transmission line, which is often matched to the source. The value of Z_o is chosen as 50 Ω in most prototyping experiments. Nevertheless, actual output impedance of a sensor node radio front-end can be different. In such situations, a matching network may be employed. The return loss is now defined as

$$\mathrm{RL} = -20 \log_{10} |\Gamma| \tag{12.3}$$

which is a positive value. Hence, higher return loss means less reflection. In generic designs, RL is kept above 10 dB; however, in high power base station applications, the targeted return loss is above 15 dB. For power limited WBAN sensor nodes, higher values of RL are highly desirable.

12.1.3 *Efficiency*

Efficiency of an antenna is defined as the ratio between input power and the radiated power. This ratio is determined by the losses incurred in the antenna system. Useful power of an antenna is the amount that is actually radiated to the far-field region. All the other losses are

- Mismatch losses due to impedance differences between the transmission line and the antenna
- Losses due to finite resistance of the conductive elements of the antenna
- Losses due to dielectric loss tangent of substrate, superstrate, or filling material (in case of dielectric filled horn antennas)
- Near-field losses due to coupling between surrounding conductive or dielectric material. Absorption losses of body tissue, device casings, and radomes are some examples for this category

In WBAN, the most prominent is the radiation absorption by the tissue. As we shall see later, proximity of lossy material to an antenna can severely degrade its efficiency. Without partial or full insulation, especially for antennas on implanted devices, sustainability of radiation is nearly impossible.

12.1.4 *Reciprocity*

The concept of reciprocity is often applied to antenna characterization. In general, all of the aforementioned parameters are valid for both transmit and receive cases. Nevertheless, this is not the case for ultra-wideband (UWB) antennas analyzed in the time domain. For wideband operation, antennas are better characterized using transfer functions or impulse responses. Transmit impulse response $h_t$of the antenna is defined as

$$E(\theta, \phi, t) = \frac{\mu}{4\pi r} I(t) \times h_t(\theta, \phi, t) \tag{12.4}$$

where E is the electric field and I is the excitation current. The receive impulse response is defined as

$$V_{rec}(t) = E(t) \times h_r(\theta, \phi, t) \tag{12.5}$$

Consequently, the reciprocity in the time domain is

$$\frac{1}{2} h_t(\theta, \phi, t) = \frac{\partial}{\partial t} h_r(\theta, \phi, t) \tag{12.6}$$

Note that it is still possible to characterize UWB antennas with radiation patterns at several frequencies inside the operating band, where the reciprocity is valid. This is what we may find in many publications. The main reason is the unavailability of resources to measure antennas in time domain. However, transfer function approach is the most effective in optimizing radiation in different directions, such as pulse shaping [1, 2]. It is also convenient to link antenna impulse response with the channel impulse response to model propagation scenarios [3].

12.2 Miniaturized Antennas

On-body or in-body devices, which are essentially the nodes of a WBAN, are no more than the size of a mobile phone. The size of a wireless endoscope can be even smaller, of the size of a size 000 capsule. This leaves the antenna designer with very little physical volume to work with. Fortunately, the problem has been in discussion over the last decade among engineers, who have developed antennas for mobile phones. The knowledge can well be used in designing antennas for all types of WBAN devices, especially those for biomedical applications [4–6].

12.2.1 *Planar Inverted-F Antennas*

A planar inverted-F antenna (PIFA) is essentially an evolution of the microstrip patch antenna. Additionally, PIFA has a shorting pin, which reduces the size of the microstrip patch antenna by half (Fig. 12.1). This miniaturization makes it very popular among small antenna designers. The location of feed with respect to shorting pin/strip can be used to tune the antenna. However, PIFA is relatively narrowband and various techniques have been employed to overcome this deficiency. One such technique is capacitive loading [7]. In addition, various other techniques employed in microstrip antenna bandwidth enhancement such as thick, low permittivity substrates with capacitive feeds can also be employed in PIFA design. Miniaturizing techniques such as meandering are also employed in advanced designs of PIFA antennas (Fig. 12.2) [8].

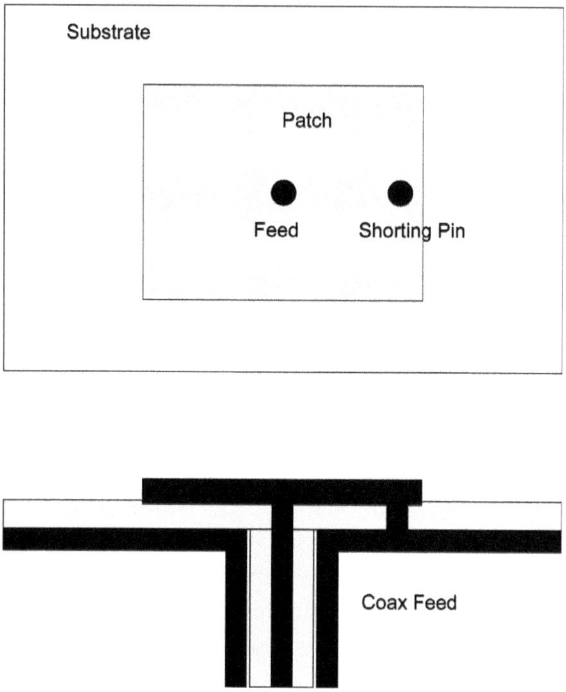

Figure 12.1. Planar inverted-F antenna configuration.

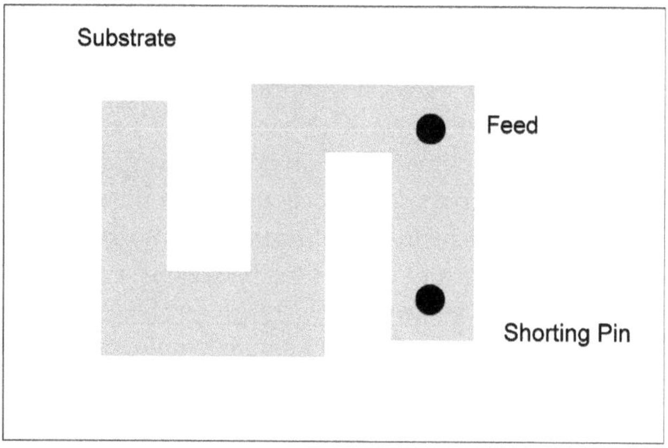

Figure 12.2. Meandered PIFA antenna.

12.2.2 *Planar Monopoles and Dipoles*

Planar monopoles discussed here are the printed version of the quarter-wave monopole antenna or the half-wave dipole antenna. Such antennas have gained popularity as compact, integrated components in mobile handheld devices. The ease of fabrication dominates the list of advantages in planar antennas. However, the tuning of quarter-wave printed monopole antennas depends on the supporting ground plane. As shown in reference [9], the finite size of the ground plane reduces the bandwidth and the frequencies of operation. Several modifications have been suggested such as adding additional strips to the ground plane. One of the challenges in integrating printed monopoles or dipole antennas in WBAN devices is retaining the compactness while manipulating the ground plane to tune the antenna. One such technique is the use of asymmetric feed location and the feed gap [9]. Being able to freely place the feed point without losing mechanical rigidity is an advantage of printed antennas.

The design challenge unique to the printed monopoles or dipoles in WBAN is the antenna detuning due to the proximity of body tissue. This has been shown in reference [5] using a planar compact capsule-shaped antenna in the proximity of a tissue sample. The change of matching is obvious in Fig. 12.3. When coupled to a conducting body such as a tissue sample, currents are induced in a larger volume. As a result, the effective size of the antenna rises by several wavelengths. This results in lower resonant frequencies. At first glance, this is helpful in antenna miniaturization. Nevertheless, it must be noted that tissue is a poor conductor, which brings down the radiation efficiency of the antenna. Furthermore, properties of the tissue are spatial and temporal variables. Therefore, usually it is not recommended that the presence of tissue is used as a miniaturization technique.

The coupling effect is prominent with the strong electric fields at the feed gap of dipole or monopole antennas, as shown in Fig. 12.4. Therefore, it is important when designing and placing planar dipole and monopole antennas in WBAN applications, the feed gap stays clear of conductive and dielectric parasitic elements or there is sufficient shielding [10], either conductive or dielectric.

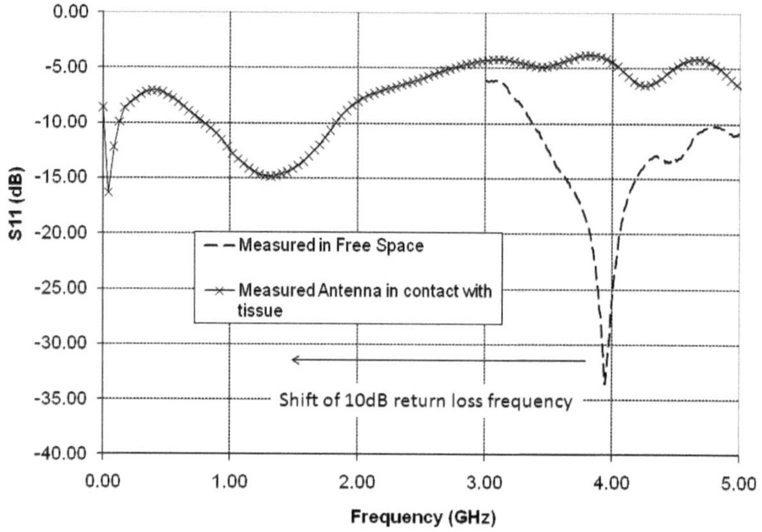

Figure 12.3. Change of measured return loss characteristics due to the presence of tissue sample [5].

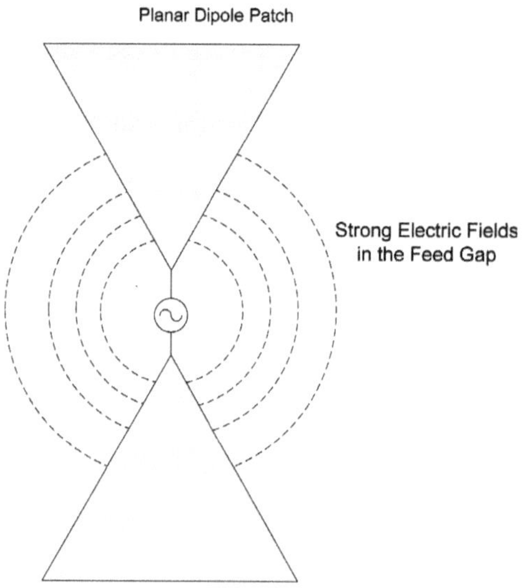

Figure 12.4. Feed gap electric fields.

12.2.3 *Planar Slot Antennas*

The slot antennas are also known as magnetic antennas. The loop antennas fall in the same general category [11]. These antennas have the advantage of loosely coupling to nearby conductive structures. The selection of slot antennas not only permits greater freedom in tuning antennas for packaged devices, but also delivers unchanged pattern stability as discussed in reference [12]. The dielectric loaded antenna presented by us, which will be discussed in the next section, was a compact wide-slot as well. The annular slot antenna presented in reference [3] proves the same since the shape of the radiated pulses remained unchanged with the distance between the antenna and the body. The power of the received pulse changed proving that the closer the antennas to the body, the lower the efficiency. However, this effect is weak in slot antennas compared with monopoles or dipoles. Use of dielectric [13] or conductive shielding can further improve the antenna performance as shown in reference [10]. Shielding also reduces the radiation absorption by the tissue. Absorption is proportional to the electric field components inside the tissue. It changes with the magnitude and the direction of the near-field patterns [14]. Although it is possible to design to minimize the absorption by manipulating the near field, total shielding is an easier, somewhat straightforward, solution. There is, however, one main disadvantage in adding a reflector to otherwise omnidirectional antenna such as a wide slot. The confinement of the fields between the shield and the antenna (or return currents in another perspective) results in high Q, and in case of UWB antennas as shown in reference [10], it causes the reduction of the matched bandwidth of the antenna. That is why in narrowband WBAN applications, PIFA with ground plane is essentially the favorite as mentioned in Section 12.2.1.

12.3 Implanted Antennas

We have seen that on-body antenna design can closely follow the principles of small antenna design for wireless handsets near lossy dielectric tissue. In contrast, the implanted antennas are totally surrounded by lossy dielectrics. This makes the antenna design, especially sustaining the required efficiency, very difficult. This challenge has been addressed in both narrowband and wideband communication by many research groups, including that of the author.

A highly cited theoretical analysis of implanted antennas can be found in reference [15]. A dipole inside a head model has been analyzed using finite difference time domain (FDTD) and dyadic Green's function (DGF) methods. The FDTD simulation has used a detailed body model with 30 different tissue types, whereas the DGF uses spherical model with six concentric layers of different tissue types. The FDTD simulation used a detailed human phantom model and, hence, delivered more accurate and realistic results. Such simulations are possible with most of the commercial electromagnetic simulation software as they all provide detailed full-body phantom models, thanks to Visible Human Project sponsored by US National Library of Medicine. However, computing resource availability may limit the simulation to only a portion of the phantom. The microstrip and PIFA antennas presented in reference [15], with 0.16% and 0.25% efficiency, respectively, have been simulated with simplified phantom models. It shows that the use of simplified models does not compromise the accuracy of return loss characteristics. After all, only the near-field electromagnetic interactions decide the return loss characteristics of an antenna. Therefore, it is possible to first simulate the antenna itself in a simplified model and then use the resulting electric and magnetic fields with a phantom model to calculate far-field electric fields. This approach allows a fine mesh for the small antenna and a coarse mesh for the large phantom model [16] when using FDTD method. Such multilevel meshing techniques are built-in for commercial software packages such as Microwave Studio by CST.

Although sophisticated modeling of antennas with phantom models is possible with modern electromagnetic solvers [17, 18], the precise modeling of the changing properties of the body tissue is nearly impossible. There is no guarantee that tissue properties remain the same from person to person or time to time. Physiological conditions can change the water content of tissue and, hence, the conductivity and permittivity. Typically, there is high contrast between the electromagnetic properties of adjacent tissue layers. For example, the relative permittivity of muscle is 64 at 434 MHz, whereas that of fat is 5 at the same frequency. One cannot expect consistency in placing implanted devices in practice. As such implanted device may end up in a fat layer or muscle. Dimensions of the antenna have to be significantly different between these two situations.

One of the methods to address this design challenge is making the antenna less sensitive to surrounding tissue properties. Note that almost all implanted antennas have some type of superstrate, which prevents it from touching the lossy tissue [15, 18–20]. The permittivity of the superstrate has been chosen similar to that of the substrate or in some cases determined by the type of biocompatible material used. We have shown that superstrate can be made to act as a shield between the antenna and the surrounding tissue properties. In other words, the permittivity of the superstrate can be selected that the sensitivity of the return loss to tissue properties can be minimized [6, 21].

12.3.1 *Dielectric Loaded Matching of Implanted Antennas*

Dielectric loading of antennas in dissipative medium is a technique that has been used for underground and underwater antenna applications [22, 23]. The objective was to completely shield the currents on the antenna from touching the conductive medium. The experiments in reference [24] have been conducted in salt water, which is very similar to an antenna in a tissue simulating medium. Recent work on insulated dipoles in dissipative medium [25] has brought to light some interesting results, which we have explored in reference [21].

It can be shown that, using expressions for power reflection coefficient at the insulator and tissue medium interface, if the dielectric constant of the insulator is sufficiently high and remains close to an average tissue dielectric constant, the antenna matching can be achieved over a wide range of tissue properties. The preferred value for dielectric constant of the insulator in the experiments conducted in reference [6] is 50. Figure 12.5 shows the antenna configuration. The antenna has the shape of a size 000 capsule. It has two distinctive regions: the grounded coplanar waveguide

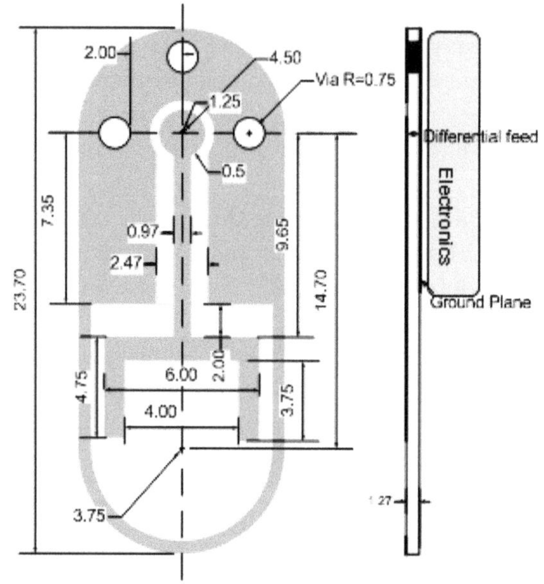

Figure 12.5. Insulator loaded printed slot antenna configuration. The shaded area is metalization.

and the wide slot, printed on a Rogers TMM10i substrate. Rogers provide a range of laminates that have very low dielectric losses. For antenna applications, it is important that high frequency laminates are chosen. Usually designers use FR4 laminates, which are widely used in printed circuit board design. But the significant dielectric losses cannot be tolerated, where high losses in propagation through tissue prevail. The TMM10i has the highest dielectric constant in the range. Furthermore, high dielectric constant of the substrate reduces the guided wavelength at a given frequency. Dimensions of the antenna and the transmission lines are determined by the guided wavelength. For example, half-wave printed dipole means the antenna length is actually half of the guided wavelength as opposed to free space wavelength.

By selecting high dielectric constant substrate and even higher dielectric constant insulator, we have achieved very high level of compactness. Thus the antenna can be planted inside a capsule-shaped enclosure. Note that the portion of the antenna with

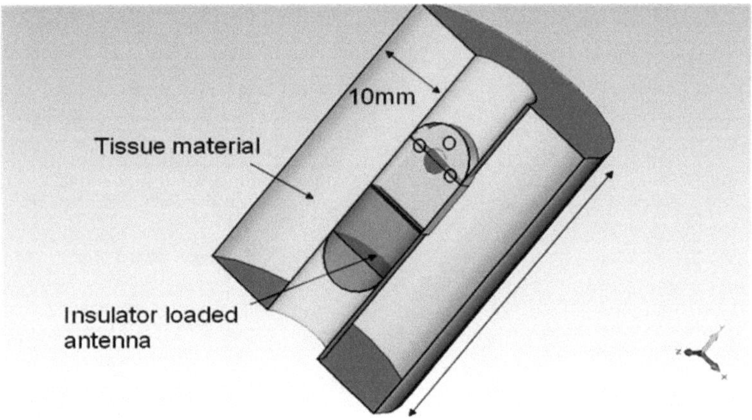

Figure 12.6. The loaded antenna inside a tissue simulating cylinder in a CST simulation. See also Color Insert.

the coplanar waveguide can indeed support other electronic components. It is essential that antenna designers must consider the effects of other components of the implanted device. We have achieved this by introducing a bulk conductor to the upper half of the antenna, and SMA connector in experiments. Figure 12.6 shows the antenna in a tissue-simulating cylinder with its insulator loading. The properties of the tissue cylinder have been changed to match various tissue types. The results are shown in Fig. 12.7. It confirms that the antenna is matched well within the expected bandwidth off 1 GHz, suitable for UWB applications.

12.3.1.1 Biocompatibility of dielectric loaded antenna

Although it is found that dielectric constant of the insulator has to be close to 50, matching the permittivity of the biocompatible material of the device casing to that value is a manufacturing challenge. As far as we are aware, the engineers developing biocompatible material have very little concern about the electromagnetic properties of those. Therefore, it is expected that the antenna with its insulator has a casing of biocompatible material such as silicone. Properties of this material definitely affect the antenna performance. The appendices of this chapter include a set of MATLAB codes

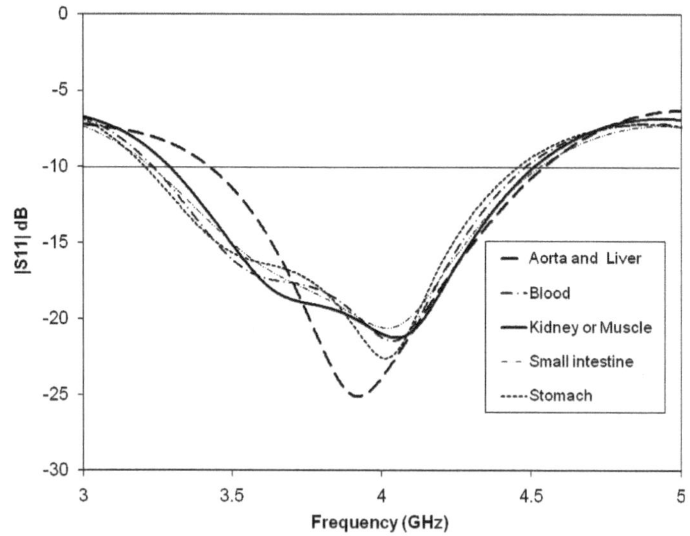

Figure 12.7. Return loss characteristics of the antenna shown in Figs. 12.6 and 12.5 for various tissue types.

for calculating the modified reflection coefficient after adding a thin shell of low dielectric constant biocompatible material as a modification to the problem analyzed in reference [25], as shown in Fig. 12.8.

As an example, Table 12.1 shows the reflection coefficient at the insulator-shell interface (note that the reflected wave inside the insulator affects the antenna return loss characteristics) for insulator dielectric constant of 50, shell dielectric constant of 10, frequency of 3 GHz, and insulator radius of 5 mm. The shell thickness

Table 12.1. Reflection coefficient at the insulator-biocompatible shell interface for fixed electromagnetic properties of the insulator shell. The surrounding media is muscle tissue.

Reflection coefficient (dB)	Shell thickness (mm)
−30	0
−37	0.1
−19	1
−5	10

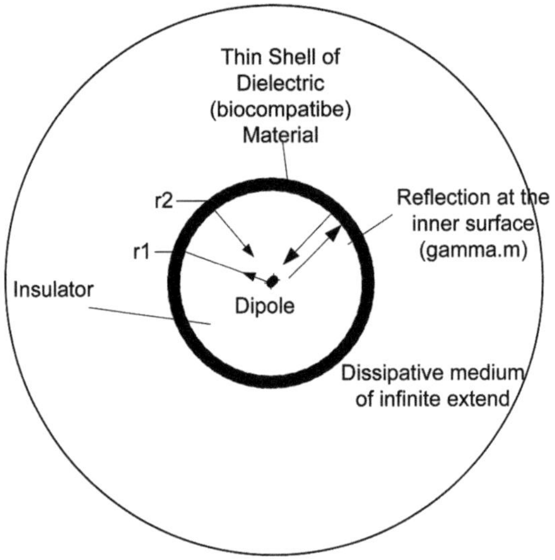

Figure 12.8. Insulated Hertzian dipole with a dielectric shell in a dissipative medium.

is varied from 0 mm to 10 mm. The implant is assumed to be inside muscle tissue. The frequency-dependent tissue properties have been determined using fourth-order Cole–Cole equation (tissue_prop.m).

It is obvious from the results in the above table that with thinner biocompatible shells, the relatively small dielectric constant of the biocompatible material does not make much difference to the reflection coefficient. Note that insulator and the biocompatible material were assumed lossless.

12.4 Volume Conduction Antennas

The antennas we have discussed so far are physically separate from the human body. Any currents induced on body tissue due to the antenna's operation, although unavoidable, are responsible for significant losses in the system operation. Thus we take measures to minimize these currents. All tissue types have increasing conductivity with frequency. At microwave frequencies, this phenomenon is a problem in terms of propagation and antenna matching. However,

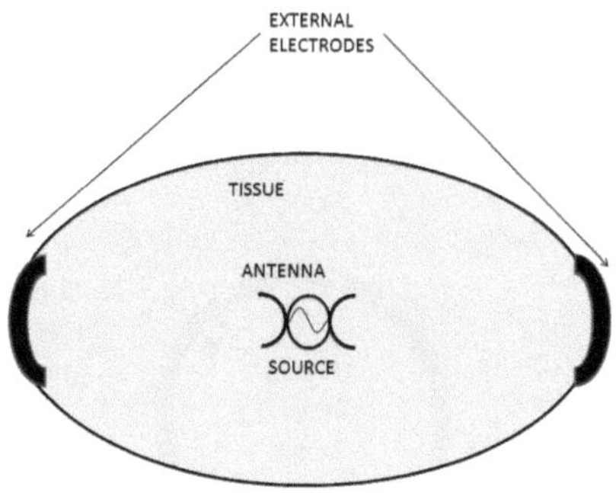

Figure 12.9. The concept of x-antenna.

there is a class of antennas suitable for WBAN communication that uses tissue conductance for its operation. They are known as volume conduction antennas [26–29]. In this technology, tissue is in direct contact with the feed probe such that the former acts as an antenna or a transmission line. Systems based on volume conduction principle use internal electrodes and skin electrodes. Although signal propagation (i.e., currents) is completely inside the tissue with nonlinear characteristics, at macro level the reciprocity theory holds [26].

The x-antenna presented in reference [29] (Fig. 12.9) has shown promising characteristics as a volume conduction antenna. At 1 kHz for an input power of 1.3 μW, 0.05 mV signal has been generated on an electrode 15 cm away. Further improvement of this method can deliver not only volume conduction, but also surface conduction communication at higher frequencies.

12.5 Summary

In this chapter, we have discussed antenna elements that are suitable for WBAN. The main focus is on designing miniaturized antennas and taking the effect of human body into consideration. The most popular antennas that have been proposed for WBAN devices are PIFA, planar monopoles and dipoles, printed slot antennas, and volume conduction antennas. Variants of these antennas can be used in either in-body or on-body devices. The newly introduced technique of dielectric loading to impedance match the implanted antennas has been discussed. As a further improvement to make the suggested method industrially viable, addition of thin shell of biocompatible material is considered. A set of MATLAB codes for theoretical analysis of such modification is provided here. Readers may modify these codes to optimize antennas for different implanted applications. For example, it is possible to calculate the optimum thickness of the superstrate layers of a dielectric-loaded PIFA by breaking the antenna currents into dipole-like elements and determining optimum distances from each of them to the insulator-tissue interface.

Appendix A

A.1 *Function Calculating the Reflection Coefficient*

```
%This is the reflection coefficient at the interface
%between the first insulator and biocompatible material.
%r1 is the inner radius of the thin shell and r2 is the
%outer radius of the thin shell. X111 and X514 are
%defined in X111.m and X514.m, respectively. k1,k2, and
%k3 are calculated using function k.m
function gamma=gamma(k1,k2,k3,r1,r2)
x1=x111(-k1,r1);
x2=x111(k1,r1);
x3=x111(-k2,r1);
x4=x111(k2,r1);

x5=x514(-k1,r1);
x6=x514(k1,r1);
x7=x514(-k2,r1);
x8=x514(k2,r1);

x9=x111(-k2,r2);
x10=x111(k2,r2);
x11=x111(-k3,r2);

x12=x514(-k2,r2);
```

```
x13=x514(k2,r2);
x14=x514(-k3,r2);

p1=x3*(x11*x13-x14*x10)/(x13*x9-x12*x10)+x4*
            (x11*x12-x14*x9)/(x10*x12-x13*x9);
p2=x7*(x11*x13-x14*x10)/(x13*x9-x12*x10)+x8*
            (x11*x12-x14*x9)/(x10*x12-x13*x9);

gamma=(x1*p2-p1*x5)/(x6*p1-x2*p2);

end
```

A.2 *X111 and X514*

```
function x111=x14(k,r)
x111=deli(k,r)*k^2;
end

function x514=x514(k,r)
x514=deli(k,r)/r+k^2*del(k,r);
end
```

A.3 *Function calculating K vectors*

```
%This function generates the k vectors using material
%properties and frequency. Frequency is an input
%variable to this function. On top of that, in case of
%biological tissue epsr and sigma are functions
%of frequency and they will be calculated by the
%function cole4.m. Usually mur=1.
function k=k(epsr,mur,sigma,freq)

eps0=8.85e-12;
mu0=4*pi*1e-7;

eps=epsr*eps0;
mu=mur*mu0;
```

```
omega=2*pi*freq;

kr=omega*sqrt(eps*mu/2)*sqrt(sqrt(1+sigma^2/
               eps^2/omega^2)+1);
ki=omega*sqrt(eps*mu/2)*sqrt(sqrt(1+sigma^2/
               eps^2/omega^2)-1);

k=kr-i*ki;
end
```

Appendix B: Calculating Frequency-Dependent Tissue Properties

A.4 *Cole–Cole function*

```
%This function delivers relative permittivity and
%conductivity of biological tissue at a given
%frequency, when all the parameters for 4th order
%Cole-Cole equation are given. Cole-Cole parameters
%for 24 tissue types are given in Table 3.1 of Peter S.
%Hall and Yang Hao, "Antennas and Propagation for
%Body-centric Wireless Communications",Artec House,
%2006,.

function [epsr,sigma]=cole4(f,einf,sig,alpha1,alpha2,
alpha3,alpha4,de1,de2,de3,de4,tau1,tau2,tau3,tau4)

eps0=8.85e-12;

ehat1=de1/(1+(j*2*pi*f*tau1*1e-12).^(1-alpha1));
ehat2=de2/(1+(j*2*pi*f*tau2*1e-9).^(1-alpha2));
ehat3=de3/(1+(j*2*pi*f*tau3*1e-6).^(1-alpha3));
ehat4=de4/(1+(j*2*pi*f*tau4*1e-3).^(1-alpha4));

ehat=einf+(ehat1+ehat2+ehat3+ehat4)+sig/
         (i*2*pi*f*eps0);
```

```
epsr=real(ehat);
sigma=-2*pi*f*imag(ehat)*eps0;
end
```

A.5 *Calculating the properties*

```
%This function delivers epsilon and sigma of a given
%tissue based on Cole-Cole equation. There is a text
%file "tissue.txt", which has all the Cole-Cole
%parameters for 24 tissue types (as given in Table 3.1
%of "Antennas and Propagation for Body-centric Wireless
%Communications" by Hall and Hao). The function reads
%from the text file and returns the permittivity and
%conductivity at a given frequency.
function [epsr,sigma]=tissue_prop(tissue_type,f)
switch tissue_type
    case 'Aorta'
        row=1;
    case 'Bladder'
        row=2;
    case 'Blood'
        row=3;
    case 'Bone(cancellous)'
        row=4;
    case 'Bone(cortical)'
        row=5;
    case 'Brain'
        row=6;
    case 'Breast'
        row=7;
    case 'Cartilage'
        row=8;
    case 'Spinal Fluid'
        row=9;
    case 'Cornea'
        row=10;
    case 'Eye Tissue'
```

```
        row=11;
    case 'Fat'
        row=12;
    case 'Bladder'
        row=13;
    case 'Heart'
        row=14;
    case 'Kidney'
        row=15;
    case 'Liver'
        row=16;
    case 'Lung'
        row=17;
    case 'Muscle'
        row=18;
    case 'Skin(dry)'
        row=19;
    case 'Skin(wet)'
        row=20;
    case 'Small Intestine'
        row=21;
    case 'Stomach'
        row=22;
    case 'Testis'
        row=23;
    case 'Tongue'
        row=24;
    otherwise
        disp('Tissue Not in Library');
end

   M=dlmread('tissue.txt');
   einf=M(row,1);
   sig=M(row,2);
   alpha1=M(row,3);
   alpha2=M(row,4);
   alpha3=M(row,5);
   alpha4=M(row,6);
```

```
    de1=M(row,7);
    de2=M(row,8);
    de3=M(row,9);
    de4=M(row,10);
    tau1=M(row,11);
    tau2=M(row,12);
    tau3=M(row,13);
    tau4=M(row,14);

    [epsr,sigma]=cole4(f,einf,sig,alpha1,alpha2,alpha3,
        alpha4,de1,de2,de3,de4,tau1,tau2,tau3,tau4);

end
```

A.6 *Function for optimization*

```
%\\A Hertzian dipole antenna embedded in tissue like
%material is surrounded by a pherical insulator and an
%outer layer of thin sperical shell of bio-compatible
%material.
%\\This function calculates the reflection coefficient
%at the inner surface of the thin shell.
%\\in_tissue (string) is the type of tissue surrounding
%the implanted antenna, selected from a set of tissue
%types as defined in the function tissue_prop(), lines
%9 to 58.
%\\sigma1 and sigma2 are conductivity of the insulator
%and the shell, may be considered as zero.
%\\2<epsr1<100 and 2<epsr2<10 are relative permittivity
%of the insulator and the shell, respectively.
%\\100e6<f<3e9 (Hz) is the frequency.
%\\0.1<r1<5 is the radius of the insulator.
%\\0.1e-3<thickness<5e-3 (m) is the shell thickness
function ref=reflection(x)
in_tissue='Muscle';
sigma1=0;
sigma2=0;
epsr1=x(1);
```

```
epsr2=x(2);
f=x(3)*1e9;
r1=x(4)*1e-3;
thickness=x(5)*1e-3;

[epsr3,sigma3]=tissue_prop(in_tissue,f);
k1=k(epsr1,1,sigma1,f);
k2=k(epsr2,1,sigma2,f);
k3=k(epsr3,1,sigma3,f);
ref=10*log(abs(gamma(k1,k2,k3,r1,r1+thickness)));
end
```

References

1. Dissanayake T., and Esselle, K. P. (2006) Waveform Optimization of UWB Radio Systems Over Range of Directions, *Electronic Letters*, **42**, pp. 384–385.
2. Dissanayake T., and Esselle, K. P. (2006) Correlation Based Pattern Stability and a Figure of Merit for UWB Antennas, *IEEE Transactions on Antennas and Propagation*, **54**, pp. 3184–3191.
3. Klemm, M., and Troester, G. (2006) Textile UWB Antennas for Wireless Body Area Networks, *IEEE Transactions on Antennas and Propagation*, **54**, pp. 3192–3197.
4. Furse, C. M. (2009) Biomedical Telemetry: Today's Opportunities and Challenges, in *IEEE International Workshop on Antenna Technology: Small Antennas and Novel Metamaterials*, Santa Monica, CA, pp. 389–392.
5. Dissanayake, T., Yuce, M. R., and Ho, C. (2009) Design and Evaluation of a Compact Antenna for Implant-to-Air UWB Communication, *IEEE Antennas and Wireless Propagation Letters*, **8**, pp. 153–156.
6. Dissanayake, T., Esselle, K. P., and Yuce, M. R. (2009) Dielectric Loaded Impedance Matching for Wideband Implanted Antennas, *IEEE Transactions on Micorwave Theory and Techniques*, **57**, pp. 2480–2487.
7. Corbett, R. W., and Ross, D. M. (1996) A Capacitively Loaded PIFA for Compact PCS Handsets, in *IEEE Antennas and Propagation Society International Symposium*, pp. 742–745.

8. Kim, J., and Rahmat-Samii, Y. (2006) *Implanted Antennas in Medical Wireless Communications*, 1st edn, Morgan and Claypool.
9. Ammann, M. J., and John, M. (2007) Some Techniques to Improve Small Ground Plane Printed Monopole Performance, *2007 IEEE Antennas and Propagation Society International Symposium,* **1–12**, pp. 2583–2586.
10. Klemm, M., Kocs, I. Z., Pederson, G., and Troster, G. (December 2005) Novel Small-Size Directional Antenna for UWB WBAN/WPAN Applications, *IEEE Transactions on Antennas and Propagation,* **53**, pp. 3884–3896.
11. Schantz, H. G. (2003) UWB Magnetic Antennas, *IEEE International Syemposium on Antennas and Propagation,* **3**, pp. 604–607.
12. Dissanayake, T., and Esselle, K. P. (April 2008) UWB Performance of Compact L-Shaped Wide Slot Antennas, *IEEE Transactions on Antennas and Propagation,* **56**, pp. 1183–1187.
13. Pelosi, M., Franek, O., Knudsen, M. B., and Pedersen, G. F. (2009) Influence of Dielectric Loading on PIFA Antennas in Close Proximity to User's Body, *Electronics Letters,* **45**, pp. 246–247.
14. Kivekas, O., Lehtiniemi, T., and Vainikainen, P. (5 November 2004) On the General Energy-Absorption Mechanism in the Human Tissue, *Microwave and Optical Technology Letters,* **43**, pp. 195–201.
15. Kim, J., and Rahmat-Samii, Y. (2004) Implanted Antennas Inside a Human Body: Simulations, Designs, and Characterizations, *IEEE Transactions on Micorwave Theory and Techniques,* **52**, pp. 1934–1943.
16. Chirva, L. C., Hammond, P. A., Roy, S., and Cumming, D. R. S. (April 2003) Electromagnetic Radiation From Ingested Sources in the Human Intestine Between 150 MHz and 1.2 GHz, *IEEE Transactions on Biomedical Engineering,* **50**, pp. 484–492.
17. Izdebski, P. M., Rajagopalan, H., and Rahmat-Samii, Y. (April 2009) Conformal Ingestible Capsule Antenna: A Novel Chandelier Meandered Design, *IEEE Transactions on Antennas and Propagation,* **57**, pp. 900–909.
18. Xia, W., Saito, K., Takahashi, M., and Ito, K. (April 2009) Performance of an Implanted Cavity Slot Antenna Embedded in the Human Arm, *IEEE Transactions on Antennas and Propagation,* **57**, pp. 894–899.
19. Houzen, T., Takahashi, M., Saito, K., and Ito, K. (2008) Implanted Planar Inverted-F Antenna for Cardiac Pacemaker System, in *IEEE International Workshop on Antenna Technology (iWAT): Small Antennas and Novel Metamaterials*, Chiba, Japan, pp. 346–349.

20. Warty, R., Tofighi, M. R., Kawoos, U., and Rosen, A. (2008) Characterization of Implantable Antennas for Intracranial Pressure Monitoring: Reflection by and Transmission Through a Scalp Phantom, *IEEE Transactions on Micorwave Theory and Techniques,* **56**, pp. 2366–2376.
21. Dissanayake, T., Esselle, K. P., and Yuce, M. R. (2009) UWB Antenna Impedence Matching in Biomedical Implants, in 3rd *European Conference on Antennas and Propagation*, Germany, pp. 3523–3526.
22. Wu, T. T., King, R. W. P., and Giri, D. V. (July 1973) The Insulated Dipole Antenna in a Relatively Dense Medium, *Radio Science,* **8**, pp. 669–709.
23. Jacobi, J. H., Larsen, L. E., and Hast, C. T. (1979) Water-Immersed Microwave Antennas and Their Application to Microwave Interrogation of Biological Targets, *IEEE Transactions on Micorwave Theory and Techniques,* **27**, pp. 70–78.
24. Siegel, M., and King, R. W. P. (1971) Radiation From Linear Antennas in a Dissipative Half-Space, *IEEE Transactions on Antennas and Propagation,* **19**, pp. 477–185.
25. Tai, C. T., and Collin, R. E. (October 2000) Radiation of a Hertzian Dipole Immersed in a Dissipative Medium, *IEEE Transactions on Antennas and Propagation,* **48**, pp. 1501–1506.
26. Sun, M., Liu, Q., Liang, W., Wessel, B. L., Roche, P. A., Mickle, M., and Scalabassi, R. J. (2003) Application of the Reciprocity Theorem to Volume Conduction Based Data Communication Systems Between Implantable Devices and Computers, in *IEEE 25th Annual International Conference of the Engineering in Medicine and Biology Society*, pp. 3352–3355.
27. Yao, N., Lee, H. N., Scalabassi, R. J., and Sun, M. (2006) Low Power Digital Communication in Implantable Devices Using Volume Conduction of Biological Tissues, in *IEEE International Conference of the Engineering and Biological Society*, pp. 6249–6252.
28. Wessel, B. L., Roche, P., Sun, M., and Scalabassi, R. J. (2004) Optimization of an Implantable Volume Conduction Antenna, in the *IEEE 26th Annual International Conference of the Engineering in Medicine and Biology Society,* 6, pp. 4111–4114.
29. Sun, M., Wessel, B. L., Liang, W., Roche, P., Liu, Q., Mickle, M., and Scalabassi, R. J. (2003) A Volume Conduction Antenna for Implantable Devices, in the *IEEE 25th Annual International Conference of the Engineering in Medicine and Biology Society*, pp. 3356–3359.

Chapter 13

Coexistence Issues with Wireless Body Area Networks

Axel Sikora
University of Applied Sciences Offenburg,
Badstrasse 24, D77652 Offenburg, Germany
axel.sikora@hs-offenburg.de

13.1 Introduction

Coexistence issues in wireless systems are known to be critical. The wireless channel is a common, i.e., shared media, where an exclusive assignment can be performed only within the limits of time, frequency, or code multiplexing. The issues may be especially unpredictable when license-free frequency bands for industrial, scientific, and medical applications (ISM- or license-exempt bands) are used, because cooperation with extrinsic and intrinsic system influences must be supported. While doing that, the ISM systems have to observe the overall rules of the regulations for the specific ISM band.

The impacts of coexistence issues are extensively discussed for hard real-time systems from industrial automation applications. But they might also have a major impact on wireless body area networks (WBAN), where they might endanger latency (jitter) due

Wireless Body Area Networks: Technology, Implementation, and Applications
Edited by Mehmet R. Yuce and Jamil Y. Khan

ISBN 978-981-4316-71-2 (Hardcover), ISBN 978-981-4241-57-1 (eBook)
www.panstanford.com

to increased packet loss or might put additional load on the energy budget, caused by prolonged channel observation times and the need for retransmissions. Both aspects are especially troublesome for low-power WBANs, which might be fed from batteries or from energy converters, and for safety-relevant (medical) systems.

First, Section 13.2 gives an overview on the coexistence issues in low-power, low-cost wireless systems in ISM bands. Special attention is paid to interferers with high-duty cycle and with unpredictable behavior. Their impact on the communication is described is Section 13.3.

Based on these characteristics, Section 13.4 presents the currently available approaches to overcome coexistence issues. Those concentrate on those relevant to WBANs and include listen-before-talk behavior, channel monitoring and prediction techniques, back-off strategies, reservation schemes, (slow) frequency adaptation and (fast) frequency hopping, and redundancy with regard to frequencies, antennas, and coding. An outlook on wide-band and ultra wide-band (UWB) systems will be given.

In a third part, a selection of solutions will be presented. These will include, but won't be limited to developments from the author's institute and shall give an overview on the state of the art.

This part will also discuss the attempts to overcome the coexistence issue from the side of standardization.

The fourth and last part will give an outlook on future research and development directions.

It should be mentioned that this relatively short chapter can give only a limited view on the issue. No exhaustive view on the manifold literature seems to be feasible.

13.2 Analysis of Interferers

13.2.1 *Classification*

The coexistence issue for wireless ISM systems includes all kinds of interferences. It is the effect of an emission from one (subject) system, called the interferer, to a second (object) system, which is potentially affected, called the interfered station. Interferers might be classified into three groups:

- Intrinsic interference from wireless systems running the same protocol in the same or neighbored frequency band
- Extrinsic interference from wireless systems running a different protocol in the same or neighbored frequency band
- Extrinsic interference from (micro) electronic systems with electromagnetic or RF-emission (EMI, RFI)

In addition, we have to distinguish between the interference to a system (i.e., the impact to our system) and the interference to other systems.

13.2.2 *Regulation Issues*

It is anticipated within the whole chapter, that the predominant number of WBAN systems are and will be operated in the license-free frequency bands for industrial, scientific, and medical applications. Those are denominated ISM or license-exempt bands.

As for all electronic devices, the use of these frequency bands is regulated from the regional regulation body. For Europe, the rules are elaborated and harmonized by ETSI and then put into national laws for each single country. The Americas and Australia follow the rules of Federal Commission of Communication (FCC). In Japan, the Ministry of Telecommunications (MKK) regulates the Telec Radio Regulations, the ARIB for Southern Pacific regions and so on.

Those rules might limit the behavior on the physical and on the data link layer:

- On the physical layer, the regulation impacts the frequency behavior. The typical parameters here to describe the frequency behavior are
  - The spectral mask
  - The effective radiated power
  - The peak power density
  - The frequency range(s)
  - The transmitter's and receiver's spurious emissions

  This behavior has consequences on the possible modulation schemes, the achievable data rates, the channel spacing, and alike. Regulations on the physical layer might also include regulation (transmit power control, TPC) or dynamic frequency selection (DFS). The latter is sometimes also known as Dynamic Channel Selection (DCS) or Adaptive Frequency Agility (AFA).

- On the data link layer, key elements of the medium access are regulated. This might include a maximum duty cycle per basic time unit and access pertinence (listen-before-talk, LBT).

The conformance to the rules depends on the frequency band and on the region. It might be certified by a self-declaration (e.g., ETSI, 2.4 GHz ISM band) or by a certified body (e.g., FCC, 2.4 GHz ISM band).

Albeit all these rules, no prediction can be made about the density of devices within the range of the system, and thus no statement can be made on the overall load on the channel.

13.2.3 *Intrinsic Interference*

The first group of interferers consists of stations that follow the same communication protocol as the interfered station. This group is characterized by the following behaviors:

- All nodes use the same frequency characteristics, i.e., the same bandwidth and the same modulation scheme.
- The nodes may work in the same frequency band; however, they might work in the same or in different channels. If the intrinsic interferers work on the same channel as the interfered stations, they might exchange information in order to perform avoidance strategies.
- All nodes follow the same channel access mechanisms, i.e., LBT or back off strategies.
- All nodes may come with similar traffic characteristics, with regard to traffic load, traffic cycles.

For completeness sake, it is highlighted that passive interferers are disregarded within this overview. As passive interferers we understand echoes due to multipath propagation. In this case, one station does not hear only the direct signal from the sender, but also the delayed signals that are reflected at surrounding objects.

These multipath effects are disregarded as data rates for WBANs tend to be low and distances to be small.

Of course, the specialitiy of WBAN networks of course lies in the attenuation through the human body.

13.2.4 *Extrinsic Interference of RF-Stations within the Same Frequency Band*

Extrinsic interference from wireless systems within the same frequency band is a common challenge for ISM-based systems. Interferers of this group can be characterized by the following behavior:

- The nodes are operated within the same frequency band. The transmission might occur such that the interferer might have a full or a partial overlap of the frequency band of the interfered station. This overlap might be timely constant or variable.
- The frequency characteristics of the interferer might be different from the interfered station. This leads to the fact that — even if the different stations work in the same frequencies — there is no possibility to exchange information and thus to cooperate.
- The traffic characteristics might be completely different. So predictions of the future behavior are not possible. Analog systems mostly come with an extremely high duty cycle. Digital systems tend to be much more event-driven.
- If the interferer's performance is reduced by the WBAN's activity, this might still worse the problem, as retransmissions and frequency agility further reduce the available bandwidth.

Figure 13.1 shows some of the most widespread systems in the ISM band from 2.400 GHz to 2.4835 GHz (as defined by ETSI).

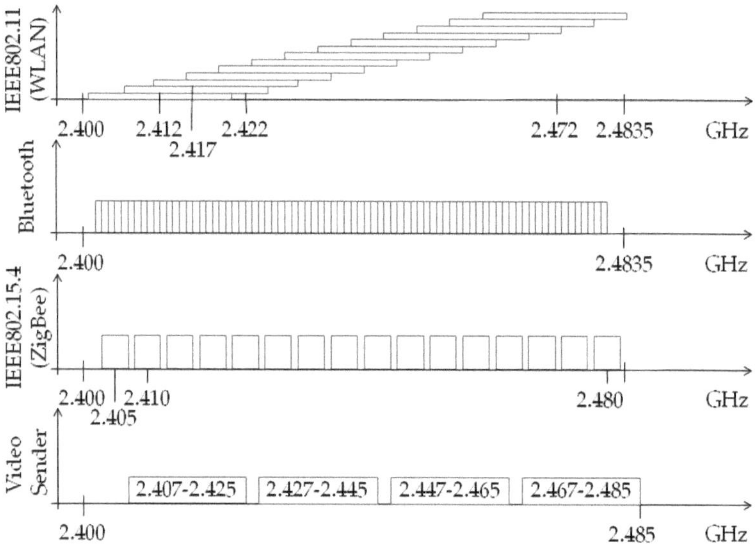

Figure 13.1. Frequency characteristics of some of the systems working in the 2.4 GHz ISM band [29].

13.2.5 *Extrinsic Interference of Other Systems within the Same Frequency Band*

All microelectronic devices, might they be dedicated for communication or not, are subject of electromagnetic interference (EMI). The characteristics are also described by regulation and should be below the so-called noise level. Practically, in all cases no limitations can be given for the timely behavior of the interferer, i.e., it may be always on. In no case, cooperation via a communication channel can be supported.

Probably, the most prominent in-band interferers for ISM band communication are microwave ovens, which are operated in the 2.4 GHz band in the whole world. For completeness sake, it should be mentioned that this frequency was chosen in the fifties as a reasonable compromise for heating food, as the lowest resonance frequency of liquid water is around 18 GHz, of gaseous water above 22 GHz.

Figure 13.2 shows the spectral characteristics of a random choice of commercially available microwave oven for home use.

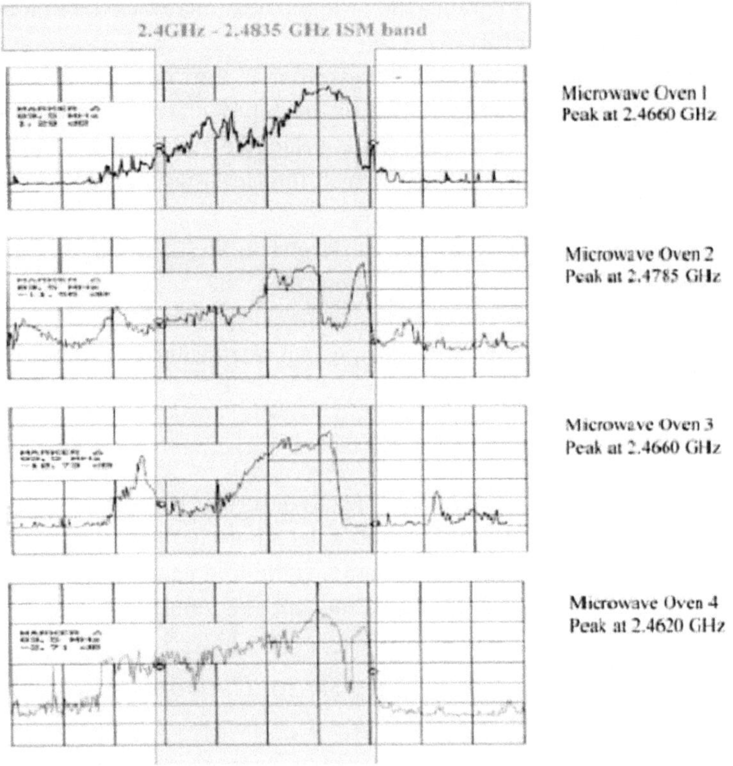

Figure 13.2. Frequency characteristics of commercially available micro ovens for home use [6]. See also Color Insert.

13.3 Effect on Transmission

13.3.1 *Fundamentals*

The required bit error rate (BER) for the communication between two nodes translates into a certain packet error rate (PER) or frame error rate (FER), which depends on the packet length. In order to achieve this BER, a certain signal-to-noise-ratio (SNR) is required, which mainly depends on the modulation scheme and on the processing gain that can be achieved, e.g., through coding or frequency spreading. In this sense, interferers contribute to a noise level, which degrades the SNR and thus increases the BER and consequently the PER.

The impact on the actual transmission shows a strong dependency from the characteristics of the input filter of the receiver. This is normally described by the selectivity, which describes the maximum power level for an intrinsic interferer working at the same channel or at neighbored channels.

The carrier-to-co-channel interference ratio (C/I ratio) or simply co-channel interference (CCI) describes the actual or required power ratio of the wanted carrier signal (P_c) with regard to the unwanted interfering signal (P_i), which is transmitted on the same channel.

$$CCI = 20 \log \left(\frac{P_c}{P_i} \right) \tag{13.1}$$

The adjacent channel selectivity (ACS) describes the same ratio with regard to an interfering signal on a neighbor channel. The first ACS describes the ratio for the directly neighbored channel, the second ACS the ratio for the second next neighbor channel, and so on.

13.3.2 *Simulation of a Dense Sensor Network (Intrinsic Interference)*

The power level of the emitted signals reduces along Frijs Free Space Equation [7], as long as line-of-sight connection in the sense of undisturbed Fresnel zones is maintained.

$$P_R = P_T \cdot \left(\frac{\lambda}{4\pi d} \right)^{\alpha} \cdot G_R \cdot G_T \tag{13.2}$$

In Eq. 13.2, P_T and P_R are the power level at the transmitter and the receiver, respectively, G_T and G_R describe the antenna gain at the transmitter and the receiver, respectively, λ is the wavelength, d is the distance between transmitter and receiver and is the coefficient which describes the attenuation. Typical real-life values are between 2 and 6 [8]. The theoretical limit is 2 and is very often used in simple model calculations for the so-called free space loss (FSL). This square dependency is predominantly calculated in logarithmic scale:

$$FSL = P_R/P_T = 20 \cdot \log(\lambda/4\pi d) + \log(G_R) + \log(G_T) \tag{13.3}$$

Doubling d, leads to the well-known reduction of FSL by – 6 dB.

Based on this square propagation law, we can easily calculate the C/I ratio under the assumption that all parameters (G_T, G_R, P_T, and λ) are identical for the two senders:

$$CI = 20\log(P_{R,c}/P_{R,i}) = 20\log(d_i/d_c) \qquad (13.4)$$

If CI is required to be larger than 10 dB (e.g., for a commercially available single chip radio transceiver at 2.4 GHz and a data rate of 250 Kbps [34]), then $d_i \geq 10^{0,5}\ d_c = 3.16\ d_c$, i.e., distance between interferer and receiver (denominated d_i) must be more than three times larger than the distance between the wanted (channel) sender and the receiver (denominated d_c).

If the first adjacent channel sensitivity is regarded, which is -20 dB, then the ratio shrinks to $d_i \geq 10^{-0.1}\ d_c = 0.1\ d_c$, i.e., the wanted channel can be 10 times farer apart from the receiver than the interferer.

Let S_1 and S_2 be two senders communicating with two receivers R_1 and R_2, respectively. Then, we can illustrate the spatial characteristics as shown in Fig. 13.3. This figure can be read as follows: Let us first assume that the CI_i must be higher than 6 dB and both station S_1 and S_2 transmit on the same channel. Left of the solid

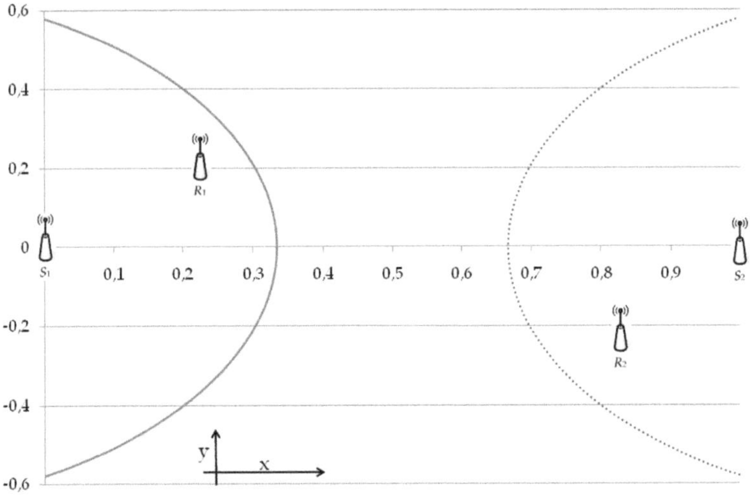

Figure 13.3. Spatial distribution of 6 dB and -6 dB C/I and ACS lines.

line, the CI_1 is above 6 dB, i.e., the receiver R_1 must be left of the line in order to successfully receive data from S_1. Respectively, the CI_2 is above 6 dB right of the dotted line, i.e., the receiver R_1 must be left of the line in order to successfully receive data from S_1.

The same kind of figure can also be used for adjacent channel sensitivity. If we assume that both stations S_1 and S_2 transmit on the neighbor channel and that $AC\,S_i$ shall be higher than −6 dB, then receiver R_1 must be left of the dotted line in order to successfully receive packets from S_1 and R_2 must be right of the solid line. In this case, we consequently have an overlap space, where both communication paths work.

As a result, we achieve a so-called space division multiple access (SDMA), which is discussed in Section 13.5.1. A notion often used in this context is the so-called spatial capacity, which is originally defined as bits/sec/cubic-meter [28]. In more recent sources, it is defined as bits/sec/square-meter, since the relevant coverage area is usually two dimensional rather than three dimensional [41].

It should be highlighted at this point that the characteristics only depend on the ratio of the distances and of C/I, but neither on the absolute distances, nor on the absolute output powers. This is shown in Fig. 13.3 for a two-dimensional case, where the axes show the relative distances between Section 13.5.1 both stations S_1 and S_2 in two directions. Increasing the output power, as it is proposed in various cases, does not help to solve the intrinsic coexistence issue. Increasing the output power even aggravates the situation, as the spatial capacity degrades. Intrinsic interference is extensively examined in the literature for mobile applications (GSM, UMTS) [2].

13.3.3 *Measurement of Real Packet Losses due to Extrinsic Interference*

Many analytical models, network simulations, and measurements in literature describe extrinsic interference behavior between WLAN and short-range wireless networks for local, home, body, or personal area networks (LAN, HAN, BAN, PAN). It is especially for the 2.4 GHz band, where a multitude of different technologies and products are available, and where SRWN are the potentially endangered, due to

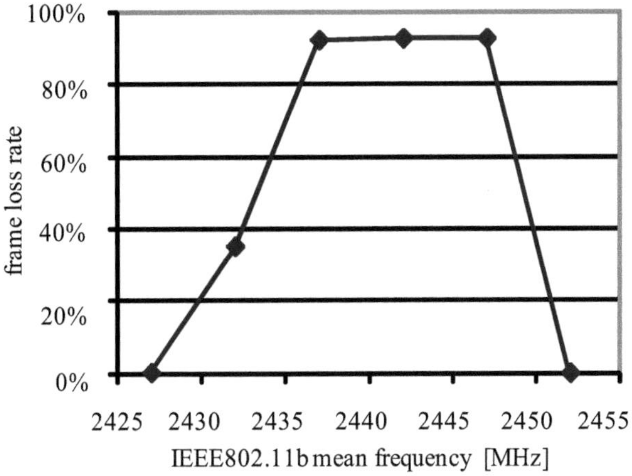

Figure 13.4. Loss of IEEE802.15.4 frames with a high activity IEEE802.11 DSSS at varying center frequencies; the 802.15.4 channel kept constant at 2440 MHz [10].

the fact that their output power may be significantly lower than that of the competing systems, e.g., WLAN.

Some representative results shall be shown here:

- Reference [18] gives a fundamental analytical estimation on the coexistence impact of an IEEE 802.15.4 network on IEEE 802.11b devices.
- Reference [35] describes lab-based measurements, where interfering signals are directly injected into the RF path.
- Reference [10] reports on real-world measurements, where IEEE802.11b- and IEEE802.15.4-stations coexist and interfere. Worst case packet loss rate for IEEE802.15.4 goes up as 92% (Fig. 13.4).

13.3.4 *Effects of Coexistence Problems*

Needless to say that packet loss is not a problem per se and of course does not necessarily lead to data loss, as within the course of reliable communication, lost packets can be retransmitted. However, these retransmissions may lead to an increased delay, may introduce delay variations (jitter), and may add to the energy consumption.

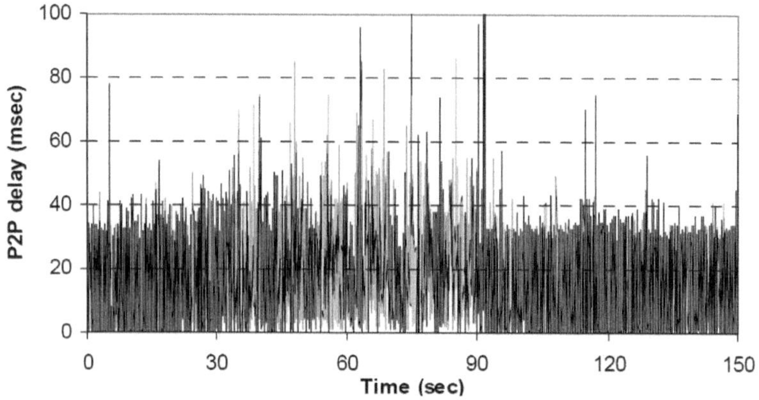

Figure 13.5. Delay of Bluetooth point-to-point UDP traffic under the influence of interference [33].

The delay and the jitter due to packet loss add to the overall system delay and jitter. There are multiple analyses on the jitter behavior of wireless communication under interference conditions. A very detailed analysis for WLAN and Bluetooth behavior can be found at in reference [33]. Figures 13.5 and 13.6 show the results for delay and jitter for UDP traffic in a point-to-point topology. In case of single coexistence with one interfering station (time 30 s to 60 s),

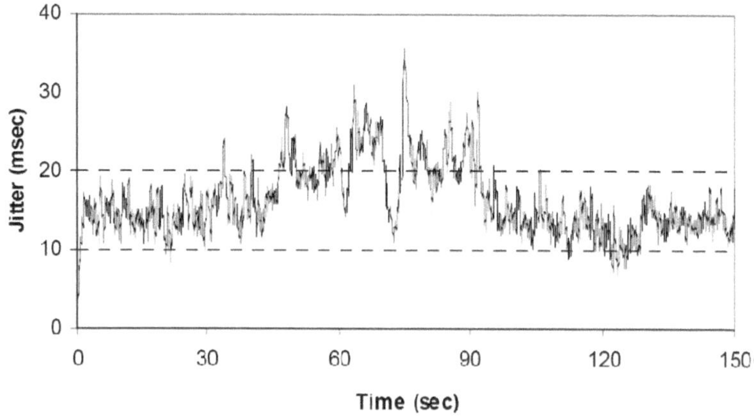

Figure 13.6. Delay of Bluetooth point-to-point UDP traffic under the influence of interference [33].

the average delay increases by 26%, whereas in double coexistence with two interfering stations (time 60 s to 90 s) by 49%.

13.4 Countermeasures — An Overview

13.4.1 *Safety Aspects*

It is one of the basic requirements for networking to provide safe operation [9]. "Safety is the condition of being protected against physical, social, spiritual, financial, political, emotional, occupational, psychological, educational, or other types or consequences of failure, damage, error, accidents, harm, or any other event .... Safety can take the form of being protected from the event or from exposure to the event [51]."

It is especially the second sentence, which describes the two main directions that

- The system either avoids to be exposed to the event. Exposure is mainly avoided by planning and/or coordination with other systems.
- Or the system attempts to be protected against the event. Protection can be achieved by redundancy and/or adaptivity.

13.4.2 *Classification*

These two directions of countermeasures can be performed on the various levels of (wireless) networking (physical, data link, network, and partially transport layer). They can be achieved through the following activities:

- **Company Policies:** This might include prohibition of some wireless products on campus, e.g., Bluetooth or 802.11b. Obviously, applications are moving away from this practice, as they are not suitable for changing topologies.
- **Regulation Bodies:** The rules from regulation bodies might include basic coexistence rules, such as LBT or TPC (Section 13.2.2)
- **Standards Bodies:** The most prominent example of a standard-based approach was offered by the legacy

IEEE802.2 workgroup with regard to the coexistence between IEEE802.11 (WLAN) and IEEE802.15.1 (Bluetooth) [43].
- **Technical Innovations:** These look for new solutions with regard to physical and data link layer protocols or with regard to system level (driver) solutions.

With regard to the collaboration with other protocols, one might also distinguish collaborative approaches, which take into the account the specific behavior or other protocols, and non-collaborative approaches, which don't.

The countermeasures on the physical layer (Section 13.5) are around the technologies to split up one medium into different channels, e.g., with space, frequency, or code division multiple access (SDMA, FDMA, CDMA). In contrast to that the countermeasures on the data link layer (Section 13.6) are built around the variations of time division multiple access (TDMA), which allows multiple stations use one channel.

13.5 Countermeasures on Physical Layer

13.5.1 *Channel Classification and Selection*

One of the basic countermeasures is the space division multiple access (SDMA) approach, already described in Section 13.3.2. Thus, the space division might be relevant not only for channel reuse, i.e., reuse of the same channel, but also for neighbor channel selection.

In order to perform the channel selection, manual frequency planning has been performed for decades, e.g., for radio and television broadcast services and for mobile (GSM) services. Of course, this manual planning has no practical importance for mobile services, such as WBANs. Automatic channel selection requires autonomous channel classification of the stations. For practical use, two flavors can be found [1]:

- Active classification can be done during the course of normal communication, or the devices can exchange dummy packets with the specific goal of building a classification list.

- Passive classification is accomplished by listening to channels. Most of today's single-chip transceivers come with two options of passive channel supervision:
  - Energy detection (ED) is an estimate of the received signal power within the bandwidth of the channel. No attempt is made to identify or decode signals on the channel [44].
  - Carrier sense reports a busy medium only upon the detection of a signal compliant with the same protocol with the same modulation and spreading characteristics of the PHY that is currently in use by the device. This signal may be above or below the ED threshold [44].

In the case of active classification, it is mostly the actual channel, which is classified, whereas in the case of passive classification, the stations may scan all available channels and thus determine a spectral activity map.

Two basic characteristics add complexity to this classification:

- As the wireless signal is spatially distributed, the observation of one station has only local significance. It cannot describe the quality at any other station or any other localization. As this has only small relevance for small-scale networks, its importance raises/becomes crucial with the size of the network. It is especially critical for spatially distributed multi-hop network, where different parts suffer from different interfering influence. This case is illustrated in Fig. 13.7. The three interferers I_1, I_2, and I_3 have an impact on only a subset of the stations S_{xy}. In addition, the optimum case for one station, e.g., S_{31}, might be prohibitive for another station, e.g., S_{24}. Some solutions to this problem are proposed in reference [11].
- If nodes are extensively using power-down modes, they might not be informed about a change in frequency — and thus have to re-register. This adds additional jitter to the transport characteristics [30]. Furthermore, channel observation is impossible in power-down phases, which

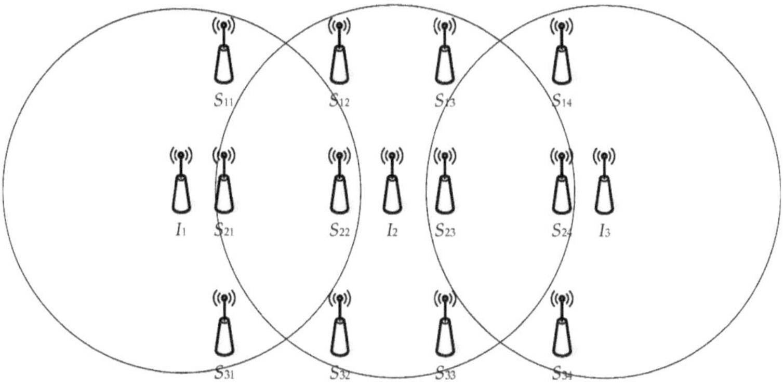

Figure 13.7. Illustration of spatial distribution in a wireless mesh network.

may cause complete communication failure when turning back to active mode.

- In addition, the classification has only limited timely relevance, too. If the channel classification is performed periodically, the channel selection can be adapted. This is performed in legacy telecommunications systems, as in DECT [46] or GSM [2].

However, many of the simpler SRWN protocols perform the channel classification only at network or connection start-up [44].

In real-life wireless protocols, adaptive channel selection is merely used as addition to basic services. For IEEE802.11, only the European *h*-extension provides the required service primitives, such as measurement request, measurement report, supported channels, and channel switch announcement (Fig. 13.8) [48].

ZigBee networks, since the 2007 specification [45], allow a slow dynamic adaptation of the used channel, which is denominated frequency agility. It provides a method for the network to change channels in the event of interference. In order to avoid the aforementioned problems of spatial distribution, it is the network coordinator that detects potential interference on a channel and directs the devices on the network to change to a better channel.

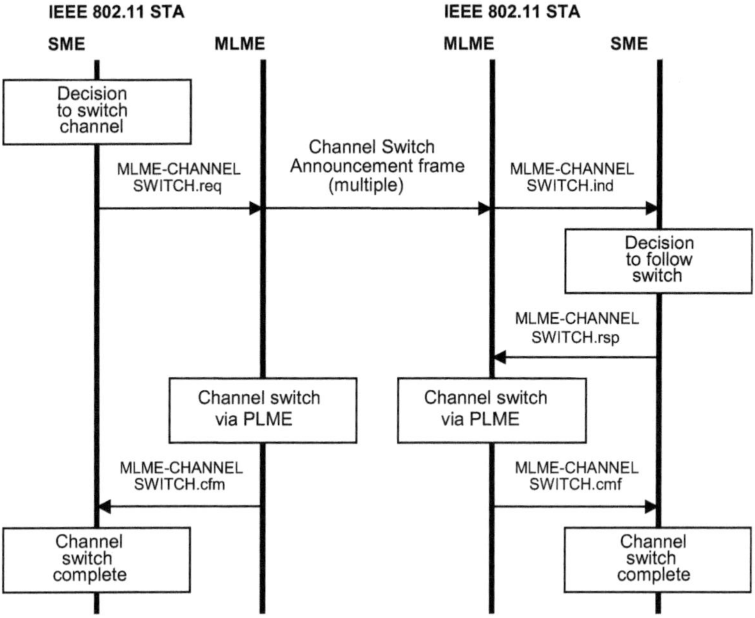

Figure 13.8. Channel switch decision in IEEE802.11h [48].

Frequency agility differs from frequency hopping with regard to the time constants of frequency changes and the adaptivity to real-world impacts.

Some vendor-specific WLAN management tools run an add-on to IEEE802.11-WLAN-Access Points, e.g., [37]. Bluetooth V1.2 [47] enables the frequency hopping (Section 13.5.2) to be adaptive (AFHSS).

13.5.2 *Frequency Hopping*

Frequency hopping spread spectrum (FHSS) is the simplest spread spectrum technique, which helps to counteract against frequency specific interference on a statistical basis. FHSS uses M different carrier frequencies that are modulated by the source signal. At one moment, the signal modulates one carrier frequency; at the next moment, the signal modulates another carrier frequency. In the long run, all M frequencies are used. The frequency pattern must be agreed between the synchronized sender and receiver.

Table 13.1. Probability of $l = 3$ subsequent collisions assuming n pairs of nodes using m channels in a random selection. $m = 58$ corresponds to one, $m = 36$ to two WLAN interferers [9]

		p	
n	$m = 80$	$m = 58$	$m = 36$
2	1,95E-06	5,13E-06	2,14E-05
3	7,81E-06	2,05E-05	8,57E-05
4	1,95E-05	5,13E-05	2,14E-04
5	3,91E-05	1,03E-04	4,29E-04
10	3,22E-04	8,46E-04	0,0035367
20	0,0025977	0,0068168	0,0285093
30	0,0087794	0,0546389	0,0963579
40	0,0208207	0,0546379	0,2285273
50	0,0406748	0,1067401	0,4464668

The stability against constant-frequency interferers is only limited to those time slots when the frequencies do not overlap (as discussed in Section 13.3.1). This also allows running several FHSS systems in parallel. However, due to non-orthogonality, the theoretical aggregate gross rate of M times of the gross rate of one channel is unachievable.

Table 13.1 describes the collision probability of multiple frequency hopping system within the same range (intrinsic interference) in the presence of WLAN systems (extrinsic interference).

In reference [13], an information-theoretic approach for analysis of various multiple access schemes is given considering various propagation factors such as path loss, shadowing, and fading.

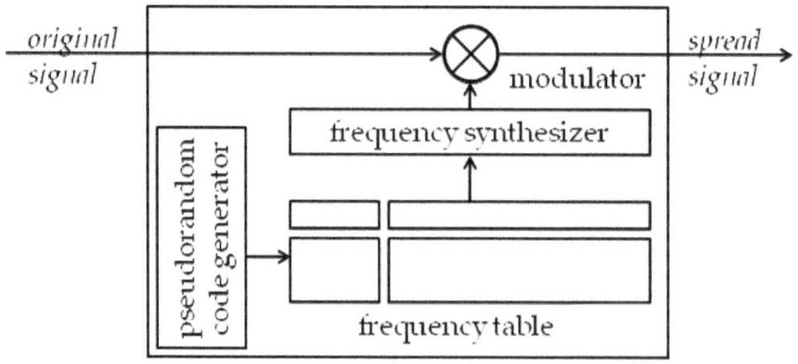

Figure 13.9. Frequency Hopping Spread Spectrum System.

However, this contribution mainly concentrates on outdoor radio channels.

Reference [14] evaluates the possible capacities of FHSS systems under indoor conditions and indicates that the number of piconets supported in an area of 20×20 m^2 varies from 15 for a fully loaded system to around 50 for a 20% load system.

Whereas FHSS systems are robust against non-hopping systems, those non-hopping systems may show a severe degradation. This problem is especially well studied for the coexistence between IEEE802.11 WLAN and Bluetooth (BT), which has been in the focus of IEEE802.15.2 work group [43] [36]. In this case, 22 out of the 79 Bluetooth frequencies overlap with one WLAN channel. In addition, WLAN packet length may be twice as long as the BT hopping time (1.293 ms for an IEEE802.11 packet with 1514 bytes at 11 MBit/s compared to 625 μs BT time constant). Furthermore, overlap times (Section 13.6.3) must be taken into account. The results [15] show the severe impact on WLAN and the relative moderate impact on BT. These results are widely validated also in recent publications, e.g., reference [16].

In general, frequency hopping is a very popular approach, as it is simple in implementation, uses narrowband transmission, and thus enables low power during operation, and delivers relatively good robustness. The major disadvantage of FHSS systems is the need of synchronization, which has to be sustained also during power-down periods. Thus, FHSS needs either a relatively long re-synchronization period after a power-down period, or a low-power timer, which is sufficiently precise to keep the node synchronized.

Bluetooth is certainly the FHSS with highest prevalence. So three of the recent developments of Bluetooth shall be mentioned here:

- In order to reduce the overall energy consumption due to synchronization times, the Bluetooth low energy technology reduces the number of synchronization channels to four. This approach is also proposed for many other frequency hopping or frequency agile systems, e.g., references [9] and [45].

- The statistical approach of plain FHSS can be extended through adaptivity, as described in Bluetooth V1.2 (adaptive FHSS, AFHSS). The adaptivity helps to blacklist a subset of frequencies. The rules, which lead to the blacklisting, are not standardized, but is obvious that — as active classification is used — it is targeted towards static interferers, e.g., IEEE802.11 WLAN.
  It should be mentioned that the blacklisting of some channels within the course of AFHSS
  - Is mainly helpful for the robustness of the WLAN transmission, as the probability that Bluetooth packets interfere with the WLAN frames on their static frequency is significantly reduced
  - Decreases the potential spatial capacity of Bluetooth
- The third approach is on the driver-side in order to coordinate the channel access of the different media. The most prominent example for this is certainly the cooperation between IEEE802.11 WLAN and Bluetooth. This cooperation on driver level has already been proposed in reference [43]. It came to life early, but only within proprietary solutions [42].

Many other protocol developments targeted towards real time and safe operation chose the FHSS as well. The most prominent examples are Wireless HART [49], a protocol for larger-scale multi-hop networks, where frequency patterns are adaptively agreed between pairs of communication partners and where synchronization is maintained through high-quality timers, and WISA [12] for short-latency transmission.

Proprietary protocols include [9] and allow up to six retransmissions on three different frequencies within 5 ms using low-cost and ultra-low energy (battery powered) transceiver and microcontroller.

13.5.3 *Frequency Spreading and Code Division Multiple Access*

The direct sequence spread spectrum (DSSS) technique also expands the bandwidth of the original signal, but the process is

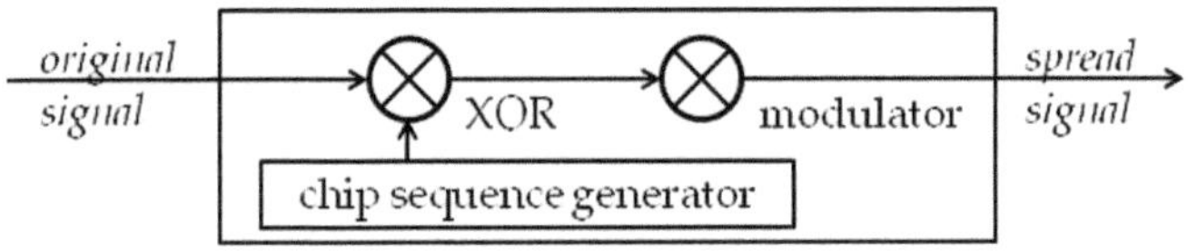

Figure 13.10. Direct Sequence Spread Spectrum (DSSS) System.

different. In DSSS, we replace each data bit with *n* bits using a spreading code. In other words, each bit is assigned a code of *n* bits, called chips, where the chip rate is *n* times that of the data bit.

At the receiver, the received signal is de-spread with the same spreading code. Due to the autocorrelation of the transmitted PN sequence with the PN sequence that the receiver believes the transmitter is using, DSSS achieves a so-called process gain, which lowers the required SNR. Thus the impact of uncorrelated interferers can be reduced.

The longer the PN sequence is, the better the achievable autocorrelation might be. As a result, the process gain increases. The process gain may be so large that code division multiple access (CDMA) can be used. However, as also the required bandwidth increases, the most protocols targeted for WBANs come with only relatively short PN sequences, e.g., between 4 (e.g., IEEE802.15.4 [44]) and 11 (e.g., IEEE802.11 [48]). The achievable process gain mainly helps to suppress deep fading in multipath environments, but does significantly improve coexistence robustness, as can be seen in Section 13.3.3.

The only exception to this, which is currently visible, is UWB.

13.5.4 *The Promise of Ultra-Wide-Band*

UWB technologies have found their place for manifold radar- and measurement applications, as well as for military communication, mainly since the eighties of the last century. It is only in this decade that UWB gained importance for communication in consumer and automation applications [3].

The basic idea of UWB transmission is the usage of an extremely broad bandwidth, so that the transmission power is distributed

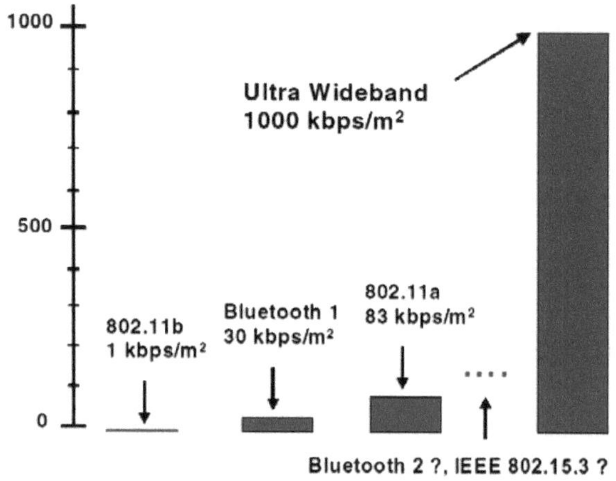

Figure 13.11. Spatial capacity comparison between IEEE 802.11, Bluetooth, and UWB [31].

over this broad spectrum. Thus the intensity of UWB-radio systems, i.e., the power per frequency (W/MHz) is correspondingly low. The demodulation and decoding of the received signal is achieved through an appropriate process gain. Hence UWB devices appear as noise to most primary users. Reference [7] gives a good overview on the challenges of UWB systems, which come with a lot of strategic advantages for sensor-actuator networks [20, 21], including the coexistence robustness. Some detect-and-avoid (DAA) strategies for UWB are discussed in reference [22]. An excellent analysis on the coexistence behavior of UWB systems can be found in reference [32].

UWB also promises to achieve an extraordinary spatial capacity, as shown in Fig. 13.11.

13.6 Countermeasures on Data Link Layer

13.6.1 *Basic Medium Access Control*

As introduced in Section 13.4, the countermeasures are built around the variations of time division multiple access (TDMA), where stations share a common physical channel. Thus the whole discipline discusses about the best moment of time for the stations to access the medium. Thus the problem is commonly denominated medium access control (MAC). For MAC algorithms, we classify centralized and de-centralized architectures. In de-centralized MAC algorithms, the coordination function is distributed among several peer stations, i.e., there exists no master node to synchronize the medium access.

For body area networks, both approaches have certain relevance. Mainly very small and simple WBANs follow a centralized approach, for example with one master data collector. For more complex and spatially distributed WBANs, more and more de-centralized algorithms are applied.

13.6.2 *Centralized Approach*

Centralized architectures, where a master device assigns time slots to its slaves, allow deterministic, i.e., hard real-time behavior, as long as the physical channel (Section 13.5) is available.

If the assignment works properly, no collisions should occur. However, centralized approaches come with two major disadvantages, which restrict their use in general sensor WBAN architectures.

- All slave stations must remain synchronized with the master, which in the general case requires precision timers and regular activity.
- In case that the synchronization is performed within the communication channel, the topology is limited to star or hierarchical star, i.e., tree topologies. This is, however, not necessarily a major restriction for WBANs. It can be overcome by synchronization methods for changing topologies. The most prominent example is certainly IEEE1588 [50], where major efforts are invested into the topology setup of a tree.

Excellent overviews of synchronization protocols for WSNs can be found in references [24], [25], and [26]. In addition, it is a relatively common practice to use side-channel synchronization, e.g., with the help of terrestrial wide area clock signals (DCF77 [38], WWV [39] or alike, or spatial synchronization signals, i.e., GPS [40].

In addition, special care must be taken for the registration of the slave at the master station. It is either registered via a (manual) side channel or during a predefined period of time, when de-centralized access is enabled. This second approach directly leads to hybrid methods.

13.6.3 *Duty Cycle Management*

A very easy approach to set up a decentralized medium access, which, however, is still commonly used in many wired and wireless systems, is the restriction of the duty cycle. If the load on the channel is low enough, the probability that two packets collide may be small enough to have no significant impact on the transmission characteristics. Duty cycle may be limited by regulation (Section 13.2.2). It can be as low as 1% (e.g., 868.6–868.7 MHz, CEPT ERC/REC 70-03, Annex 7) or even 0.1% (e.g., 869.25–869.3 MHz, CEPT ERC/REC 70-03, Annex 7). The regulation details describe the exact conditions, i.e., the related time. For example, for the above-mentioned CEPT ERC/REC 70-03, it is 1 h.

Duty cycle management is still used in many fields. It is simple and needs only (unidirectional) transmitters. However, it is of course only generally suitable for low-performance connectivity.

13.6.4 *Channel Sensing Methods*

Many regulation rules prescribe listen-before-talk (LBT), which is sometimes also called carrier sense (CS). In this case, the (bidirectional) transceiver has to supervise the channel for a certain amount of time before starting the transmission. LBT helps to avoid collision with an already active transmission and is, therefore, very helpful in reducing the collision probability.

However, two aspects restrict its use for wireless networks:

- LBT does not give a guarantee that a collision cannot occur. If two stations start their LBT simultaneously, then both stations can start their transmission without breaking the rules, but causing a collision. For a real system with propagation times and turn around times (required time to switch a half-duplex transceiver from receiver into transmitter mode), we are talking about time intervals for

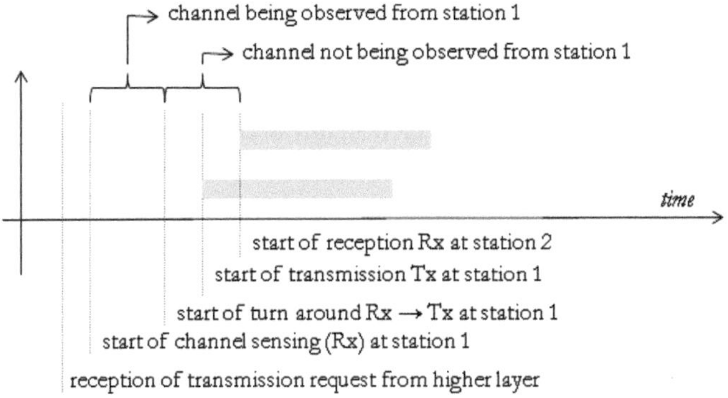

Figure 13.12. Timing behavior of channel sensing for LBT strategy, including turn around and propagation times (translated from [5]).

simultaneity. This problem is well known since decades, as it already occurred in the wired coax- or hub-based Ethernet. It is a matter of collision avoidance strategies like persistency to further reduce the collision probability. The timing behavior at one transceiver is illustrated in Fig. 13.12.

- The other restriction is caused by the spatial distribution of wireless communication and which is commonly described as hidden station problem. If it is assumed that in Fig. 13.7 two stations, e.g., S_{i1} and S_{i3}, do not hear each other, but have overlapping ranges, so that the intermediate station S_{i2} can hear both. Applying LBT at S_{i1} and S_{i3} does not remedy a collision at S_{i2}. This problem can only be solved with additional countermeasures, e.g., RTS-CTS-mechanisms.

As already described in Section 13.5.1 there are different methods of channel sensing. The impact of different channel sensing methods is, for example, evaluated in reference [17] with regard to the impact on IEEE802.15.4 performance under IEEE802.11b interference.

13.6.5 *Persistency and Collision Avoidance*

It is also the hidden station problem, which rules out the possibility to use the collision detection (CD) approach, as it is applied in

IEEE802.3 (wired Ethernet). For this, it is necessary that a sender can make reasonable assumptions about the collision (coexistence) situation at the receiver at different displacement. The CD approach has the major advantage that a transmission can be stopped at a very stage of the transmission.

Due to the hidden station problem, it is only after the end of a transmission and a missing acknowledgment (ACK) that a station learns about a lost packet. This might lead to the situation that in a busy network, multiple stations have to perform retransmissions. If all of them perform LBT directly after the medium was relinquished and use the same time constants, this necessarily leads to a collision.

Furthermore, the collision probability may be increased in case of repeated broadcast packets. If, for example, S_{21} in Fig. 13.7 sends out a broadcast packet, which is then forwarded by all directly neighbored stations (S_{11}, S_{12}, S_{22}, S_{31}, and S_{32}), this will lead to collisions despite the usage of LBT.

This problem can be improved if the stations reduce greed to access the channel, which is described by the notion of persistency. A persistent (or 1-persistent) transmits its data that is has to send with a probability of 1 whenever it finds the channel idle. If the channel is busy, the station waits until it becomes idle (greedy approach).

Non persistent behavior differs from 1-persistent, when the channel is found busy. In this case, the station deliberately waits a random period of time before sensing the channel again. This approach was first used in the legacy Aloha system [23] for retransmissions. In Aloha and in many other systems (including CSMA/CD Ethernet), a so-called binary back off algorithm is applied.

If the transmission is organized in time slots, we can define a probability p that a station transmits its data that is has to send. With a probability of $q = 1 - p$, a p-persistent station defers the transmission until the next slot and the same procedure iterates. If the channel has become busy meanwhile, the station waits a random time and starts again. If the channel is busy when first sensing it, the station waits until the next slot and repeats the procedure. The impact of the variation of p in the presence of a FHSS interferer is investigated in reference [16].

As an example, the back off mechanism of IEEE802.15.4 is shown in Fig. 13.14. Each IEEE802.15.4-compatible device shall maintain

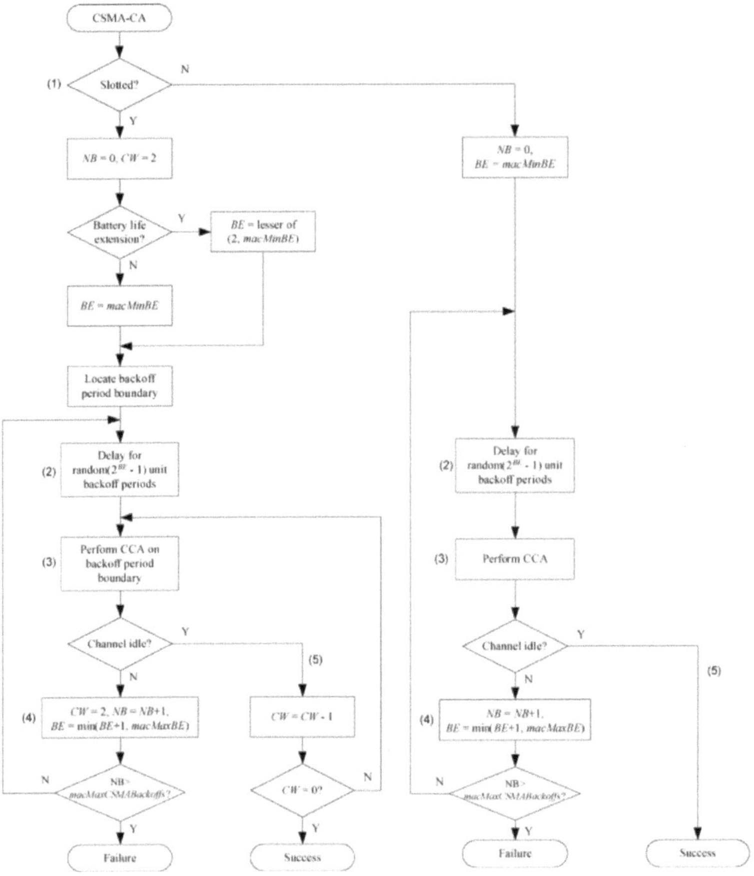

Figure 13.13. CSMA/CA mechanism including back off algorithm for unslotted operation in IEEE802.15.4 [44].

three variables for each transmission attempt: *NB*, *CW*, and *BE*. *NB* is the number of times the CSMA-CA algorithm was required to back off while attempting the current transmission; this value shall be initialized to zero before each new transmission attempt. *CW* is the contention window length, defining the number of back off periods that need to be clear of channel activity before the transmission can commence; this value shall be initialized to two before each transmission attempt and reset to two each time the channel is assessed to be busy. The *CW* variable is only used for slotted

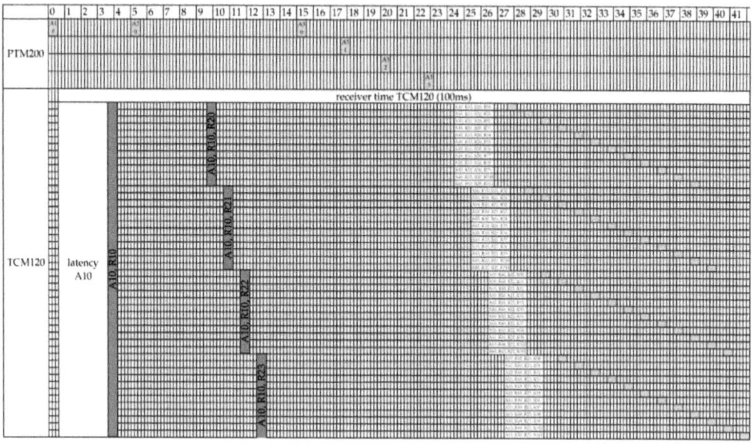

Figure 13.14. Possible delay of data frames for EnOcean Radio Protocol [own calculations].

CSMA-CA. *BE* is the back off exponent, which is related to how many back off periods a device shall wait before attempting to assess a channel. *BE* shall be initialized to the value of *macMinBE*. In slotted systems with the received BLE subfield set to one, this value shall be initialized to the lesser of two and the value of *macMinBE*. Note that if *macMinBE* is set to zero, collision avoidance will be disabled during the first iteration of this algorithm.

Although the receiver of the device is enabled during the channel assessment portion of this algorithm, the device may discard any frames received during this time.

For unslotted CSMA-CA, the MAC sublayer shall initialize *NB* and *BE* and then proceed directly to step 2. The MAC sublayer shall delay for a random number of complete back off periods in the range 0 to 2*BE* – 1 (step 2) and then request that the PHY perform a CCA (step 3), which shall start immediately.

If the channel is assessed to be busy (step 4), the MAC sublayer shall increment both *NB* and *BE* by one, ensuring that *BE* shall be no more than *macMaxBE*. If the value of *NB* is less than or equal to *macMaxCSMABackoffs*, the CSMA-CA algorithm shall return to step 2. If the value of *NB* is greater than *macMaxCSMABackoffs*, the CSMA-CA algorithm shall terminate with a channel access failure status.

It should be mentioned that persistency is not limited to use with channel sensing. Reference [27], for example, uses a random-based back off mechanism for unidirectional transmitter in order to keep the collision probability low. The usable time slots for this system are shown in Fig. 13.14. The upper part shows the timing of the three transmission attempts of a standard module, where the first two attempts are performed at fixed times. For the third attempt, one of four possible time slots is selected through a random process with equal probabilities for all of the four time slots.

The lower part of Fig. 13.14 shows the delay of a repeated frame under the condition that the repeater receives the first frame of a sender. The time slots for the repeated packets are also selected on random processes.

In addition, back off times can also be used to prioritize the traffic. This approach is mostly handled with regard to the so-called inter-frame spaces (IFS), which anyhow have to be regarded due to the physical characteristics of the channel (propagation time, echoes) and the transceivers (turn-around time). It is applied in many wireless systems in order to keep data packets and their respective acknowledgments together. In addition, it can be used to allow management or other low-latency traffic prioritized access to the channel.

IEEE802.11 defines five different IFSs to provide priority levels for access to the wireless media [48]: short IFS (SIFS), PCF IFS (PIFS), DCF IFS (DIFS), arbitration IFS (AIFS), which is used by the quality of service facility, and the extended IFS (EIFS). Figure 13.15

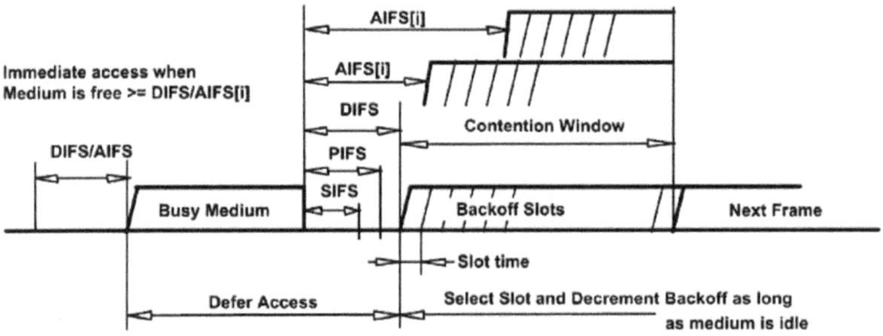

Figure 13.15. Some IFS relationships in IEEE802.11 [48].

shows some of these relationships in IEEE802.11, where the short IFS give a higher priority to the stations or the messages.

13.6.6 *Medium Reservation Methods*

The persistency and collision avoidance methods do not have any impact in case of hidden stations. This problem can be improved by the usage of ad hoc medium reservation methods. The basic and maybe mostly used mechanism is the so-called RTS-CTS handshake, as it is defined in IEEE802.11 [48, 4].

The handshake of Fig. 13.16 begins when the sender sends a short request-to-send (RTS) packet; the receiver then replies with a clear-to-send (CTS) packet.

- Only afterward, the sender begins the actual data transmission.
- A node further away from the sender may not receive the RTS during carrier sense; it receives, however, the CTS which prompts it to back off.
- If a collision occurs during RTS, the packet is deliberately made small that the collision doesn't make much damage.

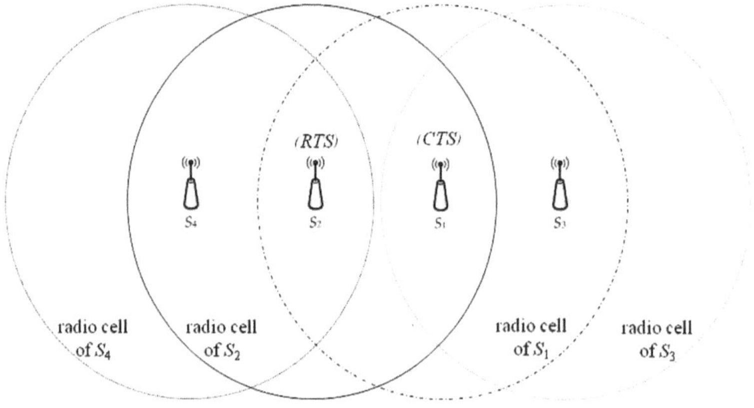

Figure 13.16. Medium reservation mechanism with RTS-CTS management frames, as used in IEEE802.11 [48], after [4].

13.7 Conclusions

Coexistence remains one of the major concerns and challenges of license-exempt bands, as they are used for WBANs. A variety of approaches has been developed, as the avoidance of coexistence impact is subject to a conflict of objectives. Therefore, no approach can be found, which is optimum for all cases, but the selection has to be performed on a case-by-case basis.

References

Book reference

1. R. K. Morrow (2004) *Wireless Network Coexistence*, McGraw Hill.
2. J. Eberspächer and H.-J. Vogel (1997) *GSM Global System for Mobile Communication* (Vogel).
3. I. Oppermann, M. Hämäläinen, and J. Iinatti (2004) *UWB — Theory and Applications*, John Wiley and Sons.
4. A. Sikora (2003) *Wireless Personal and Local Area Networks*, John Wiley and Sons.
5. G. Kupris and A. Sikora (2008), *ZigBee - Datenfunk mit IEEE802.15.4 und ZigBee*, Franzis-Verlag Poing,

Journal reference

6. C. Borean (2007) Analysis of ZigBee Interferences in WBANs, in *First European ZigBee Developers' Conference, June 07,* Munich, Germany.
7. H. T. Friis (1946) A Note on a Simple Transmission Formula, in *Proceedings of the IRE*, 34.
8. U. Walter (September 2005) Behavior of IEEE 802.15.4 Standards in the Various Frequency Bands, in *Design and Elektronik Developer Forum*, Berlin.
9. A. Sikora and D. Lill (2008) Wireless Technologies for Safe Automation – Insights in Protocol Development, in *13th IEEE International Conference on Emerging Technologies and Factory Automation (ETFA)*, Hamburg.
10. A. Sikora and V. Groza (2005) Coexistence of IEEE802.15.4 With Other Systems in the 2.4 GHz ISM Band, in *IEEE Instrumentation and Measurement Technology Conference (I^2MTC)*, Ottawa.

11. A. Sikora (2005) Adaptive Coexistence of Meshed ZigBee Networks and Other Systems in the Same ISM Band, in *Wireless Technologies Kongress*, Mannheim.
12. G. Scheible, J. Endresen, D. Dzung, and J.-E. Frey (February 2007) Unplugged but Connected: Design and Implementation of a Truly Wireless Real-Time Sensor/Actuator Interface, *IEEE Spectrum*, **44(2)**.
13. D. Verhulst, M. Mouly, J. Szpirglas (July 1984) Slow Frequency Hopping Multiple Access for Digital Radio Telephone, *IEEE Journal on Selected Areas in Communication*, SAC-2.
14. A. Kumar and R. Gupta (2001) Capacity Evaluation of Frequency Hopping Based Ad-Hoc Systems, in *Proceedings of the 2001 ACM SIGMETRICS International Conference on Measurement and Modeling of Computer Systems.*
15. N. Golmie, R. E. V. Dyck, A. Soltanian, A. Tonnerre, and O. Rébala (2003) Interference Evaluation of Bluetooth and IEEE 802.11b Systems, *Wireless Networks*, 9, pp. 201–211.
16. I. Ashraf, K. Voulgaris, A. Gkelias, M. Dohler and A. H. Aghvami (November 2009) Impact of Interfering Bluetooth Piconets on a Collocated p-Persistent CSMA-Based WLAN, *IEEE Transactions on Vehicular Technology*, pp. 4962–4975.
17. S. Y. Shin and H. S. Park (June 2008) On the Impact of Channel Sensing Methods to IEEE802.15.4 Performances Under IEEE802.11b Interference, *Journal of Communications and Networks*, 10(2).
18. I. Howitt and J. A. Gutierrez (2003) IEEE 802.15.4 Low Rate Wireless Personal Area Network Coexistence Issues, in *IEEE Wireless Communications and Networking Conference (WCNC).*
19. C. -C. Chong, F. Watanabe, and H. Inamura (2006) Potential of UWB Technology for the Next Generation Wireless Communications, in *2006 IEEE Ninth International Symposium on Spread Spectrum Techniques and Applications.*
20. G. Dolmans, O. Rousseaux, L. Huang, T. Fu, B. Gyselinkx, S. D'Amico, A. Baschirotto, J. Ryckaert, and B. van Poucke (2007) UWB Radio Transceivers for Ultra-Low Power and Low Data Rate Communications, in *IEEE International Conference on Ultra-Wideband (ICUWB).*
21. J. Zhang, P. Orlik, Z. Sahinoglu, A. Molisch, and P. Kinney (March 2009) UWB Systems for Wireless Sensor Networks, *Mitsubishi Electric Research Laboratories Report TR2009-007.*

22. S. M. Mishra and R. W. Brodersen (2007) Cognitive Technology for Improving Ultra-Wideband (UWB) Coexistence, in *IEEE International Conference on Ultra-Wideband (ICUWB).*
23. N. Abramson (1970) The ALOHA System — Another Alternative for Computer Communications, in *the AFIPS Conference Proceedings*, 37, FJCC 1970, S. 695–702.
24. F. Sivrikaya and B. Yener (July/August 2004) Time Synchronization in Sensor Networks: A Survey, *IEEE Network*, pp. 45–50.
25. S. Johannessen (April 2004) Time Synchronization in a Local Area Network, *IEEE Control Systems Magazine*, pp. 61–69.
26. B. M. Sadler and A. Swami (2006) Synchronization in Sensor Networks: An Overview, in *MILCOM 2006.*
27. F. Schmidt, G. Scholl, A. Anders, H.-J. Körber, and H. Wattar, RF-Embedding of Energy-Autonomous Sensors and Actuators Into Wireless Sensor Networks, in *Multifunctional Structures/Integration of Sensors and Antennas Meeting Proceedings, RTO-MP-AVT-141, Paper 3*, Neuilly-sur-Seine, France, pp. 3.1–3.14
28. C. Paillard and J. Wight (2006) Inside UWB Design: A Tutorial, CommsDesign, 29.3.2006.
29. C. Borean (2007) Analysis of ZigBee Interference WBANs, in *First European ZigBee Developers' Conference*, Munich, Germany.
30. G. Kupris (2008) Channel Agility and ZigBee/WLAN (IEEE 802.11 g and n) Coexistence, in *Second European ZigBee Developers' Conference.*
31. J. Foerster, E. Green, S. Somayazulu, and D. Leeper (2001) Ultra-Wideband Technology for Short- or Medium-Range Wireless Communications, *Intel Technology Journal*, Q2.
32. M. Hämäläinen (2006) Single-Band UWB Systems: Analysis and Measurements of Coexistence With Selected Existing Radio Systems, Faculty of Technology, University of Oulu, available at http://herkules.oulu.fi/isbn9514280644/isbn9514280644.pdf.
33. A. Miaoudakis, D. Stratakis, V. Zacharopoulos, and E. Antonidakis (June 2009) Co-existence Performance Evaluation of Wireless Computer Networks in a Typical Office Environment, *ComSIS*, 6(1), pp. 169–184.

Data Sheets and Application Notes

34. Nordic Semiconductor ASA, Product Specification Single Chip 2.4 GHz Transceiver nRF2401, Rev. 1.1, June 2004.

35. R. Rodriguez, MC1319x Coexistence, Freescale Semiconductor, Application Note, Document Number: AN2935, Rev. 1.2, 07/2005, http://www.freescale.com/files/rf_if/doc/app_note/AN2935.pdf.
36. M. B. Shoemake (February 2001) Wi-Fi (IEEE 802.11b) and Bluetooth Coexistence Issues and Solutions for the 2.4 GHz ISM Band, *Texas Instruments*, version 1.1, available at focus.ti.com/pdfs/vf/bband/coexistence.pdf.

Websites

37. http://www.cisco.com/en/US/products/sw/cscowork/ps3915/.
38. http://www.ptb.de/en/org/4/44/442/_index.htm.
39. http://tf.nist.gov/timefreq/stations/wwv.html.
40. http://tycho.usno.navy.mil/gps.html.
41. http://www2.rohde-schwarz.com/en/technologies/Wireless_Connectivity/UWB/information/.
42. http://www.ti.com/bluetooth80211.

Standards and Protocols

43. IEEE
802.15.2-2003 IEEE Recommended Practice for Telecommunications and Information Exchange Between Systems — Local and Metropolitan Area Networks Specific Requirements — Part 15.2: Coexistence of Wireless Personal Area Networks With Other Wireless Devices Operating in Unlicensed Frequency Band, http://www.ieee802.org/15/pub/TG2.html.
44. IEEE Std 802.15.4^{TM}-2006: Wireless Medium Access Control (MAC) and Physical Layer (PHY) Specifications for Low-Rate Wireless Personal Area Networks (WPANs), http://www.ieee802.org/15/pub/TG4.html.
45. ZigBee
46. Digital Enhanced Cordless Telecommunications (DECT); Common Interface (CI); Part 3: Medium Access Control (MAC) layer, EN 300 175-3 V1.4.2 (1999-06).
47. https://www.bluetooth.org/foundry/adopters/document/Bluetooth_Core_Specification_v1.2
48. IEEE Standard for Information technology — Telecommunications and information exchange between systems — Local and metropolitan

area networks —Specific requirements; Part 11: Wireless LAN Medium Access Control (MAC) and Physical Layer (PHY) specifications.

49. HART Field Communications Protocol Specification, Revision 7.0, August 2007.
50. IEEE P1588^{TM}/D1-E, "*Draft Standard for a Precision Clock Synchronization Protocol for Networked Measurement and Control Systems*", 18 Mar 2007.
51. IEC 61508 "*Functional safety of electrical/electronic/programmable electronic safety-related systems (E/E/PES)*"

Chapter 14

Implanted Wireless Communication Making a Real Difference

Henry Higgins
Zarlink Semiconductor, Castlegate Business Park, Portskewett, Caldicot, NP26 5YW, UK
henry_m_higgins@hotmail.com

14.1 Introduction

This chapter outlines the requirement for in-body communication with descriptions of body properties, design, testing, and ultimate manufacture of a product for widespread use.

Since the first cardiac pacemaker was implanted in Arne Larsson's chest in 1958, it has been essential to know what implants are doing and change the operation in the best interest of the patient. Fitting an implant and hoping for the best is not a way forward. The implant can be in a unique position to sense what is going on within the body without the need for invasive surgery. Operating on a patient increases the risk of infection, can be expensive, distresses him or her, and may not show what is really happening. The implant can become almost like part of the body and can feedback information when he or she is going about their normal day-to-day business. The patient does not need to be sedated or anesthetized, and the data gathering can go on almost anywhere.

Wireless Body Area Networks: Technology, Implementation, and Applications
Edited by Mehmet R. Yuce and Jamil Y. Khan

ISBN 978-981-4316-71-2 (Hardcover), ISBN 978-981-4241-57-1 (eBook)
www.panstanford.com

14.2 Why In-body Communication?

The implant may not be operating to its best, so changes in vital parameters may be necessary. This can best be done with a wireless link. It hardly needs to be said that wires passing through the skin are a source of infection, can result in injury (if pulled or knocked), and are often uncomfortable for the patient.

The communication system is an addition to the implant and is there to make it work better and provide feedback to the medical team.

14.3 Applications

Typical applications for a communication system are closely linked to the implant applications. The pacemaker has been mentioned above; other applications include the life-saving cardiac defibrillator, stimulation of muscles can bring movement back to limbs paralyzed by injury or disease, control of bowel or bladder incontinence can mean freedom for thousands of otherwise very distressed patients. The cochlear implant has brought hearing to many whom the conventional hearing aid will not help.

Many other applications are under development, such as nerve and muscle stimulation. Imaging the gastrointestinal tract has proved to be a valuable diagnostic tool where a camera pill transmits pictures of the esophagus, stomach, small and large bowel. These images are not possible to obtain even with a MRI scanner [1].

14.4 MICS and ISM Bands

There are several things to consider when choosing a frequency band, or bands, for implant communication. Tissue penetration is important as well as approved worldwide and accepted for implant use.

The band accepted worldwide for implant use is the Medical Implant Communication System (MICS) band, which is based at 403 MHz [2]. This band may be widened if the approval bodies agree. This band has the advantage of good propagation through the body, and it can be used for two-way communication. There is a restriction on the effective transmit power of 25 μW (−16 dBm); this limit applies outside of the body. The band is split up into 300 kHz channels with limits on out-of-channel interference. Use of this band within the prescribed limits does not require a license.

The Industrial Scientific and Medical Bands (ISM) are several in nature [3]. This band can be used, for example, to transmit data from a camera pill working its way through the digestive system, as this application is not considered to be an "implant." An ISM frequency can be used to wake up a MICS band circuit but not for two-way data transfer. ISM bands can also be used between non-implanted hearing aid. The permitted power levels and frequency bands vary from country to country, so it is essential to check on any national regulation that may preclude operation. One solution to this is to have country variants of the system. An important consideration when using ISM bands is that they are not universal; for example, in the United States there is a 915 MHz band, but in Europe the equivalent is 868 MHz with no overlap. If one of these bands were to be used, a variant would need to be made for each block or the implant would need to be capable of being reprogrammed in situ to change frequency. If one of these bands were to be used, then there may need to be country variants — hard wired or software controlled.

Typically, the range of operation, from patient to base-station, will be 3 m or less. If a body-worn data gathering system is used, the range could be even less.

As regulations will change with time, it is prudent to check on the current status.

A 27 MHz band has been used for control of a capsule within the gastrointestinal tract, but the antenna will be very much smaller than 5 m, which would be the case for a dipole in free space and, therefore, will be of poor efficiency.

14.5 RFID Techniques

Communication with electromagnetic induction similar to Radio Frequency Identification(RFID) has been successfully used for years, where only a small data transfer is needed. Wherever practical, this approach can be useful due to the big advantage of not needing a battery or for not draining the existing battery. RFID communication cannot, however, be initiated by the implant but needs to be started externally.

RFID is useful where an implant may be in place for a very long time without recourse to a battery change or where the data transfer is small. Applications for this include cranial pressure sensing, which is performed at regular intervals, or interrogating the status of an aortic arterial aneurism (AAA) repair [4].

14.6 Propagation Through the Body, Changes in Body Shape and Posture

Body tissue is variable in its electrical properties and also varies with frequency as Table 14.1 shows. The relative dielectric constant, ε_r, affects the wavelength of an RF signal; in free space, $\varepsilon_r = 1$ and the wavelength at 403 MHz is 74.4 cm, but when traveling through muscle ($\varepsilon_r = 58$), this will be reduced to 9.76 cm. The conductivity, σ, will affect how the tissue absorbs the signal. For air, σ is almost zero. The characteristic impedance, Z_r, is an issue when a signal goes from one medium to another as there will be a reflection at the boundary as shown in Eq. 14.1. For air, $Z_r = 377\ \Omega$, sometimes denoted Z_0.

$$\Gamma = \frac{Z_a - Z_r}{Z_a + Z_r} \tag{14.1}$$

Table 14.1. Dielectric constant, conductivity, and characteristic impedance of muscle and fat

	Muscle			Fat		
Frequency	ε r	6	Zr	ε r	6	Zr
100	66.2	0.73	31.6	12.7	0.07	92.4
400	58.0	0.82	43.7	11.6	0.08	108
900	56.0	0.97	48.2	11.3	0.11	111
MHz		$S.m^{-1}$	Ω		$S.m^{-1}$	Ω

Source: William Scanlon Queens University, Belfast.

where Γ is the reflection coefficient, or how much of the signal power is reflected at the boundary,

Z_a is the characteristic impedance of first medium Ω.
Z_r is the characteristic impedance of second medium Ω.

Much has been written about propagation of RF (radio frequency) signals through a human body. The depth of the implant is important, as is the posture. Not only the amplitude but also the polarization of the signal will be a function of the size, shape, and posture of the body and implant depth. It should be borne in mind that a surgeon will fit an implant where it is best for the patient and the function and not necessarily for the best RF propagation. As a patient lives with an implant, so he/she may put on weight or exercise and loose fatty tissue so the propagation will continue to change. Posture will also change the propagation as Johansson has shown [5].

Calculation of body propagation can be made, but this will have limitations as not every body is the same, a good reference is [6].

14.7 Antennas

The implant antenna is a very important part of the link. Because of the wavelength of a 403 MHz signal and that it is immersed in a weakly conductive medium, any antenna will appear poor compared to a full size dipole in air. The wavelength of an RF signal is found from Eq. 14.2.

$$\lambda = 300/f\text{m} \qquad (14.2)$$

where f is frequency in MHz.

This is reduced by being immersed in a dielectric material such as muscle as Eq. 14.3

$$\lambda_{\text{m}} = \lambda/\surd(\varepsilon_r)\text{m} \qquad (14.3)$$

where λ_{m} is the wavelength in medium.

Figure 14.1 shows the wavelength vs frequency for operation in air and in muscle ($\varepsilon_r = 58$).

There are several options for antenna, but the configuration will depend on the implant size and where it is to be located. A patch

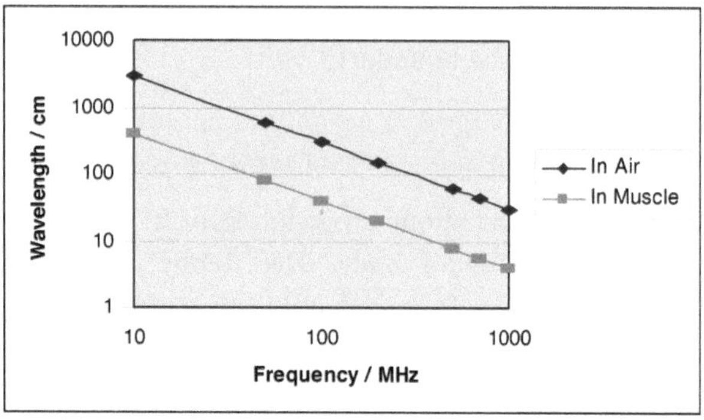

Figure 14.1. Wavelength in air and typical muscle vs frequency. Note: Log scales. See also Color Insert.

antenna is a flat structure that can be made as part of the implant case or can be attached to a convenient surface. It needs a layer of dielectric between the active surface and the 0 V ("ground") plane. The size of the patch is usually defined by the available flat surface. At 403 MHz, even in a body, any patch antenna will be electrically small and so it will have low gain. A typical example is shown in Fig. 14.2, which has been used for both 403 MHz and 2.45 GHz.

Figure 14.2. Silver on alumina patch antenna mounted on test implant case. See also Color Insert.

The impedance in air is almost entirely reactive but when measured in contact with skin and body tissue, the impedance becomes more resistive. It is essential that any antenna be measured when in contact with a body or immersed in body phantom liquid. The point of measurement needs to be clearly defined at either the antenna input or at the end of a length of coupling cable.

Small antennas can be made to be electrically larger by meandering the structure [7] or employing a fractal [8] pattern. A loop antenna will tend to be inductive and a monopole will be capacitive. Any antenna will be electrically small even within the body. Typically the antenna designer has to make the best of the volume allocated. A good theoretical reference for antennas is Kraus [9].

The measurement of the impedance of an electrically small antenna can be challenging as a network analyzer may not give the most accurate result, so other methods should be employed. Details of the measurement techniques and the hardware used can be found in Yang [10]. Fujimoto has written extensively on the design and use of small antennas [11].

Three-dimensional simulation software can be used in the design process where it is possible to simulate an antenna within a medium that resembles a typical human body. This will enable the effect of changes in body tissue electrical parameters to be estimated. Simulation of a 3D structure can take several hours and is dependent on good initial design. Software is available from Agilent (ADS), CST, and Ansoft.

14.7.1 *Use of Smith Chart in Coupling Network Design*

In the design of a coupling network between the antenna and the transmitter or receiver, calculation can be difficult; so a graphical technique can be used. This can be done either by using a software package or by hand using a preprinted chart, pencil, and a calculator.

Philip Smith designed the chart, thus bearing his name. More detail can be found in his book [12]. The concept is to give a graphical representation of any complex impedance showing clearly the resistive and reactive elements. There are two basic versions of the Smith Chart: impedance and admittance; these can be overlaid to make a powerful graphical tool. Detailed use of the Smith Chart in the design of RF networks can be found Lee [13], so it will not be repeated here.

The impedance chart comprises circles of constant resistance with their center on a horizontal line; the outer circle has 0 Ω resistance, with smaller circles indicating progressively higher resistance. The reactive (imaginary) part is represented by partial circles with inductance above the center line and capacitive below. The center line represents pure resistive impedance.

Often a Smith Chart is shown normalized where the impedance is divided by the characteristic impedance (50 Ω for example) and the admittance multiplied by the characteristic impedance. Figure 14.3 shows a typical normalized impedance only chart available from the University of Berne [14].

As well as a visual representation of lumped impedance and admittance, the chart also represents adding transmission line. Various versions of the chart are available from Internet sources.

To design a network, the antenna impedance must be known and plotted on the chart. The destination impedance must be defined. Then by adding series and parallel reactive components, including the impedance of the IC and possibly line lengths, the

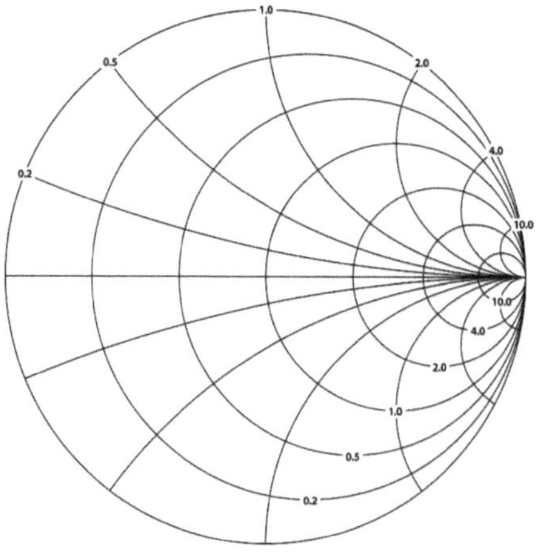

Figure 14.3. Blank normalized impedance Smith Chart (University of Berne).

target impedance at the transmitter/receiver integrated circuit (IC) can be reached.

It is useful to design a network using a paper Smith Chart and calculator as it gives the designer a better feel of what difference adding a component will make. The next stage is to design the PCB tracking for the coupling network. It is suggested the PCB comprise signal layer on top and a 0 V ("ground") plane such that it is possible to have transmission line of known impedance. Using surface-mount components, it is possible to make this very small so the track effect will be minimal at 403 MHz. The coupling to the antenna should also be known or the impedance measured where it enters the PCB such as the end of a short length coax. When the above are known and the paper initial design completed, simulation is recommended, which will include the PCB track and dielectric material with real models of the passive components; these are often available from the manufacturers or can be derived from the published data on self-resonant frequency, tolerance, and Q. Simulation of the entire network should be close to, but not the same as, the initial design. Simulation can be run to see the effect of component tolerance and potential yield given the nominal values.

Note: When operating within the body, there is no "ground" and the whole implant, along with monitor and stimulation wires, will be floating.

14.7.2 *Design of Antenna Coupling Networks*

Assuming an available transceiver is to be used, most of the RF system hardware design has already been done leaving the PCB layout, software, and antenna coupling network. The antenna coupling network is important as it needs to present the transmitter and receiver with the correct impedance, conjugately match the antenna and provide filtering. The coupling network may also be required to separate a 403 MHz data signal from a 2.45 GHz wake up and prevent each branch from loading each other. A typical example of a communication IC that uses 2.45 GHz for wake up and 403 MHz communication is the Zarlink ZL70101 [15].

The impedance required by a transmitter may be different to that required by a receiver. For optimum power, the transmitter impedance is found from Eq. 14.4

$$R_l = \frac{(0.7071 \cdot V_{pp})^2}{P_o} \tag{14.4}$$

where R_l is the optimum load resistance, P_o is the required power output, and V_{pp} is the maximum voltage swing of the transmitter output.

The receive impedance may be optimized for noise as well as gain. This may require a more complex network and possibly include a switch to present different impedances for transmit and receive. Often an optimum value is specified by the IC manufacturer.

Below are examples of the design of coupling networks:

14.7.2.1 Design example 1

Consider a simple example at 403 MHz, where the antenna impedance is 10 + j20 Ω; normalized, becomes 0.2 + j0.4 Ω, is required to be transformed to an impedance of 200 Ω, normalized 4 Ω. See Fig. 14.4. Adding a series 10 nH transforms the impedance to 10 + j48 Ω, normalized 0.2 + j0.96 Ω. Adding a parallel capacitance of 8.3 pF will produce an impedance of 200 + j0. Part of that 8.3 pF could be the tuning capacitance of the IC. See Fig. 14.5. As has been seen, the only part of the IC input impedance that is considered is the capacitance, as a transmitter or receiver will be required to work into a resistive load that is often in the 100–500 Ω range, and the real part of the IC impedance is ignored.

14.7.2.2 Design example 2 (SAW filter)

In many countries, the emergency services use a communication network called Tetra. One of the Tetra bands is 308 to 400 MHz, which is very close to the MICS band. The transmit powers are high with fixed base-stations closely spaced to provide wide coverage. This high power signal could overwhelm the sensitive input of a MICS implanted or base-station receiver as the front-end filtering may not be able to provide enough selectivity. Additional filtering to the RF path can be provided with a SAW filter centered on the MICS band. These filters have a narrow-pass band that severely attenuates

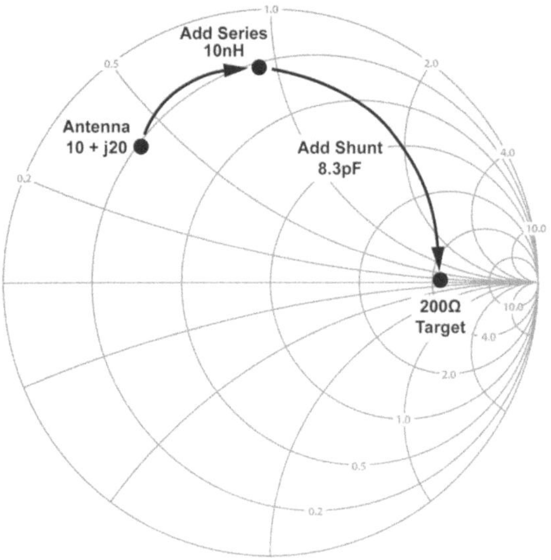

Figure 14.4. Simple matching antenna uses chart normalized to 50 ω (Software University of Berne).

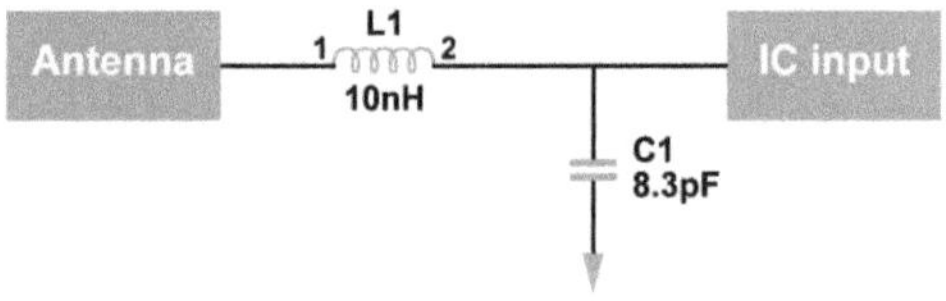

Figure 14.5. Matching schematic from Fig. 14.4. See also Color Insert.

out of band signals with relatively low attenuation of the in-band signal.

These filters are available in small size, which can be included in an implanted design. They can be single sided at both ports or single sided on one port and double sided on the other. Typically, the impedance of the ports is nominally 50 Ω on one side and 200 Ω on the other. The impedances SAW filters tend to be capacitive. This reactance needs to be conjugately matched as part of the coupling network design.

The filter can be put into the receive path or in a common transmit and receive paths as part of the antenna coupling network.

A typical example is the SF16-0403C5UB02 from AVX/Kyocera. A SAW filter in a common transmit and receive path will not only reduce out-of-band signals overloading the receiver but will also attenuate any out-of-band signals in the transmitted signal.

If a differential SAW filter is used with a single-ended RF IC, then a balun needs to be designed. A balun will transform a differential signal to single sided using capacitors and inductors. The first step is to conjugately match the SAW differential output such that it presents a non reactive impedance of known value. Next design the balun transforming the differential SAW resistance to the required impedance at the RF IC.

Consider the example of a nominal 200 Ω SAW (AVX/Kyocera SF16-0403C5UB02) coupling to a 400 Ω impedance at an IC input. Consider also that the IC input has a 2 pF internal capacitance. Each differential port of the SAW has an impedance of 35 Ω + 5.7 pF. This is shown on the normalized Smith Chart of Fig. 14.6 along with the locus of the SAW input impedance. Adding a 36 nH inductor to each differential output port to 0 V will conjugately match the capacitive element and result in an impedance that is almost a pure resistance of 183 Ω on each port giving a differential impedance of 366 Ω.

A balun to convert to 400 Ω can be designed using the circuit of Fig. 14.7 and Eqs. 5 and 6.

$$L = \surd(R_l \cdot R_S/2)/\omega \qquad (14.5)$$

where R_s is single-sided impedance, 400 Ω in this example; R_l is differential impedance, 360 Ω in this example.

$$\text{At 403 MHz, } L = 100 \text{ nH}$$

Also

$$C = 1/[\omega \cdot \surd(R_l \cdot R_S/2)] \qquad (14.6)$$

This gives C = 1.5 pF.

The conjugate matching and impedance transformation can be rationalized by adding the parallel inductor/inductor and inductor/capacitor combinations to a single inductor on each SAW port to 0 V.

The output of the balun is a pure 400 Ω resistance, but the IC input has a capacitance of 2 pF. A 77 nH inductor to 0 V from the

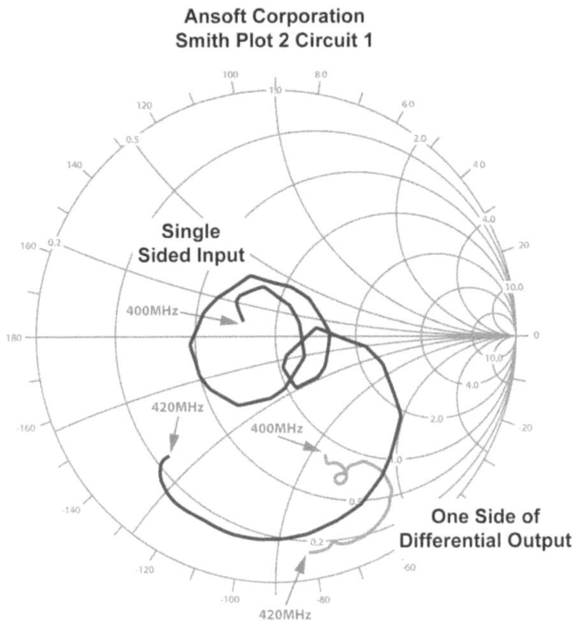

Figure 14.6. SAW filter input and (differential) output; 400 to 420 MHz; Ansoft SV software. See also Color Insert.

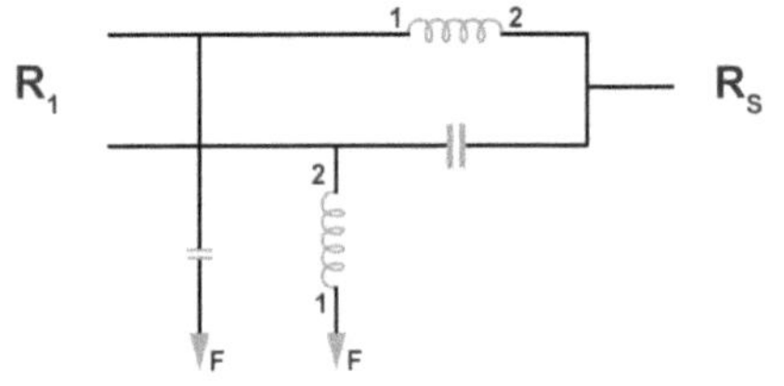

Figure 14.7. Inductor/capacitor balun.

IC pin will conjugate match this capacitance. The path from the differential ports of the SAW to the IC now looks like Fig. 14.8.

The single-ended port now needs to be matched to the antenna remembering that the SAW has a capacitive impedance. For example, if the SAW filter were to match to an antenna of impedance 10 + j20 Ω (normalized 0.2 + j0.4), then the impedance of the SAW needs to be transformed to produce a conjugate match of the antenna.

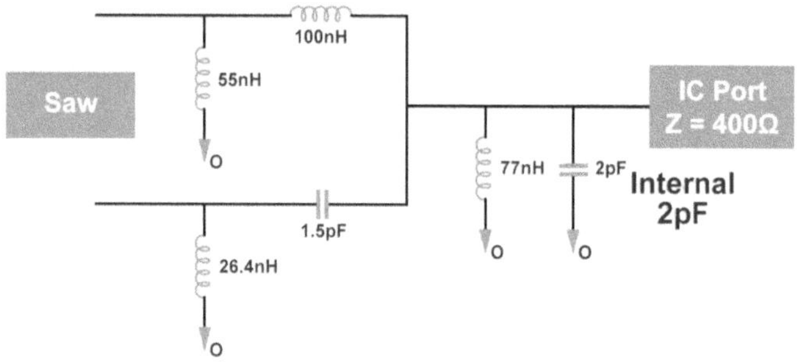

Figure 14.8. Complete differential SAW to IC port network; includes balun and tuning.

The conjugate match has the same real part but the negative of the imaginary part. In this example, the SAW impedance needs to be transformed to 10 – j20 Ω (normalized 0.2 – j0.4). Adding the balun and coupling network to the differential output changes the single-sided input to 40 – j6.2 Ω (normalized 0.8 – j0.124); this is almost a pure resistance as Fig. 14.9 shows. Adding a parallel 15 pF capacitor to 0 V transforms the impedance to 10 – j17.5 Ω (normalized 0.2 – j0.35). Adding a series 150 pF capacitance will produce the required 20 – j20 Ω conjugate match of the antenna (Fig. 14.9).

SAW filters are also available with both single-ended input and output such as the AVX/Kyocera SF16-0403C5UU01.

The schematic is shown in Fig. 14.10.

If a 2.45 GHz wake-up is employed using a single antenna, then the signals need to be split as close to the antenna as possible. This can be done with an inductor in the 403 MHz path and a small capacitor in the 2.45 GHz path.

An alternative approach can be found in Yang [10].

When the design is complete and the PCB is laid out, the whole circuit, including tracking, should be simulated using realistic models of the passive components and the SAW filter. Real components will have a specified value, but there will be a tolerance and the parameters will change with frequency that may change the response from that predicted.

The above examples used pure capacitors and inductors with no PCB allowed for. Real passive components will have

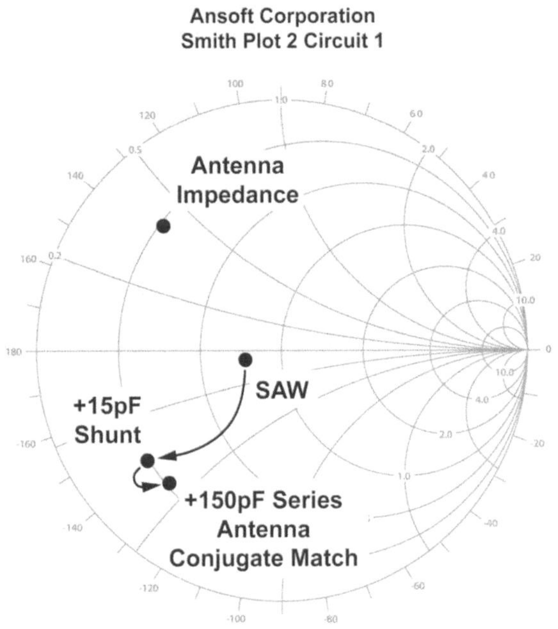

Figure 14.9. Single-sided SAW to antenna matching; normalized plot; Ansoft software.

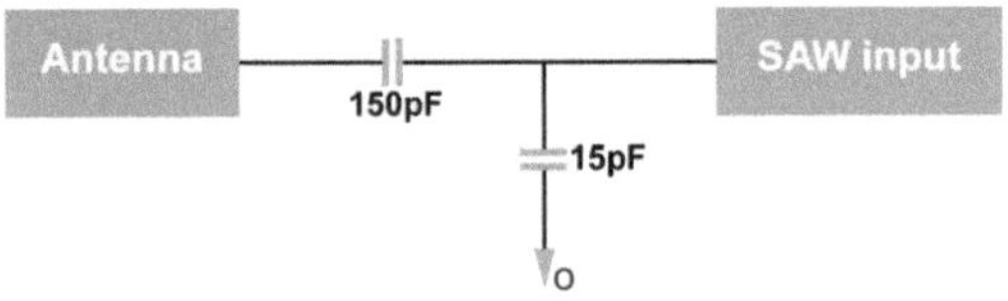

Figure 14.10. Single-sided SAW input to antenna coupling network.

parasitic elements such that a capacitor will have a typical equivalent circuit for a 10 pF surface-mount capacitor is shown in Fig. 14.11.

This will result in an impedance that will change from capacitive to inductive (and vice versa for an inductor); a typical 10 pF surface-mount capacitor impedance and series resistance plot is shown in Fig. 14.12 (courtesy AVX).

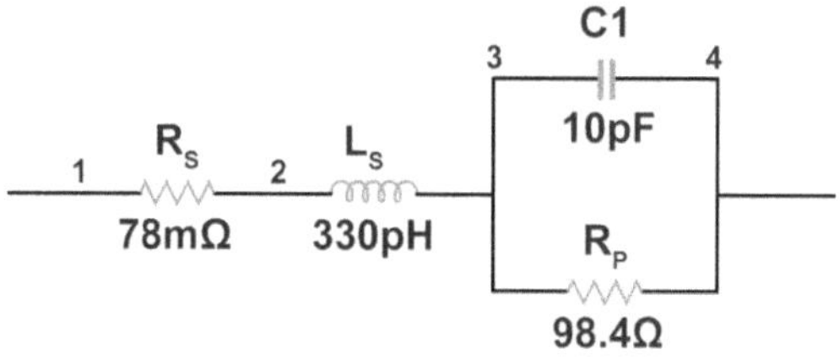

Figure 14.11. Typical capacitor equivalent circuit.

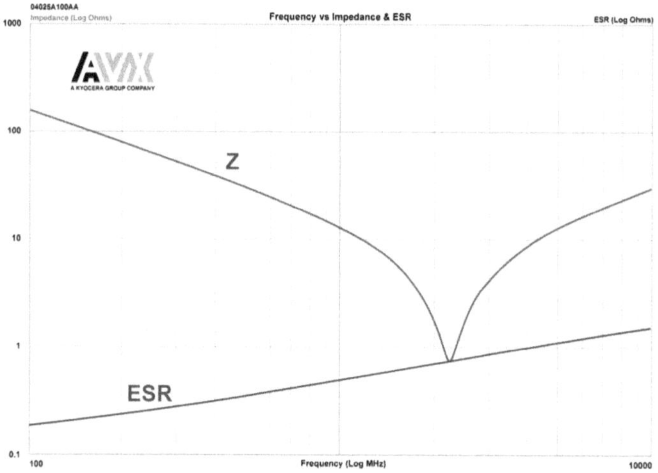

Figure 14.12. Impedance and series resistance for 10 pF 0402 capacitor (courtesy AVX). See also Color Insert.

14.7.2.3 Use of simulation for antennas and design of coupling networks

There are several software packages that will simulate a PCB with passive components. These include ADS from Agilent and Microwave Office from AWR and free software such as AppCad from Agilent, SmithCalc from the University of Berne, and Ansoft Student Version (SV). To give accurate results, the PCB and dielectric constant should be precisely defined. The models for the passive components must include the parasitic preferably using manufacturers' models.

14.7.3 *Physical Body Simulator*

Before any live trials can be considered, it is essential that the communication system be proven in an environment that resembles the electrical properties of a human body. The body phantom is a useful tool as a battery-powered implant can be placed in it at various positions and depths. This enables the effectiveness of the antenna in a body medium to be assessed and the reliability of the link transferring data.

Testing an implant in dead tissue, such as meat from a butcher, does not have the same electrical properties as the living does.

A body phantom comprises a 76 cm high, 30 cm diameter Perspex tank filled with a liquid having electrical properties close to that of living tissue. There are recipes for the liquid from Wojeck [16] that cover various frequencies as Table 14.2 shows. The implant must be water tight, battery powered, and suspended in the liquid attached to a PTFE (polytetrafluoroethylene) (or similar nonconductive) support.

A system will be used in a doctor's office where the signal will be reflected off of various surfaces causing multipath fading. To test implant to base station link without the complication of reflections, an anechoic chamber can be used. This is a chamber that has walls, floor, and ceiling covered with absorbent material, thus eliminating stray reflections. Using a body phantom in an anechoic chamber will show the link can work and the ideal data rate for given.

14.7.4 *Body Simulator Measurements and Sample Results*

Testing in an anechoic chamber will require an implant and base station as well as measurement antennas and other equipment.

Table 14.2. Body phantom liquid (* HEC = hydroxyl ethyl cellulose)

	% by weight	
Frequency GHz	0.1 to 1.0	1.5 to 2.5
Water	52.4	45.3
Suger	45.0	54.3
NaCl	1.5	0.0
HEC*	1.1	0.4

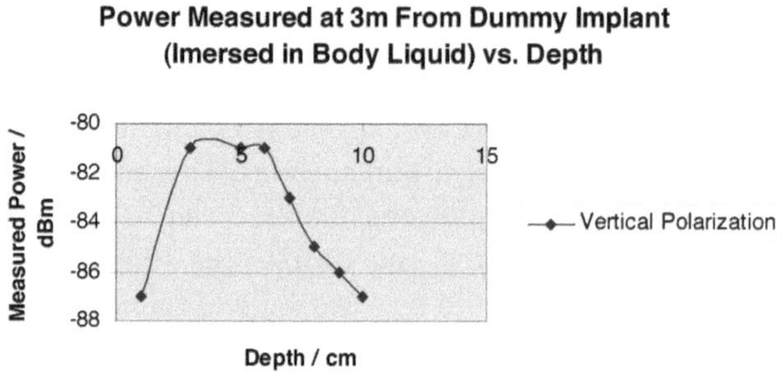

Figure 14.13. Power measured from implant in tank of body phantom liquid vs depth (transmit power = −16 dBm).

Measurement can be made with the implant and measurement antenna at various polarizations. Test software has to be written, which will enable the implant to transmit a continuous wave signal for a period of time that will enable measurements to be made. Also it must be possible to set it into the receive mode record and transmit back the RSSI (received signal strength indicator) level measured when the base station is transmitting a CW signal.

Using a body simulator tank, and a battery-powered implant with a patch antenna, measurements were made of signal levels and error rate. Signal levels were measured using a calibrated log-periodic antenna and spectrum analyzer.

Figure 14.13 shows how the signal from the implant varied with depth. Here "depth" is the distance from the front of the tank to the implant antenna.

Figure 14.13 shows the measured power level increases with depth up to 6 cm after which it declines. This is thought to be due to the body tissue acting as a parasitic antenna, which improves propagation. After 6 cm, the body phantom attenuates the signal as it is weakly conductive. Measurements made of data rate also confirm this result.

14.7.5 *The Role of Automatic Antenna Tuning*

When implanted, an antenna will present a different impedance from when it was in air. The implanted environment will not be predictable and may change with time, so it is essential that the implanted system be capable of automatically tuning the antenna coupling network for optimum performance without external intervention. This feature may be built into the RF IC along with a routine that will activate on wake up. If this feature does not exist, then external varactor diodes may need to be added and a routine written that will optimize both the receive signal and transmit power.

14.8 Implant Power Constraints and Battery Considerations

For long-term use, battery drain must be kept to a minimum as the communication system is in addition to the primary implant function. Typically, a pacemaker will be expected to last over seven years before requiring replacement, so the overhead that the communication system adds must be minimal. With implants having more features and pressure to make them smaller, the battery size will also be constrained. A smaller battery will hold less charge, so it is necessary to only power up the RF system when it is to be used.

For a new application, it should be remembered that the voltage of a primary (nonrechargeable) battery will decline gradually, whereas a secondary (rechargeable) will show a much steeper drop when a critical point is reached. The RF communication system can feed back the battery status.

14.9 Error Correction

When packets of data are transmitted, error can be caused by noise, interference, or multipath fading. To recover from this, the system, if the receiving end detects an error, can invoke error correction. This involves an overhead on the data transmitted but may enable a corrupt packet of data to be corrected. If the error is such that correction does not work, then the receiving end may request the whole packet be re-sent. Using such techniques, the data integrity is maintained and the final error rate is minimized.

14.10 RF Circuit Hardware Options

RF transmitters will need supply current related to the output power — that is fairly obvious. For a receiver to operate on a low-level signal, it will require careful design for gain and low noise. Typically, the more sensitive a receiver is, the higher the supply current will need to be. Receivers can take as much current as transmitters. There are several ICs available that operate both in the MICS and some of the ISM bands. To choose an IC for implant use, first consider the following:

- Required data rate.
- Whether the implant is going to be on permanently or intermittently
- Available supply voltage and current.
- Available volume
- Space for antenna
- Whether the IC is available in implant quality
- Whether wakeup is needed?

Several ICs have sleep mode options, which will reduce the average current drain significantly. Waking up the IC from sleep mode can be done either by transmitted signal in the band of use or by a separate ISM band. Table 14.3 shows several ICs that cover the MICS band. Most ICs will have an SPI (serial peripheral interface) to an application processor.

Other manufacturers include Lime, Maxim, and Zlinx.

When choosing an IC, it is necessary to be sure what the additional components are as they can significantly increase the required volume.

Typically, the RFIC will be controlled by a separate microcontroller, which may be from a different manufacturer or may be the controller for the primary function. This will involve setting and reading register values, which can be a considerable task; so the use of an evaluation board is recommended. These are available from most RFIC manufacturers and will enable basic functions to be run and allow access to individual registers for more advanced operation.

Table 14.3. Example values

IC	Maximum date rate	Sleep current	Wake mode	On current	Minimum frequency	Maximum frequency
Melexis TH7122	115	50	Software	6.1(Rx) 15(Tx)	27	930
RFM TRC105	200	100	Software	2.7(Rx) 16(Tx)	300	610
Semtech SX1211	200	100	Software	3(Rx) 15(Tx)	863 and 902	870 928
TICC1101	500	9800	Wake on radio, poll every second	15(Tx0dBM and Rxmode)	315	915
Ti Toumaz CC981H (Sensium)	50		None	4.5	868 and 902	870 928
Zarlink ZL70101	800	100	2.45GHz	5	402 and 433.5	405 434.4
	kb.s	nA		mA	MHz	MHz

Refer to data sheet for more details.

Some RFICs also include a processor that could be used to control the primary function if it is adequate for the purpose. Microcontrollers will also take current, which must be included in the power budget. As with RFICs, some have low power modes, which can significantly reduce current drain. The TI MSP430 series is considered to be among the lowest power products currently available.

14.11 Base Station

When designing an implant, it is also necessary to design the external base station to control the communication and handle the data. Often it is possible to use the same IC for both the implant and the base station. Power and space are not so much of a problem as it is for the implant; so enhancement of the ICs performance is possible by adding a pre-amplifier in the receive path and using a better antenna system.

In a real environment, RF signals will reflect off of walls, furniture, equipment, etc. Also polarization will depend on the implant position and patient posture. This can result in constructive interference, which will improve the RF link but can also result in

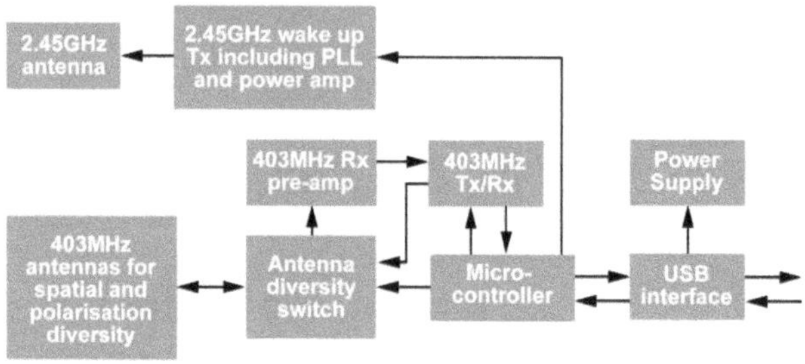

Figure 14.14. Typical base station block diagram.

destructive interference, which could break the link. This can be countered by employing an array of antennas to provide spatial and polarization diversity. The antennas can be selected with an RF switch. This will increase the size of the base station, but additional antennas could be integrated into the enclosure.

A typical base station will have an RF IC, microcontroller, pre-amplifier, antennas, antenna switch, and a USB interface to a PC as shown in Fig. 14.14. A USB interface can also be used to provide power. This includes a separate wake-up transmitter working on an ISM band. The wake transmitter will need to have the capability of operating over different frequency limits and power levels depending on the country of operation. Frequency hopping may be required for the wake transmitter, which will require reprogramming of the frequency source (usually a crystal oscillator, phase locked loop (PLL) and voltage-controlled oscillator (VCO)), allowing time for it to settle before transmission. Care must be taken to be sure that there is no transmission outside of the limits when the frequency is changed. This can be done by switching off the output stage during the frequency hops.

14.11.1 *Link Budget*

A simple link from base station to implant is shown in Fig. 14.15, where the base station antenna has a gain of 0 dB, the range is 3 m, the body loss is assumed to be 16 dB, and the implant antenna has a gain of −20 dB. The implant antenna gain is for a poor antenna at 403 MHz; better gains are possible.

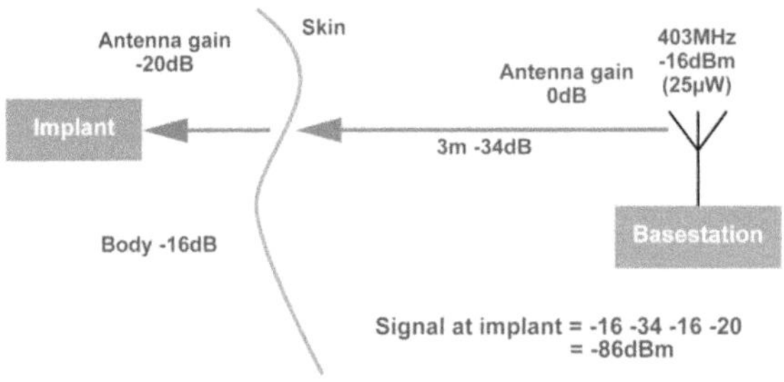

Figure 14.15. Link from base station to implant (simple example).

For the base station to implant path, the radiated power is 25 μW (−16 dBm). The path loss in air (assuming no reflected signal path) is found from Eq. 14.7 (Friis equation).

$$P_s = G_t(\lambda/(4\pi R))^2 \tag{14.7}$$

where P_s is signal at skin surface, λ is wavelength in meters, G_t is antenna gain, and R is range in meters.

The signal at the implant receiver is found from adding the transmit power antenna and path gains (losses) as Eq. 14.8.

$$\begin{aligned} P_i &= -16 - 34 - 16 - 20 \text{ dBm} \\ &= -86 \text{ dBm (or 2.5 pW)} \end{aligned} \tag{14.8}$$

where P_i is the signal delivered by the implant antenna

The implant to base station path has the advantage that the 25 μW limit is outside of the body; so the transmit power can be higher as long as the SAR (specific absorption rate) limit is not reached, and the battery can supply the current. In this example, the transmit power is 1 mW (0 dBm); with antenna and body losses, the power at the body surface is −36 dBm; this is within the capability of several ICs. Figure 14.16 shows a simple example. Adding the gains (losses) gives a base station signal of −70 dBm, or 100 pW, which is much higher than the base station to implant signal.

In a real environment, there will be stray reflections that can interfere constructively or destructively at the receiver. The signal at the receiver must be factored into a signal-to-noise ratio to access the viability of the link. There may also be background noise sources that will deteriorate the signal quality, but if the link does not work with these very simple calculations, it will not work in a real situation. See also reference [17].

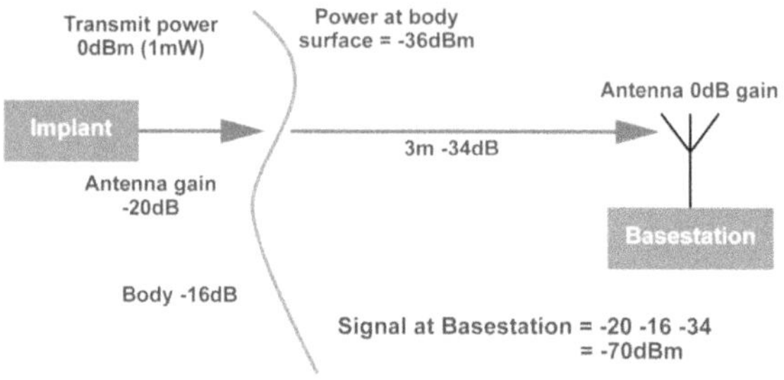

Figure 14.16. Implant to base station link (simple example).

14.12 Environment

The human body is an environment of an almost constant temperature and mostly free of mechanical shock (this is not the case for cochlear implants in children where normal rough and tumble can be the cause of significant shock). The implant will need to survive the distribution process that may involve rough treatment and large temperature swings.

An important consideration for the design of an implant is the materials that will be exposed to body liquids that can attack many materials. It is not adequate to cover a metal surface with a plastic coating as this may become porous and enable leaking of toxic materials. The metals that are considered safe for implant use are platinum, platinum/iridium, and an implant grade of stainless steel. Metals such as silver and copper are toxic and gold will be attacked by body liquids.

14.13 Manufacture

It is not too difficult to design and build a single implant for experimental use by hand adjustment of the parameters. An experimental implant may not need to operate for a lengthy time, and it may not matter if it fails. An implant that can be fitted to patients is required to work consistently, does not need physical intervention, and has to be of a very high standard. It must be possible for an implant to be purchased against a know specification.

To take the concept to a product that can be fitted to patients requires a design that can be reproduced reliably in significant volumes.

For an RF system, the design requires to be thoroughly tested in a simulated environment and may undergo live testing. The design should not require components to be selected at manufacture. Components used must come from a source that will supply to implant grade — not every manufacturer will do this. It may be necessary to change a design to enable implant grade components to be used. The printed circuit board (PCB) is an essential part of the product, and it should not be taken for granted. For consistent product manufacture, the assembly of components onto the PCB should be by machine and not by hand; even the best assembly workers will, occasionally, fit a component into the wrong location. This will result either in time consuming rework or products being scrapped. This adds to the overall cost of manufacture.

The final testing of the assembly should be to check for the part was manufactured correctly. Final testing should not be to verify the design as this should be well established along with the choice of components. Following manufacture, there may be a need for burn in testing where the product is powered for a period of time, possibly at an elevated temperature. This is to make sure that any component weakness is revealed.

14.14 Conclusions

Adding a wireless link to a human body implant can greatly improve the diagnosis of diseases and improve the quality of life of many patients. It offers the possibility of long-term monitoring without the need for a long and expensive stay in hospital. By monitoring the performance of the body and the implant, therapeutic regimes can be optimized without the need for surgical intervention. In the future for this technology will use even lower power devices with ever smaller component sizes including integrating passive components into the PCB. The PCB technology is also critical to size reduction.

Implant technology will make even more significant improvements to the lives of otherwise very ill patients.

Notation

dB: a log power ratio, $10\log_{10}(P_1/P_2)$.
dBm: a power ratio referenced to 1 mW, $10\log_{10}(P_1/1\text{ mW})$.

Rx: receiver.
Tx: transmitter
λ: wavelength in meters

References

1. Sivard, A., et al. (November 2004) Challenge of Designing In-body Communications, EE Times, http://www.eetimes.com/design/embedded/4025029/The-challenge-of-designing-in-body-communications.
2. ETSI EN3018390, Electromagnetic Compatibility and Radio Spectrum Matters (ERM); Radio equipment in the frequency range 402 MHz to 405 MHz for Ultra Low Power Active Medical Implants and Accessories; Part 1. 2006. 06921 Sophia-Antipolis, Cedex, France.
3. Loy, M., et Al. (May 2005) ISM Band and Short Range Device Regulatory Compliance Overview, *Texas Instruments Application Report* SWRA048. Dallas, Texas.
4. Finkenzeler, K (2003) *RFID Handbook*, 2nd edn, Wiley. Munchen, Germany.
5. Johansson, A. J. (2002) *Wave Propagation From Medical implants — Influence of Body Shape on Radiation*, Department of Electroscience, University of Lund, Lund, Sweden.
6. Rahmat-Samii, Y., and Kim, J. (2006) *Implanted Antennas in Medical Wireless Communications*, 1st edn, Morgan and Claypool. California USA.
7. Warnagris, T. J., and Minardo, T. J. (December 1998) Performance of a Menadered Line as an Electrically Small Transmitting Antenna, *IEEE Transactions on Antennas and Propagation*, 46(12), pp 1797–1801. www.ieee.org
8. Felber, P. (2001) *Fractal Antennas*, revised edn, Illionis Institute of Technology, Revised January 16. www.iit.edu

9. Kraus, J. D. (1988) *Antennas*, 2nd edn, McGraw Hill. Columbus OH, USA.
10. Yang, G. Z. (2006) *Body Sensor Networks*, Springer, pp 117–143. London, UK.
11. Fujimoto, K., et al. (1987) *Small Antennas*, Research Studies Press (John Wiley). Bognor Regis, UK.
12. Smith, P. (1969) *Electronic Applications of the Smith Chart*, McGraw Hill. Columbus OH, USA.
13. Lee, R. F., and Chen, W. (1997) *Advances in Microstrip Printed Antennas*, Wiley, pp. 35–70. Columbus OH, USA.
14. Dellsperger, S., University of Applied Sciences, Berne, fritz.delsperger@bfh.ch.
15. Zarlink ZL70101 Data Sheet: http://www.zarlink.com/zarlink/zweb-zl70101-datasheet-dec09.pdf.
16. Wojclk, J., et al. Tissue Recipe Calibration Requirements, SSI/DRB-TP-D01-003, Spectrum Sciences Institute RF Dosemetry Research Board, 51 Spectrum Way, Nepean, Ontario, K2R 1E6 Canada.
17. Planning For Medical Implant Communication Systems (MICS) and Related Devices, SPP 6/03, October 2003. Australian Communications Authority, Canberra, Australia. www.acma.gov.au.

Chapter 15

Wireless Power and Data Telemetry for Wearable and Implantable Electronics

Zhi Yang, Yu Han, Linh Hoang, YiKai Lo, Kuanfu Chen, Jian Lao, Mingcui Zhou, and Wentai Liu
University of California, Santa Cruz, CA, USA
wentai@soe.ucsc.edu

The recently reported power and data telemetry systems in the literature will allow wireless communication with tissue and by doing so enable implantable/portable microelectronic devices to treat presently incurable diseases, analyze biologic signals, and enhance the performance of surgical tools. While telemetry design is, in general, guided by several basic themes — heat dissipation, data communication, and integration and miniaturization — this chapter addresses the above themes through system-level modeling, analysis of critical blocks, and design examples based on dual-band scheme [1].

15.1 Introduction

Inductive links are popularly used in newer generations of biomedical microelectronic systems for providing power and data. Supported by telemetry, these systems are more appealing in terms

Wireless Body Area Networks: Technology, Implementation, and Applications
Edited by Mehmet R. Yuce and Jamil Y. Khan

ISBN 978-981-4316-71-2 (Hardcover), ISBN 978-981-4241-57-1 (eBook)
www.panstanford.com

of portability, safety, and system integration in comparison to their counterparts, those powered by a battery and commanded by signals through tether wires connected externally. Some important examples of wearable and implantable electronics are cochlear implant [2–4], retinal implant [1, 5–9], tissue imaging [10], brain-computer interface [11–13], artificial limb [14], neural recording [15–19], and health meters [20].

Functionally, a telemetry system consists of two links: a power telemetry link to deliver energy for powering electronics and a data telemetry link to communicate information between the object and an external station. To deliver power and data wirelessly for those wearable and implantable electronics, both links have two components of electronics. Those located externally or physically detached from the objects are referred to as primary side units, e.g., external battery, power transmitter, signal processing station, forward data transmitter, reverse data receiver, etc. Those located under the skin (implanted electronics) or along with the objects (wearable electronics) are typically referred to as secondary units, including power receiver, forward data demodulator, stimulator [21], signal sensor and processor [22], and reverse data transmitter, etc. To accomplish the functions of powering and communicating with secondary electronics, the telemetry system should achieve high-power efficiency to avoid heating tissue, low bit-error-rate (BER) to ensure reliable data communication, and small in size and light in weight to target the highest level of system integration and miniaturization. As will be shown in this chapter, the mentioned design goals are inherently connected; thus careful design optimization and trade-off analysis are important to achieve the optimal performance given the application.

This chapter is organized as follows. In Section 15.2, analysis and methodology to design highly efficient power telemetry link are discussed first, followed by modeling and optimization of coils, which are the discrete devices of the largest footprint size. Further, the trade-off between power efficiency and coil size is elaborated by correlating power efficiency, Q factor, coupling, and geometry parameters analytically. In Section 15.3, a review/comparison of several data communication schemes is presented based on power dissipation, design complexity, data rate, BER, and communication distance. In Section 15.4, a telemetry design example to support a 256-channel epiretinal stimulator array is given with test results. In the appendix, detailed derivations and design guidelines related to coils and power telemetry are presented.

15.2 Power Telemetry

In this section, critical components of a power telemetry link, including coupled resonant amplifier, high Q inductor, and rectifier, are introduced first, followed by detailed modeling, explanation, and implementation results.

Resonant Amplifier: As shown in Fig. 15.1, a power telemetry link used in biomedical applications involves two components: power transmitter and power receiver. The power transmitter is typically located outside the body, converting energy from a battery (Vb) to radio frequency electromagnetic (EM) wave. The receiver harvests energy from the EM wave, which further goes through rectification and voltage boosting, powering the rest electronics at the secondary side. For the convenience of discussion, the power transmitter and receiver are conceptually simplified as two isolated resonant tanks (small coupling coefficient between coils, small power load at the secondary side). The primary resonant tank has a resonant frequency at $\omega_1 = 1/\sqrt{L_1 C_{res1}}$, and the secondary resonant tank has a resonant frequency at $\omega_2 = 1/\sqrt{L_2 C_{res2}}$. If $\omega_1 = \omega_2 - \omega_0$ and the carrier frequency $\omega = \omega_0$, both primary and secondary tanks are in resonance, achieving high power efficiency. Details on modeling these two resonant frequencies will be presented in Section 15.2.1.

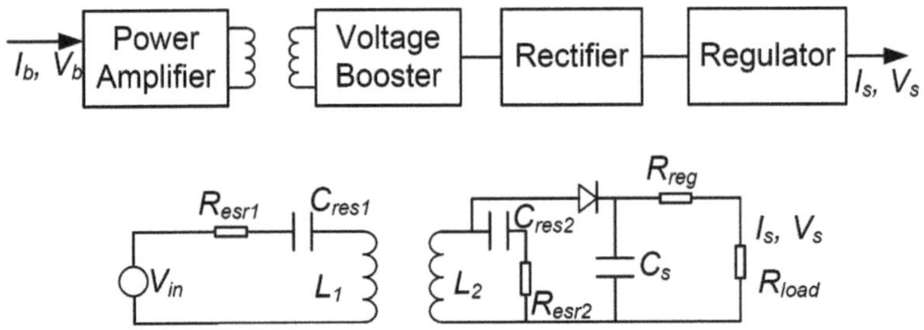

Figure 15.1. Conceptual diagrams of a power telemetry system.

Power Efficiency and Inductor Optimization: Power efficiency, defined as the ratio of power harvested at the secondary side over the power drained from the battery ($I_s V_s / I_b V_b$ in Fig. 15.1.), is an important parameter to evaluate the performance of a power telemetry link. It can be analytically approached using different simplified models, and the results show that power efficiency is a monotonously increasing function of the Q factors of coils and the coupling between them. In case the coils are weakly coupled and the power load is small, power efficiency is approximately proportional to $k^2 Q_1 Q_2$ (Q_1 and Q_2 are the Q factors of coils, and k is the coupling coefficient between the coil pair), suggesting coils to be the most critical components for power telemetry. Given the restrictions on coil size and coil separation, the achievable coupling coefficient is determined. Therefore, one of the emphases of this chapter is on improving the power efficiency by increasing the coil Q factor. Details on coil modeling and method to improve the Q at the carrier frequency are given in Sections 15.2.2 and 15.2.3.

Power Rectification and Regulation: The induced voltage at the secondary resonant tank is rectified and regulated to power the rest of the electronics. For rectifier design, particularly on chip implementation, diodes or diode-connected MOS transistors are commonly used. Figures 15.2a, b illustrate full-wave rectifier (generate two voltage supply levels) and cascoded rectifier (generate four voltage supply levels) topologies. Generally, diodes are implemented using MOS transistors, and they face two challenges. First, as shown in Fig. 15.2c, diode-connected PMOS transistor has $V_T + V_{OD}$ source to drain voltage drop (V_T is the transistor turn on voltage, V_{OD} is transistor over-drive voltage). A large voltage drop across the diode-connected transistor degenerates the power conversion efficiency.

$$\eta_{rectifier} = \frac{V_0}{V_H} = \frac{V_H - (V_T + V_{OD})}{V_H}, \tag{15.1}$$

where $\eta_{rectifier}$ is the power conversion efficiency, V_O is the rectified DC voltage, and V_H is the peak voltage in the secondary resonant tank connected to the rectifier input.

Second, substrate has to be biased to the largest positive/negative voltage level for PMOS/NMOS transistors. Otherwise,

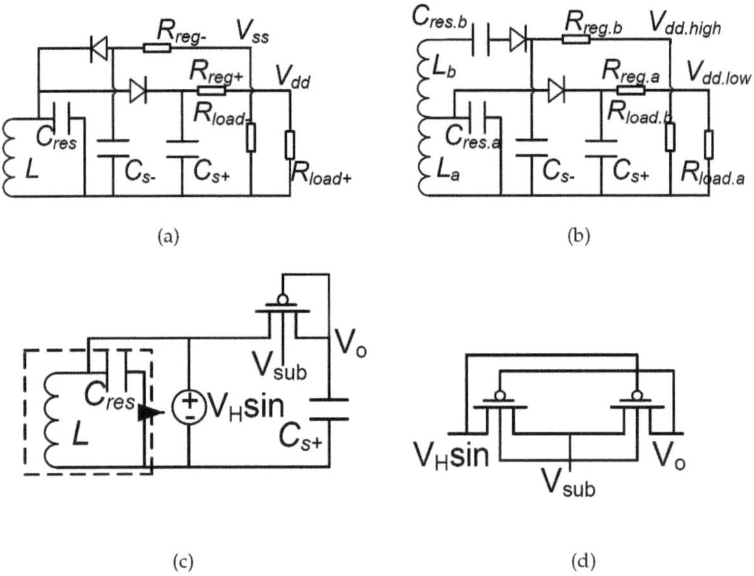

Figure 15.2. Rectifier topologies: (a) Full-wave rectifier providing V_{dd} and V_{ss}; (b) Cascoded rectifier providing four voltage supply levels: V_{dda}, V_{ddb}, V_{ssa}, and V_{ssb}, avoiding DC-DC voltage drop from V_{ddb}/V_{ssb} to V_{dda}/V_{ssa} (only positive branches V_{ddb} and V_{dda} are drawn); (c) PMOS implementation diode for rectifying positive branches; its substrate has to be biased to the highest voltage level to avoid leakage current; (d) a bias circuitry for the substrate of PMOS diode. $V_{sub} = \max(VH \sin \omega t, V_o)$.

large leakage current to substrate appears. This requirement is more complicated in biomedical applications, since the voltage in the resonant tank is AC and its amplitude appears to be the largest at the secondary side. Consequently, special bias circuitry for the substrate is required, e.g., floating well biasing, dynamic bulk switching technique. An example of bias circuitry for PMOS diode substrate is shown in Fig. 15.2d, where the substrate voltage fluctuates to track the higher one of source and drain. More discussion on the choice of rectifier structure is presented in Section 15.2.4.

15.2.1 *Mega-Hz and Sub-Mega-Hz Power*

A commonly used power telemetry architecture is shown in Fig. 15.3a, where the transmitter is adopted from Class-E power

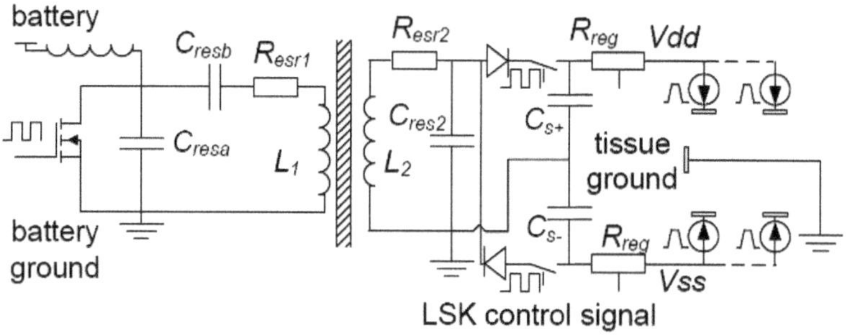

(a)

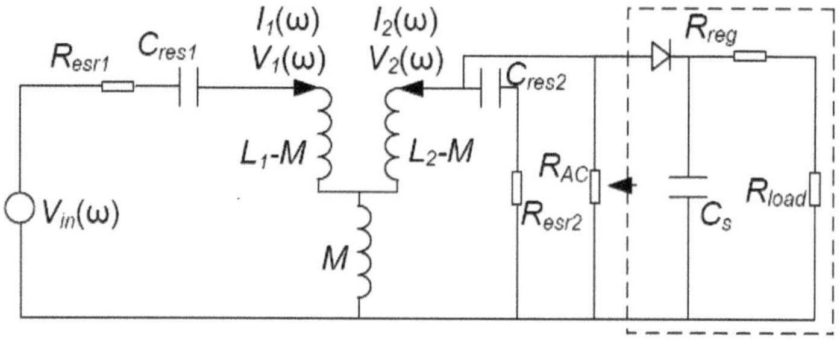

(b)

Figure 15.3. (a) Class-E amplifier based on power telemetry link, including Class-E power amplifier, power receiver resonant tank, rectifier and regulator; (b) Linearized circuit model for the power telemetry link.

amplifier, the receiver is a resonant tank followed by rectifier and regulator. The corresponding linearized circuit model of the power telemetry architecture is shown in Fig. 15.3b.

The Co-Resonance of Primary and Secondary Tanks: Given components' values and coil coupling coefficient, there are two resonant frequencies satisfying the following constraints:

$$\frac{\partial I_1(\omega)}{\partial \omega} = 0|_{\omega=\omega_{res1}}, \quad \frac{\partial I_2(\omega)}{\partial \omega} = 0|_{\omega=\omega_{res2}} \tag{15.2}$$

where ω_{res1} and ω_{res2} are the resonant frequencies of primary and secondary tanks after considering load reflection through coil coupling. At ω_{res1} and ω_{res2}, currents through the coils (I_1, I_2) reach their maximum respectively. To improve power efficiency, both primary and secondary resonant tanks are desirably to resonate at the same frequency point.

After inclusion of coil coupling and power load, ω_{res1} and ω_{res2} deviate from the ideal form. Through straightforward analysis based on linearized circuit model, V_1, V_2, I_1, and I_2 meet the following constraints:

$$I_1(R_{esr1} + 1/(j\omega C_{res1}) + j\omega L_1) + I_2 j\omega M = V_{in}, \quad (15.3)$$

$$I_1 j\omega M + I_2 \left(j\omega L_2 + \frac{R_{esr2} + 1/j\omega C_{res2}}{R_{esr2} + 1/j\omega C_{res2} + R_{AC}} R_{AC} \right) = 0 \quad (15.4)$$

$$I_1 j\omega L_1 + I_2 j\omega M = V_1, I_1 j\omega M + I_2 j\omega L_2 - V_2, \quad (15.5)$$

where M is the mutual inductance between L_1 and L_2.

Combine Eqs. 15.2–15.5, under the resonant condition, circuit components' values satisfy

$$\omega L_1 + 1/\omega C_1 + IM \left\{ \frac{\omega^2 M^2}{j(\omega L_2 - 1/\omega C_{res2}) + R_{esr2} + R_{AC.esr}} \right\} = 0, \quad (15.6)$$

$$\omega L_2 + 1/\omega C_{res2} + IM \left\{ \frac{\omega^2 M^2}{j(\omega L_1 - 1/\omega C_{res1}) + R_{esr1}} \right\}. \quad (15.7)$$

where IM{.} stands for the imaginary part, $R_{AC.esr}$ is the converted R_{AC} to series resistance in the secondary resonant tank.

Based on Eqs. 15.6 and 15.7, the optimal power carrier frequency ω_C can be represented as a function of L_1, L_2, C_{res1}, and C_{res2}.

$$\omega_C = \sqrt{\frac{\frac{R_{esr2}+R_{AC.esr}}{C_{res1}} - \frac{R_{esr1}}{C_{res2}}}{(R_{esr2} + R_{AC.ESR})L_1 - R_{esr1}L_2}}. \quad (15.8)$$

Power Efficiency: Both the primary and secondary tanks are in resonance is a necessary condition to achieve high efficiency. In this subsection, focus is made to balance undesired primary and secondary power losses.

The power drained from the external battery consists of three parts: power dissipation in the primary resonant tank (P_1), power dissipation in the secondary resonant tank (P_2), and power consumed by R_{AC} (P_d). Among these three components, P_d is the power delivered to support electronics, while P_1 and P_2 are undesirable power losses, which should be minimized.

Assume the resonant conditions of Eqs. 15.6 and 15.7 are satisfied, P_2 is approximately computed as

$$P_2 = I_2^2 R_2 = 0.5(V_r + V_T)^2 (j\omega C_{res2} + 1/R_{AC})^2 \omega L_2 / Q_2 \quad (15.9)$$

where V_r is the rectified voltage, V_T is the diode forward voltage drop when it turns on, and Q_2 is the secondary coil Q factor.

P_1 is the Ohm loss in the primary resonant tank (derived considering the energy spent in the secondary side), represented as

$$\begin{aligned} P_1 &= I_1^2 R_1 = V_{ind}^2/(k^2 \omega L_2 Q_1) \\ &= \frac{2(P_2 + P_d)^2}{(V_r + V_T)^2 (j\omega C_{res2} + 1/R_{AC})^2 k^2 \omega L_2 Q_1} \end{aligned} \quad (15.10)$$

where V_{ind} is the induced voltage on the secondary side, Q_1 is the quality factor of the primary resonant tank, and k is the coupling coefficient between the primary and secondary coils ($k = M/\sqrt{L_1 L_2}$)

Fix the power requirement of electronics at the secondary side ($P_d = P_{d0}$, $V_r = V_{r0}$); the optimization goal is to reduce the undesired power losses P_1 and P_2.

$$\arg\min \alpha P_1 + \beta P_2 |_{P_d = P_{d0}, V_r = V_{r0}, \omega = \omega_c} \quad (15.11)$$

where α and β are weighting coefficients for power loss at the primary side and the implant side, respectively.

When the carrier frequency and components' values satisfy the resonant conditions in Eqs. 1.6 and 1.7, the objective function $\alpha P_1 + \beta P_2$ is minimized to be

$$\alpha P_1 + \beta P_2 \approx 2 P_d \sqrt{\frac{\alpha\beta}{k^2 Q_1 Q_2}} + \frac{2\alpha P_d}{k^2 Q_1 Q_2} \quad (15.12)$$

It is interesting to see that this minimum undesired power loss is determined by Q and coupling coefficient alone. In the following section, focus will be on boosting the coil Q factor at the carrier frequency.

15.2.2 *Inductor Q Boosting*

When a coil operates at low frequency, its Q factor is defined as $Q = \omega L/R$, where ω is frequency, L is inductance, and R is series resistance. As frequency increases, frequency-related effects, including skin effect, proximity effect, and self-resonance modify both L and R, dramatically degenerating the Q factor. According to reference [23], a formula for predicting the Q factor is

$$Q(f) = \frac{2\pi f L}{R}\frac{1 - f^2/F_{self}^2}{1 + f^2/f_h^2} \tag{15.13}$$

where f is frequency, f_h is a parameter to quantify the impact from proximity effect (skin effect), and f_{self} is the coil's self-resonant frequency; f_h and f_{self} are expressed by geometry and physical parameters, briefly

$$f_h \propto \frac{1}{r_s^1 \sigma \sqrt{N_1 N_s}},\ f_{self} \propto \frac{1}{\sqrt{a \ln(a/r_1) \sum C_{p,k}(k-p)^2}} \tag{15.14}$$

as shown in Fig. 15.4. r_s, r_t, and a are the radii of a strand, a turn, and the coil loop (usually litz winding is used, where one turn contains multiple strands; otherwise, $r_s = r_t$). N_s and N_t are the number of strands per turn and the number of turns of the coil winding. σ is the metal conductivity, $C_{p,k}$ is the parasitic capacitance between turn p and turn k whose expression is shown in the appendix.

Importantly, for any given coil, there is an optimal frequency, f_{peak}, that has the maximal Q

$$\frac{\partial Q}{\partial f} = 0|_{f=f_{pwK}} \tag{15.15}$$

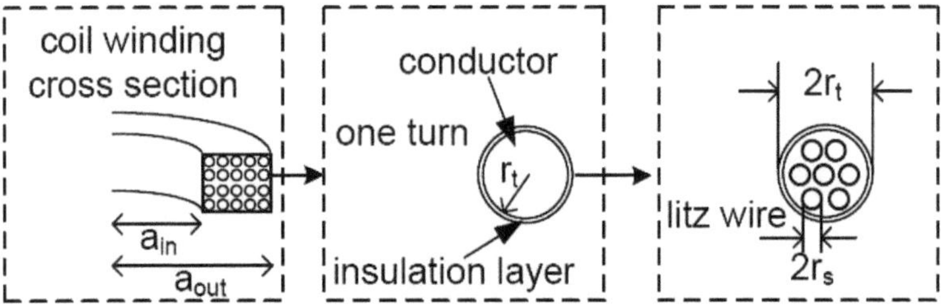

Figure 15.4. Coil geometry illustration. Left: A multiple-turn, multiple-layer coil winding. Middle: Cross section of one turn, single-strand wire. Right: Cross section of one turn, multiple-strand wire (litz wire).

To maximize the power efficiency, f_{peak}'s of the coil pair should be designed in accordance with the power carrier frequency. Based on Eqs. 15.13–15.15, the analytical form of f_{peak} is

$$1/f_{peak}^2 \approx 1/f_h^2 + 3/f_{self}^2 \tag{15.16}$$

Equation 15.16 represents a key design equation and a closed form analytical solution for f_{peak}. With this single equation of merit, the maximum Q, as well as the maximum efficiency of the telemetry system, can be determined. By changing the design parameters, one may tune f_{peak} close to the target frequency and maximize the power efficiency of the telemetry.

Example a. Inner diameter: To increase the self-resonant frequency of a coil and thus f_{peak}, an effective and practical method is to increase the separation between layers. As shown in Fig. 15.4., fixing outer radius a_{out} and reducing the inner radius a_{in}, an increased layer separation can be achieved.

According to Eq. A.7, a coil's total parasitic capacitance is

$$C_{self} = \left[C_b(l-1)m + c_m \sum_{i=1}^{l} (2i-1)^2 (m-1) \right] / N_t^2. \tag{15.17}$$

where C_b is the parasitic capacitance between two nearby turns in the same layer and C_m is the parasitic capacitance between different layers, as shown in Fig. A.3 in the appendix.

For a tightly wound coil, the parasitic capacitance between two nearby turns is

$$C_b = C_m = \varepsilon_0 \varepsilon_r \int_0^{\pi/4} \frac{\pi D_i r_0}{\varsigma + \varepsilon_r r_0 (1 - \cos\theta)} d\theta \tag{15.18}$$

For coils with spacing between layers, the parasitic capacitance is

$$\begin{cases} C_b = \varepsilon_0 \varepsilon_r \int_0^{\pi/4} \dfrac{\pi D_i r_0}{\varsigma + \varepsilon_r r_0 (1 - \cos\theta)} d\theta \\ C_m = \varepsilon_0 \varepsilon_r \int_0^{\pi/4} \dfrac{\pi D_i r_0}{\varsigma + \varepsilon_r r_0 (1 - \cos\theta) + 0.5 \varepsilon_r h} d\theta \end{cases} \tag{15.19}$$

where h is the separation between two layers. The calculation results using Eqs. 15.17–15.19 under different layer separation is shown in Table 15.1 (insulation layer thickness: $\varsigma = 3$ um, dielectric constant $\varepsilon_r = 3$), suggesting tightly wound coil is not necessarily the best choice for achieving high Q.

Table 15.1. Comparison of coils with and without interlayer separation

Coil	D_{in}	C_b	C_m	L	C_{self}	f_{self}	f_{peak}
h=0	2.76 cm	15 pF	15 pF	85 μH	17 pF	4.1 MHz	1.6 MHz
h=r_0	2.66 cm	15 pF	1.0 pF	77 μH	1.5 pF	15 MHz	2.5 MHz

Example b. The number of strands: If the cross section of the winding is fixed, reducing the number of strands is an effective way to increase f_{peak} and the Q at higher frequencies for two reasons. First, f_{self} increases due to the increased separation between nearby turns. Second, f_h increases since the total number of strands in the winding decreases. Therefore, both f_{peak} and the Q are larger at higher frequencies.

As a numeric example, comparisons of several coils with different geometry parameters are given in Table 15.2. Specifications of coils under comparison: $D_{out} = 3.2$ cm, $D_{in} = 2.7$ cm, $b = 2.5$ cm, $t = 2.5$ mm, $d_s = 50$ um, $N_t = 36$, $\varsigma = 3$ um, $\varepsilon_r = 3$, wire strand size = *American wire gauge* (AWG) 44.

15.2.3 *Rectifiers and Regulators*

On-chip rectification is commonly implemented by diode-bridge configured in a half-wave rectifier scheme for simplicity or a full-wave scheme to acquire a smaller output voltage ripple and higher conversion efficiency. However, the use of on-chip PN-junction diode and simple diode-connected transistor in rectifier circuitry usually leads to a voltage drop of ~1 V (diode turn on voltage plus overdrive voltage, as shown in Fig. 15.5a), resulting in a power efficiency degradation, which is more severe in low-voltage applications. Furthermore, the diode substrate has to be properly biased to avoid leakage current through the parasitic BJT.

Table 15.2. Comparison of coils of different number of strands

Coil	N_s	R_{DC}	f_{self}	f_{peak}	Q_{peak}
IV	3	9.5 ohm	26 MHz	8.5 MHz	198
V	7	4.1 ohm	19 MHz	4.2 Mhz	215
VI	15	1.9 ohm	14 MHz	2.1 MHz	222
VII	30	0.95 ohm	5.6 MHz	1.1 MHz	219

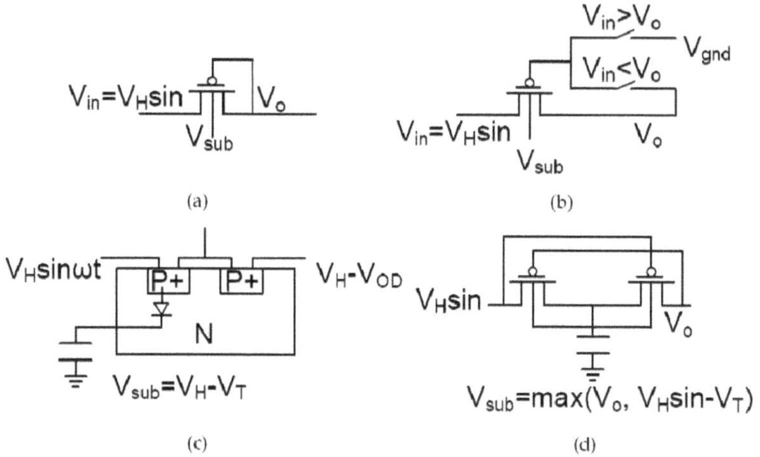

Figure 15.5. (a) Diode-connected transistor as a rectifier; (b) Synchronized rectifier; (c) Floating well substrate biasing; (d) Dynamic switching substrate biasing (V_{in} is the input voltage to the rectifier, V_o is the output voltage from the rectifier).

To enhance the conversion efficiency, active diode configuration has been re- ported and its generic type is shown in Fig. 15.5b. The basic principle of this circuitry is to mimic an ideal diode with zero drop-output voltage. The transistor MP is used as a switch, and its gate-control voltage is derived from the rail-to-rail output voltage of a comparator and accordingly, a smaller forward voltage drop across the diode can be achieved since it operates in triode region. By carefully controlling the turn-on time of the transistor, reverse current leakage that comprises power efficiency is prevented.

To avoid leakage current through the parasitic BJT, bulk can be biased using floating well biasing (the substrate V_{sub} is biased to $V_{peak} - V_T$, as shown in Fig. 15.5c), or dynamic bulk switch technique (the substrate is biased to the larger one of source and drain $V_{sub} = \max(V_{sour}, V_{Drain})$, as shown in Fig. 15.5d).

15.3 Data Telemetry

In applications involving stimulation, a data telemetry link transfers commands and stimulation patterns from the primary station to the secondary electronics. In some other applications involving signal sensing and processing, a data telemetry link transfers those recorded data at the secondary side to the primary station for further processing. Depends on applications, a set of desired performance metrics and constraints may vary, such as data bandwidth, implementation complexity, power consumption, system size, and reliability. In this section, several commonly used communication schemes are discussed for different applications.

Amplitude-Shift Keying: Amplitude-shift keying (ASK) is one of widely used modulation techniques for data communication for wearable and implantable electronics [5, 24, 25]. Under ASK scheme, binary data are transmitted through modulating the envelope of carrier wave. One prominent advantage of ASK-based system is that the circuit blocks for both modulation and demodulation are relatively efficient in terms of design complexity and integration. In reference [5], a neuro-stimulus chip with ASK telemetry units for retinal prosthesis has been reported. The reported chip can support a 10 by 10 electrode array, providing up to 250 Kbps data rate. In reference [26], Z. Tang et al. proposed a load shift keying (a special type of ASK) (LSK) modulator for data recording, called LSK circuit configuration modulator (LSK-CCM). The system modulates the secondary load with digitized recorded data, which induce voltage fluctuation across the primary side inductor for data demodulation. The design can provide 36 Kbps data with modulation index down to 0.02.

Despite the efficiency in implementation, one of the challenges an ASK-based data link faces is the susceptibility to the signal amplitude fluctuation, which can be severe if the object is moving [27] and produce erroneous bits.

Frequency-Shift Keying: Frequency shift keying (FSK) is another commonly used modulation technique where carrier frequency acts as the data-carrying entity [28–30]. Three FSK designs for biomedical applications have been evaluated in reference [28], where measurement results show that the differential FSK design achieves relatively better performance: maximum 4 MHz data rate, 0.29 mm^2 die area, and power consumption of 0.38 mW at 5/10 MHz carrier frequencies.

Compared with ASK, FSK has decreased susceptibility to signal amplitude fluctuation, achieving more robust data communication. However, there are two major drawbacks. First, FSK signal spectrum is relatively wide and cannot benefit from a high Q inductive-capacitive tank at the receiver side. Consequently, it is more inefficient to recover power from FSK signal compared with ASK signal with small modulation index. Second, FSK receiver circuitry is more difficult to design; for example, the receiver requires synchronization circuitry, since the internal clock cannot be simply derived from the carrier.

Phase-Shift Keying: The drawbacks of ASK and FSK derive from fluctuations in signal magnitude and frequency spectrum. Such fluctuations are suppressed in phase-shift keying (PSK). Instead of using carrier amplitude and frequency to represent digital data, PSK modulates carrier waves through changing their initial phases. A binary PSK (BPSK) demodulator that incorporates a digital phase-locked loop (PLL) has been reported in reference [31]. The design can support 100 Kbps forward data rate at 13.56 MHz carrier frequency. Their design has bidirectional telemetry units: BPSK-based forward data transmission and multilevel LSK-based reverse data transmission. In a different implementation of BPSK-based on COSTAS loop [32], the measured data rate is 1.12 Mbps with 0.7 mW power consumption.

Featuring stable signal amplitude and carrier frequency, however, BPSK has to use coherent demodulation method in the receiver circuitry due to its sensitivity to phase information. A phase synchronization circuit with high precision is particularly important.

Differential Phase-Shift Keying: In some biomedical applications, for example retinal prosthesis, power and data required to operate the implant goes up to a few hundred mW and a few Mbps [33]. The power telemetry design must take the coil size, tissue magnetic field exposure, and power efficiency into consideration, while the data telemetry should provide a high data rate that can control a large number of stimulation pixels at real time. In a single-band approach, where power and data are transmitted through the same carrier, high power efficiency and high data rate are two contradicting specifications [5]. Dual-band telemetry is a better approach [28, 34] because it transmits power and data at different frequencies. However, the power-to-data interference in the dual-band scheme can potentially corrupt the data demodulation. As a result, it is important to incorporate a dual-band approach with a power interference cancelation scheme.

A noncoherent differential phase-shift keying (DPSK) scheme, where "1" is coded as a phase shift of 180° and "0" is coded as no phase shift [35], can be used to reduce the power interference. By assigning data rate equal to the power carrier frequency, two consecutive symbols are under the same power interference, which is eliminated by differentiation. The reported DPSK data link in reference [1] is based on subsampling scheme [36], where sampling frequency is lower than the data carrier frequency. The adoption of the subsampling scheme is based on two reasons. First, a low frequency operation reduces demodulator power. Second, it samples the band-pass signal without down mixing, which involves the phase-locked loop circuitry. In this sense, the DPSK technique can be regarded as a noncoherent technique, and the DPSK receiver design is much simpler than BPSK receiver since the phase information is not precisely required. Measurement results in reference [1] show that the DPSK data link can provide up to 2 Mbps data with total power consumption of 6.2 mW/5 V, when tested under 12 dB interference-to-signal ratio.

Ultra-Wideband Telemetry: Very recently, ultra-wideband (UWB)-based data link has been reported in biomedical applications to transmit data. A UWB design can offer low power consumption, low signal power spectrum density, high data rate, immunity against power interference and noise [37]. UWB radio is defined by FCC to support high data rate with a fractional bandwidth f_c greater than 0.2 or a bandwidth of at least 500 MHz at all times of transmission [37–39]. The fractional bandwidth is defined by the following expression:

$$f_c = 2\frac{f_H - f_L}{f_H + f_L} \tag{15.20}$$

where f_H and f_L are the upper and lower frequencies measured at the -10 dB emission points, respectively.

Instead of using continuous carrier waves as in common communication systems, UWB generally uses very low duty cycle impulses to transmit data. Such efficient transmission mode highlights UWB to transfer high data rate in a relatively large distance (100 Mbps, several meters). Further, UWB pulse can be extremely narrow, thus robust to multipath fading effects. Though UWB data transmission features high data rate and low power consumption, there still exist many challenges, such as antenna design for wearable and implantable applications, high-peak power pulse generation, and interference with other devices in the overlapping spectrum. These challenges should be carefully considered before implementation.

BER Performance Analysis: BER performance reflects the reliability of data transmission. To compare the mentioned modulation techniques, simulated BER curves based on theoretical analysis in reference [35] are shown in Fig. 15.6. It is shown that PSK and DPSK biomedical systems are superior to those of ASK and FSK systems in terms of BER performance.

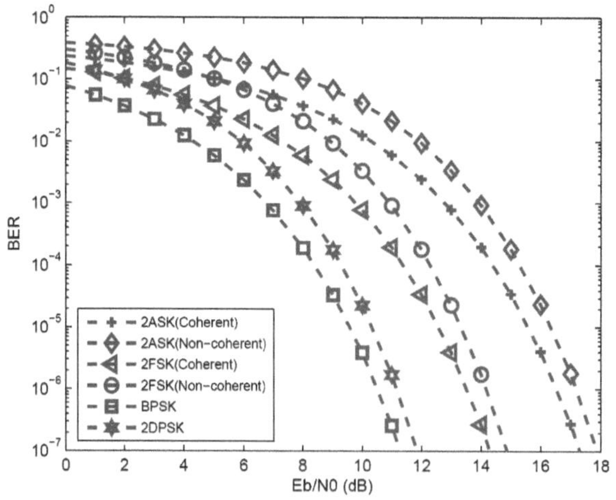

Figure 15.6. Simulated BER curves of different modulation schemes.

15.4 Design Example

A 256-channel epiretinal prosthetic system is illustrated in Fig. 15.7. Video images are captured by an external camera and encoded into bitstreams by an external signal-processing unit. The bitstreams are transferred to the implant. At the receiver side, data are demodulated and distributed to control the 256-channel stimulator at 60 frames per second.

In this epiretinal prosthetic system, the power and data telemetry links are realized under dual-band architecture, where power and data are transferred at two different frequency carriers. The 2 MHz carrier power telemetry link, including a Class-E-based power amplifier, inductively coupled coils, rectifiers, and regulators,

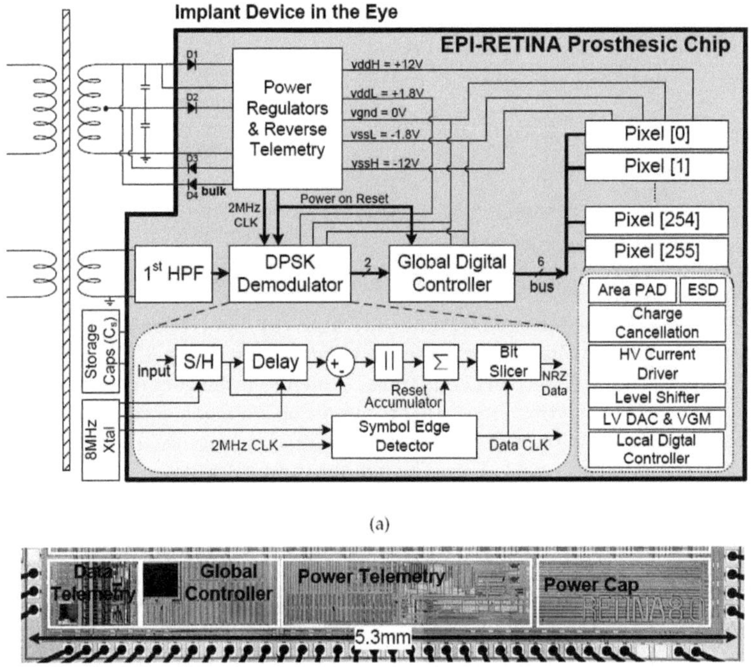

Figure 15.7. (a) Block diagrams of power and data telemetry in a 256-channel epiretinal prosthetic system [33]. (b) Chip photo of the power and data receivers.

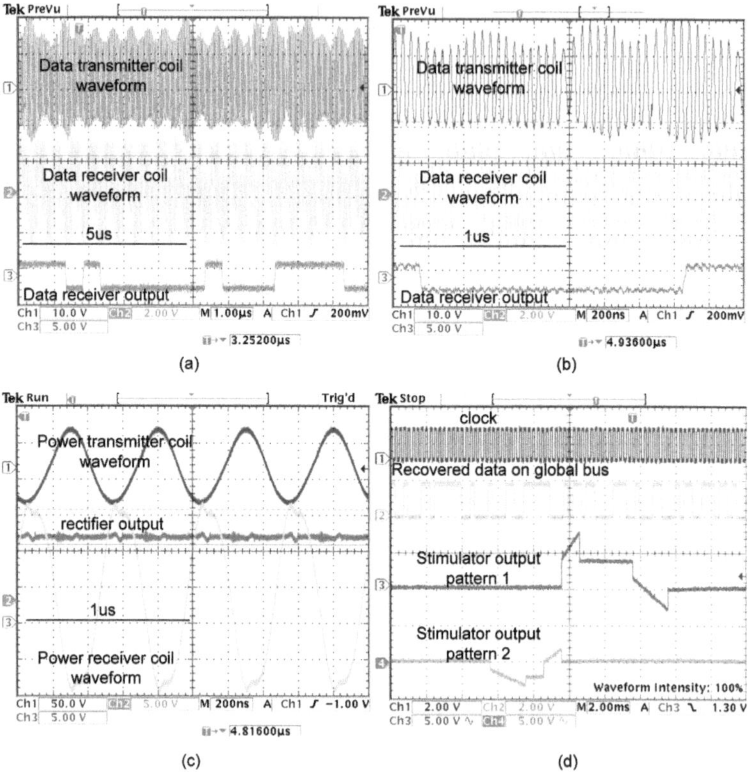

Figure 15.8. Measured results. (a) and (b) Data telemetry. Trace 1: the measured voltage across the transmitter coil; Trace 2: the measured voltage across the receiver coil; Trace 3: the measured data receiver output. (c) Power telemetry. Trace 1: the measured voltages across the power transmitter coil; Trace 2: the measured voltage across the power receiver coil (input to rectifier); Trace 3: rectifier output for powering secondary electronics; (d) output of two stimulator drivers that are powered and controlled by the power and data telemetry [33]. See also Color Insert.

generates in four supply voltages of ±12 V and ±1.8 V, in total 100 mW for the implant. The 22 MHz carrier data telemetry link is implemented based on a DPSK modulation scheme, which can tolerate power-to-data interference. The power regulators and the DPSK data receiver are integrated on a chip in 0.18 µm 32 V CMOS process and tested on the bench top. The bench-top test setup includes: an NI data acquisition card (Model: NI USB-6259), used as the signal source to generate 2 Mbps image bitstream for the full path demonstration; an FPGA (Model: Altera Flex10k), receives bitstream from the NI card and produces the DPSK data; and a 2MHz clock for external power and data transmitters. Power and data transmitters induce RF 2 MHz power carrier and modulated 22 MHz data carrier, which are inductively coupled to the implant side. A power receiver that consists of a tapped coil, resonant tank, rectifiers, and regulators harvests the RF energy and generates DC voltages. A data receiver that consists of a data coil, a DPSK demodulator, and an 8 MHz crystal oscillator decodes the 2 Mbps image bitstream from the modulated 22 MHz signal and further feeds to digital controllers. The 256-channel stimulator initiates stimuli based on decoded bitstream accordingly.

Figures 15.8a,b show the measured results of the data telemetry subsystem. Figure 15.8c illustrates the measured waveforms of the power telemetry subsystem. Both power and data coil pair are coaxial. Waveforms are measured at a coil separation of 2 cm. In Fig. 15.8d, traces 1 and 2 present the distributed system clock and the image bitstream on the global bus, while traces 3 and 4 show the output waveforms of two arbitrarily selected channels (the stimulator array is powered by the power telemetry and commanded by the data telemetry. $R_{elec} = 10\ \text{k}\Omega$ in series with $C_{elec} = 100$ nF are used to mimic the electrode-tissue interface in this test). The signal-to-interference ratio and BER of the DPSK demodulator are measured as -11 dB and 10^{-5} [33], respectively.

Acknowledgement

The authors thank funding provided by NSF through BMES-ERC, and DOE. The authors thank Dr. Guoxing Wang and Dr. Mehmet Yuce for technical discussions.

Appendix

A.1 *Equivalent AC Resistance*

The resistance and power dissipation of any conductor carrying DC current is usually calculated with the assumption of a uniform current density distribution. Any nonuniform current distribution, which is induced by the imposed magnetic field perpendicular to the current flow, can significantly increase the power dissipation. The term "skin effect" applies to nonuniformity in a single conductor, caused by the magnetic field induced by itself. "Proximity effect" is a nonuniformity caused by the magnetic fields of other conductors. In this section, we give formula to calculate power loss from skin effect and proximity effect for popular circular shape conductors.

$$P_{DC} + P_{SK} = \frac{P_{DC}\gamma}{2} \frac{ber\gamma\, bei'\gamma - bei\gamma\, ber'\gamma}{(ber'\gamma)^2 + (bei'\gamma)^2} \tag{A.1}$$

$$P_{PRO} = -\frac{2\sqrt{2}\pi r_s}{\sigma\delta} \frac{ber_2\gamma\, ber'\gamma + bei_2\gamma\, bei'\gamma}{(ber\gamma)^2 + (bei\gamma)^2} H^2_{peak}, \tag{A.2}$$

where P_{SK} and P_{PRO} are induced power dissipation from skin effect and proximity effect, ber_i and bei_i are the real and imaginary parts of a ith Bessel function, $\gamma = j\sqrt{j\mu\sigma\omega r}$ (ω is the angular frequency of the current), μ is the permeability of the conductor, σ is the conductivity of the conductor, H_{peak} is the magnitude of H field imposed by nearby conductors.

In the case that the radius of the cylindrical conductor is smaller or comparable to the skin depth, an approximation of Bessel functions allows the calculation of the AC resistance of one strand/turn

$$R_{SK} = 0.021(r_s/\delta)^4 R_{DC}, \tag{A.3}$$

$$P_{PRO} = \frac{\pi^3 r_s^4 \mu_0^2 \sigma f^2}{2} H^2_{peak}, \tag{A.4}$$

where r_s is the radius of an individual strand, R_{DC} is the DC resistance, and $\delta = \sqrt{2/\mu\sigma\omega}$ is the skin depth.

Depending on the number of turns, and how tightly the coil is wind, the contribution from skin effect and proximity effect to the total power varies. For coils of multiple turns and strands ($N_t N_s \gg 1$), proximity effect contributes much more power than skin effect does. Based on Eqs. A.3 and A.4, the AC resistance of a coil can be represented as

$$R_{AC} = R_{DC}[1 + 0.125 N_t N_s \eta\, \beta(\pi r_s^2 \mu_0 \sigma)^2], \tag{A.5}$$

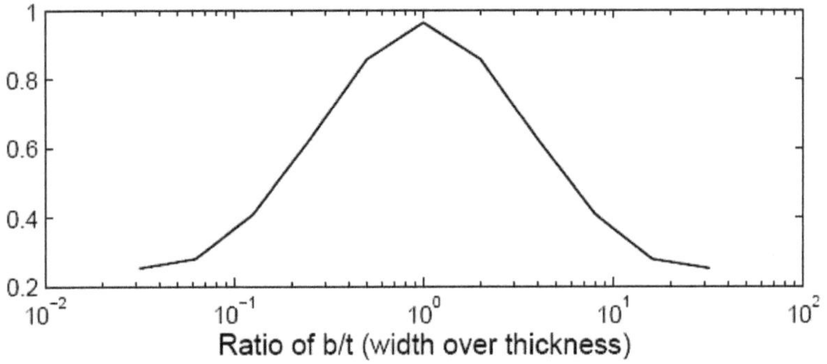

Figure A.1. Numeric values of η for coils with different cross sections. b and t are the coil's width and thickness.

where η is a parameter relates to coil winding cross section geometry (shown in Fig. A.1), β is the area efficiency defined as the ratio of the total conducting area over the cross section of the winding.

A.2 *Coil Model*

As illustrated in Fig. A.2, a coil is modeled as a distributive RCL network, where inductive and capacitive couplings between turns

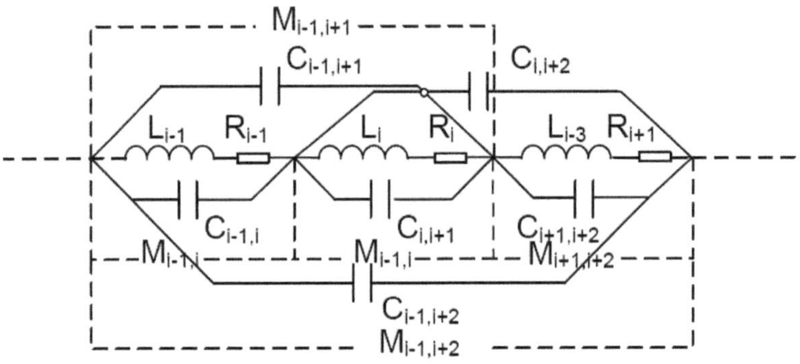

Figure A.2. Distributive equivalent model of a coil. I_{Li} represents the current going through the inductive branch L_i, R_i is the equivalent ESR of the inductive branch L_i, $I_{C\,p,k}$ denotes the mesh current through the parasitic capacitance $C_{p,k}$, I_e is the external driving current, and $M_{p,k}$ is the mutual inductance between turn p and turn k.

(p and k) are modeled as mutual inductance $M_{p,k}$ and parasitic capacitance $C_{p,k}$.

The inductive and resistive elements in Fig. A.2 can be described based on the former appendix and physics formulae

$$R_i = R_{i.DC}[1 + 0.125N_t N_s \eta\, \beta(\pi r_s^2 \mu_0 \sigma)^2]$$
$$L_i = 0.5\, \mu_0 D_i \ln(D_i/OD)$$
$$M_{p,k} = \int (\cos\theta/l) ds ds'$$

where D_i is the diameter of the conductor loop, OD is the diameter of a single turn, θ is the angle of inclination between the two loop elements ds and ds', and l is the radius vector between them.

The calculation of parasitic capacitance between turns is performed using the model shown in Fig. A.3. The turn-to-turn parasitic capacitor is formed by the insulation layers of turns and nonconductive air gap. Per unit angle, the turn-to- turn parasitic capacitance value is

$$C_t = \frac{0.5C_{gap}C_{insulate}}{C_{gap} + 0.5C_{insulate}}. \tag{A.6}$$

Under the condition that the insulation layer is thin compared with the turn radius, the parasitic capacitance between two turns is

$$C_{p,k} = \varepsilon_0 \varepsilon_r \int \frac{\pi D_i r_0}{\varsigma + \varepsilon_r r_0 (1 - \cos\theta) + 0.5\varepsilon_r h} d\theta, \tag{A.7}$$

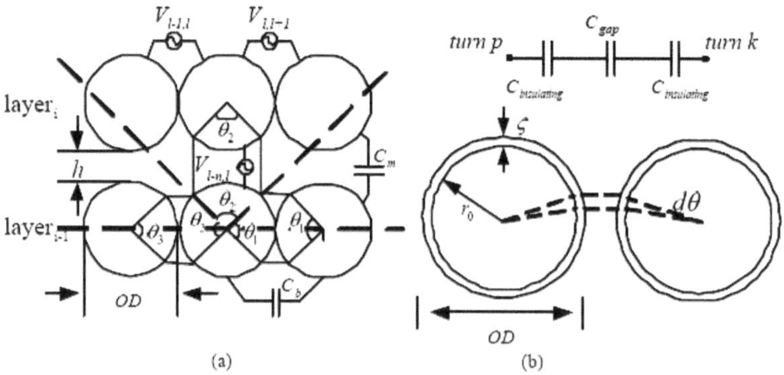

Figure A.3. Turn-to-turn parasitic capacitance. (a) Illustration of the parasitic capacitance in a coil winding. (b) Equivalent model to calculate the turn-to-turn parasitic capacitance.

where θ_e is the effective angle between turn p and turn k, and h is the separation of two turns. As a first order approximation, assume that each turn except those on the perimeter of a coil is surrounded by four turns and $\theta_1 = \theta_2 = \theta_3 = 90^\circ$.

Through analysis as shown in reference [23], the distributive coil model can be simplified as a lumped capacitor C_{self}, in parallel with and inductor, as

$$C_{self} = \sum_{p<k} C_{p,k}(k-p)^2/n_t^2 \tag{A.8}$$

Consequently, the equivalent impedance and Q factor of the coil is

$$Z_e = (j\omega L + R_{AC})||(1/j\omega C_{self}) = \frac{R_{AC} + j\omega L}{(1-\omega^2 LC_{self}) + j\omega R_{AC} C_{self}}, \tag{A.9}$$

$$Q(f) \approx 2\pi f L(1 - f^2/f_{self}^2)/R_{DC}(1 + f^2/f_h^2), \tag{A.10}$$

where $f_h = 2\sqrt{2}/\left(\pi r_s^2 \mu_0 \sigma \sqrt{\eta \beta N_t N_s}\right)$, $f_{self}^{-1} = 2\pi\sqrt{LC_{self}}$, and $R_{DC} = N_t(D_{out} + D_{in})/2\sigma\pi r_s^2 N_s$, as defined in previous sections.

References

1. Zhou, M., Yuce, M. R., and Liu, W. (2008) A Non-Coherent DPSK Data Receiver With Interference Cancellation for Dual-Band Transcutaneous Telemetry, *IEEE Journal of Solid-State Circuits*, **43(9)**, pp. 2003–2012.
2. Buchegger, T., Oberger, G., Reisenzahn, A., Hochmair, E., Stelzer, A., and Springer, A. (2005) Ultra-Wideband Transceivers for Cochlear Implants, *EURASIP Journal on Applied Signal Processing*, **2005(18)**, pp. 3069–3075.
3. Spelman, F. A. (May 1999) The Past, Present, and Future of Cochlear Prostheses, *IEEE Engineering in Medicine and Biology Magazine*, **18(3)**, pp. 27–33.
4. McDermott, H. (July 1989) An Advanced Multiple Channel Cochlear Implant, *IEEE Transactions on Biomedical Engineering*, **36(7)**, pp. 789–797.

5. Liu, W., Vichienchom, K., Clements, M., DeMarco, S. C., Hughes, C., McGucken, E., Humayun, M. S., de Juan, E., Weiland, J. D., and Greenberg, R. (2000) A Neuro-Stimulus Chip With Telemetry Unit for Retinal Prosthetic Device, *IEEE Journal of Solid-State Circuits*, **35(10)**, pp. 1487–1497.
6. Weiland, J. D., Humayun, M. S. (July 2008) Visual Prosthesis, in *Proceedings of IEEE,* **96(7)**, pp. 1076–1084.
7. Theogarajan, L., Wyatt, J., Rizzo, J., Drohan, B., Markova, M., Kelly, S., et al. (February 2006) Minimally Invasive Retinal Prosthesis, *IEEE ISSCC Digest of Technical Papers*, **2(5)**, pp. 99–100.
8. Ortmanns, M., Rocke, A., Gehrke, M., Tiedtke, H. J. (December 2007) A 232-Channel Epiretinal Stimulator ASIC, *IEEE Journal of Solid-State Circuits*, **42(12)**, pp. 2946–2959.
9. Rothermel, A., Liu, L., Aryan, N. P., Fischer, M., Wuenschmann, J., Kibbel, S., et al. (January 2009) A CMOS Chip With Active Pixel Array and Specific Test Features for Subretinal Implantation, *IEEE Journal of Solid-State Circuits*, **44(1)**, pp. 290–300.
10. Schwartz, G. D., and Barkin, J. S. (2007) Small-Bowel Tumors Detected by Wireless Capsule Endoscopy, *Digestive Diseases and Sciences*, **52(4)**, pp. 1026–1030.
11. Song, Y., Borton, D. A., Park, S., Patterson, W. R., Bull, C. W., Laiwalla, F., Mislow, J., Simeral, J. D., Donoghue, J. P., and Nurmikko, A. V. (2009) Active Microelectronic Neurosensor Arrays for Implantable Brain Communication Interfaces, *IEEE Transactions on Neural Systems and Rehabilitation Engineering,* **17(4)**, pp. 339–345.
12. Chestek, C. A., Gilja, V., Nuyujukian, P., Kier, R. J., Solzbacher, F., Ryu, S. I., Harrison, R. R., and Shenoy, K. V., Hermes, C. (August 2009) Low-Power Wireless Neural Recording System for Freely Moving Primates, *IEEE Transactions on Neural Systems and Rehabilitation Engineering*, **17**, pp. 330–338.
13. Lin, C. T., Ko, L. W., Chiou, J. C., Duann, J. R., Huang, R. S., Liang, S. F., Chiu, T. W., and Jung, T. P. (July 2008) Noninvasive Neural Prostheses Using Mobile and Wireless EEG, *Proceedings of the IEEE,* **96(7)**.
14. Low, K. S., Lee, G. X., and Taher, T. (2009) A Wearable Wireless Sensor Network for Human Limbs Monitoring, *IEEE Instrumentation and Measurement Technology Conference*, pp. 1332–1336.
15. Harrison, R., Kier, R. J., Chestek, C. A., Gilja, V., Nuyujukian, P., Ryu, S., Greger, B., Solzbacher, F., and Shenoy, K. V. (August 2009) Wireless

Neural Recording With Signal Low-Power Integrated Circuit, *IEEE Transactions on Neural Systems and Rehabilitation Engineering*, **17**, pp. 322–329.

16. Akin, T., Najafi, K., and Bradley, R. M. (January 1998) A Wireless Implantable Multichannel Digital Neural Recording System for a Micromachined Sieve Electrode, *IEEE Journal of Solid-State Circuits*, **33(1)**, pp. 109–118.
17. Mohseni, P., and Najafi, K. (2003) A Wireless FM Microsystem for Biomedical Neural Recording Applications, in *Proceedings of IEEE Southwest Symposium on Mixed-Signal Design*, pp. 217–222.
18. Chae, M., Yang, Z., Yuce, M. R., Hoang, L., and Liu, W. (2009) A 128-Channel 6 mW Wireless Neural Recording IC With Spike Feature Extraction and UWB Transmitter, *IEEE Transactions on Neural Systems and Rehabilitation Engineering*, **17**, pp. 312–321.
19. Sodagar, A. M., Wise, K. D., and Najafi, K. A. (2009) Wireless Implantable Microsystem for Multichannel Neural Recording, *IEEE Transactions on Microwave Theory and Techniques*, **57**, pp. 2565–2573.
20. Chow, E. Y., Ouyang, Y., Beier, B., Chappell, W. J., and Irazoqui, P. P. (2009) Evaluation of Cardiovascular Stents as Antennas for Implantable Wireless Applications, *IEEE Transactions on Microwave Theory and Techniques*, **57**, pp. 2523–2532.
21. Kelly, S. K., Shire, D. B., Doyle, P., Gingerich, M. D., Drohan, W. A., Rizzo, J. F., Theogarajan, L. S., Chen, J., Cogan, S. F., and Wyatt, J. L. (2009) The Boston Retinal Prosthesis: A 15-Channel Hermetic Wireless Neural Stimulator, 2nd *International Symposium on Applied Sciences in Biomedical and Communication Technologies (ISABEL 2009)*, pp. 1–6.
22. Sivaprakasam, M., Liu, W., Wang, G., Zhou, M., Weiland, J. D., and Humayun, M. S. (2006) Challenges in System and Circuit Design for High Density Retinal Prosthesis, *IEEE/NLM Life Science Systems and Applications Workshop (LSSA)*, pp. 1–2.
23. Yang, Z., Liu, W., and Basham, E. (October 2007) Inductor Modeling in Wireless Links for Implantable Electronics, *IEEE Transactions on Magnetics*, **43**, pp. 3851–3860.
24. Yu, H., and Najafi, K. (2003) Low-Power Interface Circuits for Bio-Implantable Microsystems, *IEEE ISSCC Digest of Technical Papers*, **1**, pp. 194–487.
25. Wise, K. D., Anderson, D. J., Hetke, J. F., Kipke, D. R., and Najafi, K. (2004) Wireless Implantable Microsystems: High-Density Electronic

Interfaces to the Nervous System, *Proceedings of the IEEE*, **92(1)**, pp. 76–97.

26. Tang, Z., Smith, B., Schild, J. H., and Peckham, P. H. (1995) Data Transmission From an Implantable Biotelemeter by Load-Shift Keying Using Circuit Configuration Modulator, *IEEE Transactions on Biomedical Engineering*, **42(5)**, pp. 524–528.
27. Young, D. J. (2009) Wireless Powering and Data Telemetry for Biomedical Implants, in 31st *IEEE EMBS Annual International Conference*, pp. 3221–3224.
28. Ghovanloo, M., and Najafi, K. (2004) A wideband Frequency-Shift Keying Wireless Link for Inductively Powered Biomedical Implants, *IEEE Transactions on Circuits and Systems*, **51(12)**, pp. 2374–2383.
29. Jung, L. H., Byrnes-Preston, P., Hessler, R., Lehmann, T., Suaning, G. J., and Lovell, N. H. (2007) A Dual Band Wireless Power and FSK Data Telemetry for Biomedical Implants, in 29th *IEEE EMBS Annual International Conference*, pp. 6596–6599.
30. Troyk, P. R., and DeMichele, G. A. (2003) Inductively Coupled Power and Data Link for Neural Prostheses Using a Class-E Oscillator and FSK Modulation, in 25th *IEEE EMBS Annual International Conference*, **4**, pp. 3376–3379.
31. Xu, W., Luo, Z., and Sonkusale, S. (2009) Fully Digital BPSK Demodulator and Multilevel LSK Back Telemetry for Biomedical Implant Transceivers, *IEEE Transactions on Circuits and Systems*, **56(9)**, pp. 714–718.
32. Hu, Y., and Sawan, M. (2005) A Fully Integrated Low-Power BPSK Demodulator for Implantable Medical Devices, *IEEE Transactions on Circuits and Systems*, **52(12)**, pp. 2552–2562.
33. Chen, K., Yang, Z., Hoang, L., Weiland, J. D., and Humayun, M. S. (September 2010) An Integrated 256-Channel Epiretinal Prosthesis, *IEEE Journal of Solid-State Circuits*, **45(9)**, pp. 1946–1956.
34. Wang, G., Liu, W., Sivaprakasam, M., and Kendir, A. (October 2005) Design and Analysis of an Adaptive Transcutaneous Power Telemetry for Biomedical Implants, *IEEE Transactions on Circuits and Systems*, **52(10)**, pp. 2109–2117.
35. Proakis, J. G. (1995) *Digital Communications*, McGraw Hill, Singapore.
36. Vaughan, R. G., Scott, N. L., and White, D. R. (September 1991) The Theory of Bandpass Sampling, *IEEE Transactions on Signal Process*, **39(9)**, pp. 1973–1984.

37. Buchegger, T., Ossberger, G., Reisenzahn, A., Hochmair, E., Stelzer, A., and Springer, A. (2005) Ultra-Wideband Transceivers for Cochlear Implants, *EURASIP Journal on Applied Signal Processing*, **18**, pp. 3069–3075.
38. Porcino, D., and Hirt, W. (2003) Ultra-Wideband Radio Technology: Potential and Challenges Ahead, *IEEE Communications Magazine*, **41**, pp. 66–74.
39. Siwiak, K. (2001) Ultra-Wide Band Radio: Introducing a New Technology, in *IEEE VTS 53rd Vehicular Technology Conference*, **2**, pp. 1088–1093.

Chapter 16

Ultra Wideband for Wireless Body Area Networks

Mehmet Rasit Yuce[a,b] and Ho Chee Keong[b]

[a]*Department of Electrical and Computer Systems Engineering, Monash University, Clayton, VIC 3800, Australia*

[b]*School of Electrical Engineering and Computer Science, The University of Newcastle, Callaghan, NSW, 2308 Australia*

mehmet.r.yuce@ieee.org

Wireless technologies based on narrow bands, especially 2.4 GHz ISM, have been used in medical monitoring widely. This chapter discusses the use of ultra-wide band technology (UWB) for medical monitoring systems. UWB presents some unique benefits when compared with the narrow-band systems: (i) low-power transmitter design, (ii) low radio frequency (RF) and electromagnetic interference (EMI) effects in medical environment, (iii) small-size antenna, and (IV) high data rate. Especially the UWB has efficiently been used for simultaneous monitoring of many continuous physiological signals such as ECC (electrocardiogram) /EEG (electroencephalogram) and neural signals for brain-computer interfaces because of its high data rate capability.

Wireless Body Area Networks: Technology, Implementation, and Applications
Edited by Mehmet R. Yuce and Jamil Y. Khan

ISBN 978-981-4316-71-2 (Hardcover), ISBN 978-981-4241-57-1 (eBook)
www.panstanford.com

16.1 Introduction

Currently, there is no wireless standard for a wireless body area network (WBAN). Devices designed by different manufacturers utilize different frequency bands and protocols. These resulted in a lack of interoperability among the different WBAN solutions. Therefore, standardization is required to ensure interoperability. In November 2007, IEEE 802.15 TG6 was formed to develop a communication standard optimized for body area network [1]. The goal of this task group is to standardize both the physical (PHY) and media access control (MAC) layers for a short-range, low-power, and highly reliable wireless communication scheme to operate in or near human body. Several wireless bands such as MedRadio, Medical Implant Communication Service (MICS), industrial, scientific, and medical (ISM), Wireless Medical Telemetry System (WMTS), and UWB are considered for WBAN standardization. Low data rate UWB technology can find applications in biomedical monitoring especially for multichannel continuous signal monitoring such as EEG, ECG, EMG (Electromyography), and neural signals. Herein we will take a closer look at how UWB can be applied to some of WBAN applications.

UWB is a narrow pulse-based transmission system whose spectrum is spread across a wide range of frequencies. Low data rate UWB is attractive for vital signs monitoring systems because its transmission power is lower than that of WLAN, Bluetooth, and ZigBee systems, which is less likely to affect human tissue and cause interference to other medical equipment. Furthermore, it is able to transmit higher data rate, which is suitable for real time continuous monitoring of many physiological signals. In this part of the book, we will discuss the design of sensor node electronics using the low data rate UWB transmission and its characteristics when operating in or around human body.

16.1.1 *Background of UWB*

In 2002, the Federal Communications Commission (FCC) approved the use of UWB wireless scheme. UWB is a narrow pulse baseband system, whose spectrum is spread across a wide range of frequency.

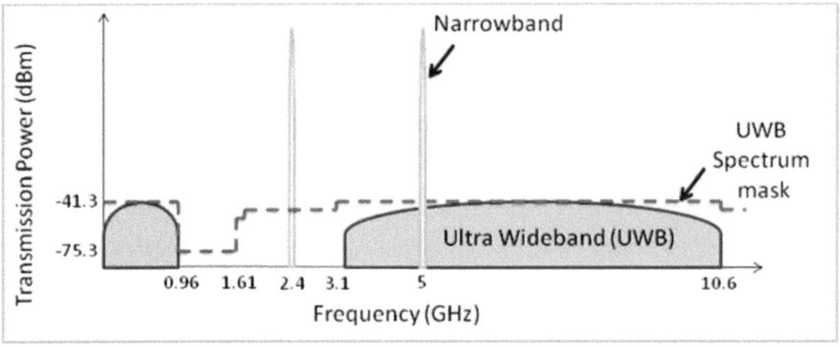

Figure 16.1. UWB and narrowband spectrum. See also Color Insert.

A generic definition of UWB is that the emitting signals have a fractional bandwidth larger than 0.2 or at least 500 MHz. The fractional bandwidth is defined as the bandwidth between the −10 dB lower and upper corner frequency. UWB is allowed to operate in 0–960 MHz and 3.1–10.6 GHz bands, but the effective isotopic radiated power has to be kept below −41 dBm/MHz [2]. The major difference between an UWB signal and a narrowband signal is that an UWB signal occupies a large bandwidth with a transmission power below the noise floor of narrowband system. Figure 16.1 shows the spectrum mask of an UWB signal that highlights the major differences between a wideband and a narrowband system.

Basically, there are two types of UWB system: the multiband carrier-based UWB system and the impulse radio UWB system (IR-UWB). A multiband system is supported by WiMedia, and it uses the OFDM (orthogonal frequency-division multiplexing) modulation scheme with the spectrum divided into bands of 528 MHz. The power consumption of multiband UWB system is high due to the complex real-time signal processing required for the OFDM modulation and demodulation. The power consumption of WiMedia compatible chip sets is typically more than 500 mW; therefore, these are not suitable for battery-powered devices. Impulse radio systems send very short pulses and use simpler modulation schemes such as pulse-position modulation (PPM) and on-off keying (OOK), which result in significant power saving and, therefore, more suitable for battery-operated WBAN applications.

In March 2007, a low data rate UWB standard (IEEE 802.15.4a) based on impulse UWB was approved [3]. This standard is developed mainly for high precision ranging application. Therefore, it is not suitable to be directly applied to a WBAN application. Modifications are required in the PHY and MAC layers in order to match the WBAN requirements.

A WBAN system requires a cross layer solution. In order to design an efficient WBAN system, it is important to consider the hardware design, PHY layer, and the MAC protocol. In this chapter, the advantages and limitations of UWB, UWB sensor node design, UWB PHY, and UWB MAC layers developments for WBAN applications will be covered.

16.2 Advantages and Limitations of UWB for WBAN

In a WBAN system, real-time data acquisition and analysis is needed; therefore, the ability to sustain higher data rate and provide an efficient communication protocol is essential. A WBAN system is not only limited to medical usage, it can also support personal entertainment applications. Therefore, a scalable data rate from 10 bps to 10 Mbps is required. A WBAN sensor node for medical usage can be wearable or implantable. Therefore, ultra-low power sensor node is extremely important, especially for the case of an implantable node. As WBAN system is designed to operate near or in human body, special absorb ratio (SAR) is an important factor to be considered. This will translate to a need for using lower transmission power. The basic requirements for a WBAN are as follows [4]:

- Supports different types of sensors, both inside and outside body
- Supports scalable data rates from 10 bps to 10 Mbps
- Data from the sensor nodes need to be secure and reliable
- Ultra-low power nodes
- Coexistence with existing networks
- Low transmission power
- Small form factor

16.2.1 *Favorable Factors for Use of UWB in WBAN Applications*

Table 16.1 shows the power consumption of various wireless sensor platforms used for a WBAN application. Energy per bit for UWB platform is much lower as compared to the other narrowband wireless platforms. Therefore, it is an advantage to use the UWB wireless scheme for WBAN applications considering the transmitter power consumption.

Table 16.2 shows the parameters of the commonly used wireless schemes for WBAN applications. A WBAN system needs to support scalable data rates up to 10 Mbps; only the UWB and WLAN wireless schemes can meet this requirement. UWB is preferred to WLAN, as a WLAN system is much more power hungry as compared to UWB.

In terms of coexistence, Bluetooth, ZigBee, and WLAN operate in the 2.4 GHz ISM band, which is overcrowded and, therefore, has a higher probability of interfering with the existing wireless

Table 16.1. Power consumption for various wireless sensor platforms [4]

Model	Company	Frequency	Data rate	Current	
				Tx	Rx
UWB	University of Newcastle	3.1–10.6 GHz	10 Mbps	15 mA	–
Mica2 (MPR400)	Crossbow	868/916 MHz	38.4 Kbps	27 mA	10 mA
MicAz	Crossbow	2.4 GHz	250 Kbps	17.4 mA	19.7 mA
Mica2DOT	Crossbow	433 MHz	38.4 Kbps	25 mA	8 mA
CC1010	Texas Instrument	300–1000 MHz	65 Kbps	26.6 mA	11.9 mA
CC2400	Texas Instrument	2.4 GHz	1 Mbps	19 mA	23 mA

Table 16.2. Wireless scheme comparison table

Wireless scheme	Max data rate	Frequency	Max EIRP
Ultra Wideband	> 10 Mbps	3.1–10.6 GHz	−41dBm
Bluetooth	3 Mbps	2.4 GHz	0 dBm
ZigBee	250 Kbps	2.4 GHz	0 dbm
WLAN (802.11b)	11 Mbps	2.4 GHz	30 dBm
MICS	300 kbps	402–405 MHz	−16 dBm

infrastructure. UWB and MICS wireless scheme are less likely to cause interference to the existing wireless infrastructure as UWB operates below the noise floor of existing wireless system, while MICS is a dedicated frequency band for medical usage.

From Table 16.2, UWB's max effective isotropically radiated power (EIRP) is the lowest as compared to other wireless schemes. The low transmission power reduces the chances of heating effects on human tissue and increases the difficulty level for unauthorized signal tapping.

Another added advantage of an UWB system is hardware simplicity for the design of UWB transmitter, which in turn will lead to a more compact sensor node design. As described so far, it is obvious that there are important favorable factors for the use of UWB wireless scheme in a WBAN application.

16.2.2 *Limitations of UWB*

A major challenge faced in the design of UWB is its receiver complexity required to demodulate the received low-power level signal and the narrow pulses. In wireless communication systems, there are basically two types of receivers: coherent receiver and noncoherent receiver. Coherent receivers synchronize the receiver clock (i.e., local oscillator) with the transmitter clock and the receiver generates template waveforms locally. A noncoherent receiver does not require the synchronization between the transmitter and the receiver, and does not generate a local template waveform. For noncoherent receiver, detection is commonly achieved by means of energy detection or using the reference waveform generated from the transmitted signal. In general, noncoherent receivers have a simpler implementation, but are penalized on performance; however, this may not be the case for UWB. An UWB coherent receiver requires a very high speed ADC with a large analog input bandwidth and needs to achieve precise synchronization. These two factors are critical for a reliable operation of a coherent receiver, which is very difficult to achieve due to the extremely narrow pulse nature (typically less than 2 ns) of an UWB system. As a result of an imperfect synchronization in case of a coherent UWB receiver, significant performance degradation will occur.

The power consumption of an UWB receiver is much higher than that of a UWB transmitter whether it is coherent or non-coherent. High power consumption is not acceptable for a power limited sensor nodes.

16.3 UWB Hardware Development

The basic hardware components of any wireless scheme are the antenna, transmitter, and receiver. The large bandwidth characteristic of an UWB signal has posed a new set of problems for hardware designers, which is different from what they faced in a narrowband system. To name a few differences, the antenna requires a flat pass band of a few gigahertz, the transmitter employs direct modulation techniques, and the receiver now has to sample and synchronize very narrow pulses. In this section, we will take a look at some of the UWB hardware components developed for WBAN applications.

16.3.1 *UWB Antennas for WBAN Applications*

UWB operates in the gigahertz frequency range, which makes it easier to achieve an antenna with smaller form factor. This is due to the fact that antenna size is relatively proportional to the wavelength. As shown in Eq. 16.1, wavelength reduces as frequency increases.

$$\text{Wavelength, } \lambda = \frac{c}{f} \qquad (16.1)$$

where c is the speed of light (3×10^8m/s) and f is the frequency.

In recent years, several compact antennas that are suitable for use in WBAN applications are proposed. Textile antenna [5] and button antenna [6] are such examples that are designed to seamlessly integrate into the clothing for on-body WBAN applications. The textile antenna is only 0.5 mm thick; it uses a highly flexible and conductive metalized nylon fabric as the conductor and an acrylic fabric as dielectric substrate. The button structure used in the button antenna is constructed using polytetrafluoroethylene (PTFE) material. The PTFE button structure is fixed to a metal disc. The metal disc is connected to a 50 Ω micro-strip line with a ground plane at the bottom.

Placement of the antenna on the human body plays an important role in determining the performance of a WBAN system. Typically, an antenna is placed at least 10 mm away from the body. If the antenna is placed very close to the body, the signal will be severely attenuated. Distance between the body and the antenna only affects the signal strength; there is no significant impact on the pulse shape and does not cause pulse spreading [5].

Another factor that should be taken into account when selecting the appropriate antenna for WBAN application is the beam width. A directional antenna is able to achieve higher gain at the expense of a smaller angle of coverage, while an Omni antenna covers a wider area but with a lower antenna gain. A directional antenna with 60° beam width is sufficient, if both transmit and receive antennas are located on the torso of the body. If the transmit or receive antenna is positioned on a moving part of the body such as arm or head, an Omni antenna would be a better choice.

Antenna designed for use in an implantable device should take dielectric permittivity and conductivity of the surrounding tissues into consideration. The implantable environment will cause the antenna operating frequencies to shift and this would affect the impedance matching for the frequency band of interest.

Apart from the normal design consideration for designing a typical antenna used in free space condition, four additional factors should be considered when designing an antenna for a WBAN application.

(1) How can it be seamlessly integrated into clothing?
(2) Effect of body on the radiation pattern.
(3) What is the minimum distance between the antenna and the body?
(4) Where will the antennas be located? (i.e., whether it is mostly static or there is much movement)

16.3.2 *UWB Transmitters for WBAN Applications*

Basically there are two methods of generating an UWB pulse: transmission line method and CMOS-based methods. UWB monocycle pulse can be generated by passing a square wave through a step

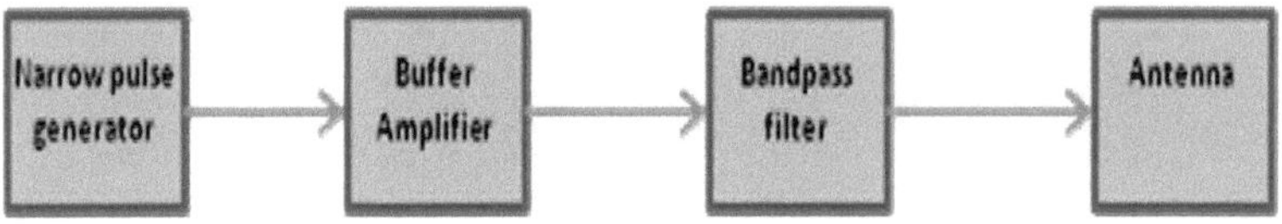

Figure 16.2. Block diagram of a carrier-less UWB transmitter for WBAN application.

recovery diode (SRD) and a pair of shorted circuit transmission line [7, 8]. But this method requires constant biasing of the diode and, therefore, is not suitable for low-power WBAN applications. Furthermore, the size of the transmitter designed using transmission line method is much larger than that of a CMOS UWB transmitter. An alternative method of implementing UWB transmitter is using CMOS integrated circuit (IC) technology. CMOS-based UWB transmitters are more energy efficient and have much smaller form factors, which is more suited for WBAN applications.

CMOS-based transmitter solutions suitable for use in WBAN have been proposed in reference [9–11]. The UWB transmitter proposed in reference [9] uses a narrow pulse generator followed by a buffer amplifier as shown in Fig. 16.2. The pulse generator is able to generate pulses with pulse width ranging from 50 ps to 500 ps. Spectral shaping is achieved by means of an external filter and an antenna. A CMOS UWB transmitter for neural recording systems has been proposed in reference [10]; the working principle of the UWB transmitter is similar to that of reference [9]. In reference [11], a carrier-based CMOS UWB transmitter for WBAN applications is proposed. The block diagram for the transmitter proposed in reference [11] is shown in Fig. 16.3. The triangular pulse determines the bandwidth, while the oscillator determines the center frequency (it is similar to a carrier frequency used in narrowband radio systems). Here the triangle-pulse generator can also be replaced with a narrow rectangular-pulse train generator. An UWB transmitter for ECG monitoring assembled using off-the-shelves CMOS components is documented in reference [12]. The working principle of the transmitter is similar to that of references [9] and [10], which uses direct modulation technique (carrier-less based).

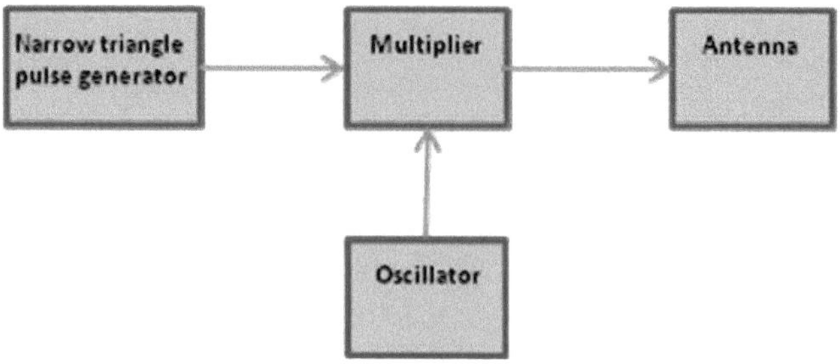

Figure 16.3. Block diagram of carrier-based UWB transmitter for WBAN application.

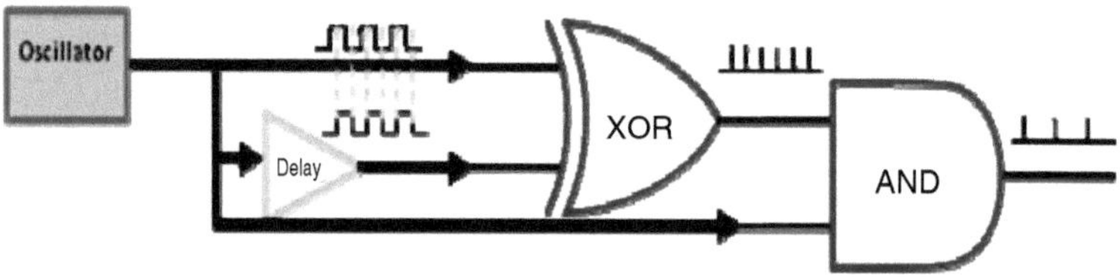

Figure 16.4. Narrow UWB pulse generation. See also Color Insert.

The basic building block for most CMOS-based UWB transmitters suitable for WBAN applications is formed using either a narrow rectangular pulse train generator or a triangle pulse generator. Figure 16.4 shows an example of how a narrow rectangular pulse can be generated in a CMOS circuit using a delay unit [12]. The AND is used to allow the narrow pulse at rising edges only. The triangular pulse can be achieved by using inverter together with NAND or NOR gates as shown in Fig. 16.5 [13]. Here the inverter causes a very short delay, which results in the formation of triangular pulse; the pulse width depends largely on the rise and fall time of the input signals.

An UWB transmitter using direct modulation does not require an oscillator, which reduces the size of the transmitter and allows lower power consumption. Therefore, it is more suitable for WBAN applications as compared to carrier-based UWB transmitter. For a carrier-less UWB transmitter, selection of the pulse width, rise and fall time will affect the performance of the system, as they determine the bandwidth and the peak and null of the UWB spectrum. Through proper selection of the pulse width, rise and fall time, it is possible

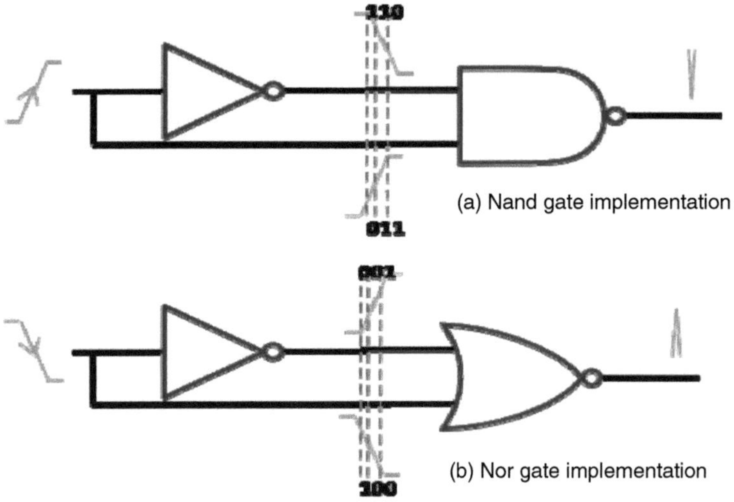

Figure 16.5. Triangle pulse generation. See also Color Insert.

to enhance the bit error rate and reduce the power consumption of an UWB WBAN system. The effects of the narrow rectangular pulse parameters on the UWB spectrum are illustrated in the next section.

16.3.2.1 Effects of pulse width on UWB spectrum

Equation 16.2 shows the Fourier series expansion of a rectangular pulse train [14]. In order to satisfy the FCC's spectrum requirement, the signal must be band limited, which can be described by Eq. 16.3 [12]. Equation 16.3 is based on an ideal pulse with zero rise and fall time. A realistic pulse with finite rise and fall time can be represented by Eq. 16.4. Finite rise time adds an additional null to the circuit as shown in Fig. 16.6. The pulse width for both spectrums in Fig. 16.6 is selected such that it is optimized to operate at center frequency of 4 GHz, with the spectrum peak at 4 GHz and nulls at 3 GHz and 5 GHz. The red spectrum has a rise time of 100 ps, while the blue spectrum has a rise time of 250 ps. It can be seen from Fig. 16.6 that the null of blue spectrum due to rise time appears at 4 GHz, which severely degrades the performance of the UWB transmitter. Null due to rise/fall time occurs every $1/t_r$, and null arising from the pulse width appears every $(\tau + t_r)$ interval.

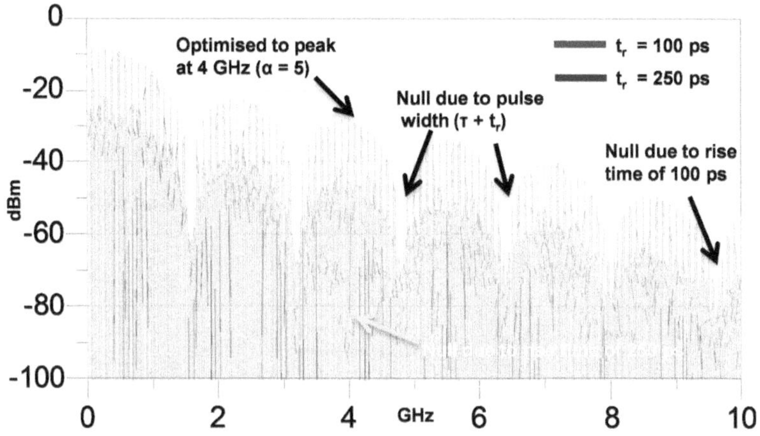

Figure 16.6. Spectrum of an UWB rectangular pulse in frequency domain and nulls in the UWB spectrum due to rise time. See also Color Insert.

Therefore, the format of the transmitted pulse should be arranged to ensure that the null does not appear at the frequency band of interest. It is desirable to keep the spectrum peak of the sinc envelope at the center frequency. The effect of null in the band of interest for a bandlimited signal is shown in Fig. 16.7. In the worst case scenario, a null at the center of the band of interest will result

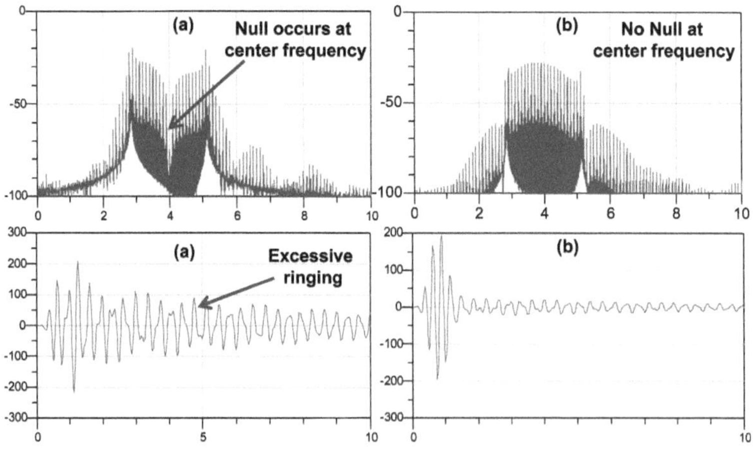

Figure 16.7. Effect of null in a band-limited signal. See also Color Insert.

in excessive ringing, which will limit the data rate and increases the noise floor.

$$x\,(t) = \frac{A\tau}{T_b} + \frac{2A\tau}{T_b} \sum_{k=1}^{\infty} \frac{\sin\,(\pi k\tau\,/\,T_b)}{\pi k\tau\,/\,T_b} \cos\,(k\omega\, t) \tag{16.2}$$

$$x\,(t) = \frac{2A\tau}{T_b} \sum_{k=n1}^{n2} \frac{\sin\,(\pi k\tau\,/\,T_b)}{\pi k\tau\,/\,T_b} \cos\,(k\omega\, t) \tag{16.3}$$

$$x\,(t) = \frac{2A\,(\tau + t_r)}{T_b} \sum_{k=n1}^{n2} \frac{\sin\,(\pi k t_r\,/\,T_b)}{\pi k t_r\,/\,T_b} \frac{\sin\,(\pi k\,(\tau + t_r)\,/\,T_b)}{\pi k\,(\tau + t_r)\,/\,T_b} \cos\,(k\omega\, t) \tag{16.4}$$

where A is the pulse amplitude, T_b is the period of the clock, τ is the pulse width, t_r is the rise time, f_c is the centered frequency, $\omega = 2\pi/T_b$, $n_1 = \omega_1/\omega n_2 = \omega_2/\omega$, ω_1 and ω_2 are the lower and upper cutoff frequencies of the band-pass filter.

It is important to select the appropriate rise time and pulse width to ensure optimized performance for an UWB transmitter. Rise time can be optimized using Eq. 16.5, where 0.5 corresponds to a 4 dB loss. The peak of the spectrum (i.e. the sinc envelope) can be put at the center frequency by using Eq. 16.6, where α represents the side lobes, "1" is the main lobe, and "3" is the first side lobe. Optimizing the nulls and peak of the spectrum enhances the performance of the transmitter, which is achieved by adjusting the pulse width.

$$f_c\, t_r < 0.5 \tag{16.5}$$

$$f_c\,(\tau + t_r) = 0.5\,\alpha, \quad where\ \alpha = 1, 3, 5, 7, 9, \ldots \tag{16.6}$$

where f_c is the center frequency, τ is the pulse width and t_r is the rise time.

Impulse UWB transmitter without carrier is the simplest and most popular form of UWB transmitter used in WBAN applications. When using this type of UWB transmitter, proper selection of pulse width and rise/fall time is required to ensure an optimized performance.

16.3.3 *UWB Receiver*

As seen from the previous section, the UWB transmitter design is simple, provides low power consumption, and is able to support a very high data rate. Typically, an UWB receiver is much more complex as compared to an UWB transmitter, and the receiver power consumption is much higher than that of the UWB transmitter. As discussed in Section 16.2, most of the favorable factors for using UWB in WBAN applications lie in the transmitter portion, while most of the limitations are found in the UWB receiver portion. In this section, we will look at the UWB receiver for use in WBAN applications.

Figure 16.8 shows the block diagram of the UWB receiver front-end designed for a WBAN application [9] using energy detection principle. The input signal is first amplified by low-noise amplifier (LNA), followed by an optional fixed-gain amplifier before the self-mixing stage. The self-mixer is a square-law device that translates the high frequencies received signal to baseband signals. The baseband signal is further amplified by a variable-gain amplifier before been fed to the integrator bank. The integrator bank is made up of eight integrators, each having a minimum integration time of 8 ns. Therefore, the maximum data rate allowed for this system is 15.625 Mb/s. Demodulation is performed at the output of the integrator bank using a comparator with a threshold set by a 5-bit digital-to-analog converter (DAC). Synchronization and bit detection are performed using the field-programmable gate array (FPGA).

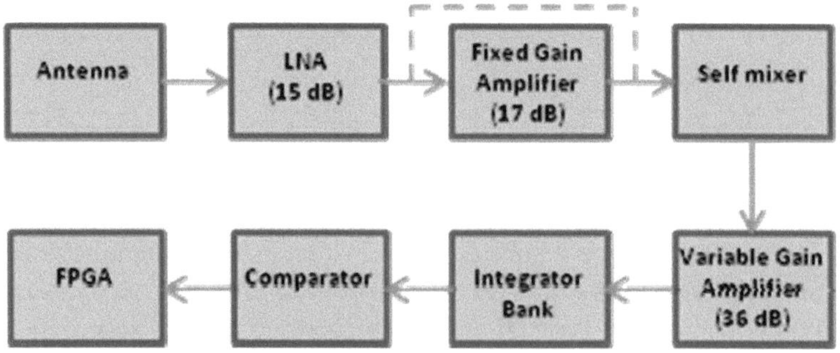

Figure 16.8. Block diagram of an UWB receiver [9].

The key objectives to be considered when designing an UWB receiver for WBAN applications are to keep the circuitry simple and the power consumption low. The most power-hungry component in an UWB receiver front-end is the analog-to-digital converter (ADC), especially for the case of a fully digital UWB receiver. For a fully digital UWB receiver, the ADC is placed after the LNA before the analog down-conversion. An ADC with sampling rate and input bandwidth of a few gigahertz is required for a fully digital UWB receiver, and these would result in significant power consumption, which is not suitable for battery-powered applications. Therefore, the receiver proposed in reference [9] chooses to place the comparator after the analog integrator bank, where the comparator is only required to sample at baseband data rate (tens of MHz). The second highest power consumption component in an UWB receiver front-end is template generation circuitry. A matching template is difficult to generate, and synchronizing the received signal with the generated template is very challenging, thus increasing the power consumption and complexity of the receiver. Therefore, self-mixing energy detection method is selected in reference [9] for implementing a suitable UWB receiver for WBAN applications.

Although a noncoherent energy detection receiver may not perform as well as a coherent UWB receiver or may not have the flexibility of a fully digital receiver, it is, however, the best UWB receiver choice for a WBAN application. The choices of antenna, UWB transmitter, and receiver have been highlighted in this section. The goal for hardware design of a WBAN application is to keep the circuit small, simple, and the power consumption low.

16.4 PHY Layer for UWB WBAN

A WBAN system can be broadly classified into two groups: wearable BAN and implantable BAN. Wearable BAN can be for both medical usage and entertainment purpose. Implantable BAN is mainly for medical applications. As the environment that wearable and implantable BAN are operating in is very different, it has been suggested to use two separate PHY layers for implantable and wearable BAN [15].

Table 16.3. UWB PHYs proposed for IEEE 802.15.6

Organization	Signal source	Modulation	Data rate	Frequency band (MHz)
ETRI-Samsung	IR	GPPM	100 k~10 Mbps	3.1~4.6 GHz 7.25~8.5 GHz
FT-CEA-Thales	IR	PPM, DBPSK	1.95~27.24 Mbps	3.1~10.6 GHz
IMEC	IR	BPM, OOK, DBPSK	100 k~27.2 Mbps	3~10 GHz
TI	IR	PPM	500 Kbps	6.4~8.9 GHz
NICT-YNU-Meiji	IR/ Chirp	BPM/ DMPSK	200 k~10 Mbps	3.1~10.6 GHz
CSEM	FM	FSK-FM	30 ~ 250 Kbps	6~9 GHz

Several UWB WBAN PHY layer solutions have been proposed to IEEE 802.15.6 task group. The proposed UWB PHYs are largely variations of IEEE 802.15.4a. Impulse radio-based UWB is the most popular choice of signal generation method [16]. Table 16.3 shows the proposed PHY specifications by the different organizations compiled by reference [16].

16.5 UWB WBAN Channel

Based on the UWB channel model provided by IEEE 802.15.4a, it can be seen that no energy is penetrating the body, instead energy is diffracting around the body. Therefore, the distance between the transmitter and the receiver is the distance along the body perimeter instead of direct distance between the transmitter and the receiver. Based on measurement data obtained from IEEE 802.15.4a, there are always two clusters of multipath components for the WBAN channel. The first cluster is the signal diffracting around the body, and the second cluster is due to the reflection of the ground. Table 16.4 shows the channel model for four scenarios defined by IEEE 802.15.6 task group for WBAN applications. Only CM3 and CM4 channel models are defined for UWB. Path loss and power delay profile (PDP) models for UWB are provided for both CM3 and CM4 scenarios. Log-normal distribution can be used to describe the path amplitude for the static case. As for the nonstatic case, where the object has large movement, the path amplitudes fit well to Weibull distribution [17]. The path arrival time is modeled by Poisson distribution.

Table 16.4. Channel models for WBAN

Scenarios	Description
CM1	Implantable to Implantable
CM2	Implantable to Body Surface/External
CM3	Body Surface to Body Surface
CM4	Body Surface to External

For the implantable case, reference [18] has developed a model based on the time domain electromagnetic simulation using the anatomy model of a human body. According to reference [18], the path loss for inside the body can be modeled as a power function as shown in Eq. 16.7. It also concludes that it is possible to improve the communication link from inside the body to body surface by ensuring the transmitted signal is polarized along the width of the body instead of polarizing along the height of the body.

$$P(d) = a\, d^{\,b} + a_1 \tag{16.7}$$

where d is the distance, $a = -2.7$ and $b = 0.7$ for the horizontally polarized antenna, and $a = -3$ and $b = 0.6$ for the vertically polarized antenna.

16.6 MAC scheme for UWB WBAN

Designing a MAC protocol for WBAN is challenging, as a single protocol needs to satisfy two different applications, each with a very unique set of requirements. Medical and personal entertainment applications have different requirements on the MAC layer. Medical applications typically have lower data rate, and the traffic is periodic. Data reliability and low latency are criteria for life criteria medical applications. Normally, the data rate for personal entertainment applications is higher and busty in nature. Data reliability and latency requirements are not as stringent as that of medical applications. In this section, we will look at a few MAC protocols designed for WBAN applications and discuss whether it is suitable for UWB PHY layer.

Most of the MAC schemes designed for WBAN are a modification from the IEEE 802.15.4 MAC scheme. An IEEE 802.15.4 network

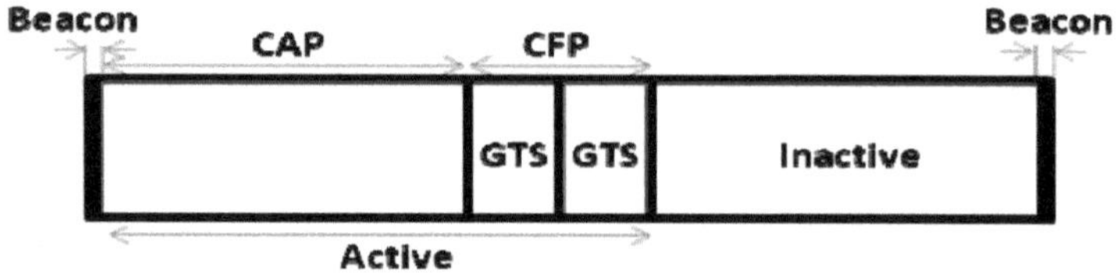

Figure 16.9. IEEE 802.15.4 super-frame.

can operate in beacon and nonbeacon modes. In the beacon mode, a periodic beacon is broadcasted by the personnel area network (PAN) coordinator. The period between two consecutive beacons is the super-frame structure. A super-frame structure is made up of an active and an inactive period as shown in Fig. 16.9. The active period consists of a contention access period (CAP) and a contention free period (CFP). CSMA-CA mechanism is used to access the channel during the CAP period, and reservation of guarantee time slot (GTS) is used for channel access for the CFP portion. In order to reserve a GTS slot, the node has to send a GTS request during the CAP period using CSMA-CA, and the network coordinator decides the GTS allocation. This MAC scheme is not entirely suitable for WBAN applications as it permits only up to seven GTS, which does not provide sufficient channel access for time critical WBAN data. Furthermore, quality of service (QoS) could not be ensured because in order for the node to obtain a GTS, it needs to be able to send a GTS request during the CAP period. Nonbeacon mode is not considered for a WBAN application because in this mode the network is always active and, therefore, is not energy saving [19].

A priority-guaranteed MAC protocol for WBAN application is proposed in reference [20]. In this scheme, there are two contention-free periods; CFP1 is for timeslot reserved for periodic traffic (TSRP) and CFP2 is for timeslot reserved for burst traffic (TSRB). TSRP is mainly used for periodic medical signal, while TSRB is catered for burst data package transmitted by personal entertainment devices. Two separate control channels AC1 and AC2 are used for medical and entertainment devices, respectively. Randomized slotted ALOHA is the mechanism used for accessing the control channel in the CAP period. The advantages of this scheme are that there are two separate control channels for medical and entertainment

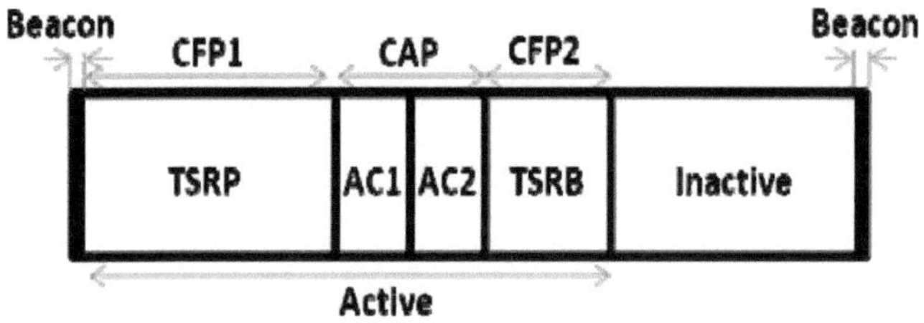

Figure 16.10. Priority-guaranteed MAC [20].

applications, therefore ensuring that the QoS of medical application is not affected during the period where there is high traffic for entertainment devices. Furthermore, all the data are transmitted in the dedicated contention-free slot and, therefore, is able to ensure low latency. In a multiuser scenario, nodes on different users are not synchronized; collision is, therefore, unavoidable, and this would significantly degrade the performance of the scheme. From an UWB hardware perspective, constantly listening for the beacons for synchronizing purposes would require extensive use of an UWB receiver, which is not preferred as this leads to higher power consumption.

A TDMA (time division multiple access)-based MAC scheme, the BodyMAC, is proposed in reference [21]. This scheme utilizes sleep mode to enhance the energy efficiency of the network. The frame structure of BodyMAC is shown in Fig. 16.11. As most of the traffic is from the nodes to the coordinator, a shorter duration is allocated for downlink. The uplink portion occupies majority part of the frame. CSMA-CA mechanism is the mechanism used for channel access in the CAP zone. Typically, nodes in a WBAN system have low duty cycle; therefore, a good sleep-mode mechanism can greatly enhance the energy efficiency of the system but at the expense of flexibility. In this scheme, in order for the nodes to enter into sleep mode, the nodes should send a request to the gateway using CSMA-CA during the CAP period. The nodes will enter sleep mode when an ACK (acknowledgment or acknowledged) is received from the gateway. In the sleep mode, the nodes will not be able to listen to beacons, broadcast information, and receive data. But these can transmit data to the gateway if a GTS has been allocated. GTS is

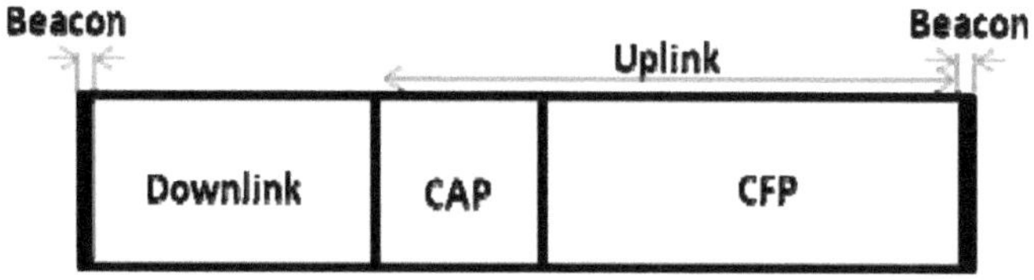

Figure 16.11. BodyMAC frame structure.

used as the synchronization mechanism when the node is in sleep mode. After the sleeping node successfully transmitted a data packet in the allocated GTS, the gateway will send an ACK back to the node containing timing information. The node will then adjust its time according to the received timing parameter. This is to ensure that the node remains synchronized when the node wakes up after a long period of sleep. Sleep mode will be terminated and re-synchronization process will be activated if the ACK is not received correctly. BodyMAC solves the problem of excessive listening, which is not preferred due to UWB hardware limitation. But the use of CSMA-CA mechanism for UWB PHY will significantly degrade the reliability of the system due to the low signal-to-noise ratio (SNR) nature of an UWB signal.

A transmit-only (Tx-Only) MAC scheme for UWB WBAN is used in reference [22]. The main goal of this scheme is to overcome the hardware limitation of UWB technology to allow higher energy efficiency and to tackle the problem faced in the multiuser scenario. In this scheme, there is no synchronization between the sensor nodes and the gateway, therefore eliminating the need for a power-hungry UWB receiver in the sensor node. The data is transmitted at a much higher data rate to achieve extremely low duty cycle, which aids in reducing the energy consumption and reduces the chances of data packet collisions. In this scheme, each sensor is allocated a fixed transmission interval as shown in Fig. 16.12. Through careful selection of the transmission interval (T_{int}) and transmission timeslot (T_{slot}), it is possible to minimize the collision. The probability of collision is deterministic although the transmission start time is a random variable. The main factors affecting the probability of collision (P_C) are T_{slot}, T_{int}, and the number of sensors on a user. The average collision probability for

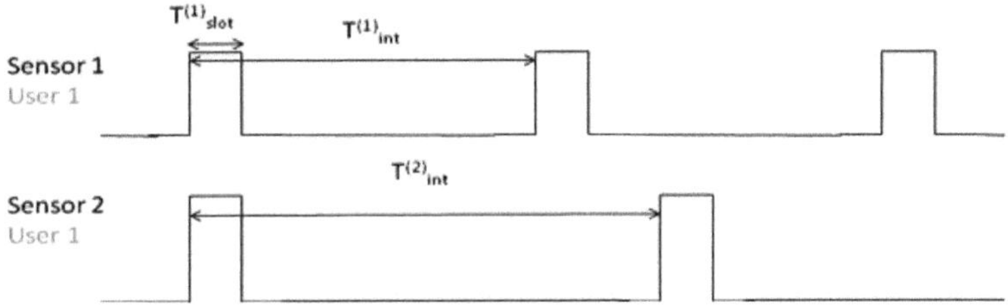

Figure 16.12. Tx-Only transmission interval.

sensor M, as shown in Eq. 16.8, is given in reference [23].

$$P_c(M) = 1 - \prod_{j \in m \setminus \{M\}} \left(1 - \frac{T_{\text{slot}}^{(j)}}{T_{\text{int}}^{(j)}}\right) \tag{16.8}$$

The Tx-Only MAC scheme also deals with a multiuser scenario by using different pulse rate for different user as shown in Fig. 16.13. As shown in Fig. 16.13a and Fig. 16.13b, sensors on user 1 uses the same pulse rate, while user 2 uses a different pulse rate (Fig. 16.13c). Figure 16.13d shows the received signal seen at user 1 receiver, when both sensor 1 from user 1 and user 2 transmit at the same time. As illustrated in Fig. 16.13e, the receiver can easily differentiate signals from another user, as the sampling interval between each

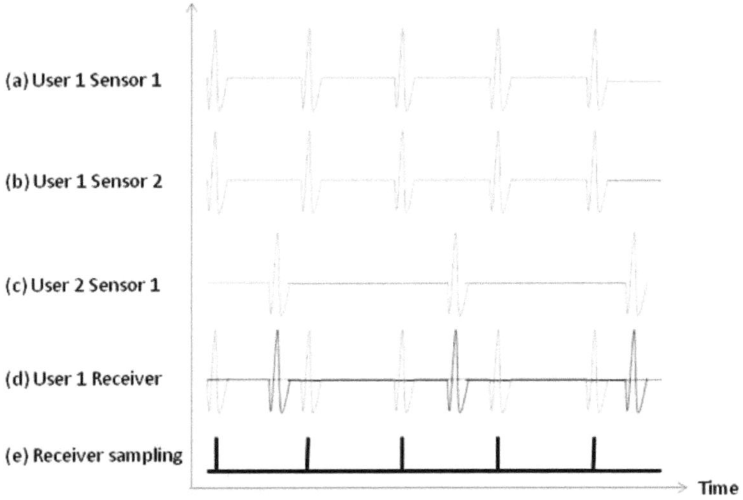

Figure 16.13. Different pulse rate for Tx-Only MAC. See also Color Insert.

pulse is known to the receiver. The receiver is usually the control unit (gateway device) carried by each user in a WBAN application.

Another characteristic of this system is that different number of pulses is allocated to the different sensors on the same user. The assignment of the number of pulses is dependent on the distance between the sensors and the receiver. Transmission power management is the main reason for assigning different number of pulses to the sensors; it is to ensure sufficient link budget is achieved. The different number of pulses per data bit can also be used as an identifier for the sensors.

The advantages of Tx-Only MAC are energy efficiency and circuit simplicity and have accounted for multiuser scenario. The major disadvantage of this MAC is that there is not feedback channel, where signal reliability could be compromised.

16.7 UWB WBAN Applications

In this section, four different medical applications for both on-body and implantable scenarios using the UWB wireless scheme will be discussed.

16.7.1 *Eight-Channel ECG (On-Body)*

An eight-channel UWB recording system for monitoring of multiple continuous ECG and EEG signals is designed in reference [12] using commercial-off-the-shelves (COTS) components. The system is depicted in Fig. 16.14. Signals from the electrodes are transferred to the transmitter through the front-end circuit and is processed by a microcontroller. The processed data is transmitted wirelessly UWB signals to the receiver (i.e., base station). A noncoherent receiver is used to demodulate the data. Figure 16.15 shows the photograph of the complete system prototype.

Typically ECG and EEG signals have amplitude of less than 500 μV with frequency less than 100 Hz. The front-end of the UWB transmitter uses an instrumental amplifier (INA321) [12]. The input ECG signal is amplified by 60 dB together with a low-pass filter with a cutoff frequency at 100 Hz.

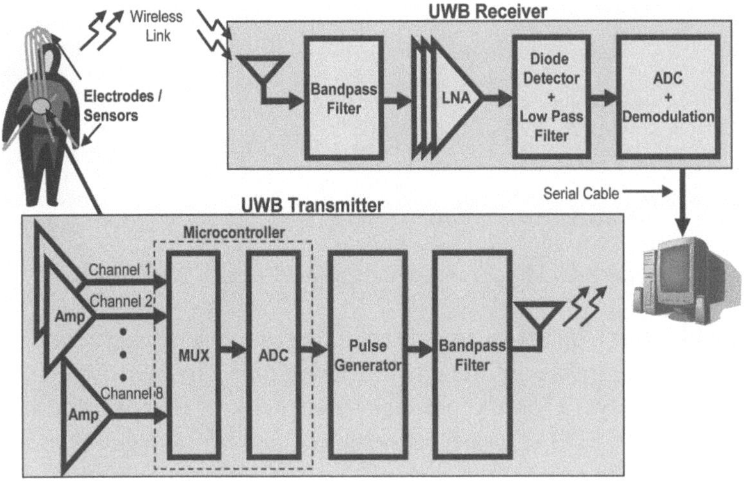

Figure 16.14. Multichannel UWB monitoring system block diagram. See also Color Insert.

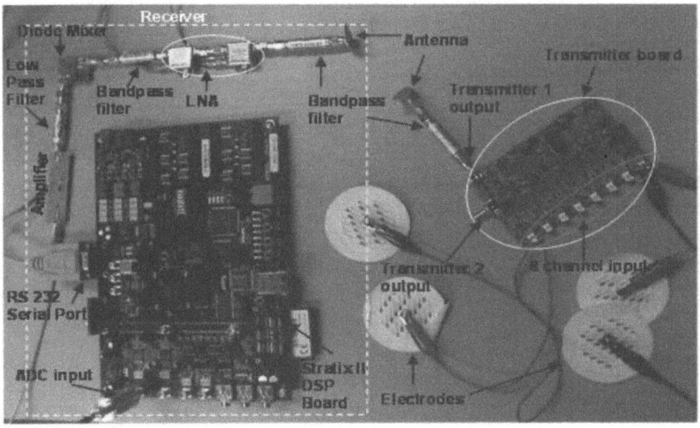

Figure 16.15. Photograph of complete system prototype. See also Color Insert.

The microcontroller operates with an external 20 MHz crystal. Each analog front-end output is connected to one of the dedicated pins for the ADC module. The 10-bit ADC is programmed to sample at 4.8 KHz with each channel sampling at 600 Hz. The 10-bit data from each sample is separated into two bytes as shown in Fig. 16.16. Low byte contains six least significant data bits (LSB) and two

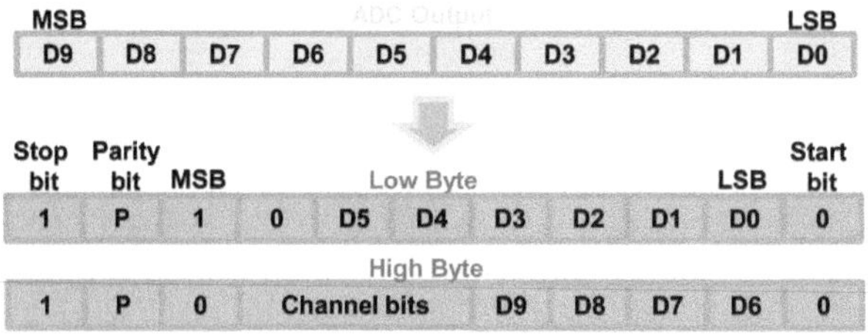

Figure 16.16. Transmit data frame. See also Color Insert.

header bits. High byte contains the four most significant data bits (MSB), three channel bits and one header bit. The MSB for the low byte is always "1" and "0" for high byte. The three channel bits range from "000" to "111", each representing one of the eight channels. High and low bytes are sent to the pulse generator through the universal asynchronous receiver/transmitter (UART) transmit pin, where a start bit, parity bit and stop bit is added. Output data rate can range from 1.2 Kbps to 1.25 Mbps. The maximum data rate is limited to 1.25 Mbps because this is the highest UART baud rate that the microcontroller can support with a 20 MHz crystal in our current prototype.

16.7.1.1 UWB pulse generators

Two UWB pulse generators for this monitoring system have been incorporated onto the PCB board. The block diagram of the two pulse generators and the generated pulses are depicted in Fig. 16.17. A programmable oscillator (LTC6905) that produces an output frequency from 17 MHz to 170 MHz with rise and fall time of 0.5 ns is a common component for both transmitters.

Transmitter 1 has a rise/fall time of 0.5 ns, which is determined by the clock generated via LTC6905. A buffer (SN74AHC1G125) with propagation delay of 8 ns is used in the delay line. LTC6905 is configured to generate a clock signal of 60 MHz, which corresponds to a half duty cycle period of 8.33 ns. The clock is adjusted slightly to account for the variation in the delay line. In the ideal case, an UWB pulse less than 500 ps will be generated, but due to finite rise/fall time, an UWB pulse between 1 to 2 ns is achieved with the current setup. As shown in Fig. 16.17c, a narrow UWB pulse

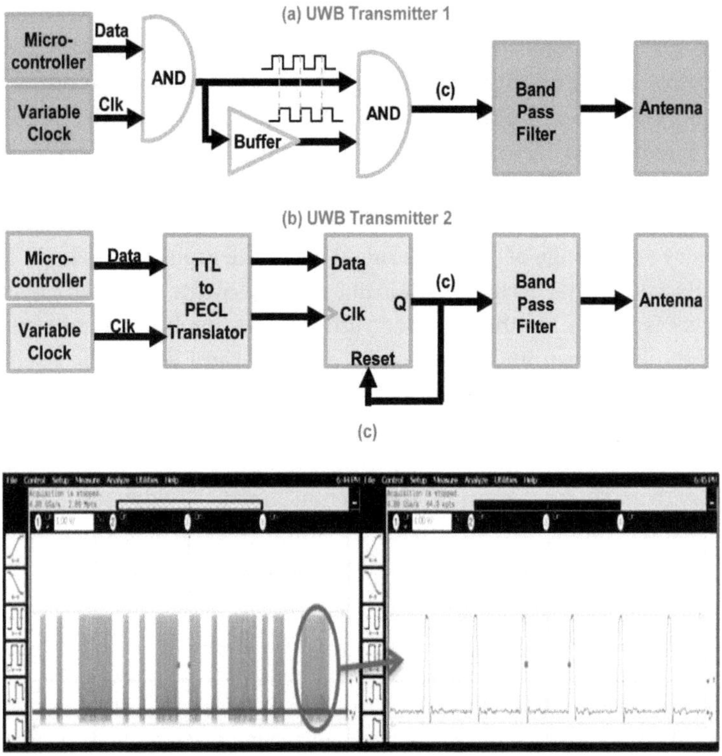

Figure 16.17. UWB transmitter block diagrams: (a) gate-based narrow pulse generation, (b) flip-flop-based narrow pulse generation, and (c) generated UWB pulse. See also Color Insert.

is generated by performing an AND (NC7SZ08) operation of the generated pulse train with its delayed version. Additional variation in the actual delay (or pulse width) may come from the delay path because of parasitic capacitances of the PCB trace or variation from the buffer propagation delay specifications. In our proposed design, the variation can be controlled by adjusting the clock rate. Another advantage of using a clock signal is that it helps to make the pulse width adjustable, which is used to optimize the spectrum shapes.

The second UWB transmitter (transmitter 2) is designed using positive emittercoupled logic (PECL), which has a rise time of 130 ps. Fast rise time is required to ensure more transmission power at high frequency as derived from Eq. 16.4. The transistor–transistor logic (TTL) signals from the microcontroller and oscillator clock are converted to PECL logic using SY10ELT22LZG and the output is fed a D-flipflop (MC10EP31) (Fig. 16.17b). The narrow UWB pulses are generated by connecting the output of the D-flipflop to the asynchronous reset pin. The propagation delay from reset to output is 400 ps, which defines the pulse width. One pulse is generated at every rising edge of the clock signal when the input data is high. The pulse repetitive period (PRF) is determined by the clock. As the PRF increases, the number of spectrum lines drops but the amplitude of the spectrum lines increases. Changing the PRF can be used to control the output power level to suit the different environment and to support a wide range of output data rate.

Both transistors use a 4 GHz bandpass filter to shape the spectrum to meet the FCC's spectrum mask. A wideband antenna designed for 3 to 5 GHz with an antenna gain of 2 dBi is used. Effective isotopic radiated power transmitted is −45 dBm.

16.7.1.2 UWB receiver front-end

The received signal is passed through a bandpass filter centered at 4 GHz with 1 GHz bandwidth to remove the narrowband interfering signals. Three LNAs, each with a gain of 14 dB is used to amplify the signal, which gives a total gain of 42 dBm. The amplified signal goes through a diode detector, which translates the high frequency components to their low frequency components. The recovered low frequency components are passed through a lowpass filter to form an envelope of the data signal. The signal is further amplified by an op-amp circuit by 60 dB before sending to the ADC. Figure 16.18 shows the recovered signal pattern from the transmitted UWB modulated signal (Note that there is a time delay between the transmitted and recovered signals due to the wireless transmission).

16.7.1.3 Data recovery

The recovered data is converted to digital signal using a 12-bit ADC embedded in the Stratix II development board. The demodulation is carried out with a FPGA chip running at 100 MHz. The receiver is

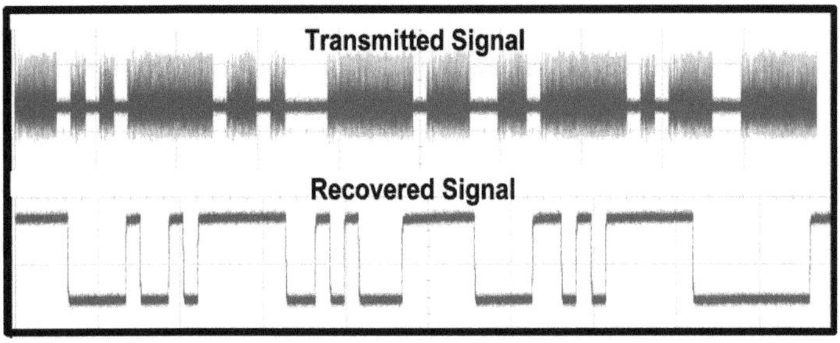

Figure 16.18. Recovered data from an OOK UWB modulated signal. See also Color Insert.

capable of demodulating data of 50 Mbps. The demodulation process is performed in two stages. In the first stage, the demodulator waits for the start signals to arrive, which is a low bit. Once the low bit is received, the demodulator samples the start bit until half of the period before sending a signal to start the demodulation clock. In the second stage, the demodulation clock goes high once the start signal is received. The first bit of data is sent to the serial port, which is the start bit. The demodulation clock runs at the same rate as the input data and samples the data at the middle of the bit. After the demodulation clock has generated 11 clock cycles, it goes low and waits for the start signal from stage 1.

ECG signals are transmitted and received wirelessly using the UWB pulses. The result is displayed using MATLAB in Fig. 16.19 on a remote computer. The signal is corrupted by the 50 Hz noise as can been seen in the waveform obtained from the oscilloscope before transmitting (Fig. 16.19a), after receiving and monitoring in MATLAB in time (Fig. 16.19b) and the frequency domain (Fig. 16.19c). The signal is passed through a 50 Hz digital notch filter designed using a MTLAB program. The 50 Hz noise is successfully removed and the ECG signal recovered. Removing the 50 Hz noise at the PC instead of the receiver helps to reduce the complexity and the programming power required at the receiver. The whole measurement has been carried out in a lab where there were other wireless standards (e.g., Wi-Fi) and equipment operating. The ECG signal has successfully been monitoring without any error.

The demodulated signal is transmitted through the serial port to the PC. A program written using Visual Basic is also developed

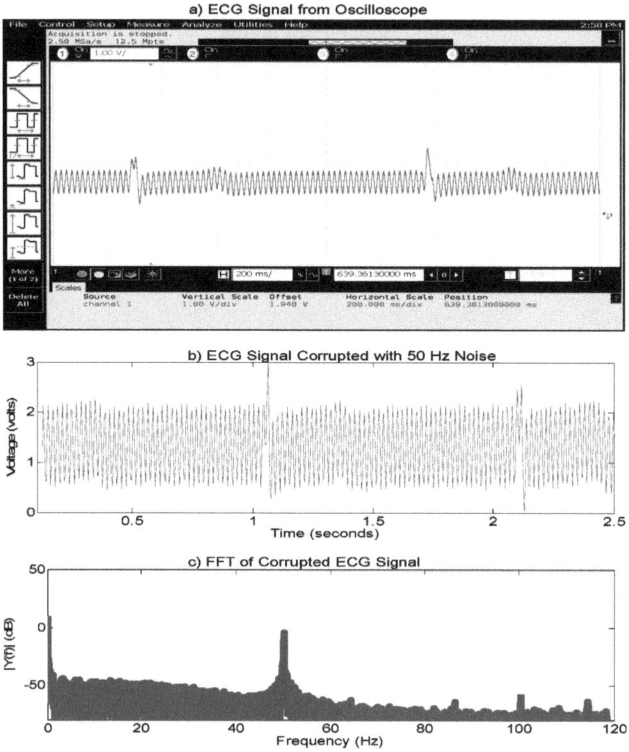

Figure 16.19. Monitored ECG waveforms wirelessly using UWB transmission. See also Color Insert.

to display the received signals on the screen. Parity bit check is performed on the received data to ensure that all data are received correctly. Once the received data is decoded, it is formatted back into a 10-bit word and separated based on the information embedded in the channel bits (Fig. 16.16). Digital filtering is performed on the received signal to remove the 50 Hz noise, which comes from the power supply. The ECG signal in Fig. 16.20 is successfully monitored. The graphical user interface program can display any eight channels by changing the button "channel selection" shown in the window.

16.7.2 *Implantable UWB WBAN*

Medical telemetry can be categorized into two groups: high data rate and low data rate systems. For example, the wireless endoscope and

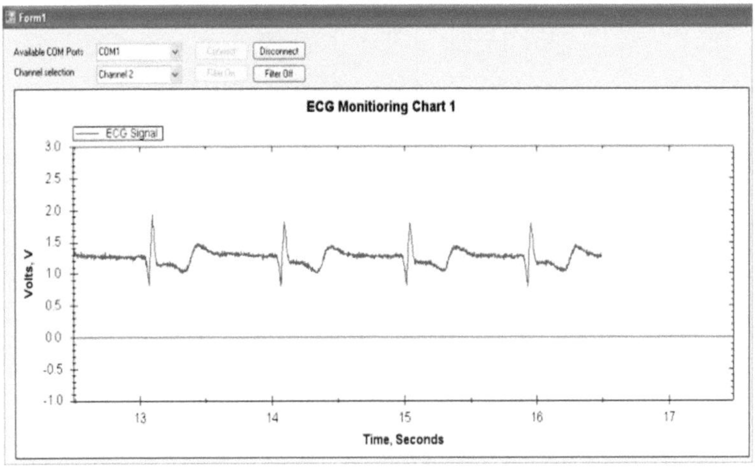

Figure 16.20. Multichannel ECG signal.

the retina prosthesis are the recent implanted systems requiring a large amount of data that should be delivered to outside or into the body. Multi implantable neural recording systems and multichannel monitoring of continuous signals such as ECG and EEG also necessitate a high data rate communication. As an example, scientists aim to achieve the recording of more than 100 channels in order to simultaneously record brain functions; a data rate more than 20 Mbps is required [14, 26]. A similar figure is also useful for wireless endoscope implants for higher resolution pictures and images. The MICS has channels with 300 KHz width and thus cannot provide such data rates.

An endoscope capsule requires high data rate to increase the resolution and image quality. For high data rate and short-range applications of this kind, UWB communication is an ideal physical layer solution by achieving a data rate equal or higher than 100 Mbps.

16.7.2.1 Multichannel neural recording systems

UWB is suitable for this application as the number of neuron channels been monitored increases, the data rate will reach a speed of 100 Mbps and higher [26]. Figure 16.21 shows the block diagram of a multichannel neural recording system. A general multichannel

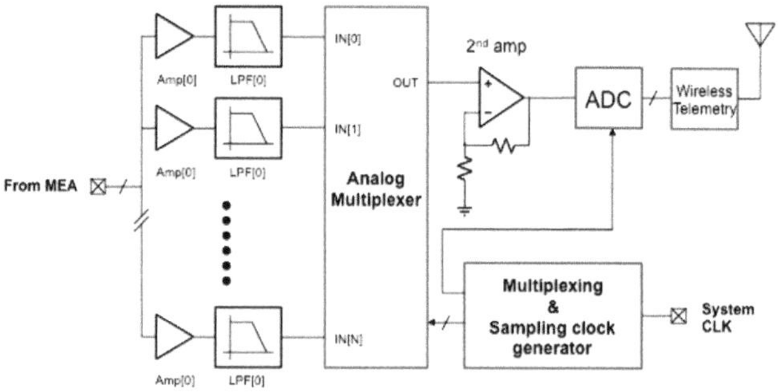

Figure 16.21. A multichannel neural recording system with UWB transmitter [14]. See also Color Insert.

wireless neural recording system consists of preamplifiers, filters, an analog multiplexer, a 2nd amplifier, an ADC, and a wireless telemetry.

16.7.2.2 Electronic pills (wireless endoscope)

A small miniaturized electronic pill can reach areas such as small intestines and deliver real-time video images wirelessly to an external console. Figure 16.22 shows a wireless endoscope (i.e., electronic pill) for a medical monitoring system. The device travels through the digestive system to collect image data and transfers them to a nearby computer for display with a distance of 1 m or more. A high resolution video-based capsule endoscope produces a large amount of data, which should be delivered over a high capacity wireless link.

Table 16.5 summarizes the commercially available electronic pill technologies that are already been used in clinical environments. One of the current state-of-the-art technologies for electronic pills is the commercially available PillCam by the company "Given Imaging" [27]. The pill uses Zarlink's RF chip [28] for wireless transmission based on the MICS band. The allowable channel bandwidth for this band is only 300 kHz. It is difficult to assign enough data rate for the high quality video data at the moment for real-time monitoring. It is quite obvious that there is a need for higher bandwidth data transmission for electronic pills that could facilitate a better diagnosis.

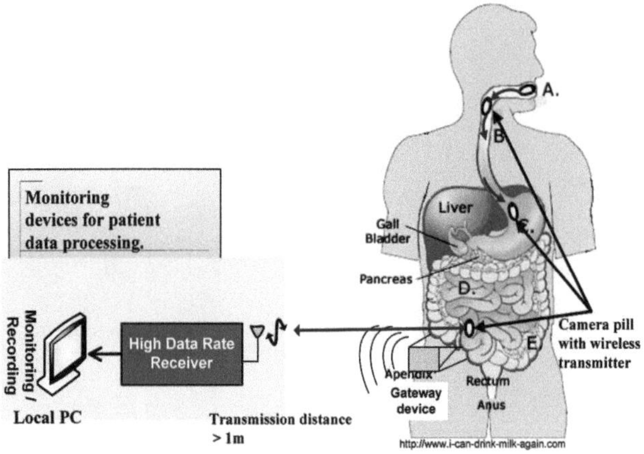

Figure 16.22. A wireless endoscope monitoring system. See also Color Insert.

Table 16.5. Comparison of hardware designs from vendors

Model	Company	Camera (sensor)	Freq. (MHz)	Data Rate	Image rate and resolution
PillCam (SB)	Given Imaging	Micron, CMOS	402–405 and 433 (Zarlink)	800 Kbps	14 images per, second or 2,600 color images
EndoCapsule	Olympus Optical	CCD camera, 1920 × 1080	—	—	2 images per second
MiroCam®	IntoMedic	320 × 320 pixel	—	—	3 images/second
SmartPill	Smartpill Corp.	acidity (pH), press., temp.	—	—	Only sensor discrete data

Wideband technology (UWB communication) is an ideal physical layer solution that achieves a data rate equal or higher than 100 Mbps. Its current applications are mostly for in-door entertainment, radar, and imaging. Due to high losses in body tissue at high frequencies many are skeptical about using UWB for implanted and, moreover, ingested devices. This challenge can only be addressed by combined design of antennas, transmitters, and receivers as described in reference [24]. Figure 16.23 shows an endoscope application using UWB proposed in reference [24].

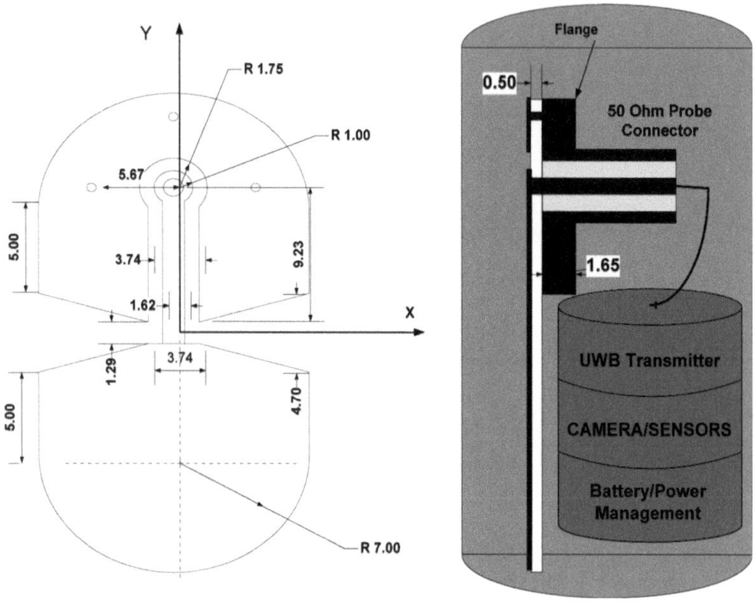

Figure 16.23. A wireless electronic pill system with main blocks identified [24]. See also Color Insert.

A wideband electronic pill can transmit raw video data without any compressing, resulting low-power, less delay in real-time and increased picture resolution. With a high definition camera such as 2 megapixels, UWB telemetry can send up to 10 frames per second (fps). When wideband technology is used with electronic pills, it will provide better diagnostic systems for medical professionals to analyze real-time video and image data wirelessly as a less invasive method. Although it is known that tissue imposes strong attenuation at higher frequencies, it may, however, be possible to send higher power levels than the regulated UWB signal level [10, 24].

Small batteries are used or being considered to supply energy to electronic pills applications, which take a considerable amount of space. A wireless power can substantially reduce the overall implant size and weight. For implantable systems like the cochlear implant and the bionic eye, the inductive link has been used to transfer power from the external unit to inside the skin to supply power for electronics. In order to transfer enough energy, inductive links

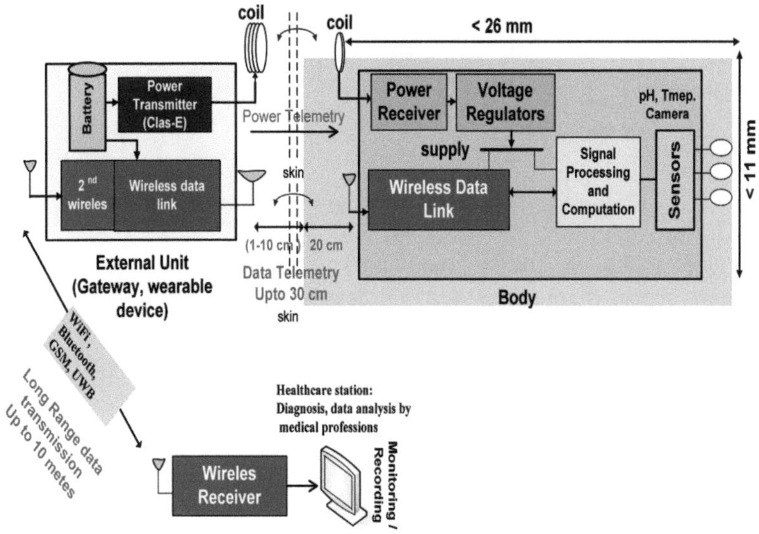

Figure 16.24. An electronic pill technology with wireless power embedded and wireless gateway for long-range data transmission. The available electronic pill devices have a dimension of 26 mm × 11 mm. See also Color Insert.

require the external unit to be very close to the implanted unit. Ideally, the external unit should be placed on the body for an efficient wireless power. A wireless power mechanism can also be used to recharge the battery of the electronic pill if a rechargeable battery is used.

Figure 16.24 shows the basic building blocks of an electronic pill with wireless power mechanism. The power signal is received and then regulated to provide power supply for the wireless data link, the sensors, and the signal acquisition unit, which processes and amplifies the sensors' data. The external unit contains a second wireless link that controls the communication between the monitoring station and the implant devices. Overall length and width of the commercially available electronic pill is usually 26 mm × 11 mm.

In this system, the patient will wear the external unit, which will be wearable on the body close to skin and, therefore, to the swallowed electronic pill. A remote wireless receiver will be used to receive data from the external unit (i.e., gateway) for monitoring,

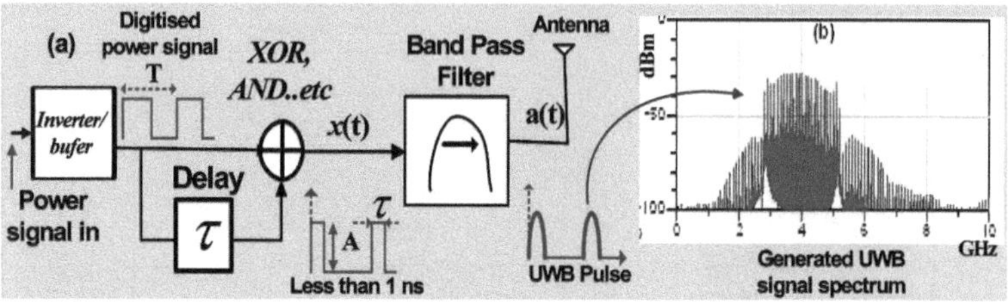

Figure 16.25. An UWB telemetry link: (a) UWB pulse generation from the power signal, (b) the UWB spectrum. See also Color Insert.

analyzing, and displaying the data. As the required specifications are not restricted for this link, commercial wireless devices could be used to undertake this duty. This additional wireless link will provide patient freedom in a room in the hospital environment. With this scenario, it is also possible for a health professional to see the data online through the Internet when the patient is home.

An interesting UWB signal generation technique for implantable electronics having a wireless power link is described in Fig. 16.25. The UWB telemetry link is based on generating the narrow UWB pulses by manipulating the received power signal. The analog power signal can be converted to a square wave using an inverter. Then the UWB narrow pulse can be generated in a similar fashion explained in Section 16.3.2. In this example, the digitized power signal and the delayed replica are passed through the XNOR gate or an AND gate to obtain a narrow square pulse. The narrow square wave is then passed through a BPF to fit the spectrum into the UWB mask (Fig. 16.25b). The selected bandwidth of the BPF is 3.5–4.5 GHz, which avoids the narrowband systems operating in the ISM bands. This UWB telemetry link eliminates the need of a reference clock at the implant site, which results in extremely low for such a high-capacity telemetry link.

16.8 Design and Implementation of an UWB-WBAN System

Figure 16.26 shows the photo of the UWB sensor node designed for a transmitter-only UWB WBAN system described earlier. The

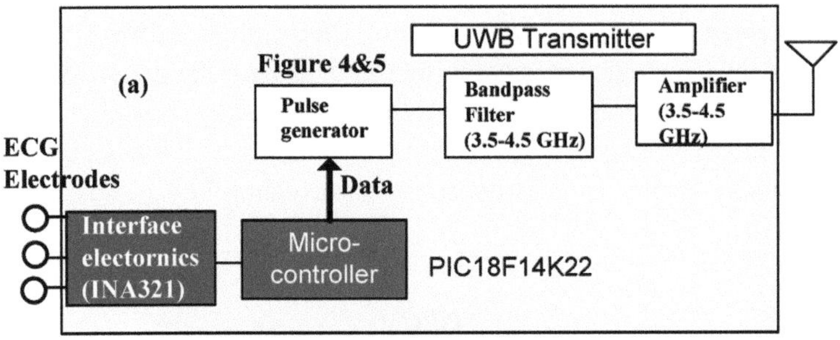

(b)

Figure 16.26. UWB-WBAN sensor node: (a) block diagram of an UWB-based WBAN sensor node, (b) photo of UWB sensor node (2.5 cm by 2.7 cm). See also Color Insert.

sensor nodes are assembled on a four layers printed circuit board with dimensions of 2.5 cm by 2.7 cm (Figure 16.26b), which is sufficiently compact for use in a WBAN application. The block diagram of the sensor node is shown in Fig. 16.26a. The narrow pulse generator is formed using a variable voltage controlled oscillator (VCO), delay circuitry, and a XOR gate. The details of UWB pulse generation were explained with Figs. 16.4 and 16.5 earlier. The sensor board is able to produce narrow pulses ranging from 0.5 ns to 2 ns, with a variable pulse repetitive frequency between 17 MHz to 170 MHz. The generated narrow pulse passes through a pulse-shaping filter (band-pass filter), which shapes the pulse spectrum to conform to the UWB requirement. After passing through the filter, the UWB pulse is amplified using a wideband LNA to meet the −41.3 dBm transmission power level. The microcontroller performs the analog-to-digital conversion, determines the transmission format, modulation scheme, and sets the data rate. The UWB pulse rate is independent of the data rate, therefore enabling the system to vary the number of UWB pulses per data bit. The logic gates combine the data with the generated UWB pulse before transmitting the information using a wideband UWB antenna. This sensor node is designed to operate in the 3.1 to 5 GHz frequency band, using direct modulation technique (i.e., no carrier is used).

16.8.1 *UWB Receiver Circuitry*

The UWB sensor node communicates with an UWB receiver. The receiver circuitry is assembled using off-the-shelves modules. The receiver antenna is placed on body, and the receiver circuitry is placed off body. The block diagram of the receiver circuit is shown in Fig. 16.27. The signal entering the receiver passes through a 3 to 5 GHz band-pass filter to eliminate the unwanted out-of-band signals. The filtered signal is amplified by 48 dBm using three wideband LNAs before down converting to baseband signal using a mixer and a 4 GHz VCO. The baseband signal passes through a low-pass filter with 100 MHz bandwidth before going through the baseband amplification stage.

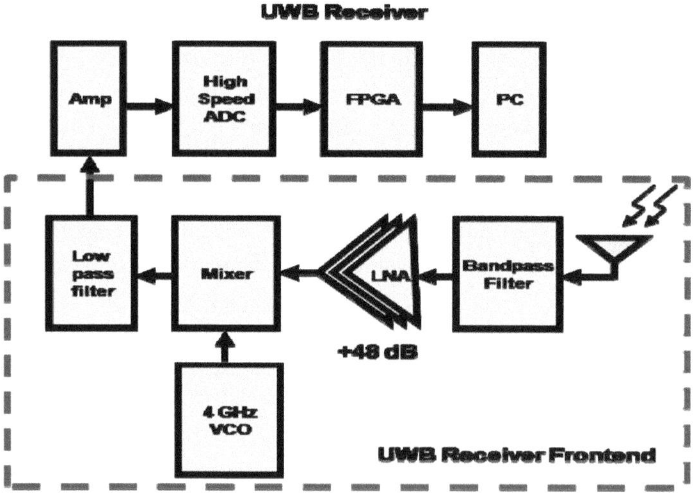

Figure 16.27. UWB receiver block diagram. See also Color Insert.

The recovered UWB pulse is digitalized using a high-speed ADC and processed and demodulated by the FPGA before transferring the data to the laptop using a serial cable. The role of the FPGA is to process the received multiple UWB pulses and determine whether it is bit "1" or bit "0" before sending to the laptop and set the appropriate baud rate for data transfer to the laptop. Preprocessing is necessary as the data transfer rate via the serial cable is much lower than the data rate the sensor nodes is sending to the UWB receiver using the gating method. After FPGA processing, the data is transferred to the laptop based on the baseband data rate. The sensor node and the receiver in Fig. 16.27 are used to maintain the best position of sensors on the body when UWB communication is used. The BER performance of the WBAN system is computed in the laptop using MATLAB program based on a known pseudorandom data sequence that the transmitter is sending [29].

16.8.2 *Experimental Setup and Measurement Result*

Table 16.6 shows the parameters used for the experimental setup. Four locations commonly used for physiological signal monitoring are chosen to evaluate the performance of UWB WBAN at different receiver positions. Figure 16.28 shows the sensors and receiver position evaluated in this measurement. Sensor 1 is at the ECG position, sensor 2 at the EEG position, sensor 3 on the wrist, and sensor 4 on the waist. All the readings are taken in a controlled environment where the subject under test remains static throughout each measurement. For each sensor and receiver pairs, five sets of readings are taken. Each set of readings consists of 10,000 data bits.

Table 16.6. Experimental parameter

Parameter	Values
UWB pulse rate	25 MHz
UWB pulse width	2 ns
UWB frequency band	3.9–4.5 GHz
Data rate	1 Mbps
Modulation	OOK
Maximum transmissin Peak power	−24.4dBm

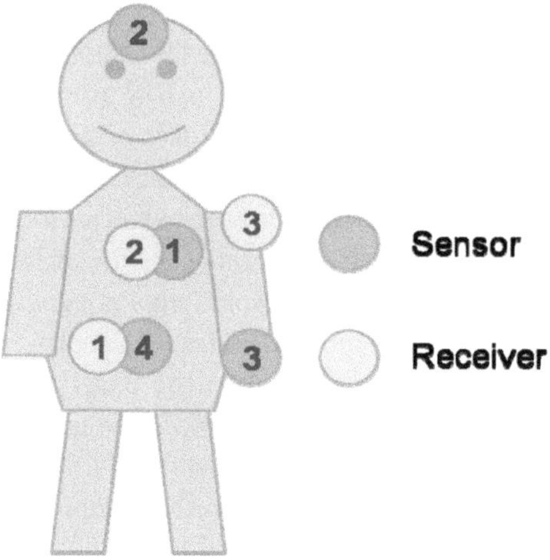

Figure 16.28. Sensor and receiver locations. See also Color Insert.

All the readings are taken using antenna orientation that gives the strongest signal strength.

Figures 29–32 show the BER plots obtained from four sensor positions across the body. Most of the commonly monitored physiological signals are located on the front upper half of the body. Therefore, we have selected four commonly used sensor locations that are well spread throughout the upper body to perform this

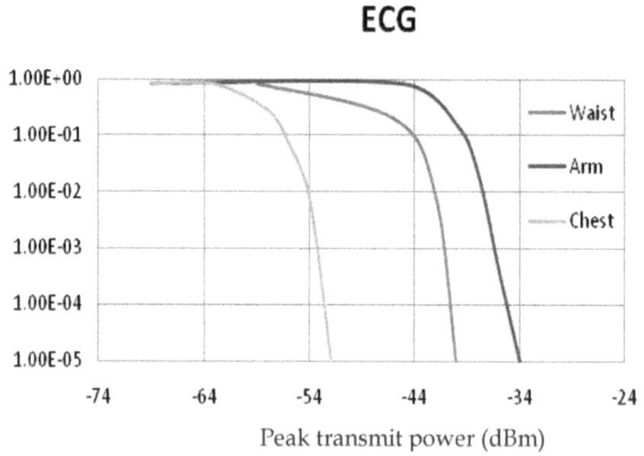

Figure 16.29. BER of ECG sensor for different receiver positions [29]. See also Color Insert.

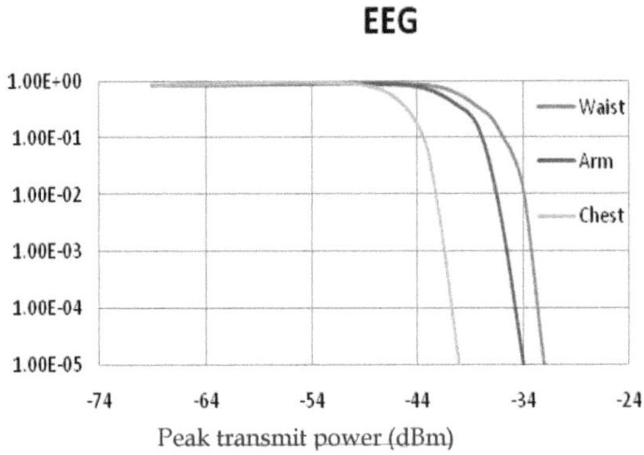

Figure 16.30. BER of EEG sensor for different receiver positions [29]. See also Color Insert.

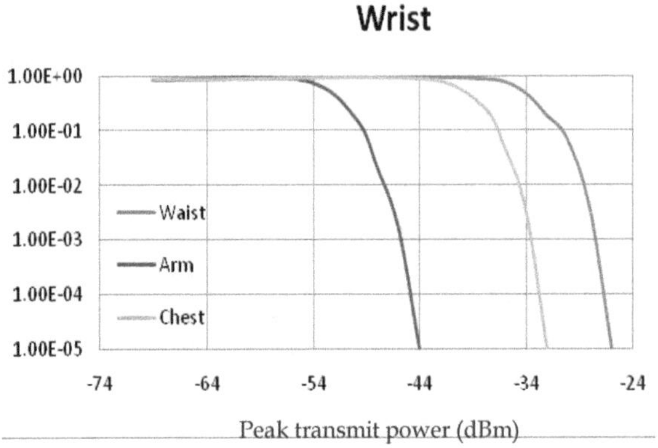

Figure 16.31. BER of sensor on wrist for different receiver positions [29]. See also Color Insert.

evaluation. All the sensor nodes and receiver are placed 10 mm away from the body in order to reduce the effect of the body on the antenna.

These experiment results show that by carefully selecting the receiver position, it is possible to reduce the amount of transmission power to up to 20 dB. There are numerous configurations for the sensor nodes in a WBAN system; therefore, there is not one receiver position that is optimized for all configurations. However, the receiver position at the center of the chest is the preferred choice for most of the configuration especially when ECG monitoring is used.

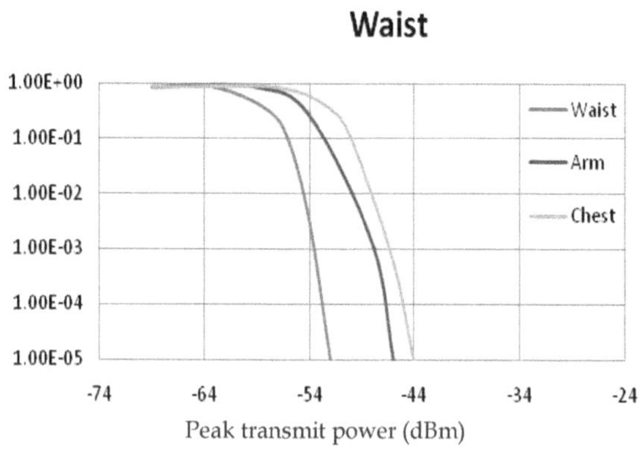

Figure 16.32. BER of sensor on waist for different receiver positions [29]. See also Color Insert.

The optimum receiver position can be selected based on the minimum total transmit power required to attain a reasonable BER performance for all the sensor nodes in a particular configuration. As an example for EEG sensor, the best receiver position is on the chest, and the worst receiver position is on the waist. For a BER of 10^{-3}, there are only 8 dB differences between the best receiver and the worst receiver positions [29].

16.8.3 *Summary*

In this section, we have looked at four different medical applications using UWB wireless schemes. The feasibility of UWB in medical WBAN applications can be clearly seen in this section. It is possible to design an UWB transmitter using COTS parts or CMOS implementation. Actual hardware measurement result shows that UWB is a promising option for both on-body and implantable applications.

One big advantage of UWB wireless technology is that its high data rate ranges from 850 Kbps to 20 Mbps, which can be used for simultaneous monitoring of many continuous physiological signals such as ECC, EEG, and EMG. In addition, UWB wireless technology also does not present an EMI (electromagnetic interference) risk to other narrowband systems and medical equipment in health care as its transmitter power is quite low and the frequencies used are at very high frequencies (>3.5 GHz). The biggest drawback for the use of the UWB technology is that the UWB chips are not available commercially to apply in a telemedicine system at the moment.

16.9 Conclusion

The UWB WBAN system is indeed a multilayer problem. In order to design an efficient UWB WBAN system, the designer has to consider the hardware level, PHY layer, and MAC protocol. UWB is a promising wireless technology that could give WBAN a breakthrough, but more research and testing on the hardware and system level is required before it can be successfully implemented. An interesting application will be to integrate UWB sensor nodes together with narrowband wireless technologies form a WBAN application for health care systems.

UWB-based WBAN sensor nodes can be applied to both non-invasive and invasive monitoring systems in medical applications such as multichannel EEG, tracking patients, and neural recording, also in gaming and entertainment industry where a high data rate transmission is required without affecting the operation of other wireless systems in the environment.

References

1. http://www.ieee802.org/15/pub/TG6.html, 2010.
2. FCC 02–48 (First Report and Order), 2002. http://transition.fcc.gov/Bureaus/Engineering_Technology/Orders/2002/fcc02048.pdf.
3. http://standards.ieee.org/getieee802/download/802.15.4a-2007.pdf.
4. L. Huan-Bang and R. Kohno (2007) Introduction of SG-BAN in IEEE 802.15 With Related Discussion, in *IEEE International Conference on Ultra-Wide Band 2007*, pp. 134–139.
5. M. Klemm and G. Troester (November 2006) Textile UWB Antennas for Wireless Body Area Networks, *IEEE Transactions on Antennas and Propagation,* **54(11)**, pp. 3192–3197.
6. B. Sanz-Izquierdo, J. C. Batchelor, and M. I. Sobhy (2007) Compact UWB Wearable Antenna, in *Loughborough Antennas and Propagation Conference 2007 (LAPC 2007)*, pp. 121-124.

7. L. Jeong Soo, C. Nguyen, and T. Scullion (2001) New Uniplanar Subnanosecond Monocycle Pulse Generator and Transformer for Time-Domain Microwave Applications, *IEEE Transactions on Microwave Theory and Techniques,* **49**, pp. 1126–1129.
8. H. Jeongwoo and N. Cam (2002) A New Ultra-Wideband, Ultra-Short Monocycle Pulse Generator With Reduced Ringing, *IEEE Microwave and Wireless Components Letters,* **12**, pp. 206–208.
9. D. Simic, A. Jordan, R. Tao, N. Gungl, J. Simic, M. Lang, L. V. Ngo, V. Brankovic (1–5 July 2007) Impulse UWB Radio System Architecture for Body Area Networks, in 16th *IST Mobile and Wireless Communications Summit 2007*, pp. 1–5.
10. M. R. Yuce, H. C. Keong, and M. S. Chae (October 2009) Wideband Communication for Implantable and Wearable Systems, *IEEE Transactions on Microwave Theory and Techniques,* **57(10)**, pp. 2597–2604.
11. J. Ryckaert, C. Desset, A. Fort, M. Badaroglu, V. de Heyn, P. Wambacq, G. van der Plas, S. Donnay, B. van Poucke, and B. Gyselinckx (December 2005) Ultra-Wide Band Transmitter for Low-Power Wireless Body Area Networks: Design and Evaluation, *IEEE Transactions on Circuits and Systems I,* **52(12)**, pp. 2515–2525 .
12. H. C. Keong and M. R. Yuce (2008) Low Data Rate Ultra Wideband ECG Monitoring System, in 30th *IEEE-EMBS Annual International Conference* 2008.
13. K. Hyunseok, J. Youngjoong, and J. Sungyong (2005) Digitally Controllable Bi-Phase CMOS UWB Pulse Generator, in *IEEE International Conference on Ultra-Wideband*, pp. 442–445.
14. M. R. Yuce, W. Liu, M. S. Chae, and J. S. Kim (24–26 September 2007) A Wideband Telemetry Unit for Multi-Channel Neural Recording Systems, in *IEEE International Conference on Ultra-Wideband,* pp. 612–617.
15. R. Kohno, K. Hamaguchi, H. -B. Li, and K. Takizawa (10–12 September 2008) R&D and Standardization of Body Area Network (BAN) for Medical Healthcare, in *IEEE International Conference on Ultra-Wideband 2008*, **3**, pp. 5–8.
16. C. Lee, J. Kim, H. S. Lee, and J. Kim (28–30 September 2009) Physical Layer Designs for WBAN Systems in IEEE 802.15.6 Proposals, in 9th *International Symposium on Communications and Information Technology 2009*, pp. 841–844.

17. H. Viittala, M. Hamalainen, J. Iinatti, and A. Taparugssanagorn (24–27 November 2009) Different Experimental WBAN Channel Models and IEEE802.15.6 Models: Comparison and Effects, in *2nd International Symposium on Applied Sciences in Biomedical and Communication Technologies 2009,* pp. 1–5.

18. A. Khaleghi and I. Balasingham (26–29 April 2009) On the Ultra Wideband Propagation Channel Characterizations of the Biomedical Implants, in *IEEE Vehicular Technology Conference 2009,* pp. 1–4.

19. C. Li, L. Wang, J. Li, B. Zhen, H. –B. Li, and R. Kohno (13–16 September 2009) Scalable and Robust Medium Access Control Protocol in Wireless Body Area Networks, in 20th *IEEE International Symposium on Personal, Indoor and Mobile Radio Communications,* pp. 2127–2131.

20. Y. Zhang and G. Dolmans (23–29 August 2009) A New Priority-Guaranteed MAC Protocol for Emerging Body Area Networks, in 5th *International Conference on Wireless and Mobile Communications 2009,* pp. 140–145.

21. G. Fang and E. Dutkiewicz (28–30 September 2009) BodyMAC: Energy Efficient TDMA-based MAC Protocol for Wireless Body Area Networks, in9th *International Symposium on Communications and Information Technology 2009,* pp. 1455–1459.

22. K. H. Chee and M. R. Yuce (2009) Analysis of a Multi-Access Scheme and Asynchronous Transmit-Only UWB for Wireless Body Area Networks, in *IEEE Engineering in Medicine and Biology Conference 2009,* pp. 6906–6909.

23. M. Weisenhorn and W. Hirt (2005) Uncoordinated Rate-Division Multiple-Access Scheme for Pulsed UWB Signals, *IEEE Transactions on Vehicular Technology,* **54**, pp. 1646–1662.

24. M. R. Yuce, T. Dissanayake, and H. C. Keong, (October 2009) Wireless Telemetry for Electronic Pill Technology, in *IEEE Conference SENSORS.*

25. M. R. Yuce, T. N. Dissanayake, and H. C. Keong (2009) Wideband Technology for Medical Detection and Monitoring, in *Recent Advances in Biomedical Engineering,* (ed. G. R. Naik), ISBN: 978-953-307-004-9, IN-TECH, available from http://sciyo.com/articles/show/title/wideband-technology-for-medical-detection-and-monitoring.

26. M. Chae, Z. Yang, M. R. Yuce, L. Hoang, and W. Liu (August 2009) A 128-Channel 6 mW Wireless Neural Recording IC With Spike Feature Extraction and UWB Transmitter, *IEEE Transactions on Neural Systems and Rehabilitation Engineering,* **17**, pp. 312–321.

27. http://www.givenimaging.com/2010.
28. P. Bradley (2006) An Ultra Low Power, High Performance Medical Implant Communication System (MICS) Transceiver for Implantable Devices, in *IEEE Biomedical Circuits and Systems Conference*, **16**, pp. 158.
29. H. C Keong, M. R. Yuce, T See, and T. M. Chiam (2010) On-Body Evaluation of UWB Receiver Position for Wireless Body Area Networks, in *IEEE International Conference on Ultra-Wideband* 2010.

Index

Color Insert

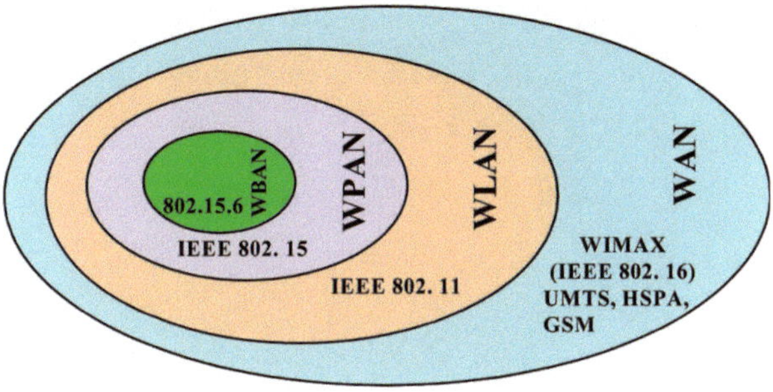

Figure 1.2. Wireless standards.

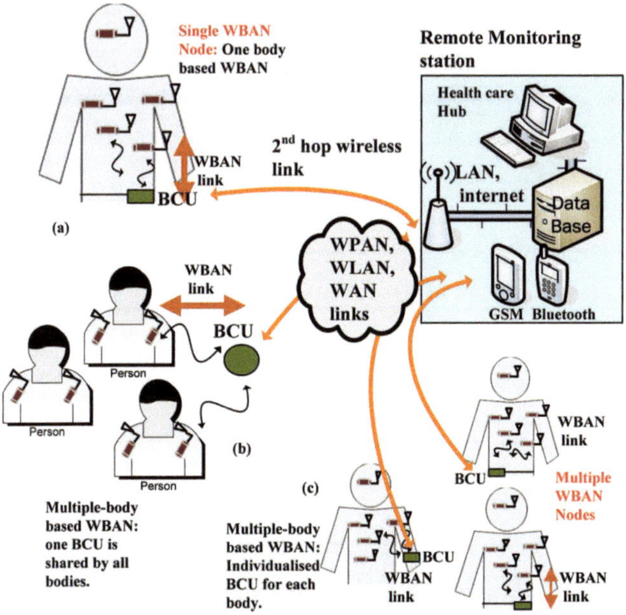

Figure 1.3. Wireless body area network implementation scenarios: (a) single WBAN node, (b) a WBAN node with multiple bodies, and (c) multiple WBAN nodes.

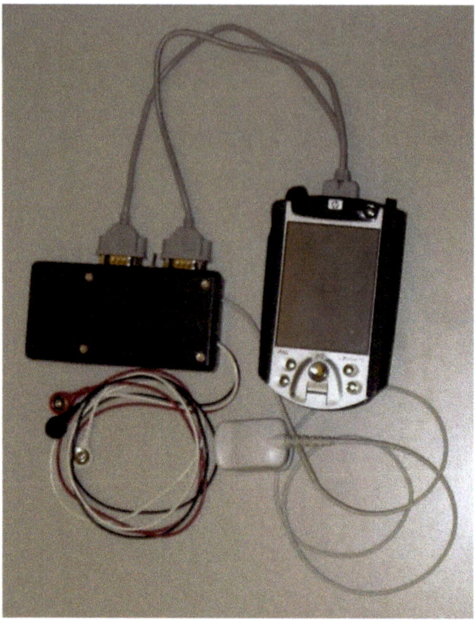

Figure 2.1. Patient PDA with ECG leads and SpO_2 sensor.

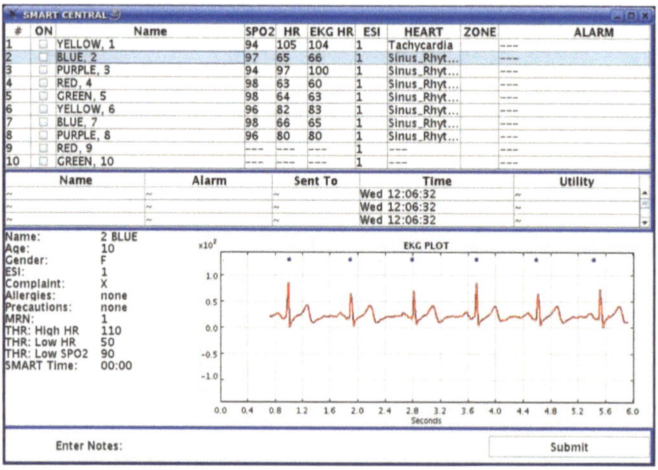

Figure 2.4. SMART central main GUI.

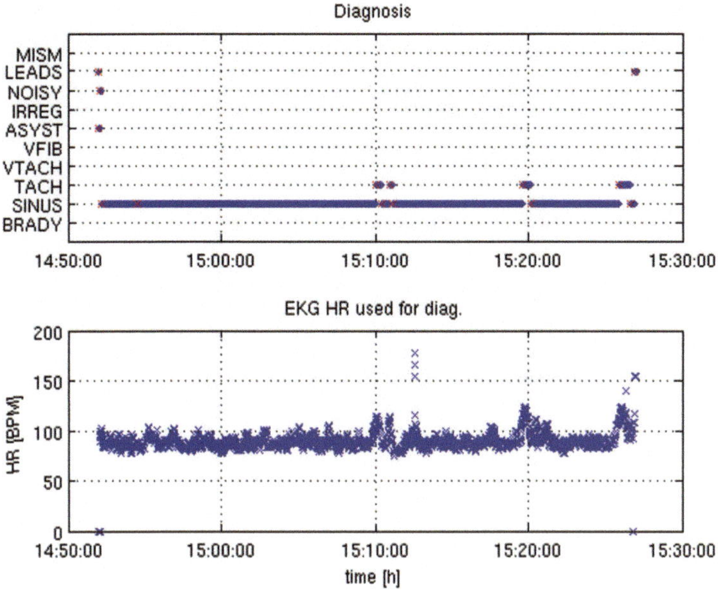

Figure 2.7. Diagnosis estimation and ECG-computed HR of an average patient.

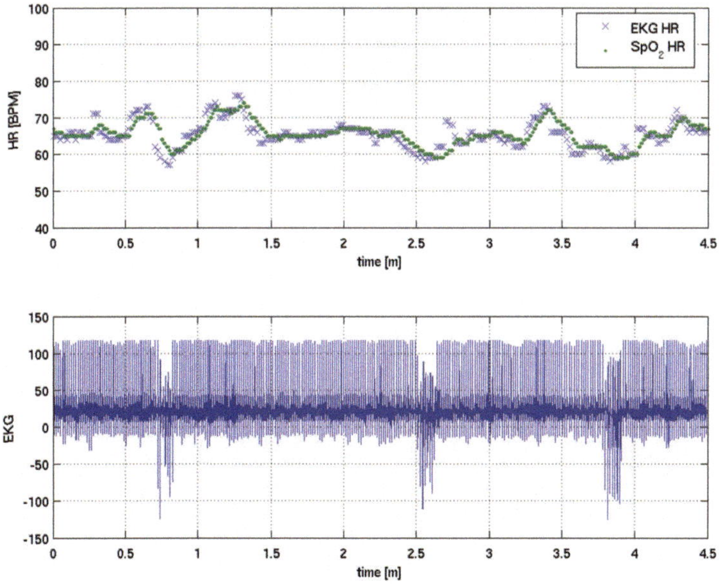

Figure 2.8. ECG – SpO_2 HR comparison and ECG pacemaker activity when patient HR dropped below 60 BPM.

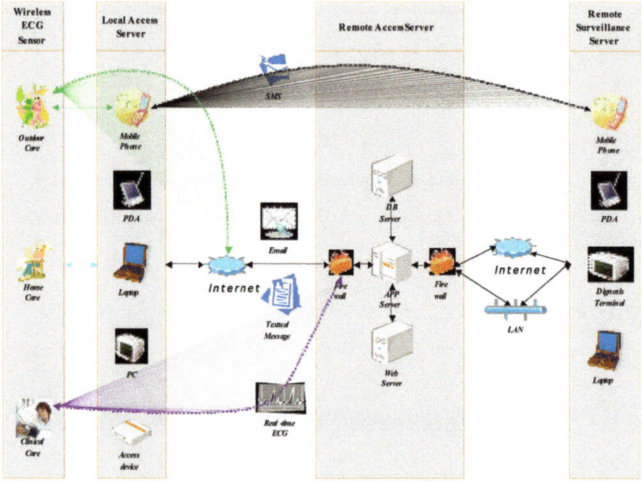

Figure 3.1. Architecture of PCC system.

Figure 3.2. Wireless ECG sensor.

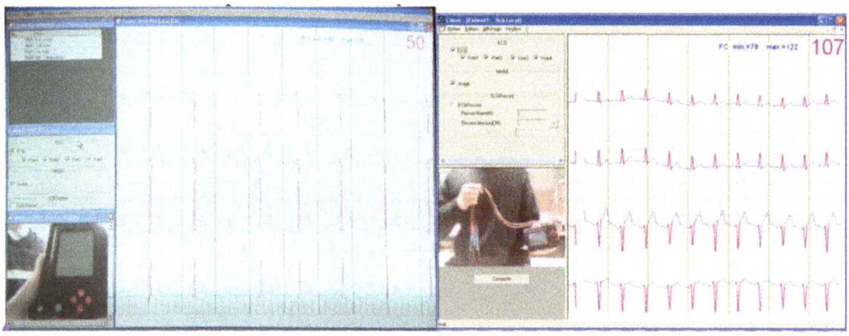

Figure 3.3. Remote surveillance server.

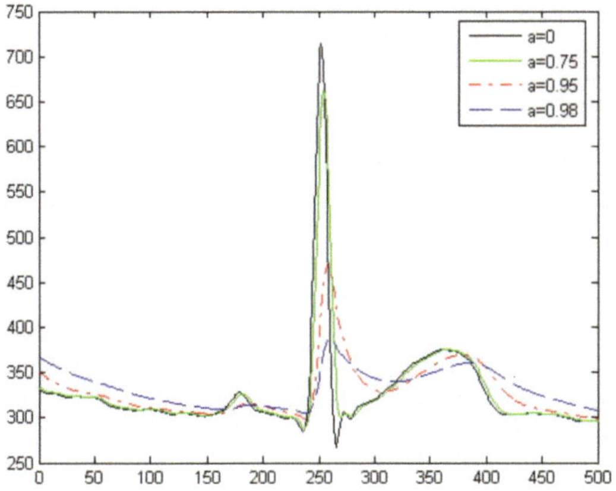

Figure 3.8. QRS complex waveform after different impact factor in the adaptive filters.

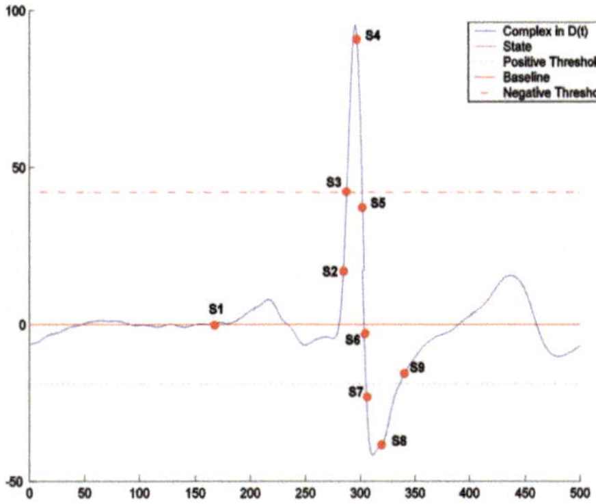

Figure 3.11. Positive states of a QRS complex in $D(t)$.

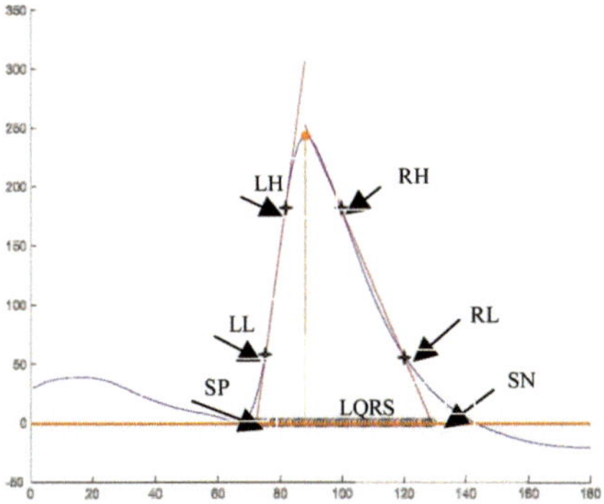

Figure 3.12. Illumination of geometric analysis method.

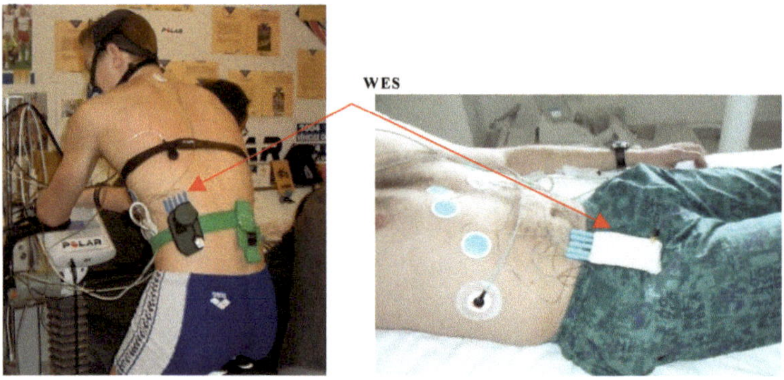

Figure 3.13. Application example of PCC system.

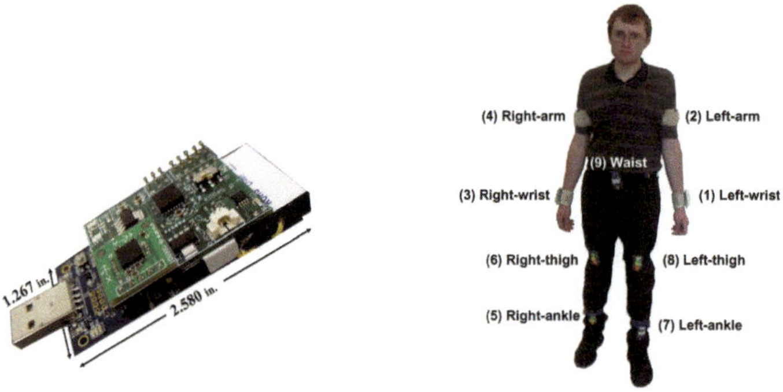

(a) Sensor node

(b) Subject

Figure 4.1. (a) A sensor node composed of processing unit and custom-designed sensor board. The motion sensor board has a triaxial accelerometer and a biaxial gyroscope Sensor node and subject (b) A subject wearing nine sensor nodes.

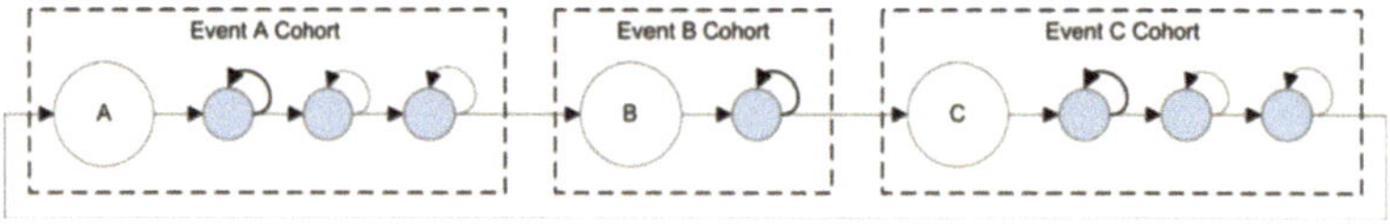

Figure 4.4. HMEM model and structure.

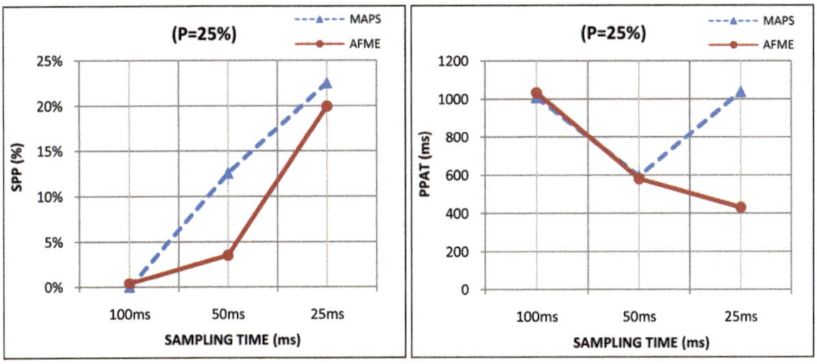

(a)

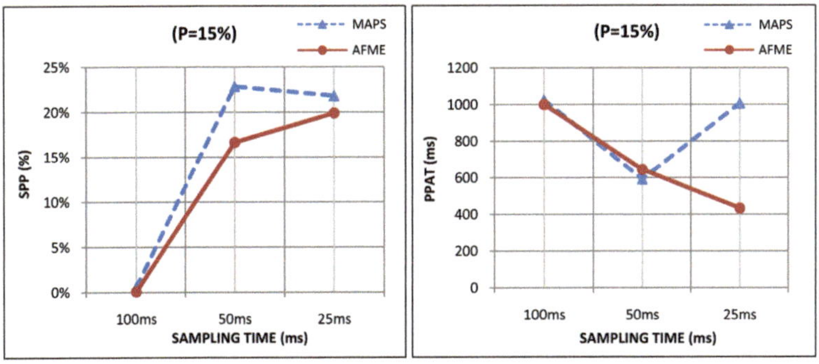

(b)

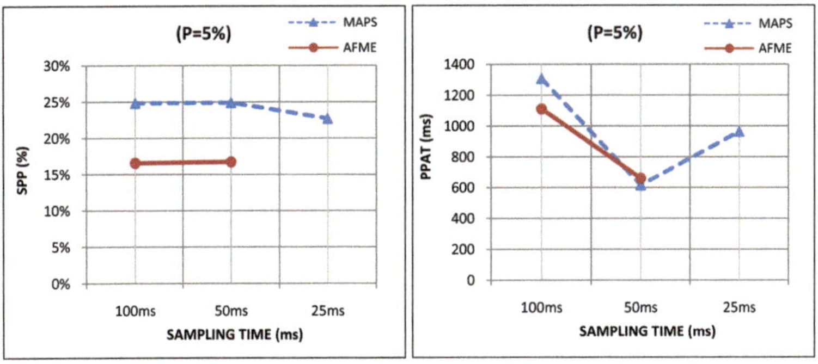

(c)

Figure 5.11. Analysis of PPAT and SPP of the sensor agents with W = 20, S = 10, and P = (a) 25%, (b) 15%, (c) 5%.

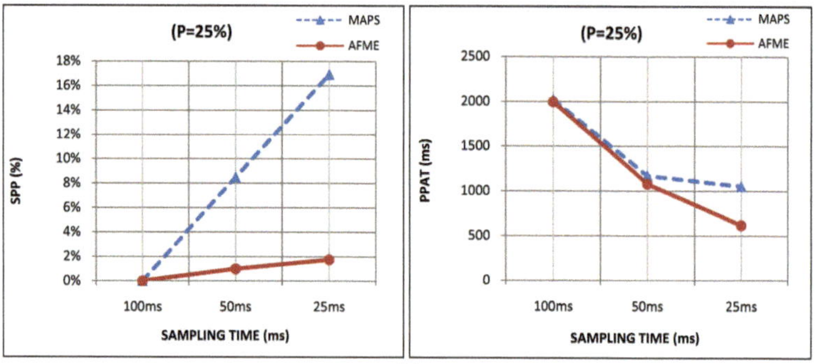

(a)

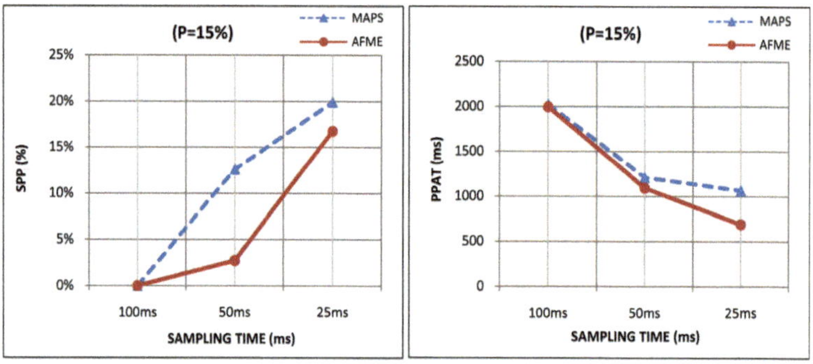

(b)

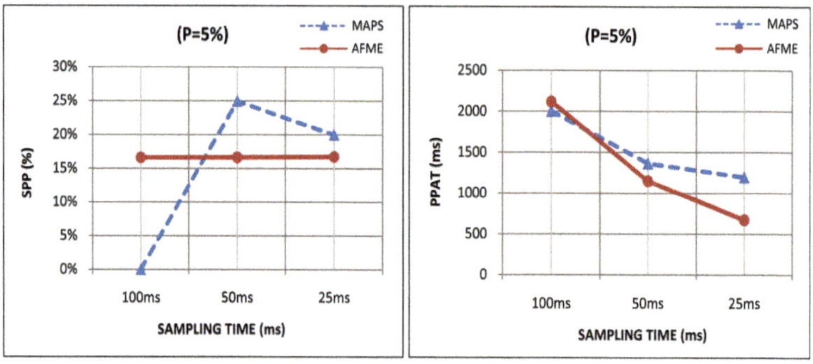

(c)

Figure 5.12. Analysis of PPAT and SPP of the sensor agents with W = 40, S = 20, and P = (a) 25%, (b) 15%, (c) 5%.

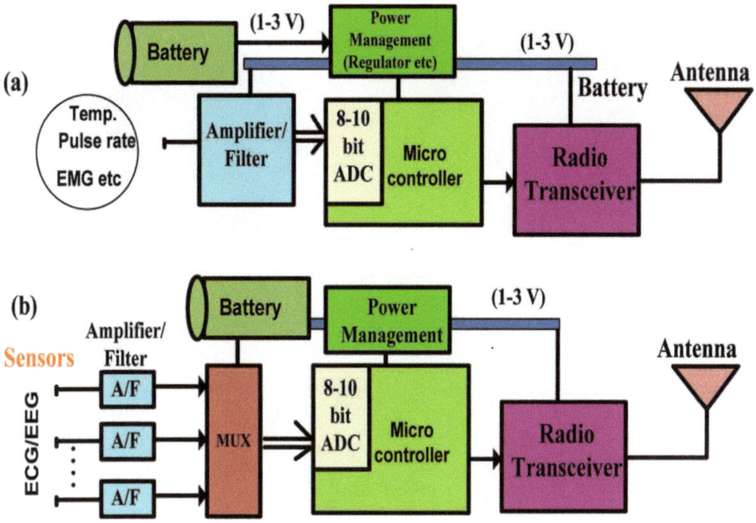

Figure 6.2. An example of sensor node hardware: (a) single wireless sensor node, (b) multi-channel wireless sensor node.

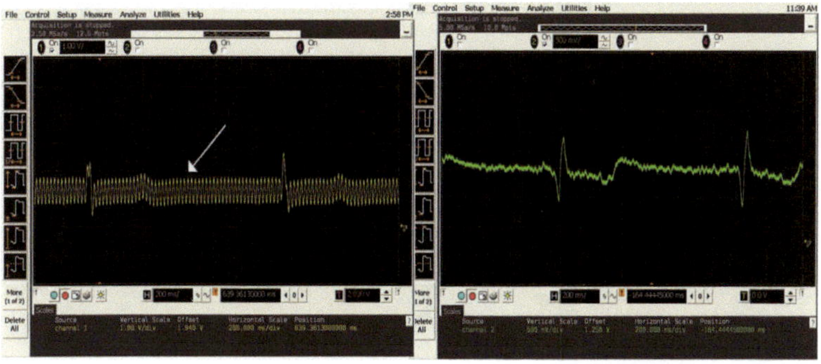

Figure 6.5. ECG with 60 Hz (or 50 Hz) noise (a) and without noise (b).

Figure 6.6. A Mica2DOT board (taken from www.xbow.com).

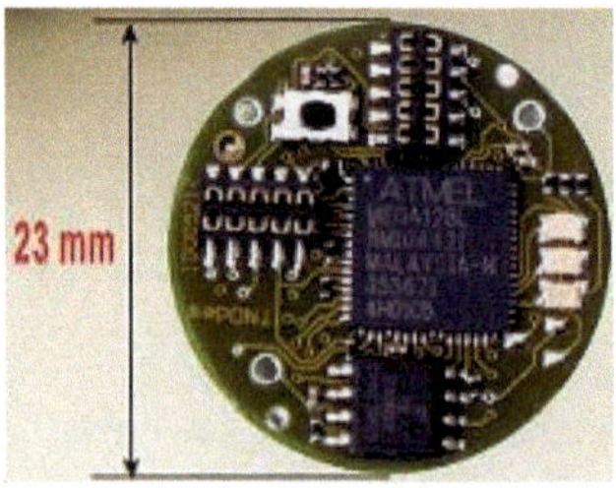

Figure 6.7. A T-node sensor node [16].

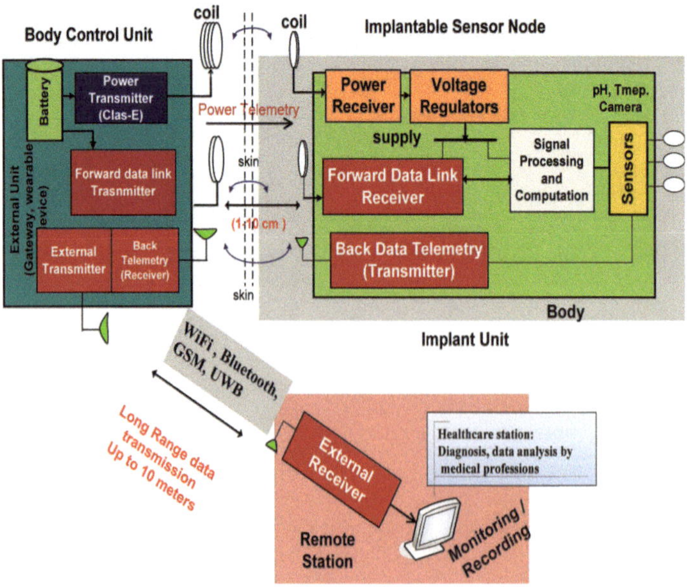

Figure 6.8. An implanted sensor node with the external unit and wireless data links.

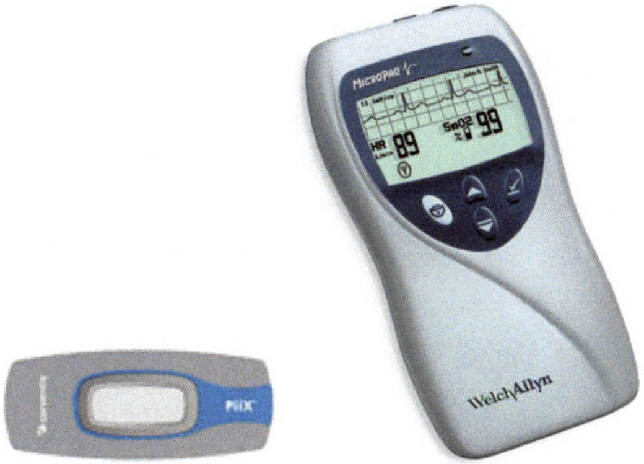

Figure 6.9. Figure PiiX sensor node from Corventis [38] and micropaq® Wearable Monitor from WelchAllyn [37].

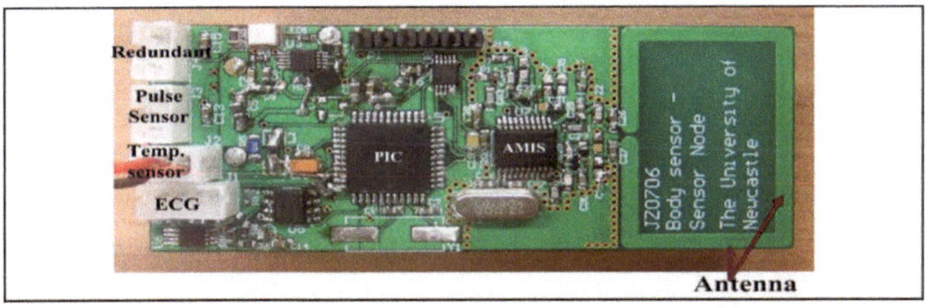

Figure 6.11. A 4-channel sensor node.

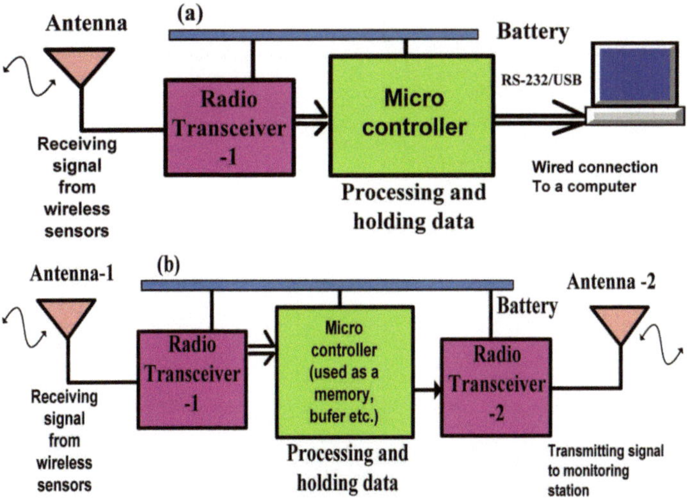

Figure 6.12. Block diagram of BCUs: (a) wired BCU, (b) portable BCU.

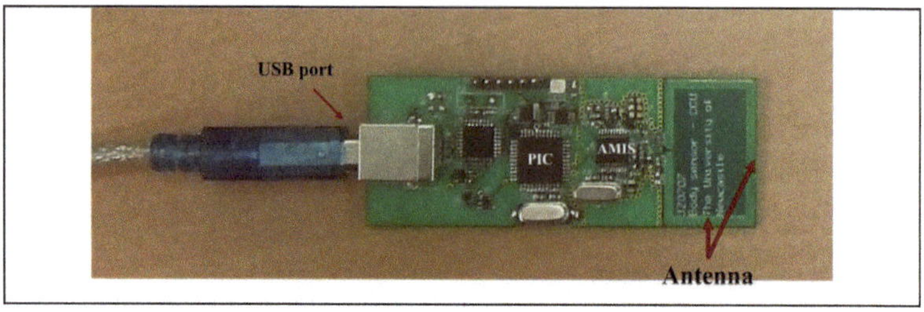

Figure 6.13. A wired BCU (BCU-1).

Figure 6.14. Intermediate body control unit (BCU-2). This device is shared by more than one patient and is portable. It contains dual wireless transceiver to support two directional wireless links.

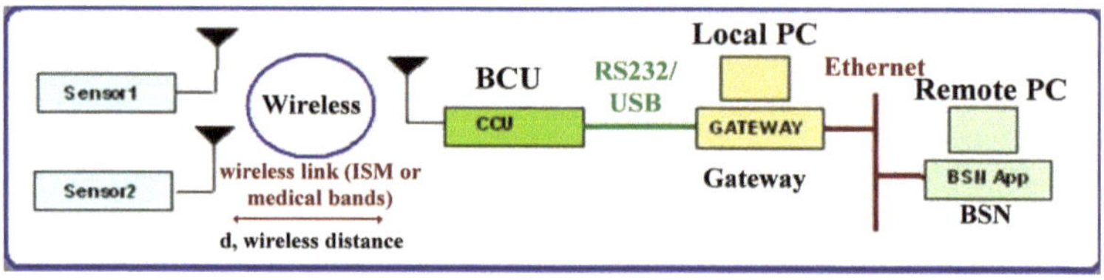

Figure 6.17. Software programs for the proposed WBAN application.

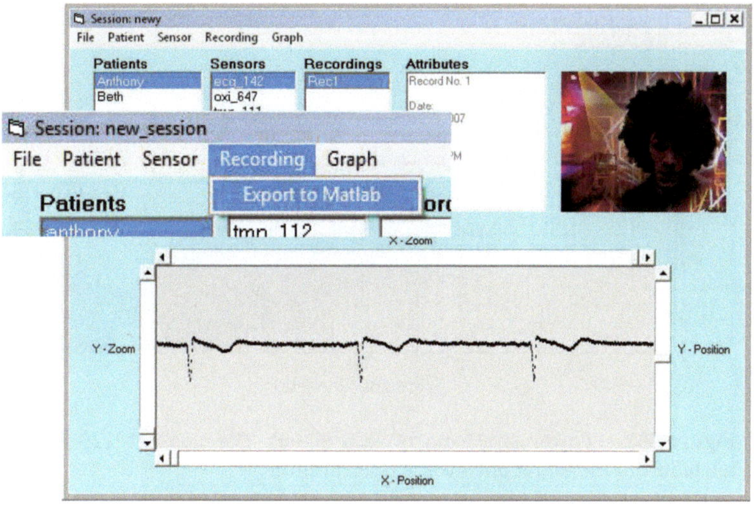

Figure 6.20. An example of ECG Monitoring.

Figure 7.1. The Berkeley Tricorder

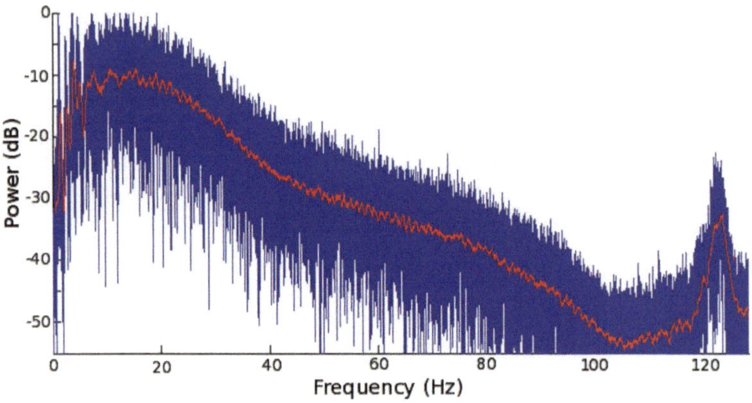

Figure 7.2. Power spectrum of ECG signal. The peak at 120 Hz is a harmonic of the 60 Hz power-line interference.

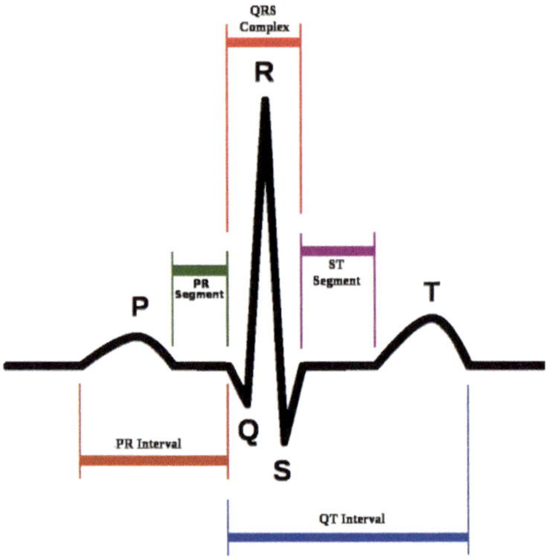

Figure 7.6. ECG waveform.

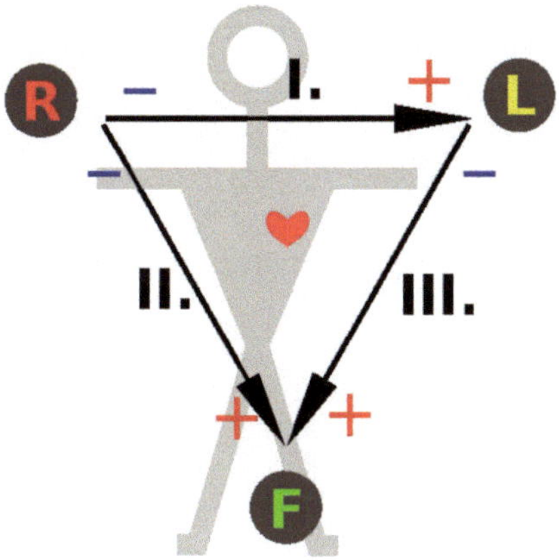

Figure 7.7. Einthoven ECG lead locations (from wikiCommons).

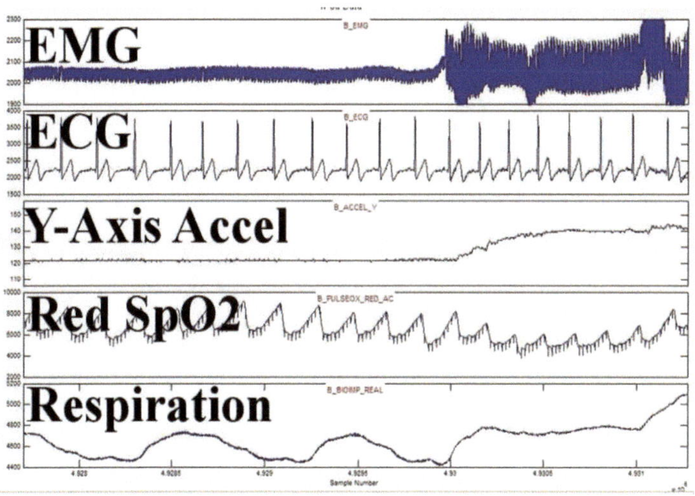

Figure 7.15. Sample data from the Berkeley Tricorder.

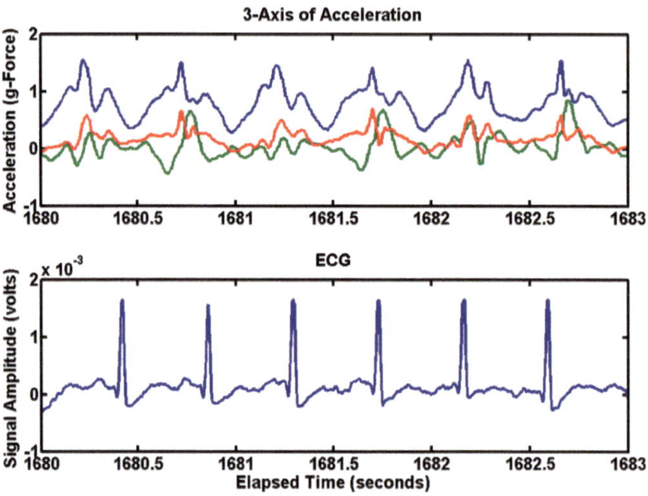

Figure 7.16. ECG signal (lower plot) captured while subject in motion (upper plot).

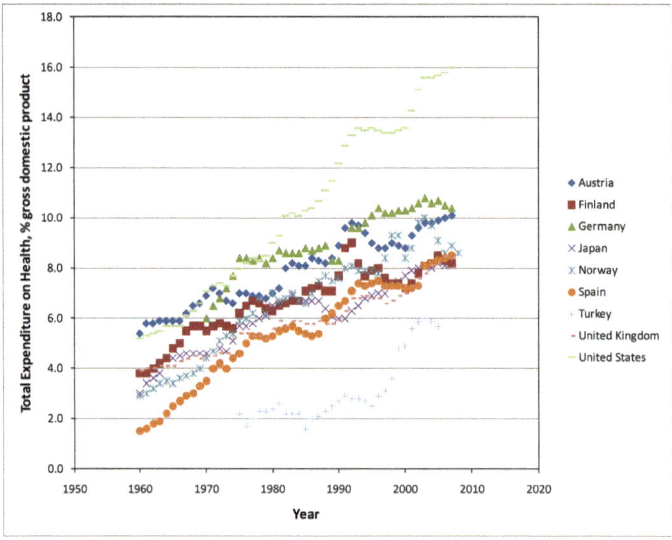

Figure 8.1. Change of total expenditure on health (percentage of gross domestic product) over years.

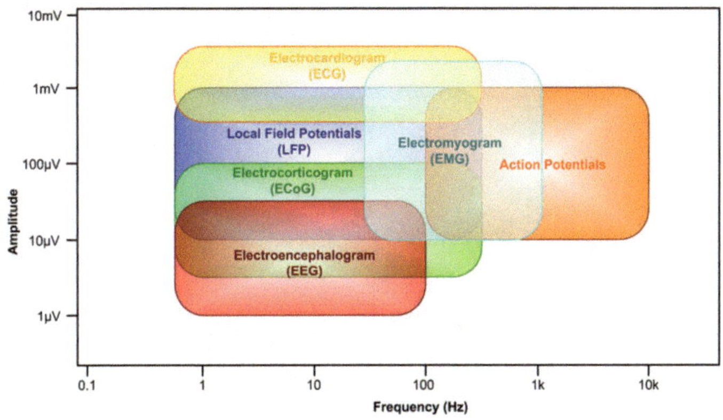

Figure 8.3. Frequency and amplitude characteristics of biopotential signals — most commonly monitored signals in medical practice [3].

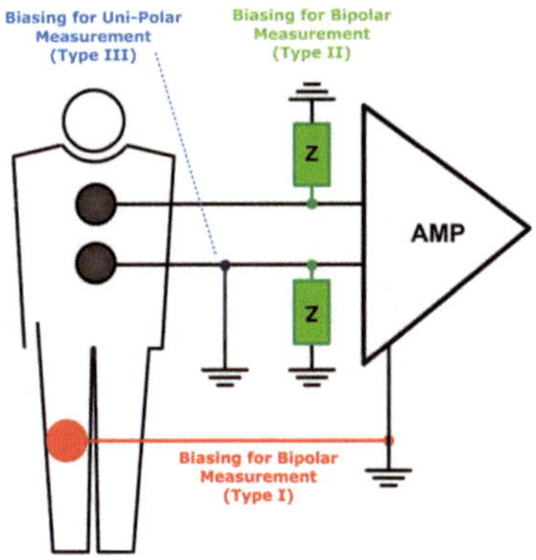

Figure 8.4. Different measurement and biasing methods for biopotential signal acquisition.

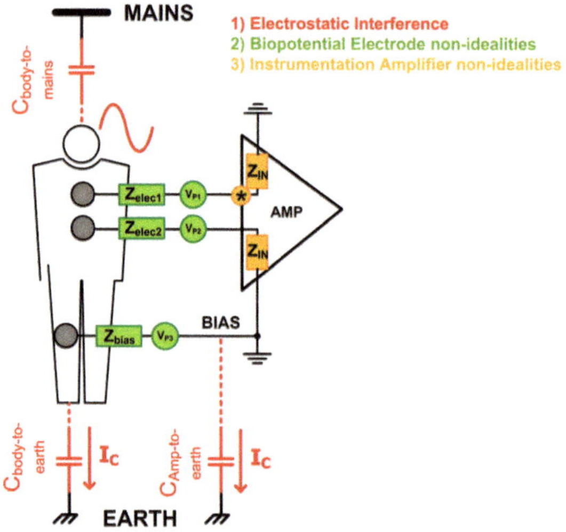

Figure 8.5. Measurement of biopotential signals (Type I measurement) introducing the nonidealities from different sources.

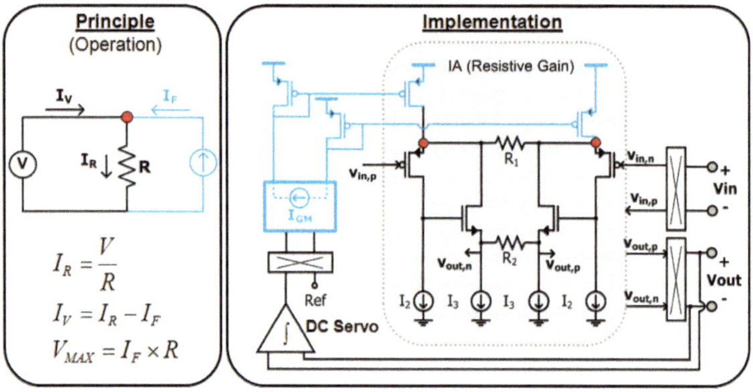

Figure 8.11. The complete architecture of a compensated instrumentation amplifier using a DC servo loop for implementing high-pass filtering characteristics [10 ,11].

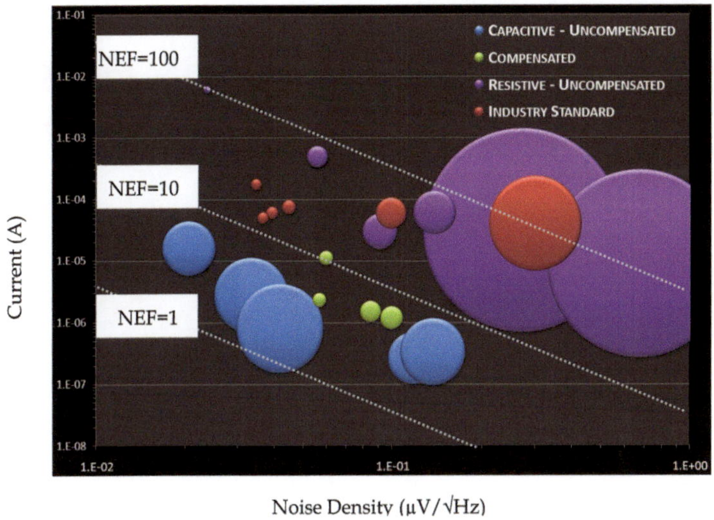

Figure 8.13. Comparison of different instrumentation amplifiers in terms of their power efficiency. Y-axis indicates the current consumption of the instrumentation amplifier and x-axis indicates the noise density of the instrumentation amplifier. Dashed lines indicate the constant contours of noise efficiency factor (NEF) lines [8]. The diameter of a circle indicates the total integrated noise of the instrumentation amplifier.

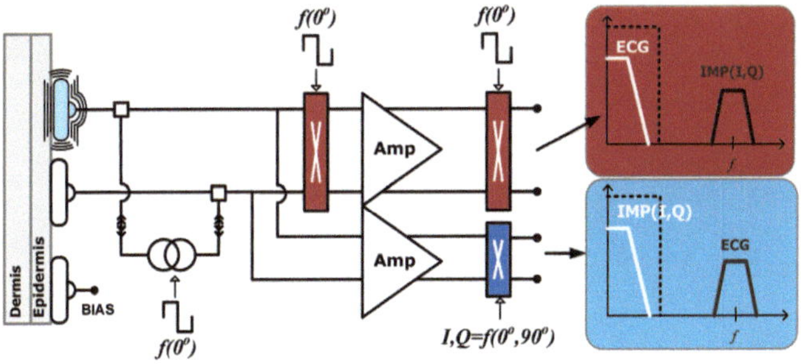

Figure 8.16. The implementation of electrode-tissue impedance measurement circuit operating simultaneously with the biopotential measurement circuit [28]. Instead of sinusoidal current sources and analog modulators/demodulators, chopper-modulation technique has been employed.

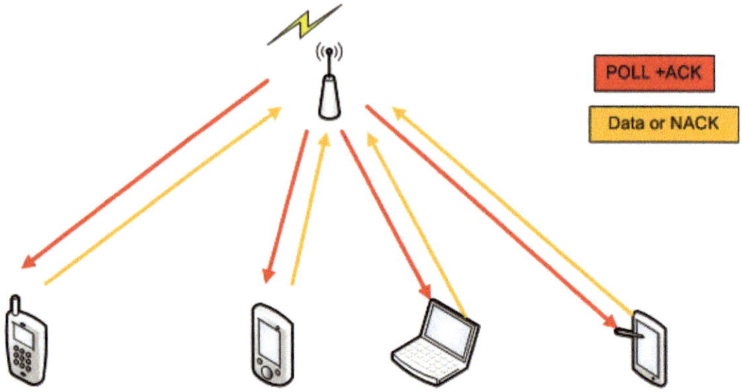

Figure 9.4. A wireless polling network showing the polling sequence.

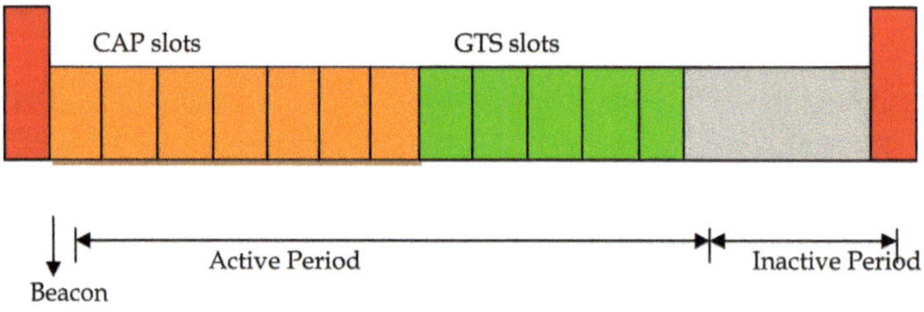

Figure 9.5. IEEE 802.15.4 MAC super frame structure.

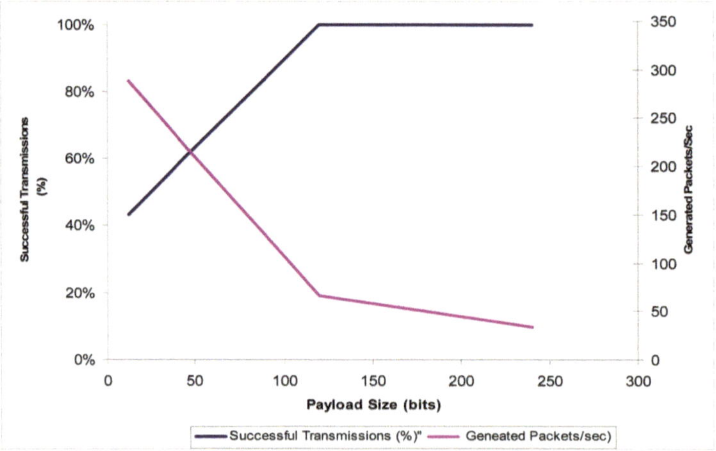

Figure 9.9. WBAN packet loss and packet generation rate for different data aggregation rate/payload size.

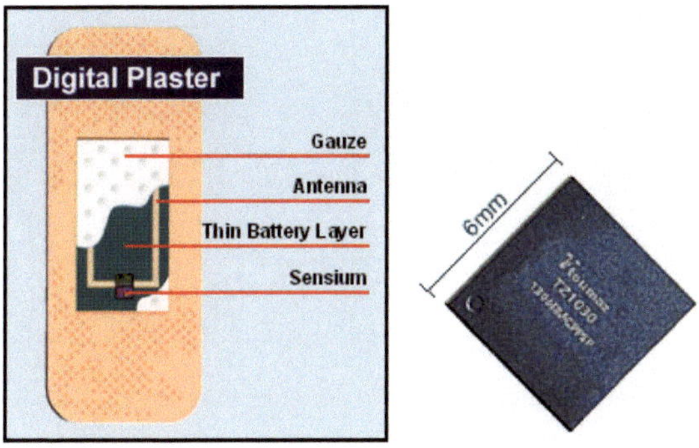

Figure 10.1. Toumaz Sensium^{TM} digital Plaster.

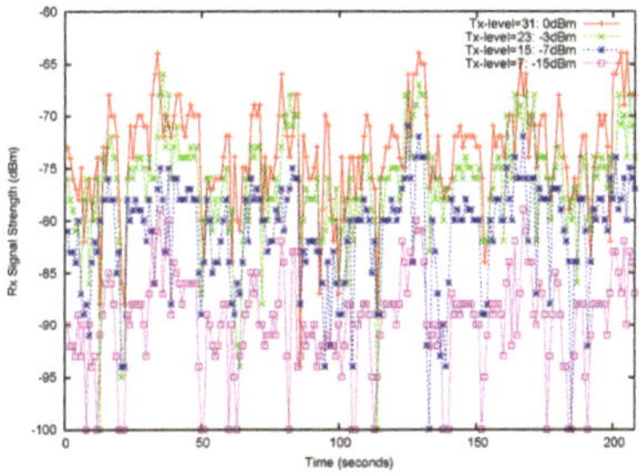

(a) Normal Walk

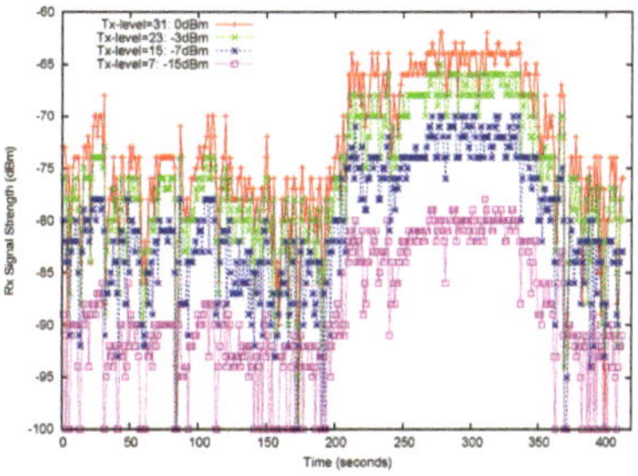

(b) Slow Walk

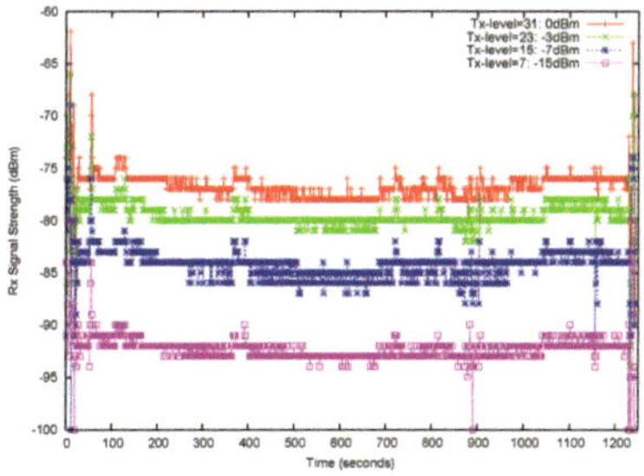

(c) Resting

Figure 10.3. RSSI vs. time for various patient scenarios.

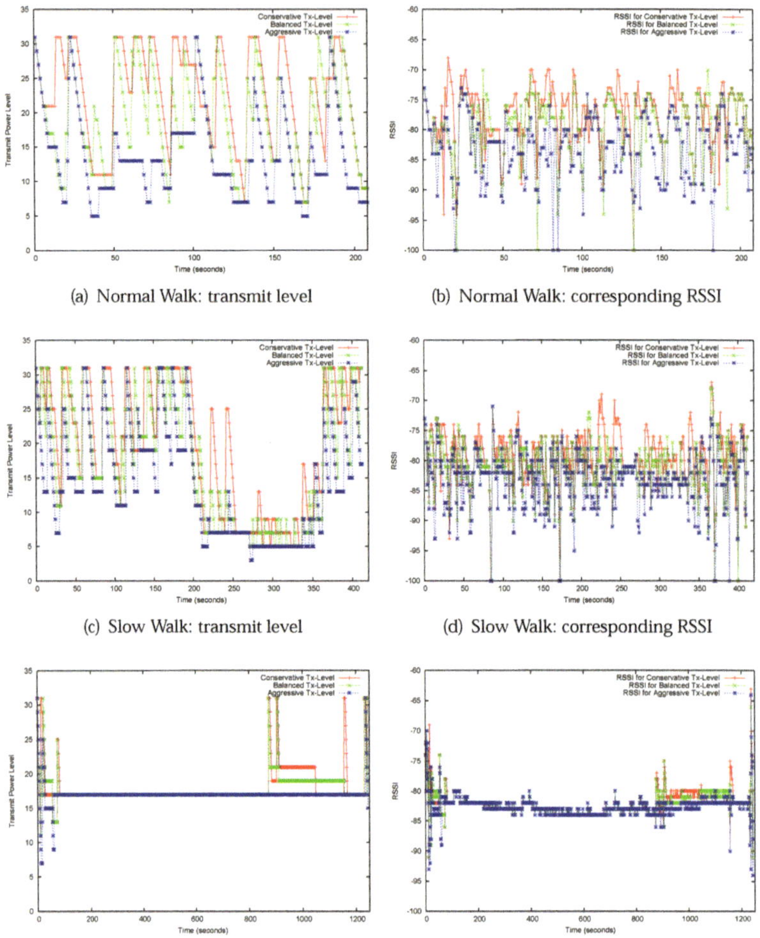

(a) Normal Walk: transmit level

(b) Normal Walk: corresponding RSSI

(c) Slow Walk: transmit level

(d) Slow Walk: corresponding RSSI

(e) Resting: transmit level

(f) Resting: corresponding RSSI

Figure 10.4. Optimal transmit power and associated RSSI for a normal walk, slow walk, and resting position.

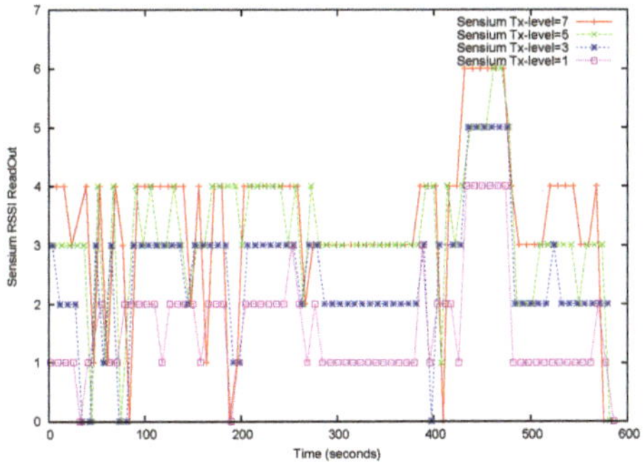

Figure 10.8. RSSI vs. time using the Sensium™ platform.

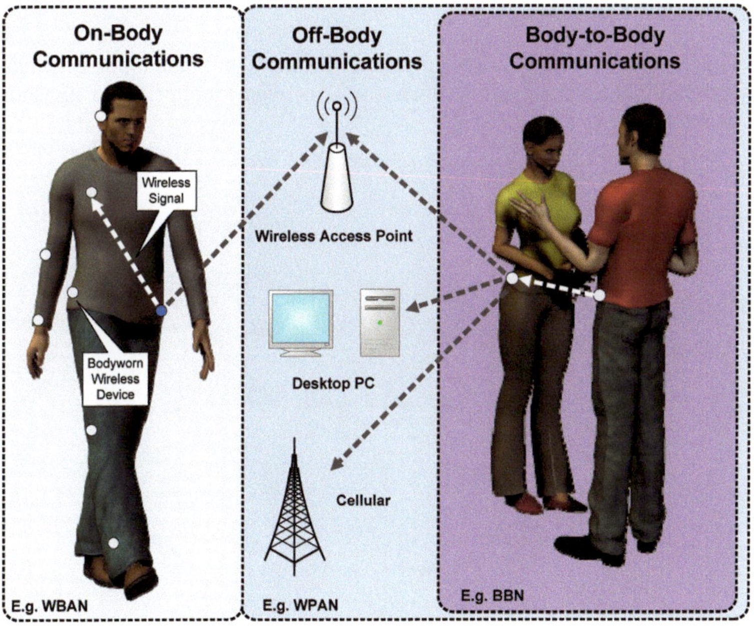

Figure 11.1. Overview of on-body, off-body, and body-to-body communications.

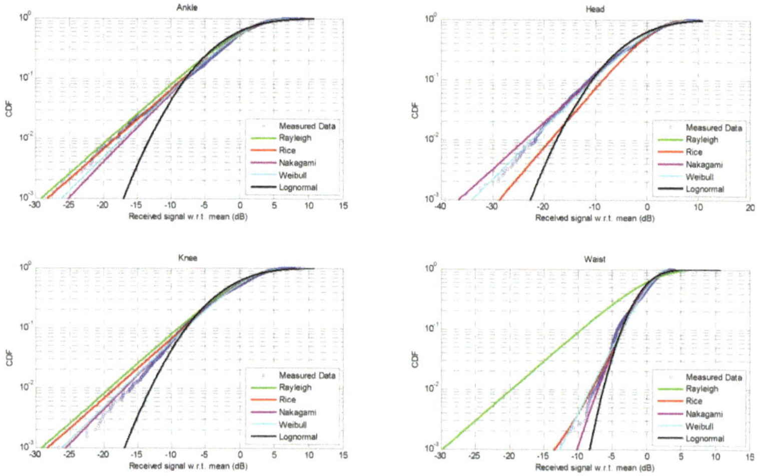

Figure 11.4. A selection of empirical CDF plots and maximum likelihood estimated theoretical plots while the user was mobile in the anechoic chamber.

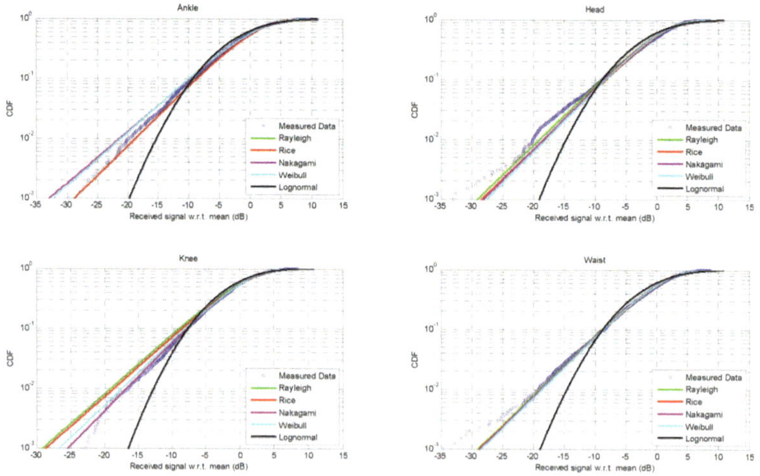

Figure 11.5. A selection of empirical CDF plots and maximum likelihood estimated theoretical plots while the user was mobile in the reverberation chamber.

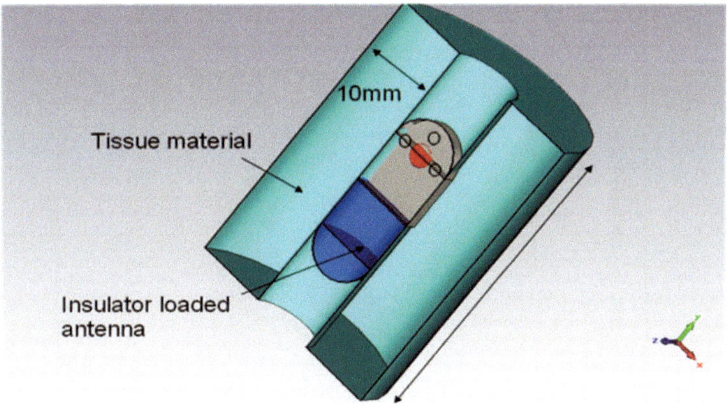

Figure 12.6. The loaded antenna inside a tissue simulating cylinder in a CST simulation.

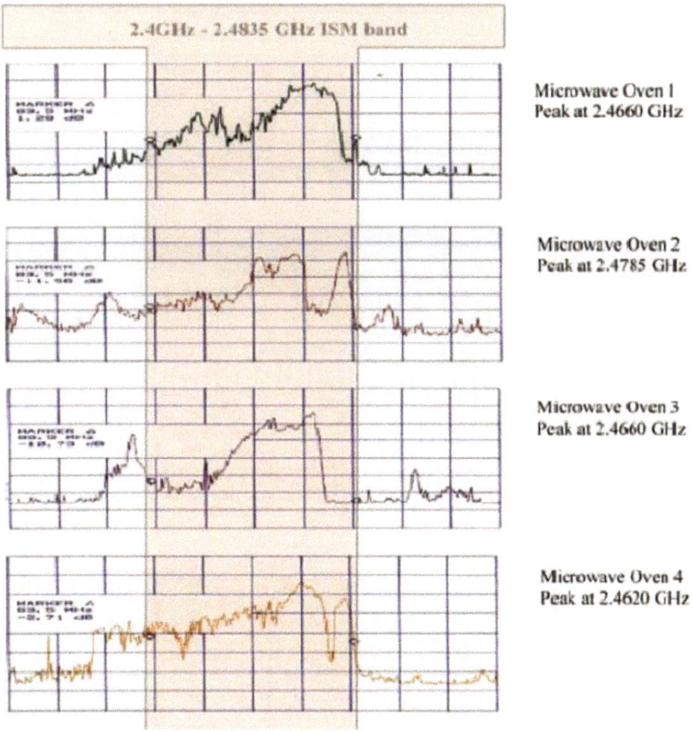

Figure 13.2. Frequency characteristics of commercially available micro ovens for home use [6].

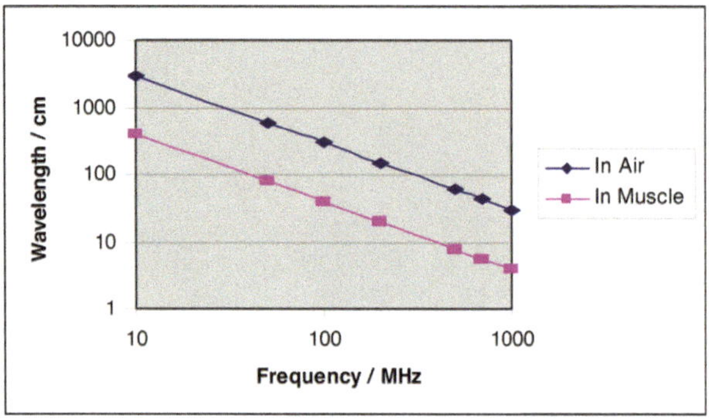

Figure 14.1. Wavelength in air and typical muscle vs frequency. Note: Log scales.

Figure 14.2. Silver on alumina patch antenna mounted on test implant case.

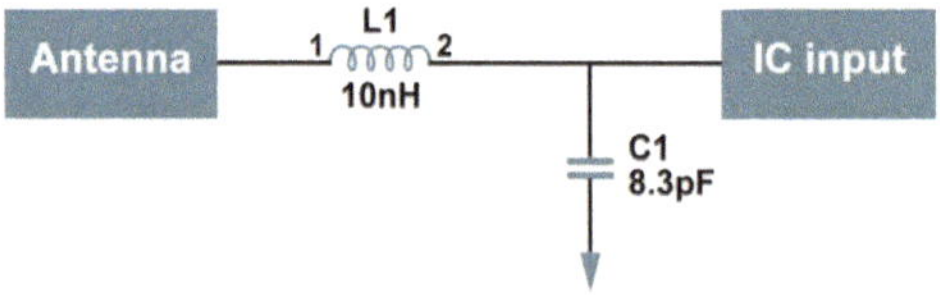

Figure 14.5. Matching schematic from Fig. 14.4.

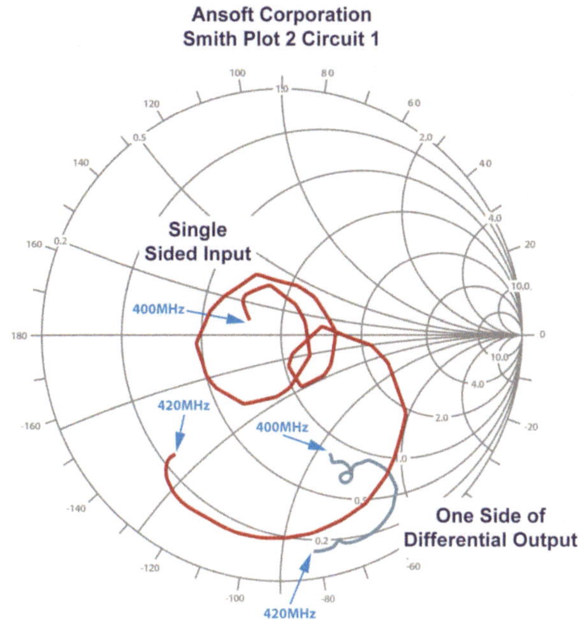

Figure 14.6. SAW filter input and (differential) output; 400 to 420 MHz; Ansoft SV software.

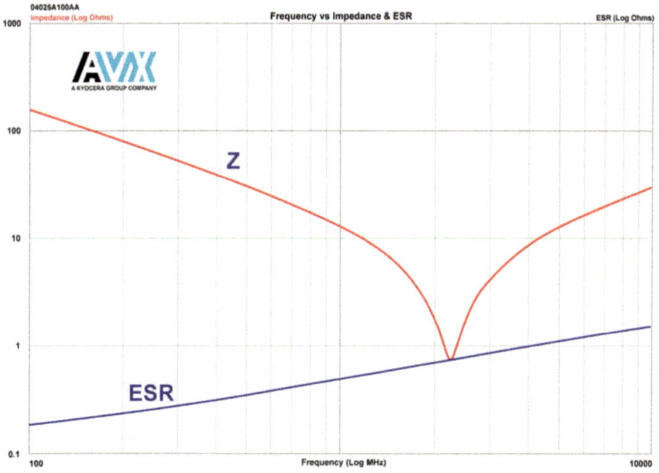

Figure 14.12. Impedance and series resistance for 10 pF 0402 capacitor (courtesy AVX).

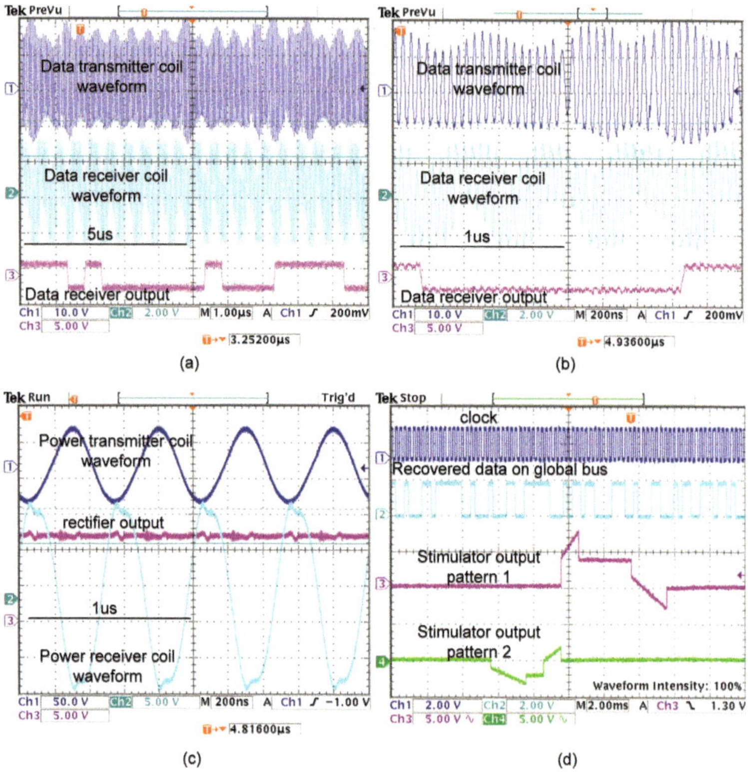

Figure 15.8. Measured results. (a) and (b) Data telemetry. Trace 1: the measured voltage across the transmitter coil; Trace 2: the measured voltage across the receiver coil; Trace 3: the measured data receiver output. (c) Power telemetry. Trace 1: the measured voltages across the power transmitter coil; Trace 2: the measured voltage across the power receiver coil (input to rectifier); Trace 3: rectifier output for powering secondary electronics; (d) output of two stimulator drivers that are powered and controlled by the power and data telemetry [33].

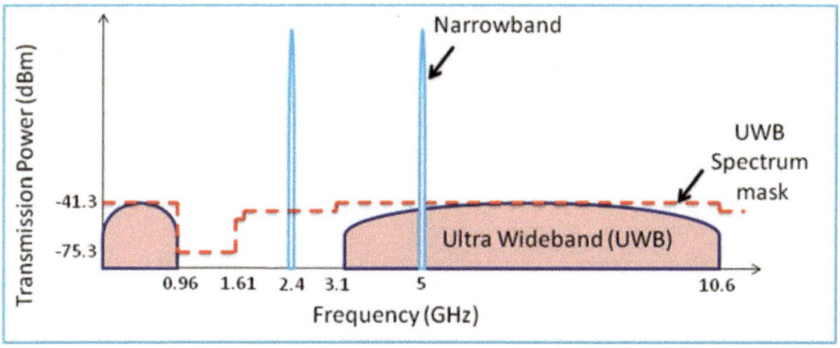

Figure 16.1. UWB and narrowband spectrum.

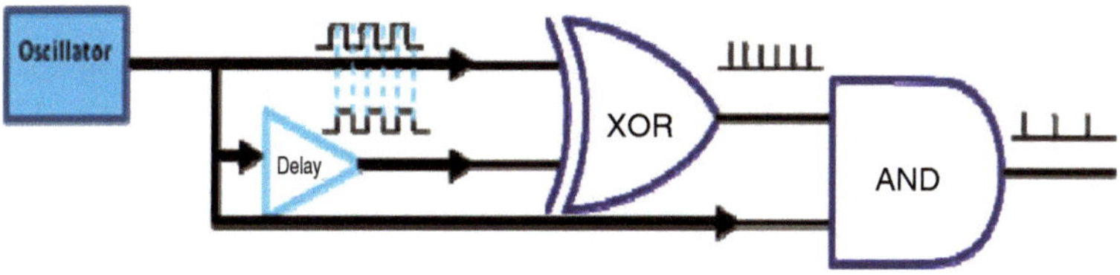

Figure 16.4. Narrow UWB pulse generation.

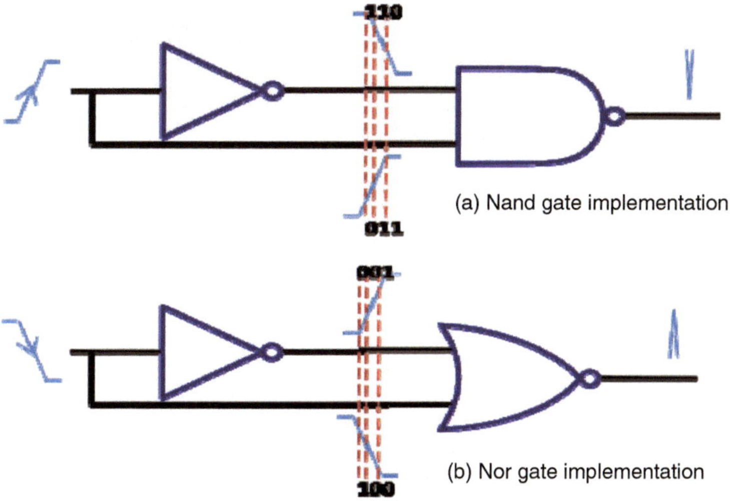

Figure 16.5. Triangle pulse generation.

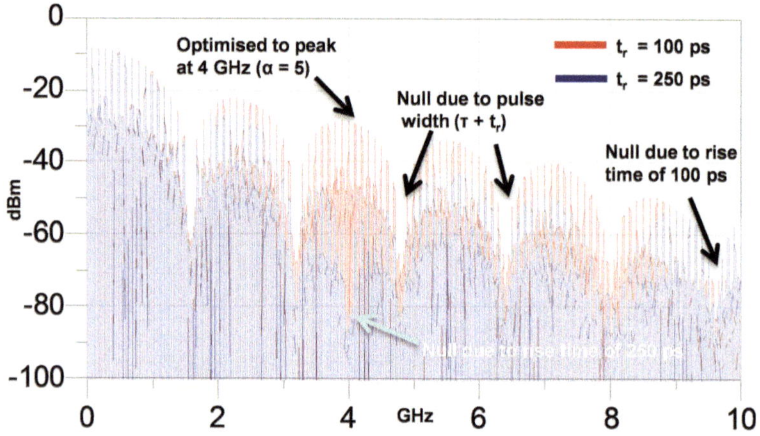

Figure 16.6. Spectrum of an UWB rectangular pulse in frequency domain and nulls in the UWB spectrum due to rise time.

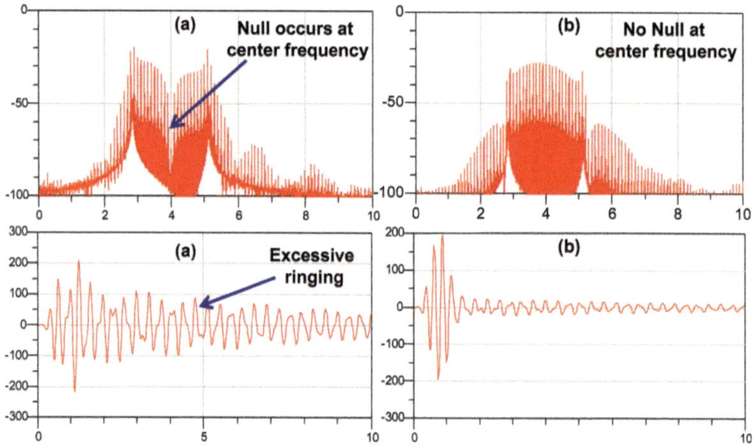

Figure 16.7. Effect of null in a band-limited signal.

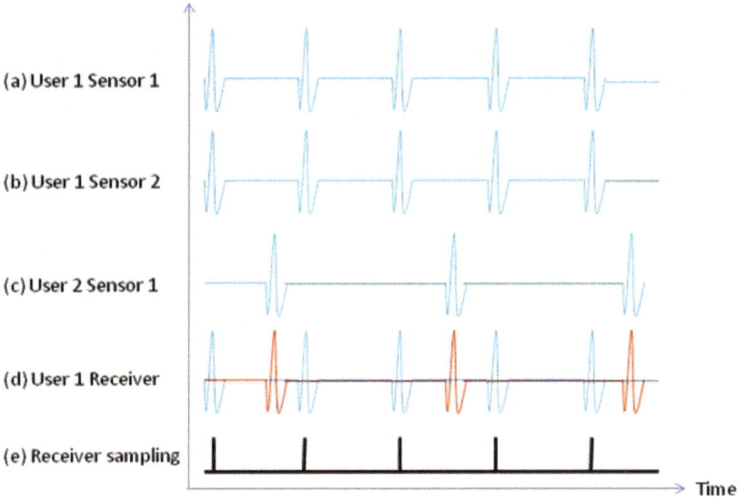

Figure 16.13. Different pulse rate for Tx-Only MAC.

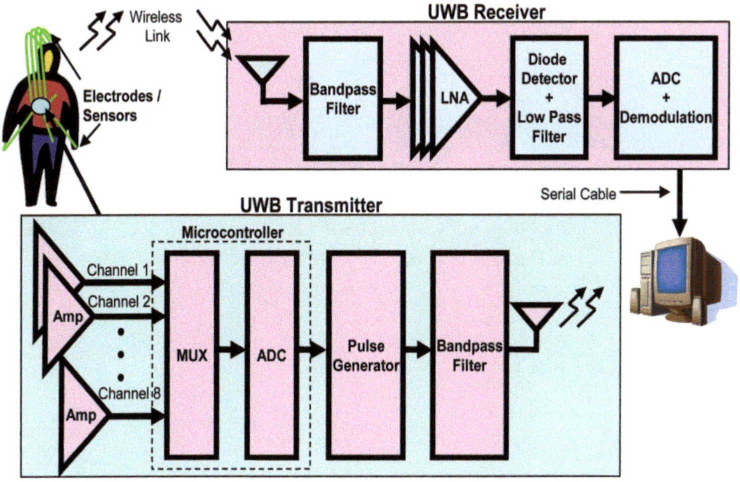

Figure 16.14. Multichannel UWB monitoring system block diagram.

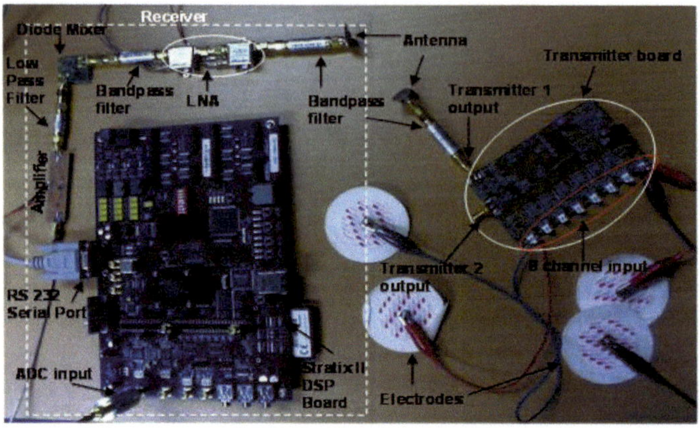

Figure 16.15. Photograph of complete system prototype.

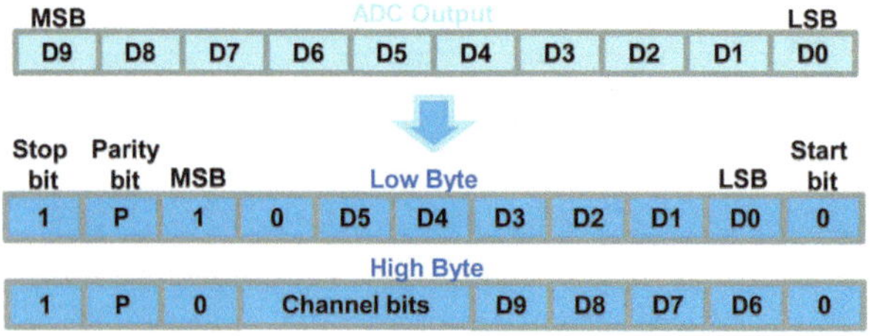

Figure 16.16. Transmit data frame.

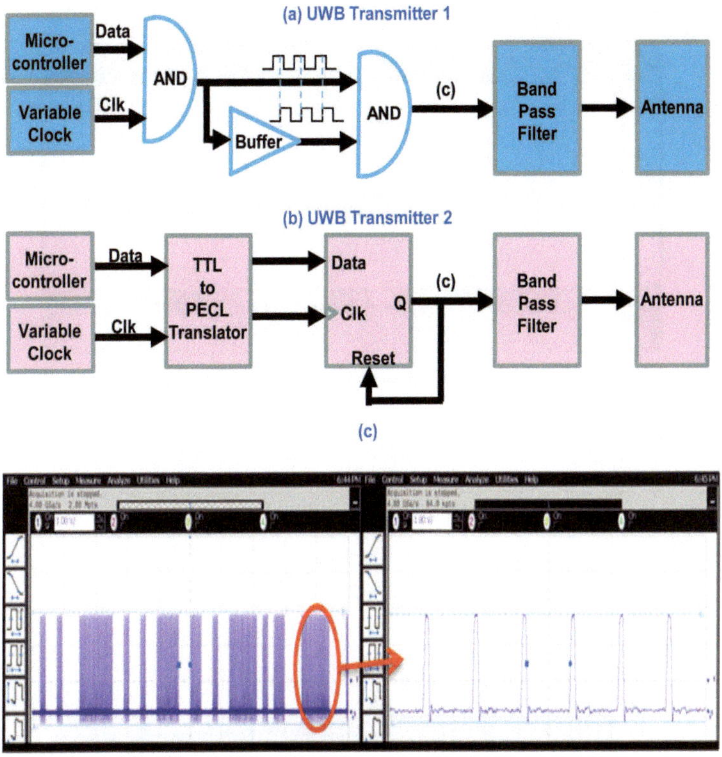

Figure 16.17. UWB transmitter block diagrams: (a) gate-based narrow pulse generation, (b) flip-flop-based narrow pulse generation, and (c) generated UWB pulse.

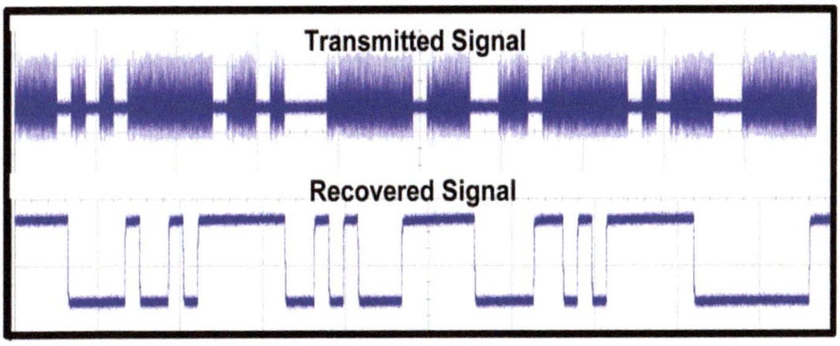

Figure 16.18. Recovered data from an OOK UWB modulated signal.

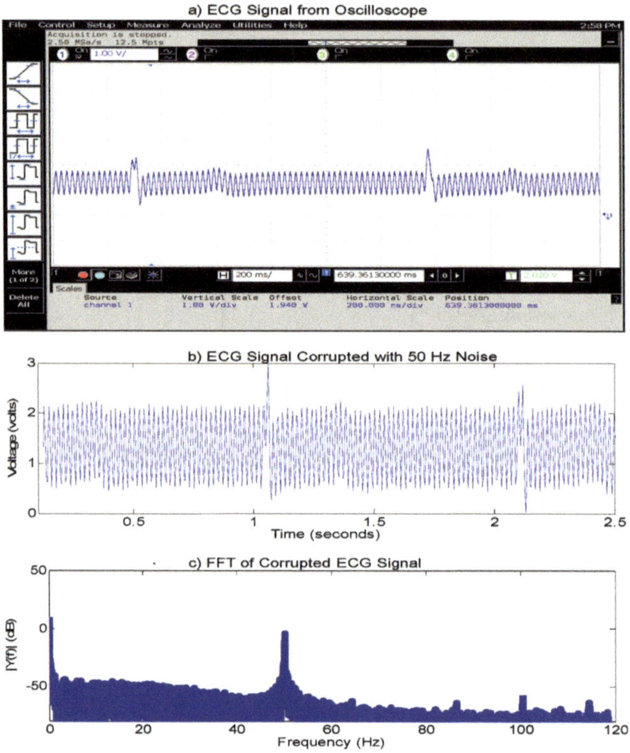

Figure 16.19. Monitored ECG waveforms wirelessly using UWB transmission.

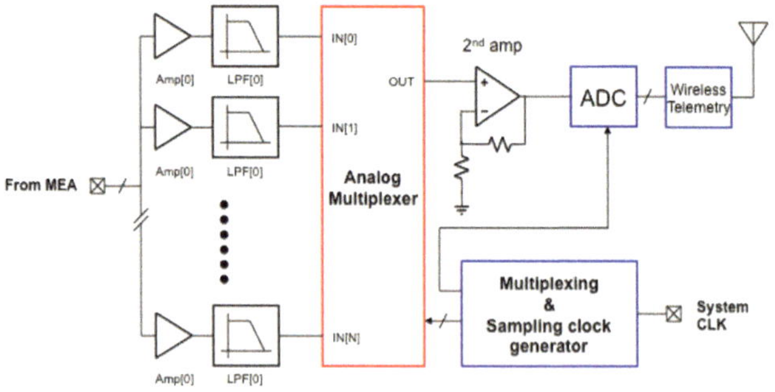

Figure 16.21. A multichannel neural recording system with UWB transmitter [14].

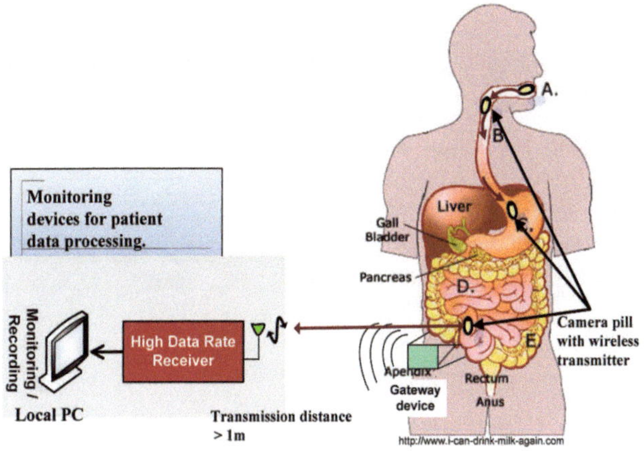

Figure 16.22. A wireless endoscope monitoring system.

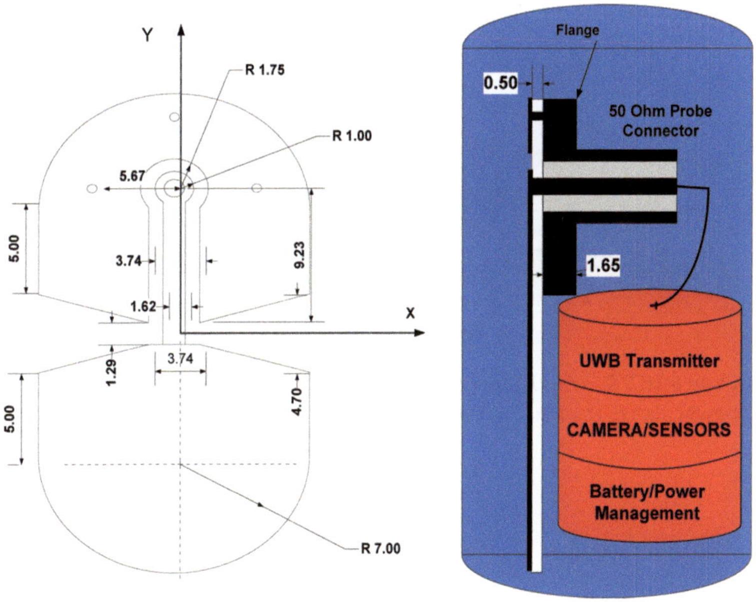

Figure 16.23. A wireless electronic pill system with main blocks identified [24].

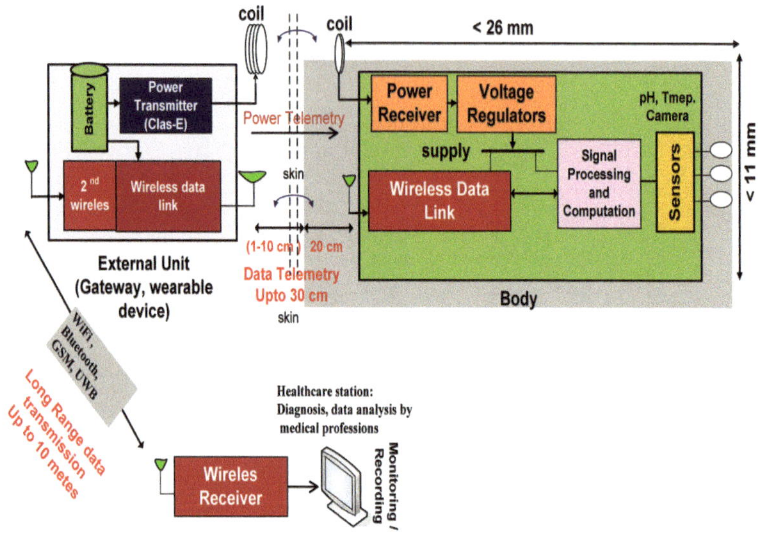

Figure 16.24. An electronic pill technology with wireless power embedded and wireless gateway for long-range data transmission. The available electronic pill devices have a dimension of 26 mm × 11 mm.

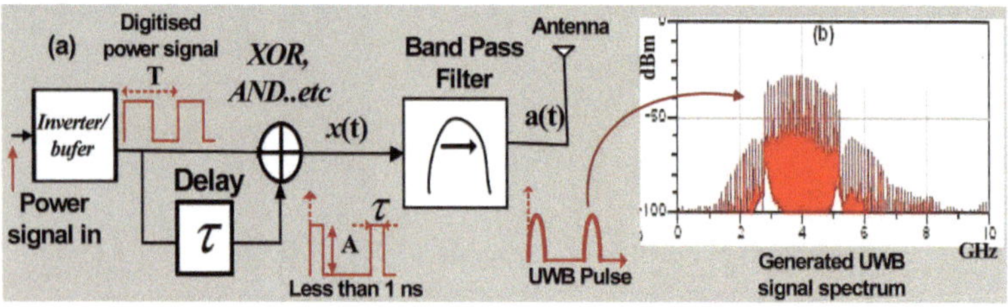

Figure 16.25. An UWB telemetry link: (a) UWB pulse generation from the power signal, (b) the UWB spectrum.

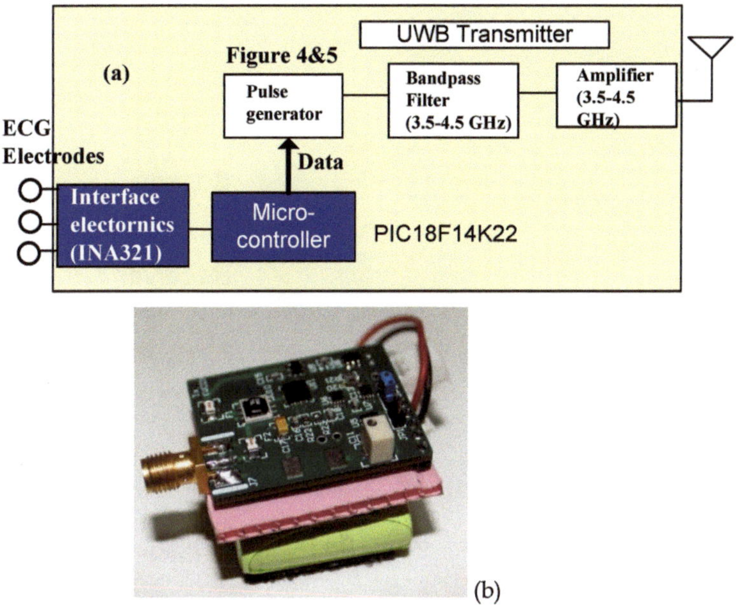

Figure 16.26. UWB-WBAN sensor node: (a) block diagram of an UWB-based WBAN sensor node, (b) photo of UWB sensor node (2.5 cm by 2.7 cm).

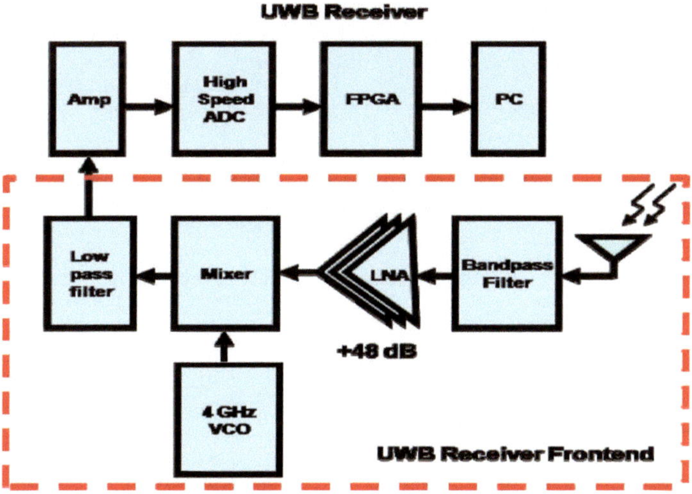

Figure 16.27. UWB receiver block diagram.

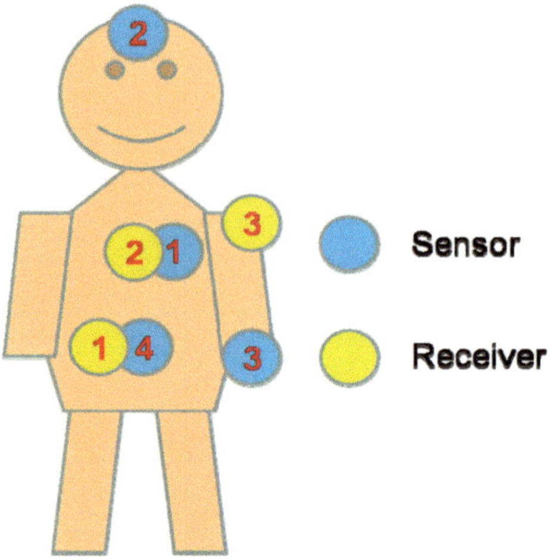

Figure 16.28. Sensor and receiver locations.

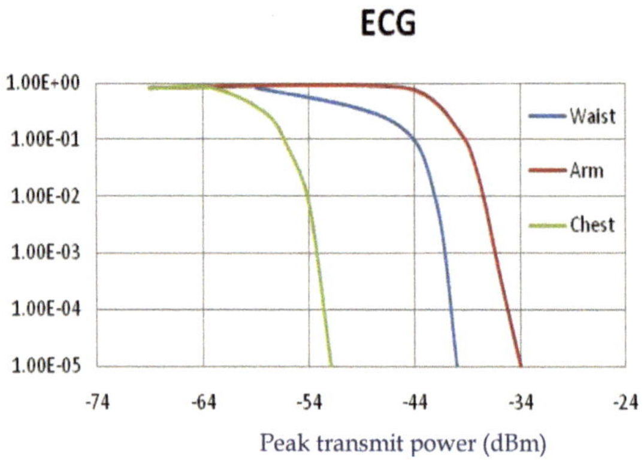

Figure 16.29. BER of ECG sensor for different receiver positions [29].

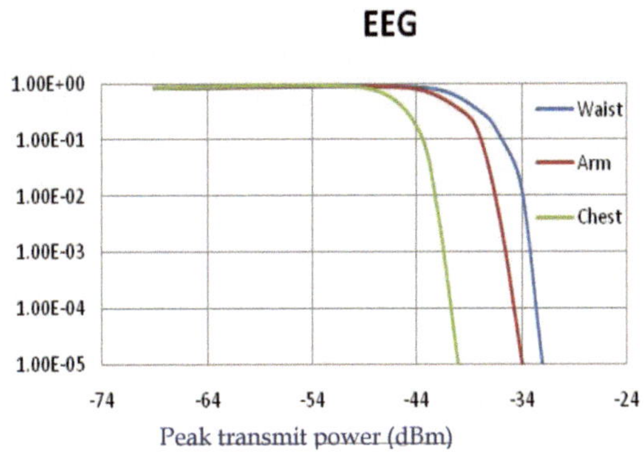

Figure 16.30. BER of EEG sensor for different receiver positions [29].

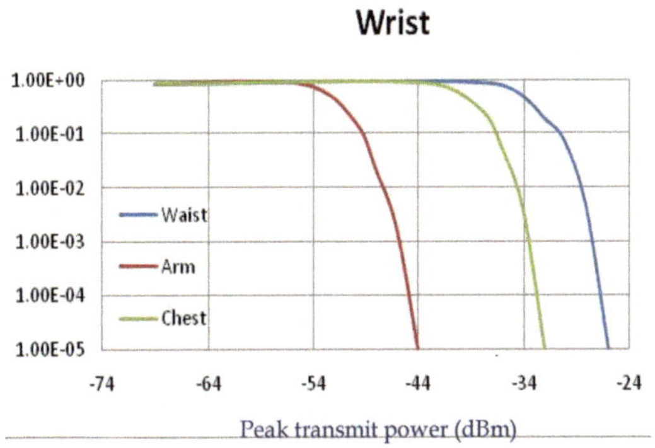

Figure 16.31. BER of sensor on wrist for different receiver positions [29].

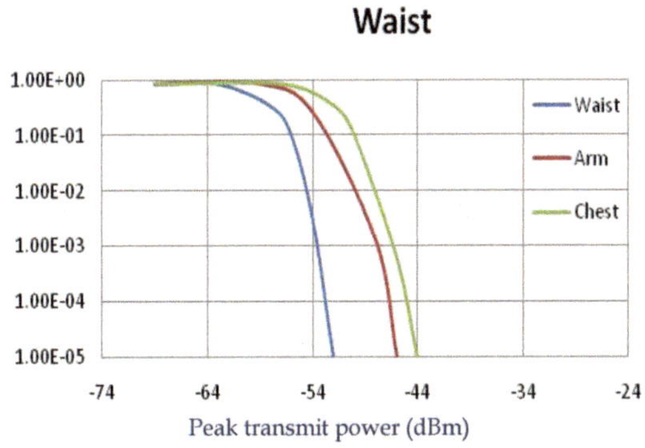

Figure 16.32. BER of sensor on waist for different receiver positions [29].